THIRD ASIAN-PACIFIC REGIONAL MEETING OF
THE INTERNATIONAL ASTRONOMICAL UNION

Third Asian-Pacific Regional Meeting of the International Astronomical Union

September 30 – October 5 1984, Kyoto, Japan

Part 1

Edited by

M. KITAMURA
Tokyo Astronomical Observatory

and

E. BUDDING
Carter Observatory, Wellington

Reprinted from
Astrophysics and Space Science, Vol. 118, Nos. 1/2

D. REIDEL PUBLISHING COMPANY

A MEMBER OF THE KLUWER ACADEMIC PUBLISHERS GROUP

DORDRECHT / BOSTON / LANCASTER / TOKYO

Library of Congress Cataloging-in-Publication Data

CIP

International Astronomical Union. Asian-Pacific
Regional Meeting (3rd : 1984 : Kyoto,
Japan) Third Asian-Pacific Regional
Meeting of the International Astronomical Union.

Bibliography: p.
1. Astronomy–Research–Asia–Congresses. 2. Astronomy–Research–Pacific Area–Congresses. 3. Astrophysics–Research–Asia–Congresses. 4. Astrophysics–Research–Pacific Area–Congresses. I. Kitamura, Masatoshi, 1926- . II. Budding, E., 1943- . III. Title.
QB61.I56 1984 520 86-469
ISBN-13:978-94-010-8503-8 e-ISBN-13:978-94-009-4496-1
DOI: 10.1007/978-94-009-4496-1

Published by D. Reidel Publishing Company
P.O. Box 17, 3300 AA Dordrecht, Holland.

Sold and distributed in the U.S.A. and Canada
by Kluwer Academic Publishers
190 Old Derby Street, Hingham, MA 02043, U.S.A.

In all other countries, sold and distributed
by Kluwer Academic Publishers Group,
P.O. Box 322, 3300 AH Dordrecht, Holland.

TABLE OF CONTENTS

Part 1

SECTION 1: A VIEW OF ASIAN-PACIFIC ASTRONOMY

SECTION 2a: THE SUN

SECTION 2b: SOLAR-STELLAR CONNECTIONS

SECTION 3: STARS

SECTION 4a: COMPACT STARS, AND GALACTIC STRUCTURE

SECTION 4b: GALACTIC STRUCTURE

SECTION 5a: GALACTIC STRUCTURE AND COSMOLOGY

SECTION 5b: GALAXIES AND COSMOLOGY

TABLE OF CONTENTS

part 2

(Astrophysics and Space Science Vol. 119, No. 1)

INTRODUCTION

M. KITAMURA
Tokyo Astronomical Observatory, Japan

and

E. BUDDING
Carter Observatory, Wellington, New Zealand

The Third Asian-Pacific Regional Meeting of the International Astronomical Union was held from 30 September to 5 October, 1984, at the Kyoto International Conference Hall, Kyoto, Japan, under the auspices of the Union and the Astronomical Society of Japan with Kyoto University as host. Three hundred and twenty-seven astronomers from twenty-two countries participated at the meeting and more than two hundred papers were presented.

The aim of the meeting was not only to promote scientific developments and cooperation, but also to offer a chance for all participants to become acquainted with major astronomical projects of the Asian-Pacific Region. Therefore, two new sessions of 'A View of Asian-Pacific Astronomy' and 'Astronomical Education in the Asian-Pacific Region', which had not been undertaken in the previous two Regional Meetings, were arranged as a first trial, besides the other ordinary scientific sessions.

The Scientific Organizing Committee consisted of D. C. Morton (chairman), R. N. Manchester, S. M. Gong, K. J. Feng, C. S. Shen, J. C. Bhattacharyya, G. Swarup, B. Hidayat, H. M. K. Al-Naimiy, H. S. Yun, J. B. Hearnshaw, S. C. Wolff, I. Kawaguchi, M. Kitamura, M. Morimoto, M. Oda, and J. P. Swings (IAU, ex officio); and the Local Organizing Committee of T. Kogure (chairman), T. Ishizawa, M. Saito, R. Hirata, S. Inagaki, E. Hiei, M. Kitamura, B. Takase, N. Kaifu, H. Maehara, Y. Osaki, and A. Yamasaki.

The programme was organized into five major sessions, as indicated below, with the respective chairmen as shown in parentheses. Most of these sessions were further divided into sub-sessions a and b:

Session Ia	A View of Asian-Pacific Astronomy (J. B. Hearnshaw and M. Kitamura),
Session Ib	Astronomical Education in the Asian-Pacific Region (M. Kitamura),
Session IIa	The Sun (S. M. Chitre),
Session IIb	Solar-Stellar Connections (H. S. Yun),
Session III	Stars (Z. Kopal),
Session IVa	Galactic Structure (B. Hidayat),
Session IVb	Galactic Structure (J. A. Galt),
Session Va	Galaxies and Cosmology (R. W. West),
Session Vb	Galaxies and Cosmology (A. J. Turtle),
Session VI	Conference Summaries (D. C. Morton).

In Session Ib, on Astronomical Education, six review papers were presented from six countries. Those papers are not in this volume, but they have been published elsewhere as a separate special issue – 'Supplement to the Proceedings of the Third Asian-Pacific Regional Meeting of the International Astronomical Union'. Conference summaries are also contained in that same 'Supplement'.

The Meeting was sponsored by the International Astronomical Union. Also, other financial support was obtained from the Japan Society for the Promotion of Science, the Regional Office of UNESCO at Jakarta (Indonesia), the Japan World Exposition Commemorative Fund, the Kyoto School of Computer Science, various Scientific Foundations in Japan, and private donations from many Japanese and overseas companies. All this help is gratefully acknowledged.

Of the presented scientific papers, eighty-five are contained in this book, and the papers appear in the order in which they are given in the various Sections. The remainder are mostly expected to be published elsewhere, in this or some other international journal in the near future.

Most memorable of this Third Asian-Pacific Regional Meeting of the International Astronomical Union were so many interesting presentations and very stimulating discussions, as well as the generous hospitality of the Local Organizing Committee. It is hoped that the Proceedings published here will provide a useful permanent record of this truly successful gathering.

REMARKS BY THE CHAIRMAN OF THE SCIENTIFIC ORGANIZING COMMITTEE

DONALD C. MORTON

Anglo-Australian Observatory, Epping, Australia

(Received 7 October, 1985)

The Scientific Organizing Committee has been extremely pleased with the response to this meeting, both in the number of attendees and the number of papers offered for presentation.

Altogether there are 327 participants representing 22 countries. The excellent facilities of the Kyoto International Conference Hall easily can handle such numbers, but I must confess that we were overwhelmed by the contributed papers. Almost all were of exceptionally high standard so that we could not easily eliminate many, and we did not want parallel sessions. Consequently the two poster sessions are filled to capacity with 72 papers each and the 52 oral papers are strictly limited to 8 min.

The programme is very tight so that I have asked the session chairman to be very firm. As you see this is a very modern conference hall. I would not want any speaker to be surprised by the the giant robot arm that can appear suddenly and carry away anyone who does not stop on time!

A regional meeting like this is going against the trend towards specialist meetings on particular topics. Nevertheless I think it is good for astronomers representing a wide variety of fields to get together to meet each other and learn about disciplines outside their immediate research areas. For this reason the oral programme contains many review papers. Furthermore, the entire first day is devoted to summaries of developments in the countries in our region. We shall hear reports on new instruments, recent scientific results and astronomical education which I am sure will demonstrate how active our science is in the Asian Pacific Region.

Finally, I wish to express our thanks to the Local Organizing Committee under Professor Kogure. This committee has worked extremely hard to make all aspects of this conference a success. In particular I should mention Professor Kogure's considerable success in obtaining financial contributions from numerous Japanese sources towards this meeting. We are most grateful to all these organizations as well as to the IAU and UNESCO for their assistance with travelling and other expenses involved with this Third Asian-Pacific Regional Meeting of the IAU.

Astrophysics and Space Science **118** (1986) 3.

THE JAPANESE NATIONAL LARGE TELESCOPE (JNLT) PROJECT*

K. KODAIRA

Tokyo Astronomical Observatory, Japan

and

T. KOGURE

University of Kyoto, Japan

(Received 15 May, 1985)

Abstract. The astronomical community in Japan is promoting a national project to construct a large optical infrared telescope on Mauna Kea in Hawaii. The scientific objectives and the basic concept of the project are briefly presented, as they stand in September 1984.

Introduction

In July 1984, the committee for astronomy of the Science Council of Japan accepted the basic concept of the Japanese National Large Telescope (JNLT) project, which was proposed by the Group of Optical and Infrared Astronomers in Japan (GOPIRA). This proposal aimed at the construction of an optical infrared telescope with an aperture larger than 5 m in Hawaii.

The largest available telescope at present in Japan is the 188 cm reflector of the Okayama Astrophysical Observatory which has been in operation since 1960. After this telescope was joined by the 105 cm Schmidt telescope of the Kiso Observatory in 1974, the committee for astronomy recommended the construction of a 4 m-class optical telescope, together with a cosmic radio telescope and others. The latter were first realized as the 45 m mm-wave telescope and its interferometric systems, composed of five 10 m dishes, at the Nobeyama Radio Observatory (NRO) in 1982. A review of the optical telescope project started in 1978, and the conceptual study has been carried out since 1980. In the following, we briefly describe the main scientific objectives and the basic concept of the JNLT project, as they stand in September 1984.

2. Scientific Objectives

The JNLT was conceived as one of the basic national facilities to keep the astronomical research in this country up to the frontiers of science, and, thereby, to contribute to the international astronomical community, in close cooperation with the Kiso Schmidt telescope, the radio facilities at the NRO, and the X-ray satellites of the Institute for Space and Astronautical Sciences.

* Paper presented at the IAU Third Asian-Pacific Regional Meeting, held in Kyoto, Japan, between 30 September–6 October, 1984.

We identify three major objectives to which the JNLT is to be devoted. The first one is the understanding of the origin and evolution of galaxies, which has been the hard kernel of the unresolved problems of cosmology in this century. The cosmology of this century began with the theory of general relativity, whose relevence first became realized by Hubble's detection of the redshift law of galaxies, promoted by the detailed studies of chemical abundances of stars and galaxies and, finally, verified by the measurements of the cosmic background radiation. Although these facts fitted neatly to the picture of a uniform expanding universe, the epoch of galaxy formation remained uncomprehended. The activities of 4 m-class telescopes in the 1970's and 1980's brought additional problems into light; – the existence of missing mass or the 'dark halo' problem, superclusters of galaxies and voids. The best way to deal with these problems is just straightforwardly to penetrate deep into the Universe, to look back at the far past, and to examine the nature of galaxies and clusters of galaxies there.

Not less important than the first objective is the research on the formation of stars and planetary systems. Explorers of the solar system in these last decades revealed that the earth is the only one place where life is prospering within our planetary system. Quite natural questions are now asked with an interest stronger than ever. Is the case 'the Earth' very special? Or is our planetary system very peculiar? The answer shall be sought in careful observations of our own planetary system, as well as any other planetary systems and proto-stars. Star formation, as elementary process, is closely related to the formation of planetary systems on the one hand, and on the other hand as a collective process, related to the evolution of galaxies. In order to observe the sites of star formation, usually we have to penetrate into the depth of thick interstellar matter.

In contrast to these classical subjects of astronomy (in a modern version), there is a new class of subject which has appeared in the window opened up by high-energy astrophysics. This environment of ours, the Universe, has turned out to be more violent and dynamical than the human being used to imagine. Physical processes and the nature of matter under extreme conditions in the cosmic environment, such as seen in flares, pulsars and quasars, just to mention a few of them, are most challenging subjects for human intelligence to understand. We need to observe these peculiar objects in detail and, at the same time, to collect widespread data in order to draw general conclusions. The latter task often requires us to approach the center of the galaxy where the star density is much higher than in the solar neighborhood.

All of these three major objectives require high-resolution imaging and a field of view wide enough to secure efficient observations of stars and galaxies fainter than the 20th magnitude. High-quality performance of the telescope is required in the optical region as well as in the infrared region.

3. Basic Concept

The working group of GOPIRA has been optimizing the basic concept of the JNLT according to the requirements from the side of the scientific objectives, and by taking into account that the 2.4 m Space Telescope (ST) will be in operation in the 1990's. The ST will be the most powerful optical telescope, whose faint-object camera can resolve two-hundredths of an arc sec in its ultimate capability in the visible region. This camera, however, has a field of view of only 4.4 arc sec square and needs to be accurately pointed at pre-selected objects. The wide-field camera of the ST, which will be always in action as a survey instrument, has a resolution of two-tenths of an arc sec and a field of view of 2.7 arc min square. This resolution, 0.2 arc sec, is comparable to that achieved by some ground-based telescopes located on good sites, such as by the CFHT on Mauna Kea. We require that the JNLT should have a high-resolving power, comparable to that of the wide-field camera of the ST, but a field of view about hundred times larger, at least on one of the available foci. This image quality, limited by the seeing and the optics, generally improves towards infrared, so far as the diffraction limit of the telescope aperture does not become effective. The ST's 2.4 m aperture reaches the diffraction limit of 0.2 arc sec around a near-infrared wavelength of 2.4 micron, though the first instrument package of the ST does not contain any specific infrared facilities. In order to make best use of the conditions available to the ground-based large telescopes, we require that the JNLT should have an aperture larger than 5 m, in complementing the function of the ST over the entire range of near-infrared, and possibly a part of the middle-infrared region.

A preliminary design study has been carried out for a single-dish optical-infrared telescope on an alt-azimuth mount. A merit-demerit study of various light-weight mirrors is underway, by applying finite-element analysis, including honeycomb, egg-crate, thin meniscus, and bored structures. The design target of maximum size of the primary mirror is provisionally set at 7.5 m. The focal ratio $F/2$ is assumed for the primary focus, $F/12$ – $F/15$ in the optical, and $F/60$ in the infrared region, for the Cassegrain and Nasmyth foci. A corrector lens system for the $F/2$ parabolic mirror is being designed to secure the required high-quality image over a 0.5 deg field of view. The infrared emissivity of the telescope should be suppressed to below 8% by introducing a removable tertiary-mirror system for the Nasmyth focus, and changeable top-rings.

4. Conclusions

In accordance with the scientific objectives and the required quality of the JNLT, we strongly wish to locate the telescope on Mauna Kea of the Hawaiian large island, which is known as one of the best astronomical sites in the world nearest to Japan. This does not mean, however, that the tasks of the telescopes that are operated on the Japanese islands (longitude 135 E) are decreasing. A remote communication link between Hawaii and Japan will help us to combine the separated facilities, and to monitor the telescope situation and observational procedure.

Optimistic Japanese astronomers are hoping to install the JNLT in the early 1990's. We believe that the JNLT on Mauna Kea, if realized in time, will surely serve to promote international collaboration, not only among those countries having facilities in Hawaii, but also among countries of the Asian-Pacific region.

SOLAR STUDIES IN CHINA*

SHI-HUI YE

Purple Mountain Observatory, Academia Sinica, China

(Received 8 May, 1985)

Abstract. After a brief historical retrospect, this paper describes the present status of solar physics in China. Besides an overall account of the observational network, instrumentation, topics of research, personnal and international cooperation, emphases are laid on the works about the maximum year of the 21st cycle of solar activity, as well as observations of the solar total eclipses in 1968, 1980, and 1983. Section 5 presents a short description of the trends of further developments.

1. A Short Historical Account

The solar studies in China may be traced back to the Xia Dynasty (21st–16th century B.C.). It is commonly acknowledged that the earliest record of a solar eclipse in the world is that of a total eclipse in the Zhong Kan reign of the Xia dynasty. It probably occurred on 22 October, 2137 B.C. From the Spring and Autumn Period (770–476 B.C.) to the beginning of the Qing Dynasty (1644–1911) there are altogether 916 items about solar eclipses recorded in the official annals of Chinese history. This is the most complete and systematic list of solar eclipses in the world. In the book *Zhuo Zhuan* there is the following statement: 'Clouds, like multitudes of red birds, carried the Sun to fly'. According to the famous British scientist Needham (1959), this may be interpreted as a description of the corona during an eclipse in 490 B.C. He also suggests that the bone inscription stating that 'three flames ate the Sun' is probably a record of solar prominences.

The earliest record of sunspots is commonly acknowledged to be that of 10 May, 28 B.C., described in Han Shu. From that time to the end of the Ming Dynasty (1368–1644) there are 112 items of sunspot records in the official historical annals. Besides, there are even more records in local chronicles. As for the phenomena of solar-terrestrial relationships, there were, in ancient China, systematic records of polar aurorae since 208 B.C.

In the realm of instruments for solar observations, it was in the Zhou Dynasty (11th century–256 B.C.) that the Chinese learned to measure time with sundials and to determine the moment of midday, as well as the seasons and festivals, with gnomons.

Since the Ming Dynasty science and technology in China gradually became backward. Modern astronomy was developed in China rather lately. It is worthy of mentioning that Chinese astronomers successfully observed the solar total eclipse of 19 June, 1936 in Hokkaido, Japan; and that of 21 September, 1941 in the Lintao County

* Paper presented at the IAU Third Asian-Pacific Regional Meeting, held in Kyoto, Japan, between 30 September–6 October, 1984.

Astrophysics and Space Science **118** (1986) 9–14.

of the Ganshu Province. The former Institute of Astronomy of the National Central Research Council made routine Hα monochromatic and sunspot visual observations. After the establishment of the People's Republic in 1949, solar studies in China entered a period of tremendous development.

2. Present Status

In China today a comparatively complete network of solar observatories has been formed. It consists of four stations, i.e., Beijing, Nanjing, Kunming, and Urumqi. The span in longitude exceeds 30°. The participating units are the Beijing, Purple Mountain, and Yunnan Observatories as well as the Urumqi Astronomical Station, all belonging to Academia Sinica; and also the Department of Astronomy of the Nanjing University. Routine observations consist of (1) the white-light photography, visual projection patrol, and magnetic field measurements of sunspots; (2) the Hα monochromatic observations of the chromosphere; (3) the multi-channel spectral observations of flares, prominences, etc., and the measurements of their velocity fields; (4) the solar radio observations from meter to millimeter wavelength regions. Most of these observations are performed with Chinese-made instruments. At present the largest optical solar telescope in China is the solar tower of Nanjing University (Fang *et al.*, 1983). The aperture of its coelostat is 46 cm, the diameters of solar images are 37.3 and 20 cm and the measured spectral resolution of the multiple wavelength spectrograph in the second order is 140 000. The best solar radio telescope is the composite interferometer of Beijing Observatory. It is composed of 4 × 16 antennae. The spatial resolution is better than 1 arc min, the sensitivity is 0.03 sfu and the working frequency is 460 MHz. A new solar magnetograph has been constructed by the Nanjing Astronomical Instruments Factory and is now being installed at the Huairou Station of Beijing Observatory. Its aperture is 35 cm and the width of the passband of the birefringent filter is 0.125 Å. The fine structure telescopes of Yunnan and Purple Mountain Observatories (aperture 25 cm and width of passband of Hα filters 0.5 Å) are now under construction at the above-mentioned factory.

On the basis of a large amount of observational material our country, in 1969, began to publish its *Solar-Geophysical Data.*

Since 1982 the Shanghai Observatory has used balloons to investigate the solar infrared radiation (Zhou, 1984). The volume of a balloon is 30 000 m^3. The height of flight is 35 km. The working wavelengths are 4.6 and 18 μ. The balloon-borne instrument is a Newtonian telescope of 15 cm. The brightness temperature of the Sun at 18 μ has been found to be 4685 $\pm$ 136 K.

There are now in China about 200 solar astronomers. In the fifties and sixties the main topics of research were the internal structure of the Sun and its energy source (Shu-Mu, 1955), the solar convection layer (Chen, 1956), the solar granulation (Chen, 1959), the spectrophotometry of prominences (Yeh, 1961), the limb darkening at the extreme limb of the Sun (Chang *et al.*, 1961), the photometry of flares and their internal motions (Zhang *et al.*, 1961), the regularity of solar activity (Chen, 1962). Since the middle of the seventies the scientific work has seen a large-scale increase. Only a few themes of research are now mentioned in the following: the solar type-I bursts in the metric wavelength band (Yin, 1978), the U-shaped spectrum of type-IV radio bursts (Li *et al.*, 1979), the dynamics of surges (Cao *et al.*, 1979, 1980), the solar wind (Wang *et al.*, 1979; Zhang *et al.*, 1982), the coronal transients (Wang *et al.*, 1982), the plasma instability in active regions (Song *et al.*, 1983, 1984), the fine structure and rotation of spot groups (Ding *et al.*, 1976), the spectral analysis of solar impulsive, hard X-ray bursts (Zhang *et al.*, 1979), the magneto-sensitive lines in the solar spectrum, the gradient of sunspot magnetic fields and the effects of Alfvén waves (Ye *et al.*, 1978, 1979, 1981), the theory of force-free magnetic fields in spots (Yang *et al.*, 1980), the slowly varying components of solar radio-emission (Zhao *et al.*, 1980; Yao, 1982), the particle acceleration in flares (Zhang *et al.*, 1982), the mechanism of heating of the chromosphere-corona transition layer (Li *et al.*, 1979), the coronal Langmuir turbulence (Song *et al.*, 1979). Moreover, the abundant records of sunspots in Chinese historical annals may be utilized in the investigation of the stability of the cycle of solar activity and the Maunder Minimum (e.g., Xu, 1977).

Much attention has been paid to application studies, and a large amount of work has been done in the forecasts of solar activity. In 1968, scientists from Beijing Observatory joined an expedition to Mount Qomolungma and for the first time carried out solar radiation measurements in the near infrared region at an altitude of 5000 m above sea level (Lin *et al.*, 1974).

In recent years the scientific intercourses of China with other countries in the field of solar research are becoming more and more frequent. Many distinguished foreign solar physicists have come to China. Some Chinese solar astronomers visited institutions of solar research in other countries and have taken an active part in some international scientific enterprises, in particular the Solar Maximum Year (1979–1981). An international workshop of solar physics was held in Kunming in November of 1983. A workshop on solar-stellar physics was set up by some Chinese and Japanese astronomers, and its two sessions were convened in Nanjing in April of 1982 and December of 1983. Under the promotion of this workshop some cooperative researches on the convective theory, dynamo theory and solar magnetic field measurements have been carried out (Unno, 1984).

3. Observations During the Maximum Year of the 21st Cycle of Solar Activity

In this maximum year a large amount of both optical and radio routine observational materials have been secured, and they have been regularly sent to the Solar-Geophysical Data Center located in Boulder, U.S.A. In the following we give short descriptions to some specific items of comparatively great interest.

(1) The fine structures of the solar microwave millisecond spike emission: such structures were recorded at 08:08 UT on 16 May, 1981 at 2.84 nc by Beijing Observatory with a specially designed apparatus. Within 1 s there may be several hundred fine spikes with flux density larger than 5×10^3 sfu. Such observations have been performed

many times in different frequency bands. Chinese radio astronomers have studied the properties and the mechanism of production of this phenomenon and some interesting results have been obtained (see, e.g., Zhao *et al.*, 1982).

(2) The SSHG observations of solar active features: the flare loop of 27 April, 1981 and some other active features were observed in the Hα line at Yunnan Observatory with a spectra-spectroheliograph (SSHG) attached to the 40 cm horizontal solar spectrograph. The observational data have been utilized in the analyses of velocity fields in these active features (Gu *et al.*, 1983).

(3) The multiple wavelength spectra of solar flares: more than 20 sets of spectrograms of flares have been obtained at the Purple Mountain Observatory with its nine-wavelength solar spectrograph during the maximum of the 21st cycle. They include the spectra of the white-light flares of 5 September, 1981 and 20 June, 1982 (Wang *et al.*, 1981).

4. Solar Eclipse Observations

The annular eclipse of 19 April, 1956 in Hainan Island (observed in collaboration with the Soviet Union), the annular eclipse of 29 April, 1976 in northern Tibet and the partial eclipse of 31 July, 1981 in Heilongjiang Province were all observed with some success. Large-scale and highly successful observations belong to three total eclipses. They are those of 22 September, 1968 in Xinjiang Province, 16 February, 1980 in Yunnan Province and 11 June, 1983 in Papua New Guinea, respectively.

The natural conditions of the 1968 eclipse in Sinkiang were bad (the duration of totality amounted only to ten or more seconds, and the height of the Sun above the horizon during the totality was only 7–10°) but some excellent materials were secured. They are mainly the photographs of a condensation region in the corona, and the records of solar radio intensities in three wavelength bands and the polarization at 3.2 cm (Solar Eclipse Group, 1976).

In the eclipse of 1980 about 30 scientific institutes and universities carried out comprehensive and coordinated observations. There were about 300 observers and 30 items which belong to the fields of optical and radio astronomy, geophysics, meteorology and physics, respectively. Almost all the observations were successful. The most outstanding ones are the photography of the chromospheric flash spectra in 4800–5960 Å as well as the records of solar radio fluxes in six wavelengths extending from metric to millimetric regions. The results of observations of all items can be found in the collection of articles (1983).

In order to observe the total eclipse of 1983 Academia Sinica sent an expedition of 14 persons to Port Moresby, the capital of Papua New Guinea. All the observations belonging, respectively, to astronomy, terrestrial magnetism and ionospheric explorations were successfully accomplished. The most important achievement was the photography of the flash spectrum in the wavelength range 3600–6600 Å. More than 180 spectrograms have been obtained. The time resolution is 3 spectrograms per second and the dispersion is about 6 Å mm^{-1}.

5. Concluding Remarks

Solar physics in China has already a sound foundation in instrumentation, observational data, topics of research and personnel. In recent years much progress has been achieved. But there are some deficiencies in our work. At first, our observational facilities and techniques are comparatively backward. We lack large and advanced instruments. At present our observations are mainly ground-based ones. Routine coronal optical observations are still a vacant area. Secondly, our field of research is comparatively narrow. It is confined mainly in the study of solar active features, in particular flares. Not enough attention has been paid to solar oscillations, the 'neutrino crisis', the basic mechanism of solar activity, the solar-stellar physics, etc. Besides, the international cooperation has to be further strengthened. We sincerely hope to develop friendship and collaboration with our foreign colleagues.

References

Cao, Tian-jun *et al.*: 1979, *Acta Astron. Sinica* **20**, 143.
Cao, Tian-jun *et al.*: 1980, *Acta Astron. Sinica* **21**, 27.
Chang, Chen-ta *et al.*: 1961, *Acta Astron. Sinica* **9**, 11.
Chen, Piao: 1956, *Acta Astron. Sinica* **4**, 31.
Chen, Piao: 1959, *Acta Astron. Sinica* **7**, 163.
Chen, Piao: 1962, *Acta Astron. Sinica* **10**, 1.
Ding, You-ji *et al.*: 1976, *Acta Astron. Sinica* **17**, 60.
Fang, Cheng *et al.*: 1983, *Acta Astron. Sinica* **24**, 189.
Gu, S. M. *et al.*: 1983, *Solar Phys.* **87**, 155.
Li, Chun-sheng *et al.*: 1979, *Acta Astron. Sinica* **20**, 153.
Li, Xiao-qing *et al.*: 1979, *Acta Astron. Sinica* **20**, 281.
Lin, Yuan-zhang *et al.*: 1974, *Acta Astron. Sinica* **15**, 149.
Needham, J.: 1959, *Science and Civilization in China*, Vol. 3, Cambridge Univ. Press, Ch. 20, 9.
Reports on Observations Made in Yunnan of China on the Total Solar Eclipse of February 16, 1980 (1983), Science Press.
Shu-Mu, Kung: 1955, *Acta Astron. Sinica* **3**, 21.
Solar Eclipse Group for Coordinates Observations: 1976, *Acta Astron. Sinica* **17**, 41.
Song, Mu-tao *et al.*: 1980, *Acta Astron. Sinica* **21**, 20.
Song, Mu-tao *et al.*: 1983, *Acta Astron. Sinica* **24**, 1.
Song, Mu-tao *et al.*: 1984, *Acta Astron. Sinica* **25**, 15.
Unno, W. (ed.): 1984, *Japan–China Collaboration on Solar Physics of Stars*, Japan Society for the Promotion of Science.
Wang, Shui *et al.*: 1979, *Sci. Sinica* **22**, 1056.
Wang, Shui *et al.*: 1982, *Sci. Sinica* **25**, 1305.
Wang, Zhen-yi *et al.*: 1981, *Astron. Circ.*, No. 11.
Xu, Zhen-tao: 1977, *Rep. Sci. Works* **8**, 63.
Yang, Hai-shou *et al.*: 1980, *Acta Astron. Sinica* **21**, 136, 143.
Yao, Jin-xing: 1982, *Acta Astron. Sinica* **23**, 309.
Ye, Shi-hui *et al.*: 1978, *Acta Astron. Sinica* **19**, 152.
Ye, Shi-hui *et al.*: 1979, *Acta Astron. Sinica* **20**, 275.
Ye, Shi-hui *et al.*: 1981, *Acta Astron. Sinica* **22**, 265.
Yeh, Shih-Huei: 1961, *Izv. Krymskoj Astrofiz. Obs.* **25**, 180, 234.
Yin, Qi-fen: 1978, *Sci. Sinica* **6**, 648.
Zhang, Chen-yue *et al.*: 1982, *Acta Astron. Sinica* **23**, 317.
Zhang, He-qi *et al.*: 1961, *Acta Astron. Sinica* **9**, 20.

Zhang, He-qi *et al.*: 1979, *Acta Astron. Sinica* **20**, 268.
Zhang, Zhen-da *et al.*: 1982, *Acta Astron. Sinica* **23**, 102.
Zhao, Ren-yang *et al.*: 1980, *Acta Astron. Sinica* **21**, 262.
Zhao Ren-yang *et al.*: 1982, *Sci. Sinica* **25**, 422.
Zhou, Hui-cheng: 1984, *Acta Astron. Sinica* **25**, 189.

NEW TELESCOPES IN CHINA*

SHU-MO GONG

Purple Mountain Observatory, Academia Sinica, Nanjing, China

(Received 10 May, 1985)

Abstract. This paper covers the main aspects of three new optical telescopes: a 1.26 m aperture one for use in the infrared, a 1.56 m aperture one for astrometry, and a 2.16 m reflector for general astrophysical work. It also briefly mentions the 13.7 m telescope designed for the mm wavelength band, the first VLBI in China and the meter wavelength aperture synthesis telescope. All these telescopes, optical and radio, are now being built in China.

1. Optical Telescopes

1.1. 1.26 INFRARED TELESCOPES

The 1.26 infrared telescope (Figure 1) is under construction in the Nanjing Astronomical Instruments Factory (NAIF), Academia Sinica. The existing 6 m dome at the Beijing Observatory and the ready-made extra-low thermo-expansion VO2 glass blank in the Xin-hu Glass Factory in Shanghai restrict the size of the infrared telescope. The project period is from 1982–1985.

(1) *Optical system.* The telescope has Ritchey–Chrétien (RC) optics with only one focus, to reduce the cost. The primary mirror, of an effective diameter of 1260 mm with a focal ratio $\frac{1}{2}$, and a secondary mirror, of an effective diameter of 110 mm, form a system having a composite focal ratio $\frac{1}{30}$.

(2) *Main mirror cell and secondary mirror assembly.* The primary mirror is a thin type, with a thickness only $\frac{1}{12}$ of its diameter. An 18 point whiffle-tree-type back support, and a mercury filled tube side support are used. The chopping of the secondary mirror is driven by a linear-motor. The displacement and speed of the secondary mirror are checked by two corresponding sensors. The vibration frequency can be adjusted from 10 to 30 Hz, with adjustable amplitude, ranging from 6″ to 40″ in the image plane.

(3) *Drives, position encoder, and control.* The telescope can rotate with variable speeds: slewing at $1\overset{\circ}{.}5$–2°/s, setting at 1′–2′/s. The tracking speed for the polar axis is 15″ $\pm 0\overset{\prime\prime}{.}2$/s. Both axes have identical worm-gear drives and encoders to read the position of the telescope axes. Rotary inductosyns of 12-inch/720-pole are used as position encoders. The outputs, $0\overset{s}{.}1$ in right ascension and 1″ in declination, from the inductosyns are shown digitally by a logic circuit. As spur gear is used as the worm gear, in order to attain the accuracy. For controlling the telescope, an on-line computer system will be used as a control unit.

* Paper presented at the IAU Third Asian-Pacific Regional Meeting, held in Kyoto, Japan, between 30 September–6 October, 1984.

(4) *Mounting*. The mount is a traditional yoke type. As the dome is a little small for the telescope, the frame must be confined to a small dimension, with the result that the tube, unable to enter the frame, leaves a somewhat large blind area about 17°. The Serrurier truss is a traditional one, but each end of the truss bar has its own screw and ball joint so that the length and the tilt of the truss can be adjusted.

1.2. The 1.5 m Astrometric Telescope

The 1.56 m astrometric telescope was designed for the exacting demands of determining trigonometric parallaxes of faint stars (Figure 2). The telescope has only one focus (the RC focus) with a 200 × 200 mm plate.

(1) *Optical system*. The design of the optical system was based on the requirement that it should reach a limiting magnitude of 17.5 in 15 min with a direct photograph. The main optical system consists of a concave hyperbolic mirror of 1560 mm in diameter and a secondary convex hyperbolic mirror of 530 mm in diameter, both these and all other mirrors being provided by the Xin-hu Glass Factory in 1979: The equivalent focal length of the main optical system is 15 600 mm and the scale is 13″.22/mm. The coma free field is ± 15′. Optical calculations showed that the maximum image disc within the field ± 10′ is less than 1″.

The clear apertures of two guiding telescopes are both 220 mm and their focal lengths 2860 mm, with a field of ± 1°. That of the finding telescope is 100 mm, with a focal

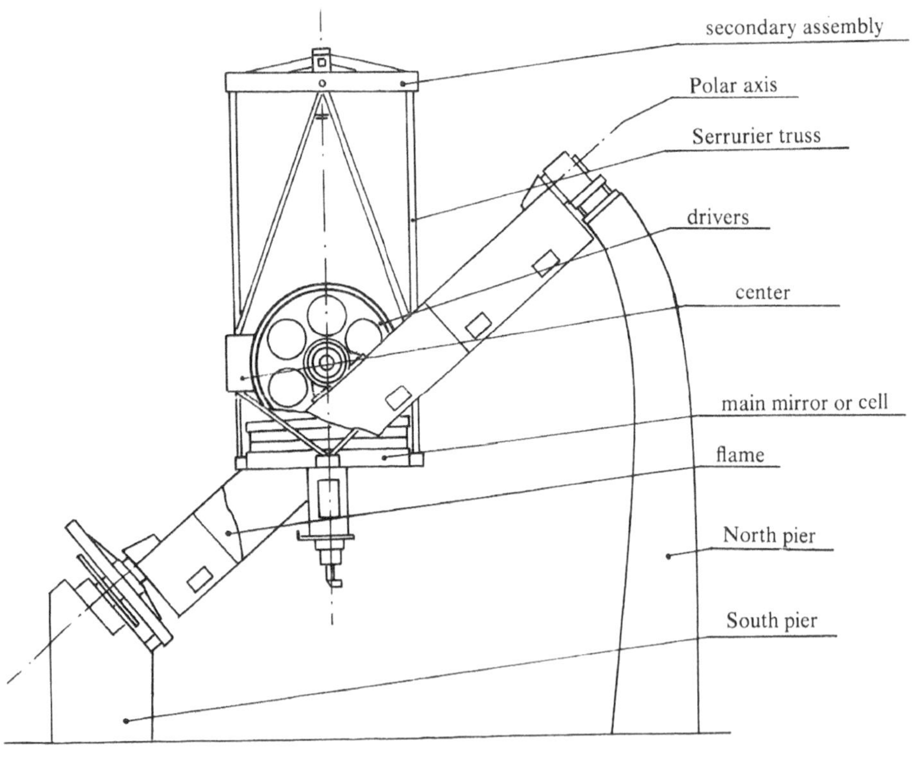

Fig. 1. Side view of the 1.26 m infrared telescope.

length of 800 mm, and a field of view of $\pm 1\overset{\circ}{.}5$. An off-set guiding system has been devised.

(2) *Mechanical assembly*. The mounting is a combination of fork and yoke. It is formed by welding two short fork arms on the upper plane of the yoke frame. This mounting exploits the advantages of the fork for freedom from a blind polar region, and of the yoke for rigidity. The northern and southern bearings of the frame are hydraulic oil pads. They have hemispherical shape, and are 1 m in diameter.

The structure of the telescope is that of a truss of Serrurier type. The primary mirror is supported in both axial and radial directions by an 18 lever counterweight mechanical system. The secondary mirror is supported in the axial direction by atmospheric pressure. Its edge support is provided by a mercury tube. A multiple exposure camera was also designed. It contains four plates, and can transport plates automatically.

For the automatic focusing device two VO2 glass rods are provided so as to keep the distance between the vertexes of the primary and secondary mirrors within the permitted tolerance. A lens is fixed on the center of the secondary mirror. It images an illuminated slit, mounted at the upper end of one rod, to a CCD element fixed on the other rod. The signal from the CCD is amplified to drive a stepping motor to return the secondary mirror to the optimum position.

The total weight rotating on the polar axis is about 32 tons. As the hydraulic oil pads are used for the bearings of the polar axis, it is possible to use a single worm wheel for both slewing and tracking in right ascension. The slewing, setting, and guiding rates are, respectively, 90°/min, 1°/min, and 1′/min. Variable tracking ranging up to $\pm 5\%$ from the sidereal rate is provided. To prevent any accident there is an overload safe device in the drive system.

(3) *Control system*. The high speed motor and several resolvers constitute a digital speed system to realize the slewing, setting, guiding, and various sidereal rates. According to the angular distance between the star and the optical axis of the telescope, the various rates of the drive system will be selected automatically in the setting process until the telescope is pointed to the star; then the telescope will track at the sidereal rate automatically. In order to prevent impacts resulting from sudden speed changes there

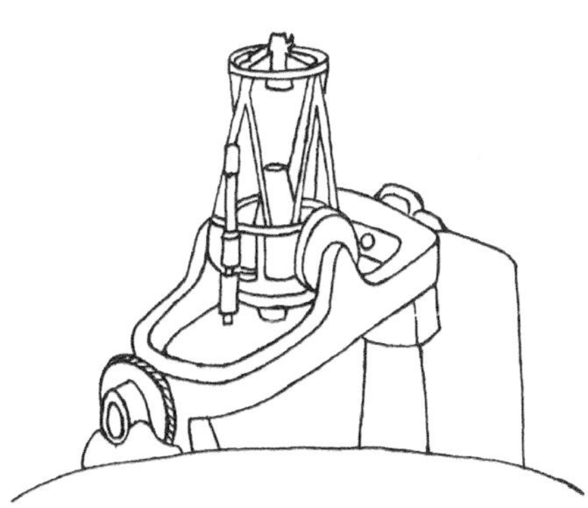

Fig. 2. 1.56 m astrometric telescope.

is a smoothing logic in the control circuit. The telescope is also provided with a photoelectric autoguide system and the dome and wind screen are also driven by servo systems.

1.3. The 2.16 m reflector

The 2.16 m telescope, being built now in NAIF, will be installed at Xin-Long station of the Beijing Observatory. At the early stage of operation, the instrument available will include a camera for direct photography at the Cassegrain focus. This will be completed in the factory, together with the telescope, in 1986, and later the grating spectrograph

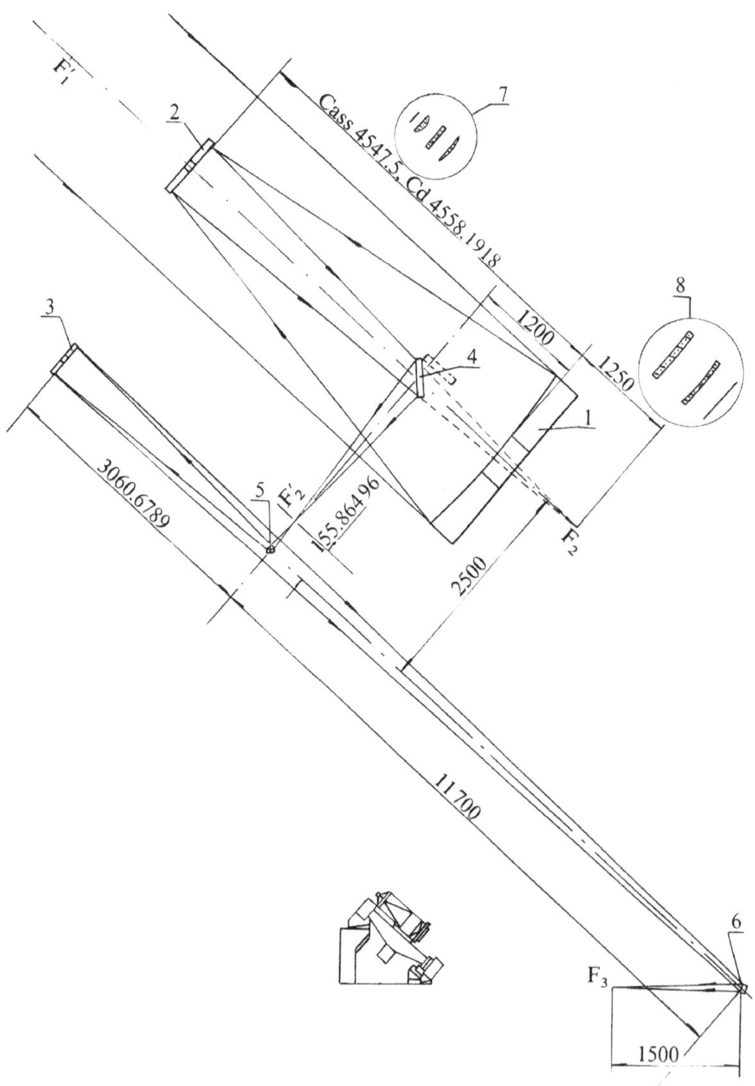

Fig. 3. Optical system of 2.16 m reflector.

for the Coudé focus. The Beijing Observatory will equip it with photoelectric photometers and a medium dispersion spectrograph.

(1) *Optical system.* The telescope has three foci, i.e., prime, Cassegrain, and Coudé focus. The Cassegrain focus is an RC system with $\frac{1}{9}$ focal ratio. With the same secondary and a concave relay mirror, the RC system is transferred to a Coudé system with *f*/45 (Figure 3). The RC system is equipped with field corrector of optics. The corrected flat field of view is 53′. In order to provide suitable wavelength coverage, two designs have been worked out for the prime focus system, though only one of these designs will be accepted.

The primary mirror and all other reflecting optical elements are made of the same VO2 glass provided by the factory as mentioned above. The aspect ratio of the primary mirror is 1 : 7.3. It is supported from the back by a mechânical floating lever system in 18 sockets acting both axially and radially. Thc sccondary mirror is also supported by a mechanical lever system. It is expected that 75% light energy concentrates within 0″.5 and 100% within 1″.2 in the optical system.

(2) *Guiding.* An off-set guiding device is equipped to the RC focus. Two refracting telescopes are mounted on either side of the main tubes. One of them is equipped with a video-unit, so that the observed area may be displayed on a TV screen in the control room.

(3) *Mounting, driving, and servo system.* The telescope uses the cross axis mounting of asymmetrical type. The polar shaft is supported by hydraulic pads. The main driving wheel for the polar shaft is a spur gear having 720 teeth and 1.95 m diameter pitch circle. That is meshed with two pinion gears locating on the east and west sides of the spur gear in an anti-backlash arrangement. The spur gear is very accurate. Its long-period error is very small, and the tooth-tooth arithmatic mean error is only 0″.39.

A torque motor drives the polar shaft with a reduction ratio of 720. The servo system can work over a wide speed range for slewing, slowing, tracking and guiding. The angular readout of polar axis is realized by using four channel rotary transformers in phase-regulated module to convert the analog signal to digital and finally displayed in degrees coarse and fine readings. The resolution is 1″.5. Besides, the servo system of the telescope also has the functions of automatic setting for observing a scheduled object and automatic setting for two objects alternately within the range of 1°.

2. Radio Telescopes

2.1. THE 13.7 m mm WAVELENGTH TELESCOPE

The 13.7 m mm wavelength telescope, being built cooperatively now by the Purple Mountain Observatory and NAIF, is a general, multi-waveband and multi-function mm wavelength radio telescope. The surface accuracy of its antenna is 0.13 mm, r.m.s. The pointing and tracking accuracy both is 3″, r.m.s., for each axis. The antenna is covered by a radome, 20.7 m in diameter, with air circulation installation to keep the temperature difference around the dish surface to less than 2°.

The first system stage of the project is equipped with quasi-optical feed system and radio receiving systems for 18–26 GHz and 80–115 GHz. The noise of the system is 1000–1800 K in a common receiving system; a low noise temperature receiving system of 500–700 K is strived for. Subsequently, an acoustic optical spectrograph with 1024 channels of bandwidth 40 MHz and a resolving power of 1 MHz and 0.25 MHz will be provided. The whole observational operation will be controlled by a PDP 11/24 computer, while a PDP 11/44 will be used for real-time data processing.

The telescope is planned to be completed in 1987 and will be installed at Dlingha station, Qinhai province in the northwest of China where the yearly average water vapor is 5 mm. It is 3200 m above sea level and its latitude is $37°15'$ N.

2.2. VLBI PROJECT

An experimental VLBI system was established in 1979 at the Shanghai Observatory. The first trans-Eurasian VLBI experiment using one element of the system, a 6 m radio-telescope, in cooperation with 100 m radio-telescope in Effelsberg, West Germany, was successfully carried out in November, 1981. The experiment lasted 54 hr, and 14 radio-sources were observed.

Presently, the Shanghai Observatory is engaged in constructing a VLBI station equipped with a 25 m radio-telescope for building up the VLBI network in China itself. The Shanghai VLBI station is the first one of the network. Another two sites are envisaged to situated at Kumming in the southwest of China and at Urumchi in the northwest of China.

The main VLBI facilities in the station are the 25 m radio-telescope, VLBI MK II and MK III data acquisition system (on order from U.S.A.), hydrogen masers and S/X band capability for compatibility in international cooperation for geodetic use. An MK II processor, of our own, is expected to become operational in 1986 mostly for domestic baselines.

2.3. THE METER-WAVE APERTURE SYNTHESIS TELESCOPE

The meter-wave aperture synthesis telescope is to be reported separately in detail by Qiu Yu-chai of the Beijing Observatory at this meeting. Whoever is interested in this work, please read Qiu's paper.

Acknowledgements

It is a pleasure to thank N. S. Hu and Z. Q. Yao of the NAIF, N. H. Zu of the Shanghai Observatory, J. H. Pan, K. R. Bao, and Z. M. Hua of the NAIF, Z. C. Xu and P. Y. Xu of the Purple Mountain Observatory, T. S. Wan of the Shanghai Observatory, and Y. H. Qiu of the Beijing Observatory for supplying material on the telescopes used in this paper.

THE EAST-SIBERIAN COMPLEX OF SibIZMIR SOLAR OBSERVATORIES

G. YA. SMOLKOV, V. E. STEPANOV, V. M. GRIGORYEV, and V. G. BANIN
SibIZMIR, Irkutsk, U.S.S.R.

(Received 8 March, 1985)

Abstract. The Sayan Solar Observatory, the Tunkinskaya Radiophysical Observatory, and the Baikal Astrophysical Observatory of SibIZMIR have been described. Some results of the investigations carried out over the past years have been expounded. Clarifying solar activity mechanisms and the nature of solar magnetic fields, active regions and flares as well as solar wind are their main problems. In the interest of these investigations the Institute has initiated and developed laboratory-scale simulation of plasma processes.

1. Introduction

Solar physics research that was begun in Irkutsk in 1958 in connection with the IGY, and placed on a broad footing after the organization of the Siberian Institute of Terrestrial Magnetism, Ionosphere and Radio Wave Propagation (SibIZMIR) in the Siberian Division of the U.S.S.R. Academy of Sciences, is currently being carried out at the facilities set up by the Institute in East Siberia, namely the Sayan Solar Observatory (SSO), the Tunkinskaya Radioastrophysical Observatory (TRAO) and the Baikal Astrophysical Observatory (BAO) (Stepanov *et al.*, 1981). They are equipped with up-to-date optical and radio telescopes designed to monitor solar activity using virtually all possible ground-based methods (Smolkov, 1982). Their major instruments, as well as most of the other facilities, have been devised and built at the Institute, while the commercially available telescopes have been upgraded significantly. These observatories serve as observational bases for a large body of SibIZMIR solar researchers. The investigations pursue a variety of challenging problems posed by solar physics and the physics of solar-terrestrial relationships, namely those of clarifying solar activity mechanisms and the nature of solar magnetic fields, active regions and flares, and solar wind. In the interests of these investigations, the Institute has also initiated and developed laboratory-scale simulation of plasma processes.

The present report is devoted to a description of the SibIZMIR solar observatories and to an exposition of some results of the investigations carried out over the past years.

2. The Sayan Solar Observatory (SSO)

The SSO is situated at an altitude of 2007 m above sea level in the East-Sayan Mountains. The observations made there are primarily directed at comprehensive

* Paper presented at the IAU Third Asian-Pacific Regional Meeting, held in Kyoto, Japan, between 30 September–6 October, 1984.

Astrophysics and Space Science **118** (1986) 21–30.

investigations of solar active region magnetic fields, the corona and the solar wind (Smolkov, 1966). The SSO is equipped with the following instruments.

(1) A horizontal solar telescope ($D = 800$ mm, $F = 20$ m) (Osak *et al.*, 1979) including a diffraction spectrograph and an echelle spectrograph, a vector magnetograph (Stepanov *et al.*, 1975; Grigoryev *et al.*, 1984) and a high-resolution panoramic magnetograph of longitudinal field (Klochek, 1975) (Figure 1). Tracking of the Sun, image quality control, selection of the time instant and execution of exposures and scanning of the target being observed, are all automatized. Data are obtained on a real time basis.

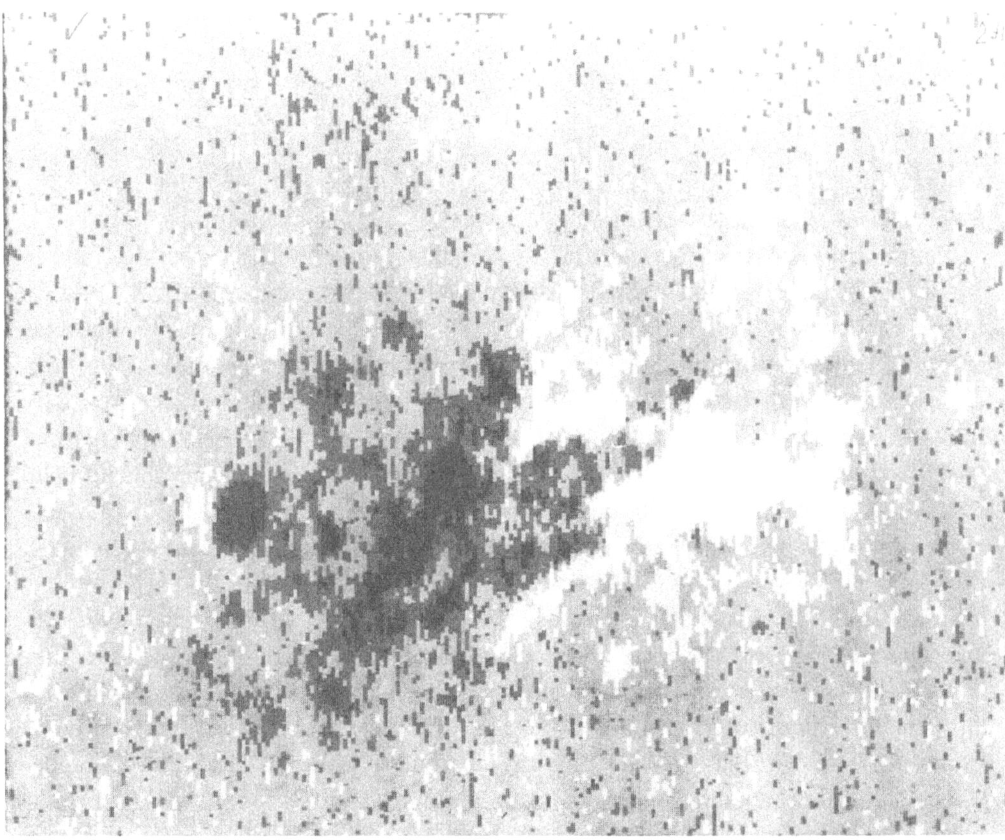

Fig. 1. Panoramic magnetogram of 13 February, 1981 active region N 11 W 28 in the λ5253 Å line of Fe I, $2 \times 4''$ resolution.

(2) A horizontal telescope ($D = 500$ mm, $F = 17$ m) including a spectrograph for auxiliary observations.

(3) A horizontal telescope for operative predictions ($D = 200$ mm, $F = 5$ m), designed to evaluate the total magnetic field of the Sun as a star, to record distributions of background magnetic fields (Figure 2) and large-scale line-of-sight velocities, and to study solar rotation and global pulsations (Grigoryev *et al.*, 1981).

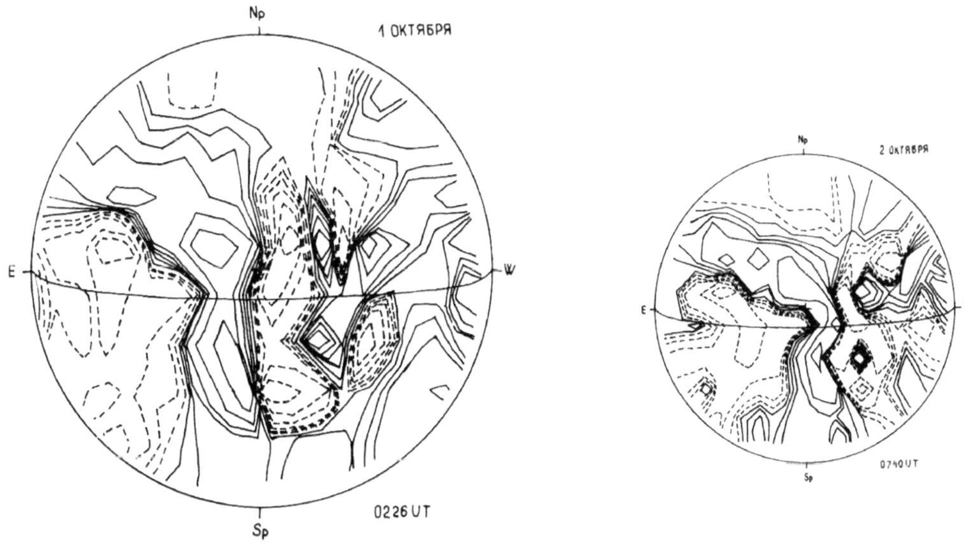

Fig. 2. Magnetogram of background magnetic fields of 1 and 2 October, 1982, 3′ resolution.

(4) A large coronograph (D = 535 mm, F = 8 m) with a Coudé focus (Nikolsky and Sazanov, 1976) for spectroscopic studies of the corona and chromospheric spicules, as well as of chromospheric structure in the 10830 Å line of He I with the aid of an electronic image converter.

(5) A similar coronagraph equipped with a birefringent filter in the Hα line and a narrow-band filter tuned in the Hβ and Ba II λ 4554 Å lines (Aleksandrovich *et al.*, 1975). It is intended to study the structure and dynamics of the active chromosphere and quasi-periodic oscillations in the active and quiet chromosphere.

(6) A chromospheric telescope (D = 180 mm, F = 5 m) for patrol observations of the whole disk of the Sun in the Hα line with the aid of a birefringent filter.

(7) A cosmic-ray spectrograph, a complex of neutron monitors situated at different altitudes above sea level (Dvornikov *et al.*, 1983).

Global characteristics of solar magnetic fields. Over the past few years, particular emphasis has been placed on the study of the total magnetic field of the Sun and background field distribution. The mean field strength of the Sun as a star is 0.5 g s, and the main period of its variation is 27^d. Also, its quasi-periodic variations, with a period of about 60 min have been found. The same periodicity is exhibited also by the variation in the magnetic flux of active regions (Grigoryev *et al.*, 1983a). Regular observations of background magnetic fields are being carried out with the purpose of investigating evolutionary properties of the poloidal and toroidal magnetic fields in the photosphere over the cycle of solar activity (Grigoryev *et al.*, 1983b). We have detected nonuniform structure of the toroidal magnetic field, and its association with Hall and anti-Hall boundaries of the large-scale photospheric field. We have also observed the appearance of a toroidal field of opposite polarity at high latitudes at the epoch of maximum activity (Grigoryev *et al.*, 1982). The detection of two modes of rigid-body rotation of background fields with periods of 27^d and 30^d led to a suggestion about the existence of two systems of magnetic field in the convection zone of the Sun (Grigoryev and Peshcherov, 1983).

Magnetic fields of active regions (AR). Extensive studies of the structure and dynamics of the magnetic field during all stages of AR development have traced in detail the picture of field emergence from below the photospheric layers (Bappu *et al.*, 1968). Following the formation of the main sunspots, field emergence proceeds impulsively during several hours. The observable net effect has the form of an oscillation of the total magnetic flux with a quasi-period of a few hours (Grigoryev and Ermakova, 1976). The character of dissipation of the AR magnetic field is reminiscent of a process opposite to field emergence, that seems to be accompanied by the submergence of strong-field ropes into subphotospheric layers (Ermakova, 1984). Certain relationships of flare activity with longitudinal field zero line structure in AR have been found (Grigoryev *et al.*, 1981).

Physics of sunspots. Spectrophotometric investigations of the dynamics of the chromosphere above the sunspot umbra in the H and K Ca II lines have shown that umbral flashes represent a particular dynamic process occurring locally (Turova *et al.*, 1983) in an optically thick medium (Firstova, 1980; Turova, 1983) and covering a greater part of the chromospheric thickness of the sunspot umbra (Turova, 1983). The difficulties in accounting for the phase relationships between the intensity and velocity, as well as for the complicated shapes of line profiles in terms of propagating or disappearing oscillations (Turova *et al.*, 1983; Teplitskaya *et al.*, 1983) led to an exotic supposition of UF as being the intrusion of material from below the visible umbral layers into the chromosphere. The nature of UF is intimately related to umbral dots (Teplitskaya and Turova, 1983). In order to create a theory of UF and form umbral models, observational estimations have been made of the height of formation of the H and K Ca II lines (Teplitskaya and Grigoryeva, 1984).

Corona. A study has been made of kinematic and dynamic properties of the corona. Coronal rotation, as determined from spectroscopic observations, differs substantially from the rotation of bright coronal features (tracers) associated with AR. The heliographical latitudes 35–40° show a drastic decrease in coronal rotation velocity associated with zonal W–E flows (Stepanov and Tyagun, 1976). We have determined that nonthermal velocities in the corona increase at the latitudes 50–60°, and also found some other regularities in the behaviour with latitude of electron density and equivalent line width (Stepanov and Tyagun, 1977). It is planned to construct a coronal polarimeter.

Solar wind. A model has been worked out for quasi-stationary high-speed streams, based on the configuration effect of coronal magnetic fields, which determine the boundary conditions of formation and outflow of the solar wind (Kovalenko, 1982). A study of the evolution of the large-scale structure of the solar wind velocity field of the heliosphere has resulted in the detection of a relationship between the global velocity of solar wind and the level of solar activity.

3. The Tunkinskaya Radioastrophysical Observatory

To diagnose the state of solar activity and to study the structure and development of active regions and flares with high two-dimensional angular resolution in the radio range, the TRAO has been equipped with the recently constructed Siberian Solar Radio Telescope (SSRT) (Smolkov *et al.*, 1983), a 256-element cross-shaped interferometer. It consists of linear 128-antenna arrays that are oriented in the E–W and N–S directions. Each parabolic dish is 2.5 m in diameter. The spacing between neighboring antennae (4.9 m) corresponds to the radio diameter of the Sun. The base length of the arrays is 622 m. The angular resolution at the zenith is about 20″ × 20″. The working wavelength is 5.2 cm. AR and flares are most pronounced around this wavelength on the average. The SSRT design took into account working wavelengths of other very large radio telescopes. Scanned radio images of the Sun are synthesized consecutively on a real time-scale as the Sun intersects the interference lobes of the beam pattern. This is achieved by means of a multi-frequency parallel recording of the unpolarized and circularly polarized emission. The N–S line has been put into operation and adjustment work on the E–W line is currently underway (Smolkov *et al.*, 1984).

Even in this stage of SSRT phase adjustment with respect to the Sun, we have obtained new data on spatio-temporal properties of the development of AR and flares (Smolkov *et al.*, 1983). These data reveal substantial variability of the slowly-varying component of the solar radio emission and, as flares are in progress, also of the structure

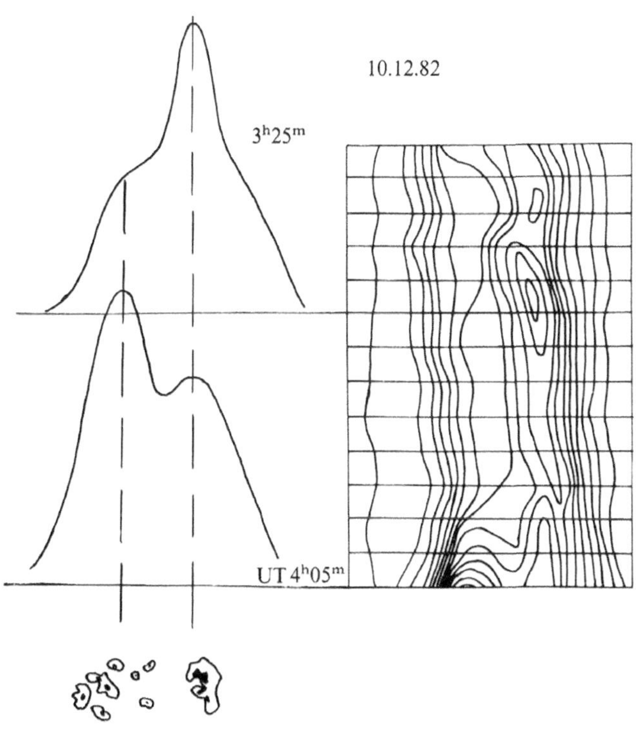

(a)

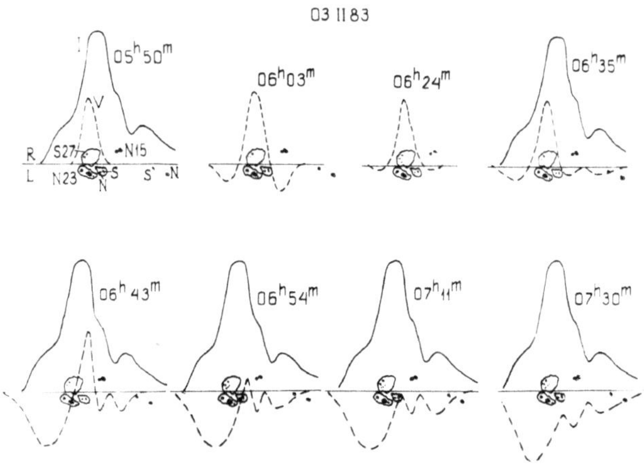

(b)

Fig. 3. (a) Variability of radio brightness distribution of microwave emission above a sunspot group on 10 December, 1982. (b) Variation in distribution, structure and of circular polarization sign of AR microwave emission in the course of development of 3 February, 1983 flare.

of local sources of microwave emission from active regions (Figure 3) (Smolkov *et al.*, 1983, 1984).

For the purposes of obtaining information about manifestations of solar activity in other ranges, the SSRT has been equipped with a photoheliograph, a radio polarimeter (λ = 5 cm), a small-baseline radio interferometer (SBRI, λ = 3.5 cm) and a radio spectrograph (70–140 MHz). It is planned to increase the number of auxiliary instruments. Investigations on the oscillations of active region microwave emission characteristics revealed that: (a) the fluctuation spectrum is discrete in character, (b) quasi-periodic components with periods of 3.5 and 5–10 min tend to persist irrespective of the state of solar activity, and (c) the enhancement of these components has a train-like character. The production of quasi-periodic fluctuations has been attributed to the back reaction of transient gas-dynamic processes brought about by the emergence of sunspot magnetic fields on the temperature plateau in the atmosphere of active regions (Zandanov and Uralov, 1983).

Theoretical investigations are aimed to elucidate the mechanisms for the radio emission of flares, the nature of the flares themselves and the association of gamma- and radio-emission of flares with the escape of protons into interplanetary space (Mazur and Stepanov, 1984; Zaitsev and Stepanov, 1983).

4. The Baikal Astrophysical Observatory

The main research area at the BAO is the study of solar activity manifestations in the photosphere and chromosphere with a resolution which should be the maximum attainable on the ground. For that purpose, the BAO is equipped with a Large Solar Vacuum Telescope (BSVT, the diameter of the achromatic objective lens is 760 mm, focal length 40 m, resolution $\sim 0\overset{''}{.}2$, and tracking with the aid of a siderostat 1500 mm in diameter) (Stepanov *et al.*, 1979), a photoheliograph (D = 180 mm, F_{equ} = 9 m) and a chromospheric telescope (full disk) (D = 180 mm), F_{equ} = 5 m, resolution ~1″). Under construction now is also a chromospheric telescope for large-scale imaging (D = 265 mm, F_{equ} = 12 m, resolution ~0″.5–1″.0).

The high quality of chromospheric pictures in the Hα line made it possible to investigate the structure and evolution of activity complexes (AC) as well as their association with flare processes. A study of an AC, which produced three proton flares in May 1981, has established the following. (i) The AC was a large-scale feature of quasi-regular shape that was pronounced against the ambient background. (ii) Its structure was like a sunspot of δ-configuration. (iii) The overall 'penumbra' identified from finestructure elements, covered the 'umbra', which was made up of 'cores' and 'hills' of opposite polarity to the magnetic field (of the sunspots and flocculi) (Figure 4) (Banin, 1984).

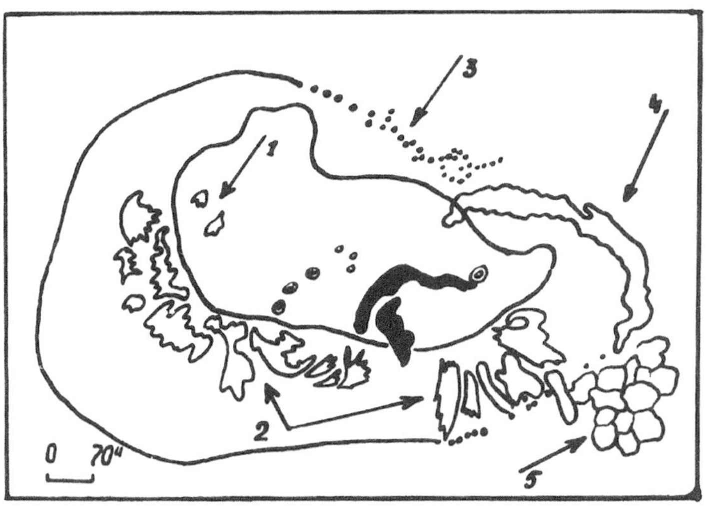

Fig. 4. Structural features in the activity complex as observed in the wing of Hα. The boundaries of AC 'umbra' and 'penumbra' are shown by solid lines; circles in the 'umbra' represent sunspots: (1) 'remnants' of cells in the 'umbra'; (2) enhanced cells; (3) cluster of dark points; (4) 'substrate' of filament; (5) cells in the quiet region; full shading: location of the flare on 16 May, 1981.

Large flares develop in the 'umbra' and were accompanied by noticeable activations of a significant part or, possibly, the entire AC. Therefore, the flares are most probably the product of the entire AC, rather than a separate AR (Banin, 1983, 1984). Activations within the entire activity complex and even in the adjacent regions evolve, mainly, along the polarity inversion line (PIL) of the longitudinal component of the magnetic field (Banin, 1983; Banin *et al.*, 1983). Therefore, the PIL system including the AC, may reflect a closed electromagnetic configuration in the solar atmosphere, and flares may reflect local manifestations of its disturbances (Banin, 1983). In the vicinity of the flare of 13 May, 1981 there were cells of size of the order of a supergranule, or multiple to

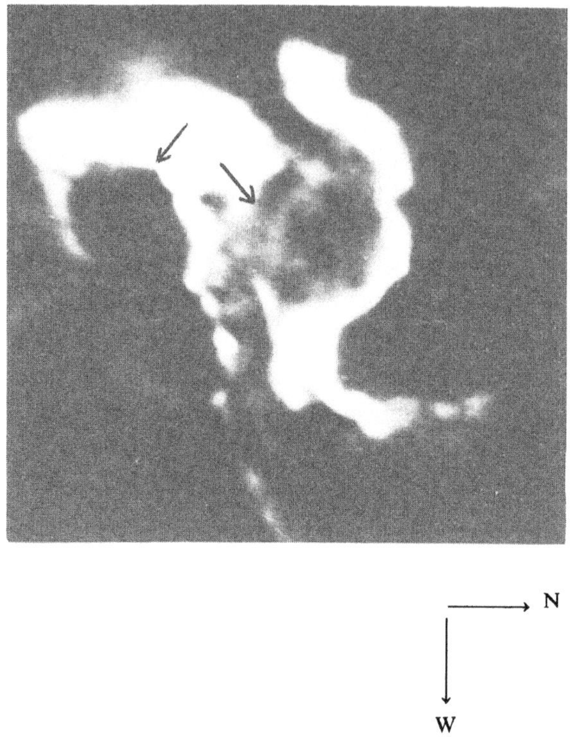

Fig. 5. A general view of 13 May, 1981 flare. Dark cells are seen – the flare does not penetrate into them.

this quantity. These represent a so called 'forbidden zone' for flares: bright ribbons stop at their boundaries, or skirt them (Figure 5) (Banin, 1984).

5. Laboratory-Scale Simulation

With laboratory devices, it is comparatively easy to attain plasma parameters inherent to flare regions on the Sun. Within the framework of a current model of the flare, the spatial scale size characterizing the energy release is represented by the current sheet thickness. The power released in the experiment, agrees satisfactorily with that of a flare. The best agreement with observations is provided by experiments with a turbulent current sheet (Altyntsev *et al.*, 1977, 1979, 1981, 1982, 1983). The evolution of current ropes produced by reconnection within a neutral sheet, has been studied experimentally (Altyntsev *et al.*, 1983). Spatially localized, directed fluxes of energetic electrons trapped by the magnetic field of a current rope, were found.

The principal problems of laboratory experiments are related to the study of a possible long-lasting and effective energy release, as well as of the influence of heat abstraction mechanisms on the effectiveness of dissipative processes.

6. Summary

Thus, the SibIZMIR solar observatories have accumulated a rather extensive experience of investigations of magnetic fields, line-of-sight velocities, various oscillatory processes of AR, solar flares and other nonstationary phenomena at all levels of the solar atmosphere. On this bases, some regularities of the development of solar activity phenomena have been defined more exactly and models for active features proposed. With the commissioning of the SSRT and BVST, the level of research will, undoubtedly, be increased in the next years. The SibIZMIR observatories have actively and successfully participated in a number of international programs, in the interests of the further development of solar physics and solar-terrestrial physics. They are interested in taking part in cooperative investigations together with other observatories located in the relevant region. Advantageous prerequisites in this regard are the coincidence of scientific objectives, overlapping or contiguous observing intervals as well as mutually beneficial supplements with wavelength ranges, methods and kinds of observations. We have already gained some experience of coordinated observations using vector-magnetographs with the Okayama observatory, as well as of joint investigations on active region birth with the Kodaikonal observatory.

References

Aleksandrovich, S. V., Domyshev, G. N., Korovkin, A. I., Sadokhin, V. P., and Skomorovsky, V. I.: 1975, *Novaya tekhnika v astronomii*, Nauka, Leningrad, p. 34.

Altyntsev, A. T., Banin, V. G., Kuklin, G. V., and Tomozov, V. M.: 1982, *Solar Flares*, Nauka, Moscow.

Altyntsev, A. T., Bardakov, V. M., and Krasov, V. I.: 1981, *ZhETF* **81**, 3(9), 901.

Altyntsev, A. T., Krasov, V. I., Lebedev, N. V., and Paperny, V. Z.: 1979, *J. Physique*, Colloque C7, Suppl. No. 7, Vol. 40, pp. C7–723.

Altyntsev, A. T., Krasov, V. I., Lebedev, N. V., Paperny, V. Z., and Simonov, V. G.: 1983, 'Electron Heating and Acceleration in "Magnetic Island" of a Neutral Sheet', *Proc. 16th Intern. Conf. on Phenomenon in Ionized Gases, Dusseldorf*, p. 340.

Altyntsev, A. T., Krasov, V. I., and Tomozov, V. M.: 1977, *Solar Phys.* **55**, 69.

Banin, V. G.: 1983, *Issled. geomagn. aeron. i fiz. Solntsa* **65**, 129.

Banin, V. G.: 1984, *Issled. geomagn. aeron. i fiz. Solntsa* **68**, 173.

Banin, V. B., Borovik, A. N., and Yazev, S. A.: 1983, *Issled. geomagn. aeron. i fiz. Solntsa* **65**, 151.

Bappu, M. K. V., Grigoryev, V. M., and Stepanov, V. E.: 1968, *Solar Phys.* **4**, 409.

Dvornikov, V. M., Krestyannikov, Yu. Ya., Sdobnov, V. E. *et al.*: 1983, *Izvestiya AN SSSR, Ser. Fiz.*, Vol. 47, 10, p. 2061.

Ermakova, L. V.: 1984, *Issled. geomagn., Aeron i fiz. Solntsa* **68**, 99.

Firstova, N. M.: 1980, *Astron. J. (USSR)* **57**, 666.

Grigoryev, V. M. and Ermakova, L. V.: 1976, *Proc. of VII Consult. Meeting on Sol. Phys. of Acad. Sci of Social. Countries*, Nauka, Moscow, p. 54.Grigoryev, V. M. and Peshcherov, V. S.: 1983, *Issled. geomagn., Aeron. i fiz. Solntsa* **65**, 3.

Grigoryev, V. M., Demidov, M. L., and Osak, B. F.: 1983, *Issled. geomagn., Aeron. i fiz. Solntsa* **65**, 13.

Grigoryev, V. M., Klochek, N. V., and Peshcherov, V. S.: 1981, *Issled. geomagn., Aeron. i fiz. Solntsa* **56**, 95.

Grigoryev, V. M., Kobanov, N. I., Osak, B. F., Selivanov, V. L., and Stepanov, V. E.: 1984, *Proc. of Huntsville Workshop.*

Grigoryev, V. M., Osak, B. F., Kobanov, N. I. *et al.*: 1981, *Issled. geomagn., Aeron. i fiz. Solntsa* **56**, 129.

Grigoryev, V. M., Peshcherov, V. S., and Demidov, M. L.: 1982, in W. Fricke and G. Teleki (eds.), *Sun and Planetary System*, D. Reidel Publ. Co., Dordrecht, p. 119.

Grigoryev, V. M., Peshcherov, V. S., and Osak, B. F.: 1983, *Issled. geomagn., Aeron. i fiz. Solntsa* **64**, 80.
Klochek, N. V., Grigoryev, V. M., and Kruglov, V. I.: 1975, *Novaya tekhnika v astronomii*, Nauka, Leningrad, p. 25.
Kovalenko, V. A.: 1982, *Solar Phys.* **73**, 383.
Nikolsky, G. M. and Sazanov, A. A.: 1976, *Astron. J. (USSR)* **43**, 868.
Osak, B. F., Grigoryev, V. M., Kruglov, V. I., and Skomorovsky, V. I.: 1979, *Novaya tekhnika v astronomii*, Nauka, Leningrad, p. 84.
Smolkov, G. Ya.: 1966, *Issled. geomagn. i aeron.*, Nauka, Moscow, p. 189.
Smolkov, G. Ya.: 1982, in W. Fricke and G. Teleki (eds.), *Sun and Planetary System*, D. Reidel Publ. Co., Dordrecht, Holland, p. 123.
Smolkov, G. Ya., Treskov, T. A., Krissinel, B. B., and Potapov, N. N.: 1983, *Issled. geomagn., Aeron. i fiz. Solntsa* **64**, 130.
Smolkov, G. Ya., Treskov, T. A., and Potapov, N. N.: 1983, *Issled. geomagn., Aeron. i fiz. Solntsa* **65**, 204.
Smolkov, G. Ya., Pistolkors, A. A., Treskov, T. A., Krissinel, B. B., Putilov, V. A., and Potapov, N. N.: 1985, 'The Siberian Solar Radio Telescope: Parameters and Principle of Operation, Objectives and Results of First Observations of Spatio-Temporal Properties of Development of Active Regions and Flares', Report at Third Asian-Pacific Regional Meeting of the IAU, 30 September – 6 October, 1984, Kyoto, Japan, *Astrophys. Space Sci.*, in press.
Stepanov, V. E., Banin, V. G., and Kruglov, V. I. *et al.*: 1979, *Novaya tekhnika v astronomii* **6**, 42.
Stepanov, V. E., Grigoryev, V. M., Kobanov, N. I., and Osak, B. F.: 1975, *Issled. geomagn., Aeron. i fiz. Solntsa* **37**, 147.
Stepanov, V. E., Smolkov, G. Ya., Kuklin, G. V. *et al.*: 1981, *Issled. geomagn., Aeron. i fiz. Solntsa* **56**, 76.
Stepanov, V. E. and Tyagun, N. F.: 1976, *IAU Symp.* **71**, 101.
Stepanov, V. E. and Tyagun, N. F.: 1977, *Issled. geomagn., Aeron. i fiz. Solntsa* **42**, 60.
Teplitskaya, R. B. and Grigoryeva, S. A.: 1984, *Issled. geomagn., Aeron. i fiz. Solntsa* **68**, 12.
Teplitskaya, R. B. and Turova, I. P.: 1983, *Solnechn. Dann.* **8**, 102.
Teplitskaya, R. B., Turova, I. P., and Kuklin, G. V.: 1983, *Publ. Debrecen Heliophys. Obs.* **5**, 267.
Turova, I. P.: 1983, *Issled. geomagn., Aeron. i fiz. Solntsa* **65**, 86.
Turova, I. P., Teplitskaya, R. B., and Kuklin, G. V.: 1983, *Solar Phys.* **87**, 7.
Zaitsev, V. V. and Stepanov, A. V.: 1983, *Solar Phys.* **88**, 297.
Zandanov, V. G. and Uralov, A. M.: 1983, *Issled. geomagn., Aeron. i fiz. Solntsa* **65**, 97, 107.

RECENT DEVELOPMENTS WITH THE ANGLO-AUSTRALIAN TELESCOPE*

DONALD C. MORTON
Anglo-Australian Observatory, Epping, NSW, Australia

(Received 2 July, 1985)

Abstract. This report describes the instruments currently available on the 3.9 m Anglo-Australian Telescope, with particular reference to the recent additions of optical fibres for multi-object spectroscopy, a low resolution spectrograph with a red-sensitive charge-coupled device and software for obtaining infrared images by scanning the telescope in synchronism with the recording of data from the InSb photometer.

The Anglo-Australian Telescope (AAT) has been operating now for more than nine years. It is funded equally by Australia and the United Kingdom to provide a major optical telescope for astronomers of both countries. It is situated at an altitude of 1165 m on the Siding Spring site of the Australian National University. This location near Coonabarabran in the state of New South Wales, Australia is about 450 km NW of Sydney and is very dark. Grubb Parsons in England figured the 3.9 m Cervit mirror and provided the telescope tube, while Mitsubishi of Japan manufactured the equatorial mount and the drive and control systems. The telescope is operated by the Anglo-Australian Observatory which also has a laboratory in the Sydney suburb of Epping on the grounds of the Commonwealth Scientific and Industrial Research Organization (CSIRO) Division of Radiophysics.

A wide range of instruments are regularly available to users of the AAT, including Boksenberg's image photon counting system (IPCS). It is used mainly at the *f*/8 Cassegrain focus on the Royal Greenwich Observatory (RGO) spectrograph which provides a remarkable range of dispersions from 2 to 156 Å mm^{-1}. For infrared observations we have a sensitive indium antimonide photovoltaic photometer built by John Barton of our staff. It can be used with a circular variable filter to give a spectral resolution of about 100. For prime focus imagery we have an RCA Charge-Coupled Device (CCD) built by the RGO, as well as photography, for which David Malin continues to find new ways to improve the techniques. Soon we expect to add a low noise General Electric Corporation (GEC) CCD to the RGO spectrograph and an infrared spectrometer with 16 InSb detectors and Fabry–Pérot etalons giving resolutions up to 60000.

Now I shall describe in more detail three recent developments in our instrumentation.

* Paper presented at the IAU Third Asian-Pacific Regional Meeting, held in Kyoto, Japan, between 30 September–6 October, 1984.

Astrophysics and Space Science **118** (1986) 31–33.

1. Optical Fibres

Peter Gray of he AAO staff has been remarkably successful in developing the capability of the AAT for multi-object spectroscopy by using optical fibres. Since 1982 we have been feeding the RGO spectrograph with a line of fibres simulating the entrance slit. Initially the aperture plate, where the fibres picked up the individual objects, was at the 12 arc min square auxiliary $f/8$ Cassegrain focus, but recently we have been able to use the full 40 arc min diameter Ritchey–Chrétien field of the AAT, by inserting a spacer box between the acquisition and guidance unit and the spectrograph. The Ritchey–Chrétien design was intended for wide field $f/8$ photography, but now it has an even more important application for multi-object spectroscopy. The present aperture plates are about 36 cm in diameter and are drilled by a milling machine which is numerically-controlled from a punched paper tape prepared with our VAX computer. A special $f/6$ collimator is fitted in the spectrograph to compensate for the beam spreading of the fibres. With the present system it is possible to change from one field of fifty objects to another in less than 5 min.

2. Faint Object Red Spectrograph

Under the direction of Peter Gillingham we have built an extremely efficient low-resolution spectrograph that mounts underneath the RGO spectrograph so that the two instruments can share the same slit. The collimatorless optical design by C. G. Wynne uses a grism and an aspheric corrector in the diverging beam, and a camera mirror which focuses the spectrum from about 5200 to 10 200 Å on a GEC CCD giving about 20 Å resolution. The read-out noise of 6.7 electrons r.m.s. and excellent charge-transfer characteristics bring this CCD very close to the regime of photo-event counting. A dichroic beamsplitter can be inserted after the slit to permit simultaneous observation of the blue light with the IPCS on the RGO spectrograph. A fibre feed also is possible with the CCD spectrograph. Recently we observed one hundred objects at once, with fifty fibres going to each spectrograph.

3. Infrared Images

Jeremy Bailey has further developed the outstanding capability of the AAT and its computer control for precise pointing, tracking and raster scanning. He has synchronised the raster scanning with the data acquisition by the IR photometer. Since this instrument has excellent DC stability, it is possible to obtain images in various infrared bands from 1.2 to 4.8 μm without frequent reference to standard sources.

For example, David Allen has obtained images of Uranus which demonstrate how easy it is to study the detailed shape of its rings from the ground at 2.2 μm where gaseous methane strongly absorbs the light from the planet itself. His observation of Venus at 2.4 μm, with John Crawford, discovered clouds on the dark side, which appear to have a rotation period of about 5.4 days compared with 3.9 days for the UV clouds seen on the bright side.

Images also have been obtained of the Galactic Centre in several IR bands resolving the central source IRS16 into three components. The middle one has CO absorption like a K giant, while those to the north–east and south–west emit Brackett γ, like H II regions. The stellar objects in the IR images have provided a link to the radio maps. At present the non-thermal radio point source appears to be offset almost 1 arc sec to the north–west of the centre of IRS16, though there are uncertainties yet in tying together the radio and optical reference frames.

GALACTIC AND EXTRAGALACTIC STUDIES WITH THE KISO SCHMIDT*

BUNSHIRO TAKASE

Kokugakuin University, Shibuya, Tokyo, Japan

(Received 15 May, 1985)

Abstract. On 1 October, 1974 the 105 cm Schmidt-telescope was dedicated at the Kiso Observatory. Since then it has been used for several galactic and extragalactic studies. The main works are described here together with some data on the telescope.

1. Basic Data on the Telescope

The apertures of the corrector plate and main mirror are 105 cm and 150 cm, respectively. The focal length of the system is 330 cm, so its focal ratio is 3.1, and the plate scale is $62\farcm5$ mm^{-1}.

Two objective prisms with vertex angles of 2° and 4° are available. The dispersions at Hγ and the A band of the former are 800 and 3800 Å mm^{-1}, while those of the latter 170 and 1000 Å mm^{-1}, respectively. The estimated limiting magnitude of the direct photograph is about 21 mag. in B. Those of the 2° and 4° prism spectral photographs are 18 and 16 mag., respectively, when the spectra are taken unwidened.

A general description of the telescope has been given by Takase *et al.* (1977).

2. Galactic Studies

2.1. Space Distribution of Red Giants and the Galactic Structure (Mikami *et al.*, 1982; Ishida, 1984)

Objective prism survey data for red giants in the Scutum region ($l = 26°–27°$, $|b| \leqq 2°$) were combined with infrared observation data (K mag. and $H - K$ colour of 469 point sources, and integrated brightness distribution in 2.4 μm) and were analyzed to determine a model of the space distribution of stars, and the stellar contents contributing to the near-infrared radiation of the Galaxy.

2.2. Star Count in the North Galactic Pole Region in the *UBV* Color Bands (Stobie, *et al.*, 1985)

Data on 19000 stars in the *UBV* bands ($12 < V < 19$), covering 24 sq. deg of the NGP region were used for counting, and a photometric reduction was made to obtain two-colour diagrams for each apparent magnitude, from which parameters of the disk and spherical components of the Galaxy were derived.

* Paper presented at the IAU Third Asian-Pacific Regional Meeting, held in Kyoto, Japan, between 30 September–6 October, 1984.

2.3. A survey of northern Bok globules and the Mon OB1/R1 association for Hα emission stars (Ogura and Hasegawa, 1983; Ogura, 1984)

An objective prism survey of 15 Bok globules has revealed 60 Hα emission stars in their vicinities, of which 80% are new. At least a half of them can be young stars associated with the globules. In the Mon OB1/R1 association 135 new Hα emission stars were detected, most of which are presumably of the T Tau type.

2.4. A study of carbon stars in the Cassiopeia region (Maehara, 1985)

From an objective prism survey for C stars with Kodak F-type plates, their subclasses can be determined. In the Cassiopeia region ($l = 115° - 130°$, $|b| \leqq 5°$), the result is that the space number density for C4–C9 stars is 200 kpc^{-3} (with the ratio of 1:0.5:<0.1 for C4–C5, C6–C7, and C8–C9, respectively), which is ~1/15 of that estimated for M giants.

2.5. A search for ultraviolet-excess objects (Noguchi *et al.*,1980, 1982; Kondo *et al.*, 1984)

From UGR three-image photographs, 588 (first catalogue) and 598 (second catalogue) UV-excess stellar objects down to 18.5 mag. were detected in each ~600 sq. deg area. According to the identifications using other catalogues, and to the follow-up spectroscopic observations, white dwarfs, subdwarfs, cataclysmic variables, and faint blue galaxies or quasars are the main constituents of these objects. A proper motion determination was made for the objects in two pilot fields.

2.6. Surface photometry of simple H II regions (Mizuno *et al.*, 1981; Mizuno, 1982; Nakano *et al.*, 1983)

Photometry in Hα and *V* bands has been made for several tens of Sharpless H II regions (which are illuminated by a single exciting star) to study their structures and to seek for any index of their evolutionary stage.

3. Extragalatic Studies

3.1. Digital surface photometry of galaxies toward a quantitative classification (Watanabe *et al.*, 1982, 1985; Kodaira *et al.*, 1983; Okamura *et al.*, 1984; Takase *et al.*, 1984)

From homogeneous photographic data of 261 galaxies ($E \sim$ Im) in the Virgo and UMa clusters, several photometric parameters (V_{26}, D_{26}, SB, and $XI(P)$ – which represent integrated magnitude, diameter, surface brightness and an index of light concentration within the isophote of $V = 26$ mag. per sq. arc sec, respectively) were derived. Correlations between these parameters and the galaxy type (T) are found to be useful for a quantitative classification of galaxies from their photometric data and for the theory of their evolution.

3.2. Surface photometry of edge-on galaxies (Hamabe, 1982)

Photometric data of nine selected edge-on galaxies ($S0^- \sim$ Scd) were analysed to discuss the global structure of disk galaxies on the basis of a three-dimensional model analysis.

3.3. A study of ultraviolet-excess galaxies based on the Kiso survey (Takase, 1980; Takase *et al.*, 1983; Takase and Miyauchi-Isobe, 1984)

So far ~ 1500 sq. deg areas have been surveyed on the UGR three-image photographs and ~ 3000 UV-excess galaxies have been detected. Their distributions in the general fields and clusters, statistics for their morphological types and degree of UV-excess have also been studied. Now follow-up spectroscopic and radio observations of these galaxies are under way.

3.4. A study of poor clusters of galaxies (Yamagata and Maehara, 1985)

Luminosity functions and number density distributions for sixteen poor clusters of galaxies have been obtained, applying a new method of discriminating images of galaxies from those of stars.

References

Hamabe, M.: 1982, *Publ. Astron. Soc. Japan* **34**, 423, and references given therein.

Ishida, K.: 1984, in M. Capaccioli (ed.) 'Astronomy with Schmidt-type Telescopes', *IAU Coll.* **75**, 257.

Kodaira, K., Okamura, S., and Watanabe, M.: *Astrophys. J.* **274**, L49.

Kondo, M., Noguchi, T., and Maehara, H.: 1984, *Ann Tokyo Astron. Obs., 2nd Ser.* **20**, 130.

Maehara, H.: 1985, *Astrophys. Space Sci.*, in press.

Mikami, T., Ishida, K., Hamajima, K., and Kawara, K.: 1982, *Publ. Astron. Soc. Japan* **34**, 223.

Mizuno, S.: 1982, *Astrophys. Space Sci.* **87**,121.

Mizuno, S., Sakka, T., Sasaki, T., and Kogure, T.: 1981, *Astrophys. Space Sci.* **78**, 235.

Nakano, M., Kogure, T., Mizuno, S., Sakka, K., Sasaki, T., and Wiramihardja, S.D.: 1983, *Astrophys. Space Sci.* **89**, 407.

Noguchi, T., Maehara, H., and Kondo, M.: 1980, *Ann. Tokyo Astron. Obs., 2nd Ser.* **18**, 55.

Noguchi, T., Yutani, M., and Maehara, H.: 1982, *Publ. Astron. Soc. Japan* **34**, 407.

Ogura, K.: 1984, *Publ. Astron. Soc. Japan* **36**, 139.

Ogura, K. and Hasegawa, T.: 1983, *Publ. Astron. Soc. Japan* **35**, 299.

Okamura, S., Kodaira, K., and Watanabe, M.: 1984, *Astrophys. J.* **274**, L49.

Stobie, R.S., Ishida, K., Yoshii, Y., and McGillivray, H.T., 1985, *Astrophys. Space Sci.* **118**, 419.

Takase, B.: 1980, *Publ. Astron. Soc. Japan* **32**, 605.

Takase, B., Ishida, K., Shimizu, M., Maehara, H., Hamajima, K., Noguchi, T., and Ohashi, M.: 1977, *Ann. Tokyo Astron. Obs., 2nd Ser.* **16**, 74.

Takase, B., Kodaira, K., and Okamura, S. (eds): 1984, *An Atlas of Selected Galaxies with Illustrations of Photometric Analyses*, Univ. Tokyo P., Tokyo and VNU Science P., Utrecht, p. 19.

Takase, B. and Miyauchi-Isobe, N.: 1984, *Ann. Tokyo Astron. Obs., 2nd Ser.* **19**, 595

Takase, B., Noguchi, T., and Maehara, H.: 1983, *Ann. Tokyo Astron. Obs., 2nd Ser.* **19**, 440.

Watanabe, M., Kodaira, K., and Okamura, S.: 1982 *Astrophys. J., Suppl.* **50**, 1.

Watanabe, M., Kodaira, K., and Okamura, S.: 1985, *Astrophys. J.* **292**, 72.

Yamagata, T. and Maehara, H.: 1985, *Astrophys. Space Sci.* **118**, 459.

COOPERATION IN ASTRONOMY BETWEEN INDONESIA AND JAPAN*

T. KOGURE

Department of Astronomy, Faculty of Science, University of Kyoto, Japan

(Received 15 May, 1985)

Abstract. The cooperation programme in astronomy between Indonesia and Japan is presented as an example of a well-coordinated bilateral cooperation between Asian countries. This cooperation has started in 1979, in the fields of galactic astronomy and stellar physics. The form of the cooperation is principally the exchange of persons between both countries for the purposes of collaborative works in some specific fields, the training of young astronomers, and technical cooperation. The main subjects of collaborative works are: (1) survey of red giants in the galactic center region, (2) survey of Hα-emission stars in some areas of the sky, (3) theoretical study of the dynamical evolution of stellar systems, (4) photoelectric observations of southern close binaries, (5) X-ray binaries and pulsars, (6) Be stars, and (7) other topics. The organization and main results of these works are presented, along with a future aspect of the cooperation.

1. Introduction

The close relationship between Indonesia and Japan in the fields of galactic astronomy and stellar physics became realized in about 1975 after the construction of two Schmidt telescopes in Japan. They are the 40/70/120-cm Schmidt at the Ouda Station of the University of Kyoto and the 105/150/330-cm Schmidt at the Kiso Observatory of Tokyo Astronomical Observatory. At the Bosscha Observatory in Indonesia, the 50/70/127-cm Schmidt telescope was constructed in 1960, and since then, observational works by this telescope have been accumulated.

Based on the support of JSPS (Japan Society for the Promotion of Science) and DGHE (Directorate General for Higher Education, Ministry of Science and Education, Indonesia) the cooperation programme has started in 1979 under the title of *Spectroscopic and Photometric Study of the Galaxy* for the first three-year programme. The second three-year programme in 1982–1984 was extended to include stellar physics and adopted the general title of *Galactic Structure and Variable Stars.*

The form of cooperation is principally the exchange of persons between both countries for the purposes of carrying out collaborative works in some specific fields, training of young astronomers, and of technical cooperation. The number of astronomers mutually sent during the two three-year programmes are 20 (short period < 2 months) and 4 (long period 2 to 12 months) from Japan to Indonesia, and 9 (short period) and 6 (long period) from Indonesia to Japan. For the training of young astronomers, a number of lectures and colloquia has been organized in both countries, besides of some fellowships for

* Paper presented at the IAU Third Asian-Pacific Regional Meeting, held in Kyoto, Japan, between 30 September–6 October, 1984.

longer study being afforded from the Ministry of Education, Science and Culture of Japan, or from other sources.

In the following sections we present main activities in the respective fields of collaborative works, and an aspect of future cooperation.

2. Galactic Astronomy

2.1. Red giant survey in the galactic center region

Red giant stars which are luminous and sufficiently numerous in the galaxy can serve as an effective probe of the galactic center region, especially when observed in near-infrared spectral regions. K. Ishida, K. Hamajima, T. Ichikawa, B. Hidayat, and M. Raharto organized an extensive survey programme using the Schmidt telescope of the Bosscha Observatory with the 6° objective prism and R and I band photography. The observations have been carried out for the region $l = 350° - 10°$ and $|b| \leqq 2°$ in 1979–1981, and several thousands of faint red stars were detected. Spectral classification and measurements of position, magnitude, and colours have been made for these stars. The main results of this work are as follows.

(a) The 2.4 μm infrared enhancement in the region $l = 355°$ and $b = -1°$ was explained as a characteristic of the galactic window (Hamajima *et al.*, 1981, 1982).

(b) A map of interstellar extinction up to about 5 kpc toward the galactic center region was derived (Ichikawa *et al.*, 1982a).

(c) Red giant stars detected in the survey were compiled to a catalogue of M-type stars by Ichikawa *et al.* (1982b) and Raharto *et al.* (1983).

(d) Two large dark cloud complexes were found in the region $l = 351°$ to 1° and $|b| \leqq 2°$ (Ichikawa, 1984).

2.2. Survey of Hα-emission stars

The survey work of Hα-emission objects by the Kiso Schmidt telescope with the 4° objective prism conducted by Maehara (1981) and by Ogura (1984) have been extended to the southern sky at the Bosscha Observatory (Maehara, 1982; Ogura and Hidayat, 1986). Ogura and Hidayat (1986) have found a weak association between Bok globules and emission-line stars in their survey of 24 regions around Bok globules.

The survey of the CMa OB1/R1 complex region has been conducted by Kogure and Wiramihardja using the Kiso Schmidt telescope with the 4° objective prism and 128 Hα-emission stars have been detected in total, among which 77 were new ones. A two-colour diagram was constructed for these stars (Wiramihardja *et al.*, 1986).

2.3. Dynamical evolution of spherical stellar systems

The dynamical evolution of spherical stellar systems such as globular clusters has been a subject of the collaborative work of S. Kato, S. Inagaki, and P. Wiyanto. They have approached this problem by numerically solving the Fokker–Planck equation and examined the following two points:

(a) Evolution of two-component stellar systems, and (b) effects of escape of stars on the evolution of stellar systems. Their results are summarized by Inagaki and Wiyanto (1984) and Wiyanto *et al.* (1984).

2.4. Other activities

Ohtani (1984) has conducted a star-counting observation of the Southern Coalsack at the Bosscha Observatory. Kogure and his group have carried out surface photometry of some small H II regions at the Kiso Observatory (Nakano *et al.*, 1983; Yoshida *et al.*, 1986). The sub-beam prism system was prepared by T. Noguchi and B. Hidayat on the Bosscha Schmidt telescope. A set of interference filters was prepared for both of the Ouda and Bosscha Schmidt telescopes by T. Tsujimura, S. D. Wiramihardja, T. Kogure, and B. Hidayat. J. Ibrahim has worked on some cosmological problems.

3. Stellar Physics

3.1. Photoelectric observations of close binaries

The photoelectric observations of various types of close binary stars has long been an important subject of the research group of M. Kitamura at the Tokyo Astronomical Observatory. The extension of this work to southern close binaries has been planned and carried out by this group and Indonesian collaborators, using the 37 cm refractor of the Bosscha Observatory.

A photon-counting photoelectric photometer was built by Hartono Partoharsodjo while he stayed at Tokyo Astronomical Observatory in 1979–1980. Then adjustments and test observations were conducted by M. Kitamura, S. Nishimura, A. Yamasaki, and A. Okazaki. From Bosscha Observatory, C. Nurwendaya, H. Luthfi, and D. N. Dawanas participated in this programme.

Since 1982, extensive observations have been carried out in both countries, using the 91 cm reflector at Dodaira Station, the 91 cm reflector at the Okayama Astrophysical Observatory, and 37 cm refractor at Bosscha Observatory. Particular subjects of observations were short-period non-contact close binaries, metallic-line eclipsing binaries, and cataclysmic binaries (Yamasaki *et al.*, 1983a, b; Kitamura and Nakamura, 1983; Kitamura *et al.*, 1983). The first work to emerge from this collaboration was that of Okazaki *et al.* (1984) on DV Aqr. They derived the orbital and physical parameters of this peculiar A + K system through extensive photoelectric observations at Bosscha Observatory.

3.2. Be/Ae binaries

From a statistical study of Be and Ae binaries, Kogure (1981) has found thet they are divided into two groups of short (orbital period $P < 30^{d}$) and long ($P > 30^{d}$) period binaries. Sutantyo (1982) has considered the effects of tidal interaction on the rotational velocities of component stars, and shown that different effects for the short and long period binaries can possibly explain the above-mentioned grouping of Be and Ae binaries.

3.3. Be STARS

Spectroscopic observations of Be stars have long been a subject of the research group of Kogure and Hirata in Kyoto (Kogure and Hirata, 1982; Hirata and Kogure, 1984). By the participation of D. N. Dawanas of Bosscha Observatory to this group in 1982, observational studies of Be stars have become an important part of the cooperation. Dawanas and Hirata have analysed the IUE spectral data on ζ Tau (Dawanas and Hirata, 1983) and HD 166596 (Dawanas and Hirata, 1986).

3.4. EVOLUTION OF NEUTRON STARS AND PULSARS

Sutantyo and Wiyanto (1984) have considered the evolution of pulsars in the *BP* diagram (B = magnetic field, P = pulse period), based on the possible interaction between stellar wind matter from the non-degenerate companion and the magnetosphere of the neutron star. They showed that this interaction could explain the evolutionary track of pulsars in the *BP* diagram.

4. Future Aspect

Most of the joint works of the present cooperation programme are still in progress and further development is highly desired from both countries. When we consider the future aspect of this cooperation, the following points may be worthy to be mentioned.

(1) The important role of moderate-size telescopes has been widely recognized in survey, patrol, and photometric observations. The construction of such telescopes for observations of the southern sky is quite desirable in the future.

(2) The significance of long lasting cooperation is emphasized in both theoretical and observational work, particularly, in the observations of variable objects.

(3) The extension of bilateral cooperation to some wider areal cooperation in the Asian-Pacific region is also desirable, not only to carry out coordinated observations, but also to promote an overall development of astronomy in this area.

Acknowledgements

The author expresses his indebtedness to JSPS and DGHE, and to all participants to the present cooperation programme, particularly, to the co-organizers, Professors B. Hidayat, W. Sutantyo of Indonesia, and Professors B. Takase, M. Kitamura, and S. Kato of Japan.

References

Dawanas, D. N. and Hirata, R.: 1983, *Astrophys. Space Sci.* **99**, 139.
Dawanas, D. N. and Hirata, R.: 1986, *Publ. Astron. Soc. Japan* (in press).
Hamajima, K., Ichikawa, T., Ishida, K., Hidayat, B., and Raharto, M.: 1981, *Publ. Astron. Soc. Japan* **33**, 591.
Hamajima, K., Ishida, K., Ichikawa, T., Hidayat, B., and Raharto, M.: 1982, *Contr. Bosscha Obs.*, No. 74.
Hirata, R. and Kogure, T.: 1984, *Bull. Astron. Soc. India* **12**, 109.

Ichikawa, T., Hamajima, K., Ishida, K., Hidayat, B., and Raharto, M.: 1982a, *Publ. Astron. Soc. Japan* **34**, 231.
Ichikawa, T., Hamajima, K., Ishida, K., Hidayat, B., and Raharto, M.: 1982b, *Contr. Bosscha Obs.*, No. 75.
Ichikawa, T.: 1984, *Publ. Astron. Soc. Japan* **36**, 525.
Inagaki, S. and Wiyanto, P.: 1984, *Publ. Astron. Soc. Japan* **36**, 391.
Kitamura, M. and Nakamura, Y.: 1983, *Ann Tokyo Astron. Obs. 2nd Ser.* **19**, 413.
Kitamura, M., Nakamura, Y., and Yamasaki, A.: 1983, *Ann. Tokyo Astron. Obs. 2nd Ser.* **19**, 361.
Kogure, T.: 1981, *Publ. Astron. Soc. Japan* **33**, 399.
Kogure, T. and Hirata, R.: 1982, *Bull. Astron. Soc. India* **10**, 281.
Maehara, H.: 1981, *Proc. Schmidt Symp. 1981*, p. 130 (in Japanese).
Maehara, H.: 1982, *Contr. Bosscha Obs.*, No. 71.
Nakano, M., Kogure, T., Mizuno, S., Sakka, K., Sasaki, T. and Wiramihardja, S.D.: 1983, *Astrophys. Space Sci.* **89**, 407.
Ogura, K.: 1984, *Publ. Astron. Soc. Japan* **36**, 139.
Ogura, K. and Hidayat, B.: 1986, *Publ. Astron. Soc. Japan* (in press).
Ohtani, H.: 1984, in M. Capaccioli (ed.), 'Astronomy with Schmidt-Type Telescopes', *Proc. IAU Colloq.* **78**, 329.
Okazaki, A., Yamasaki, A., Nurwendaya, C., and Luthfi, M. H.: 1985, *Publ. Astron. Soc. Pacific* **97**, 62.
Raharto, M., Hamajima, K., Ichikawa, T., Ishida, K., and Hidayat, B.: 1983, *Ann. Tokyo Astron. Obs., 2nd Ser.* **19**, No. 4, 469.
Sutantyo, W.: 1982, private communication.
Wiramihardja, S. D., Kogure, T., Nakano, M., Yoshida, S., and Tatematsu, K.: 1986, *Proc. Third Asian-Pacific Regional Meeting, Astrophys. Space Sci.* **118** (this issue), p. 311.
Wiyanto, P., Kato, S. and Inagaki, S.: 1984, private communication.
Yamasaki, A., Okazaki, A. and Kitamura, M.: 1983a, *Publ. Astron. Soc. Japan* **35**, 131.
Yamasaki, A., Okazaki, A. and Kitamura, M.: 1983b, *Publ. Astron. Soc. Japan* **35**, 423.
Yoshida, S., Mizuno, S., Nakano, M., Kogure, T., Sakka, K., Sasaki, T., and Wiramihardja, S, D.: 1986, *Proc. Third Asian-Pacific Regional Meeting, Astrophys. Space Sci.* **119** (in press).

NEW TELESCOPES IN INDIA*

J. C. BHATTACHARYYA
Indian Institute of Astrophysics, Bangalore, India

(Received 2 July, 1985)

Abstract. The review covers the description and present status of the two optical/infrared telescopes designed and presently near fabrication in India. The 234 cm aperture optical telescope expected to be installed at Kavalur shortly has several interesting features in its control and data acquisition systems. Design features of the 122 cm Infrared Astronomical Telescope expected to be completed next year are described. The review also covers the Ooty Synthesis Radio Telescope and the Gauribidanur Low Frequency Array operating at 327 and 34 MHz, respectively, and a 10 m aperture millimeter wave telescope being built in Bangalore. Plans and preparations for still bigger systems for observations in different bands of Electromagnetic spectra are also briefly described.

For the major part of the last hundred years, observational astronomy in India was limited to the activities of two observatories at Kodaikanal and Hyderabad. But after independence in 1947, a determined thrust has been put to build up new astronomical observatories, where front line research will be possible. There have also been advances in the field of astronomical observations with balloons, rockets, and satellite-borne instruments, probing the portions of electromagnetic spectra inaccessible from the ground. The present report covers only the ground based installations, carrying out observations in those bands which reach the Earth's surface through several transparent windows in the atmosphere.

The locations of the major installations are shown in Figure 1. Concentration is in the southern part of India, partly because almost the entire sky is accessible from these locations. The southernmost location is in Kodaikanal ($+10°14'$), where optical observations were started at the turn of the century‡ and the first radio observations of the Sun from India commenced in 1952†. The observatory now concentrates on observations of the Sun and the solar system in optical bands. A 'Solar Terrestrial Physics' unit is also located here which carries out regular observations of the geomagnetic and ionospheric variations connected with solar events.

Moving northwards, the next major installation is at Ooty ($+11°23'$) where the large steerable parabolic cylinder carries on investigations of the radio sources in the meter wavelength region. This telescope was totally designed and built in India and was commissioned in 1970. The instrument operates nominally at 327 MHz, and was originally designed for radio source studies by the Lunar Occultation technique. Recently this has been linked up with seven more smaller dishes and can now function as a synthesis telescope (Swarup, 1984).

* Paper presented at the IAU Third Asian-Pacific Regional Meeting, held in Kyoto, Japan, between 30 September–6 October, 1984.

‡ Report of the Kodaikanal and Madras Observatories for 1900–1901.

† Annual Report of Kodaikanal Observatory for the year 1952.

Astrophysics and Space Science **118** (1986) 45–49.

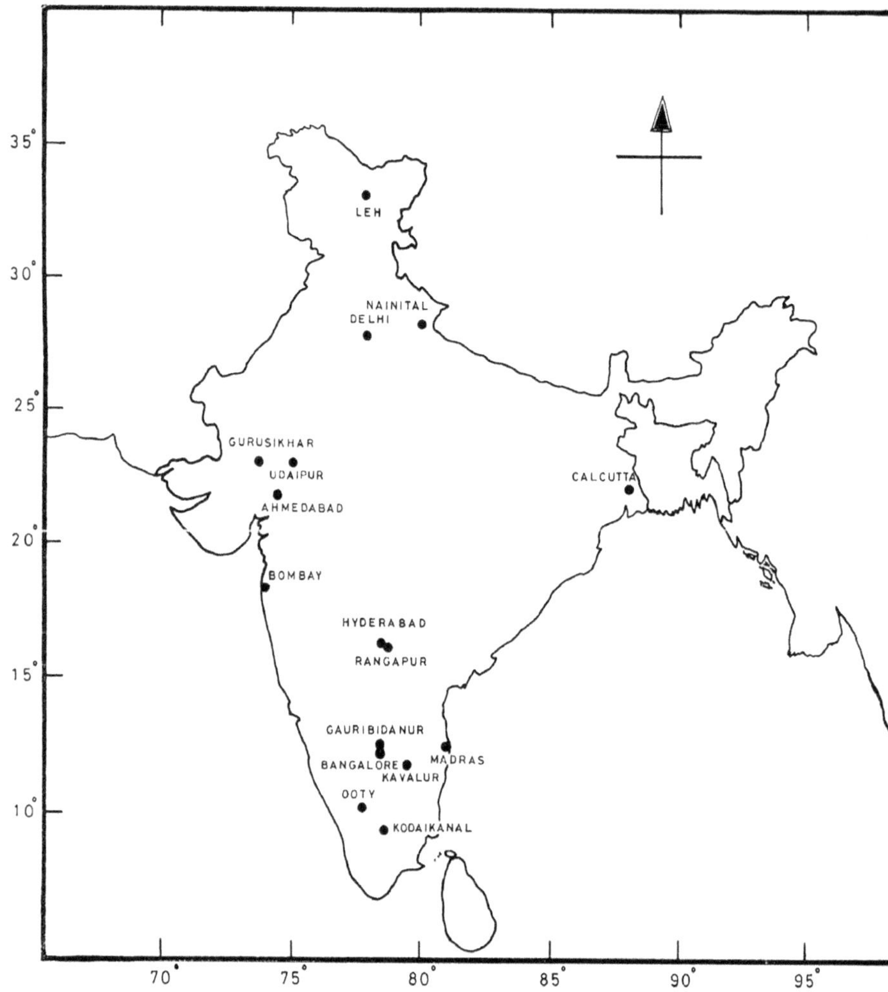

Fig. 1. Locations of the major installations.

Still further north, we come across Kavalur Observatory (+12°34′) which is the major optical observatory in India. The largest telescope at the moment is 102 cm Zeiss Ritchey–Chrétien reflector; there are two more, locally made, 75 and 38 cm reflector telescopes. In an adjacent building, installation of a 234 cm reflecting telescope has just started.

This telescope is of indigenous design and manufacture. Figure 2 shows an artist's sketch of the general arrangements of the optical elements and mounting. Equatorially mounted, with a 6 m horse-shoe as its north bearing, this is driven by large spur gears around both polar and declination axes. The driving motor set consists of two torque motors in opposition to reduce the gear backlash. Twenty bit absolute position encoders feed back position information to a set of microprocessors which control the driving rate with the help of a precision clock. A VAX 11/780 computer will exercise supervisory control over the movements, and also regulate the collection and on-line processing of data. Elaborate peripherals have been provided to receive data from two-dimensional detector arrays and process the same for a variety of experiments.

The optics has been ground and figured at the optical laboratory of the Indian Institute of Astrophysics (IIA) at Bangalore, the organization which controls both the Kodaikanal and Kavalur observatories. The optical arrangement will provide a *f*/3.25 prime, *f*/13 Cassegrain, and *f*/43.25 Coudé foci. It will immediately operate in the prime focus mode with a Wynne corrector for photography, and a photometer and image tube spectrograph. Cassegrain and Coudé instrumentation will follow in due course.

In Bangalore, besides the IIA, we have another institution, the Raman Research Institute (RRI) which has undertaken a project of constructing a millimeter wave telescope. A 10.4 cm dish has been fabricated and is presently under installation. The equipment will later be moved to a nearby location after the initial trials are over. The complete antenna and the receiver system are of indigenous design and fabrication.

100 km north of Bangalore, we come across an extended array of dipoles set up for studies of celestial radio sources in the decameter region. The array consists of 1000 dipoles and stretches in the form of a T, 1400 m in the E–W direction and 450 m in N–S direction. Studies of solar decametric bursts, and mapping of selected H II regions and extended sources in the sky have been carried out with this equipment (Sastry, 1983). This is a joint venture between the two institutions in Bangalore, IIA and RRI; several important modifications are now being carried out to refine the capabilities of the telescope.

A scheme has just been submitted for the construction of a Giant Meterwave Radio Telescope in this region. This is a very ambitious project, and when completed, will be composed of 27 large dishes located in the form of a Y in an area of 30 km across. If approved, work will start immediately, and we may have the telescope operational in the mid-nineties.

Still further north, we have the old Nizamiah Observatory at Hyderabad which is being moved to a new location in the University campus outside the city. But the observatory's main installations are at a site 60 km away at Japal–Rangapur ($+17°06'$), where a 122 cm reflector, with Nasmyth focus is regularly carrying observational programmes in variable stars. The old 10-inch astrograph which obtained all the plates of the 'Carte du Ciel' programme in the early part of this century, is also located there (Sanwal, 1975).

Following our northerly course, we will now come across two observatories almost at the same latitude in the western part of India. The first one is at Udaipur ($+24°35'$) where on an island in the midst of a lake, a small solar observatory was established in 1974. The instrument, presently in use is a solar telescope and monochromator mounted on a 3.5 m spar (Bhatnagar *et al.*, 1979). Because of the surrounding body of water, the seeing throughout the day remains quite good.

The other observatory is near the hill town of Mount Abu ($+24°36'$), a peak called Gurushikhar. It is a new site where a 1.22 m aperture telescope for IR observations will be installed by end of 1985. The design of the system has been drawn up and frozen.

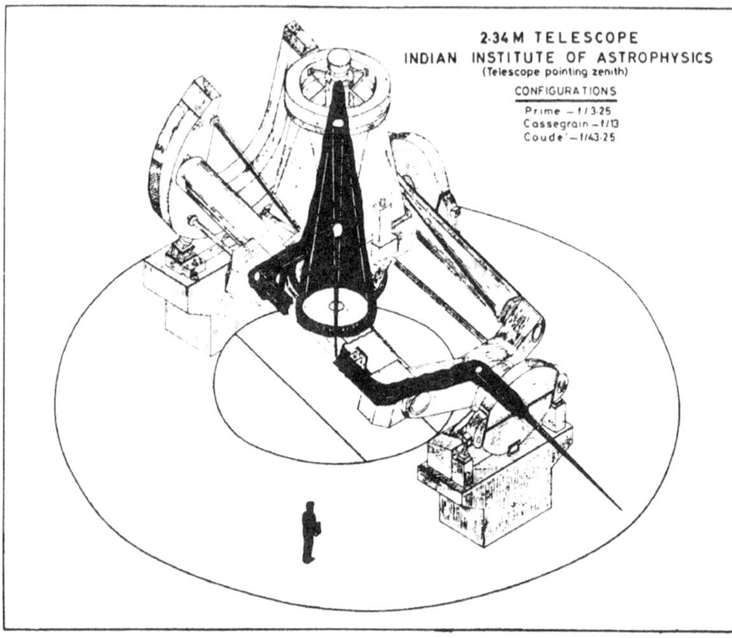

Fig. 2. General arrangements of the optical elements and mounting.

The civil works at the site are nearing completion, the mechanical mount fabrication is about to start and the optical works are well on their way to completion. This is an effort by several institutions in India and it is expected that the telescope will be completed in time.

The next optical observatory in the northern sector is in Nainital ($+29°22'$), in the foothills of the great Himalayas. The observatory was established in 1955 and now has several functioning optical telescopes. The largest one is a Zeiss 104 cm reflector, a twin of the instrument at Kavalur (Pande, 1979). There are several telescopes of the half-meter class in this observatory.

The observatory has proposed a very ambitious plan of acquiring a 4 m telescope. The details of the project are not known yet; but it is understood that they propose indigenous construction along the lines of the recently constructed 2.34 m telescope at Kavalur. To make the fabrication plan conform to existing manufacturing capabilities in the country, several changes in the design would, however, be necessary. These are now being vigorously pursued by the scientists of this observatory. Testing of several sites are going on in the nearby hills before deciding on the final location of the instrument.

Further up north, a new high altitude site is being tested thoroughly in the Ladakh region of Kashmir. The experiment will consist of the detailed recording of meteorological parameters and conducting a series of micrometeorological, seeing and transparency observations. A 50 cm aperture optical telescope has already been moved up there and parts of the photometric programmes of several observatories in India are being carried out. Humidity in that region is very low; the total precipitable water above the site is often less than 1–2 mm; the underlying idea is to locate a suitably-sized IR telescope here in future.

This sums up the current efforts in creating new major groundbased observational facilities in India. There are concentrated efforts for installing half-meter class optical telescopes in several universities where astronomy or astrophysics is being taught. The present estimate of the number of such universities is about twenty in the entire country. A totally indigenous design has been drawn up and a prototype constructed by the IIA in Bangalore. The first telescope is being installed in the campus of the Indian Institute of Science, Bangalore. A 60 cm aperture Zeiss telescope has been acquired by the Panjabi University, Patiala, which is currently in the process of installation. Several old educational institutions in the country also have small working telescopes of 15 to 25 cm aperture. Although extensively used in the past, they are now lying unused. A vigorous effort to bring them up into work is being made by the astronomical societies in the country.

References

Bhatnagar, A.: 1979, in A. Bhatnagar *et al.* (eds.), 'Solar Optical Astronomy Programmes of the Vedhshala Udaïpur Solar Observatory for SMY', *Proc. Indo–U.S. Workshop on Solar Terrestrial Physics*, Udaïpur, p. 21.

Pande, M. C.: 1979, *Twenty-five Years of Uttar Pradesh State Observatory*, Naini Tal, p. 9.

Sanwal, N. B.: 1975, in R. V. Karandikar *et al.* (eds.), 'Observations with the 8-inch Astrograph of the Nizamiah Observatory', *Optical Astronomy with Moderate-Size Telescopes*, Osmania Univ., Hyderabad, p. 107.

Sastry, Ch. V.: 1983, *Phys. News* **14**, 59, 63.

Swarup, G.: 1984, *J. Astrophys. Astron.* **5**, 139.

THE IRAQI NATIONAL ASTRONOMICAL OBSERVATORY*

HAMID M. K. AL-NAIMIY

Astronomy and Space Research Center, Jadriyagh, Baghdad, Iraq

(Received 15 May, 1985)

Abstract. Iraq is currently experiencing a praid cultural, scientific, and technical renaissance, and astronomy is a natural focus for the country's pride in the past achievements of the civilization which have flourished in Iraq. The current plans of the Space and Astronomy Research Center (SARC) include building a major observatory to work in the optical, IR and radio region of the spectrum.

The core of the optical facility will be a 3.5 m optical telescope, together with 1.25 m telescope designed for efficient performance in the IR. These telescopes will be equipped with instruments for photographic, photometric and spectroscopic observations. A 30 m dish is also being built for millimeter/radio observations.

SARC has selected an excellent observing site in the northern mountains of Iraq which has good seeing and clear dark skies. The sites selection was made with the collaboration of several leading astronomers and observatories from various countries.

1. Introduction

It is a well-known fact that in the middle ages, the world received important gifts from the Arabs, particularly from Mesopotamia. One of these gifts is the institution which we call today the 'Astronomical Observatory' – a place where groups of astronomers worked together for more than one generation, advancing common goals. In this sense, the concept of on astronomical observatory represents an invention of the Arabs and is one of their permanent contributions to human culture (Kopal, 1983).

We all know that the Babylonians were the first to name the constellations of the Zodiac, and they discovered the Saros period which they measured to be ten years and ten days.

A large number of bright stars and astronomical terms still carry their Arabic names (Abetti, 1952) the terms zenith, nadir, azimuth, alidade, almucanter are only a few examples. Names of stars such as Aldebaran, Betelgeuse, Altair, Algol are also Arabic names.

Several observatories were established in Mesopotamia, the most important are:

- Al Maamouni Observatory,
- Benou Moussa Observatory,
- Samarra Observatory.

Several great astronomers lived in Mesopotamia, one of whom Al-Battani (856–929), latinized Albategnius, carried out observations of remarkable accuracy. He noticed that the longitude of the Sun's apogee had increased by $16°43'$ since the time of Ptolemy, which implies a slow variation in the 'Equation of Time'. He determined the length of

* Paper presented at the IAU Third Asian-Pacific Regional Meeting, held in Kyoto, Japan, between 30 September–6 October, 1984

the year to be $365^d5^h46^m24^s$ (with an error of 2^m23^s).Al-Battani's work in Mesopotamia shows like a star of the first magnitude in the firmament of astronomy, illuminating a way to the future (Kopal, 1983).

With all this in the background, it is not surprising that interest in astronomy is still considered a matter of heritage in Mesopotamia. In fact, a first class observatory is being built in this area.

2. Site Selection

Several expeditions were arranged during the period 1974–1977 to select a suitable site for the observatory. These expeditions were made with help from Soviet, Egyptian, and American Astronomers. Finally, two sites were selected (Table I), and a site testing program started in both sites. The program included the monitoring of meteorological data and seeing measurements using the Polaris Trail Telescope (PTT) method developed by Prof. M. F. Walker from University of California at Santa Cruz. The programme was carried by Iraqi personnel assisted by a group from the Royal Greenwich Observatory (RGO). PTT plates were processed and evaluated by Walker.

TABLE I

	Sinjar	Korek
Altitude	1500 m	2127 m
Longitude	41°45′ E	44°28′ E
Latitude	36°28′ N	36°35′ N

TABLE II

Seeing in sec of arc	≤1″	1.1″–1.5″	1.6″–2″	>2″
Korek	18%	43%	18%	21%
Sinjar	9%	13%	15%	63%

TABLE III

Percentage of observed nights with average seeing as indicated

Location	Best seeing observed	Sec of arc				Total nights
		≤1.0	1.1 to 1.5	1.6 to 2.0	>2.0	
Junipero Serra	0″.5	26	38	13	23	558
Siero Toledo	0″.7	24	32	22	22	509
Korek	0″.5	11	29	23	37	106
Kitt Peak	0″.75	15	30	16	39	253
San Pedro Martir	0″.75	15	25	17	42	2
Piper Mountion	0″.75	9	30	20	42	164
Mt. McKinley	0″.75	2	32	20	46	50
Sinjar	0″.75	1	19	212	59	100
Flagstaff	1″.0	1	5	29	65	80

Table II shows seeing measurements for both sites (October 1977–January 1978). According to these results, Mt. Korek was selected, but site testing continued for a full year (October 1977–August 1978). Results are compared with other international sites (Table III).

3. Iraqi National Astronomical Observatory (INAO)

The INAO project is being built as a West-German joint venture, in a 'Turn-Key' contract, which includes as well civil work and a residential complex. Work started in 1981 and the last section of the project is expected to be finished in 1986. The observatory comprises the following telescopes:

- 30 m Millimeter Radio Telescope (MRT).
- 3.5 m Optical Telescope (LOT)
- 1.25 m Optical Telescope (SOT).

Other astronomical observing equipment is also included in the project.

In what follows, we give a brief description of the system:

(A) Radio telescope

The radio telescope is to be manufactured by Krupp–MAN (Ltd. – of W. Germany) and is designed along the lines of 30 m mm radio-telescope. The telescope design provides for an antenna mounted in an altazimuth system, in accordance with the turning-head principle. It is designed for operation without a radio-dome.

The *reflector surface* is to be constructed in a very special way that will minimize dead weight and thermal deformations in the panel area. The panels are constructed from an aluminium honeycomb sandwich material and are adjusted the workshop in pairs on frames which are fixed to the reflector truss-work.

The reflector panels are constructed with an r.m.s. deviation of <40 μm while the overall surface accuracy is <80 μm. This accuracy is necessary for operation in the millimetre wavelength region.

To minimize thermal effects, a thermal insulation layer is used. This forms the front enclosure of an air-circulation system around the reflector truss-work. This sytem ensures a maximum temperature difference of <1 K.

To minimize gravitational effects the homology principle is utilized, where the structure deforms under gravity at any elevation angle into another parabola, thus minimizing the effect of gravity on surface accuracy.

Working at high frequencies with large aperture puts stringent requirements on the control system, as the beam width of the telescope gets sharper. Pointing and tracking errors are less than 1 arc sec.

Extensive simulation studies proved that a conventional cascaded control system does not meet the high accuracy required, and provides insufficient damping of wind load. A more exact state controller was designed which considerably reduces pointing error.

The telescope is controlled by a VAX 11/780 computer.

(B) Receiver system

The receiver system is designed and manufactured by MBB. It consists of the following:

(1) *Front end*

The front end allow measurements in the following frequenties:

(i) 4.6 to 5.1 GHz;
(ii) 8.9 to 10.8 GHz;
(iii) 21.8 to 23.8 GHz.

Different operation modes are initiated by a switching network following the antenna feed/polarizer equipment. The switching network is controlled by the measurement control equipment.

Input signal is down-converted to 770 MHz at a bandwidth of 500 MHz for both polarizations. The IF signal is then transmitted to the control and research building.

(2) *Back End*

The back end consists of the following receivers:

(1) *The continuum polarization receiver (CPR)* processes the two polarization channels to output channels enabling the definition of the total power for each channel, and the Stokes parameters.

(ii) *The acousto-optical spectrometer (AOS)* allows spectral measurement of radio sources. The AOS allows the monitoring of 1024 channels of 500 MHz total bandwidth, with overlapping single channel bandwidths of 0.7 MHz.

(iii) *The auto-correlation spectrometer (ACS)* measures the autocorrelation function of the received signal, which is then Fourier transformed into the power spectrum. The ACS allows a maximum of 1024 channels with a channel bandwidth of 48 Hz–195 KHz selectable by the user.

(3) *Receiver System Control*

Data collection and analysis is done by a second VAX 11/780 computer.

(C) 3.5 m optical telescope (LOT)

The 3.5 m optical telescope is designed after the Max-Planck-Institute telescope at Calar Alto and will be manufactured by Zeiss. The telescope mounting is an equatorial horseshoe frame, similar to the 5 m telescope of Mt. Palomar. The telescope tube is of Serrurier type. The hour axis is driven by a single-stage spur gear at the large diameter of the horse-shoe. Eight motors are used for hour drive.

The telescope utilizes the $F/3$ prime focus, the $F/10$ Ritchey–Chrétien (RC) focus, the $F/35$ coudé focus, and an $F/35$ wobbling IR focus. It is controlled by two computers (PDP) 11/44 and PDP 11/24). It is also equipped with a 35 cm guiding telescope, a TV offset guider and TV control equipment which allow automatic guiding and remote operation at the different foci.

The telescope is provided with the following astronomical equipment:

(1) *Prime focus attachment (PFA)* is provided with a 24 × 24 cm photographic attachment which can be used with 2-lens and 3-lens correctors. Remote control and TV monitoring allows unmanned operation. A detachable cabin is also provided for manual operation.

(2) *Cassegrain spectrograph* of Boller and Chivens type 31523 is also provided, which allows dispersions in the range of 40–550 Å mm^{-1}. The spectrograph has six interchangeable gratings and is provided with an image intensification tube.

(3) *Coudé echelle spectrograph (CES)* is designed according to the white pupil principle, which allows a compact system. The spectrograph is provided with two echelle gratings 79 lines/mm and 316 lines/mm blazed at 63°.

One $F/3$ Schmidt-meniscus Cassegrain camera is provided for recording echellograms. This camera is equipped with two interchangeable cross-dispersers. In addition $F/1$ and $F/1.5$ cameras are provided for recording single-order spectra.

(D) 1.25 m OPTICAL TELESCOPE (SOT)

This telescope will also be manufactured by Zeiss and is designed after the 1.23 m telescope of Max-Planck-Institute at Calar Alto. The telescope mounting is a pole – universal type where the entire weight of the telescope floats on a spherical oil bearing. The telescope drive is controlled by two computers (PDP 11/34 – PDP 11/24).

The telescope has three foci, $F/10$ RC-Cassegrain, $F/10$ Nasmyth, and $F/35$ IR cassegrain focus. It is also provided with a 35 cm guiding telescope, TV offset guider, and TV control equipment. The telescope is provided with a fixed Nasmyth Spectrograph similar to the LOT Cassegrain Spectrograph.

(E) ADDITIONAL INSTRUMENTS

In addition to the equipment listed with each telescope, there is equipment that can be used for observations on both telescopes as follows:

(1) *UBVRI photometer* which will be used for singe-channel photometry in the five spectral bands. The photometer is equipped with two interchangeable photomultipliers for photon counting and DC-mode observations.

An LSI 11/2 computer is used for photometer control and data acquisition. The photometer has been developed by the Max-Planck-Institute at Heidelberg.

(2) *Two infrared photometer/spectrophotometers* are provided, one for the 1–5 μm spectral region and the other for the 5–14 μm region. Both photometers are identical, except for filters and detectors. Both photometers are equipped with advanced detectors and sophisticated data acquisition systems which allow background-sky limited operation.

Each photometer is cooled by two-stage closed cycle cryogenic cooling system.

Each photometer is controlled by PDP 11/23 computer which also controls the lock-in amplifier and allows automatic observation.

The photometer is provided with eight filters at selected bands and a circular variable filter (CVF) for spectrometric purposes.

The photometer control can be synchronized with telescope motion to allow automatic beam switching, raster scan and search scan.

(3) Data reduction facilities include:

(a) Perkin Elmer PDS micro 10 microdensitometer.

(b) PDP 11/44 computer.

References

Abetti, G.: 1952, *The History of Astronomy*, Henry Schuman, New York.

Kopal, Z.: 1983, in K. Goffstein (ed.), 'The Role of Astronomy in Islamic Culture', *Islamic Cultural Identity*, Nomos Verlag.

THE AUSTRALIAN RADIO-TELESCOPE*

B. J. ROBINSON

Division of Radiophysics, CSIRO, Sydney, Australia

(Received 2 July, 1985)

Abstract. A large synthesis radio telescope is under construction in Australia at a cost of $A 30.7 million. In comprises a 6 km-long 'compact array' of 22 m antenna at Culgoora and a 'long baseline array' of 319 km, achieved by adding another 22 m antenna near Coonabarabran and the existing 64 m antennae at Parkes. On occasion it will be possible to use this array with one of the deep space network antennas at Tidbinbilla to achieve a baseline of 580 km.

The arrays have been designed to provide high-resolution radio images over a wide field with high-dynamic range plus good polarization and spectral line performance. The operating frequency range is from about 327 MHz to over 50 GHz.

1. Introduction

An array of radio telescopes is being constructed in Australia to provide high-resolution radio images with a wide-field mapping capability, high dynamic range and good polarization purity. Particular emphasis has been put on spectral line performance. The projected completion date is 1988, to mark Australia's 'bicentennial year'.

The array has been subdivided into a 'compact array' of baseline 6 km and a 'long baseline array' of baseline 319 km (with extension to 580 km).

The array sizes have been chosen to provide angular resolution at radio wavelengths which will match that available, or planned, for infrared, optical, ultraviolet, and X-ray telescopes. For example, at 3 cm wavelength the synthesized beam of the 'compact array' is 1″ arc, which matches the resolution of ground-based optical telescopes such as the 3.9 m Anglo–Australian telescope close by. The long baseline array will provide angular resolution of 0.″06 arc at 13 cm wavelength, which will be complementary to the images from Space Telescope after its launch in 1986.

Higher resolution still will be made possible by:

(a) Extension of the long baseline array to Hobart, Tasmania – giving a maximum baseline of 1400 km.

(b) Operation as part of international VLBI networks, including those incorporating QUASAT – the proposed orbiting radio telescope.

2. The Compact Array

The compact array is being built at Culgoora (latitude 30° S) on the site of the CSIRO radioheliograph. It consists of six mobile 22 m antennae located on an east–west line.

* Paper presented at the IAU Third Asian-Pacific Regional Meeting, held in Kyoto, Japan, between 30 September–6 October, 1984.

Five of the antennae will be disposed at selected stations (35 in all) along a 3 km rail track. The sixth antenna will be constructed on 100 m of rail track 3 km west of the end of the 3 km rail track.

A configuration study group has established a set of station locations for 1.5, 3, and 6 km configurations that provide good 8, 12, and 16 day minimum redundancy solution made up from 4 day groupings which themselves give acceptable synthesized beams.

Tests carried out on synthetic sources with 4 day groupings predict a dynamic range of about 29 dB after CLEANing or maximum entropy restoration. These sophisticated image-processing techniques have made it possible to build a viable aperture-synthesis telescope with only a small number of antennae. Few antennae means low cost. The penalty is the many days it takes to build up an image (compared, for example to the speed of the VLA with 27 antennae). But, to exploit the richness of the southern skies, a slow radio telescope is preferable to none at all.

The six antennae of the compact array will be connected by fibre optics, carrying wide-band digital signals from the antennae to the digital correlator at the centre of the array. The local oscillator and timing signals may be distributed by fibre optics too. The operating bandwidth of the compact array is 160 MHz, which ensures that the sensitivity of the array is good despite the small antenna size. The r.m.s. noise in the synthesized maps is expected to be 15 μJy for a 4 day observation.

The frequency bands of the Australian Telescope have been chosen to contain the astrophysically most important spectral lines, in a frequency range extending from about 327 MHz to at least 50 GHz. The correlator design (to be discussed later) contains sufficient capacity to provide good spectral coverage and resolution for both galactic and extragalactic molecular line sources.

3. The Long-Baseline Array

The long-baseline array consists of one or more antennae of the compact array operating in conjunction with a seventh 22 m antenna on Siding Spring Mountain, near Coonabarabran (latitude 31°.3 S) and the existing 64 m telescope at Parkes (latitude 33° S). With this baseline of 319 km the angular resolution will be about 0".02 arc at a wavelength of 3 cm.

For part of the time the NASA 64 m telescope at Tidbinbilla (latitude 35°.5 S) will be used with the long baseline array, extending the baseline to 580 km and improving the resolution proportionately.

The possible placement of a 26 m antenna near Hobart in Tasmania (latitude 43° S) would provide a VLBI capability to the antennas of the Australian Telescope.

The antennae of the long baseline array could be linked by microwave links, fibre optics or satellite links (using AUSSAT, a national communications satellite in geostationary orbit). A choice will be made after field tests. Microwave links would be the most straightforward system, but the limited bandwidth (10 MHz per antenna per polarization) would increase the r.m.s. noise in the synthesized maps to 400 μJy.

4. Antenna Design

Detailed design of the new 22 m antennae has been carried out over the last twelve months by Sydney consulting engineers Macdonald, Wagner Pty. Ltd. Tenders for construction will be called in November, 1984.

The antennae will be fully-steerable paraboloids, alt-az mounted, with an overall diameter of 22 m. The field of view will be 40′ arc at a frequency of 1420 MHz. The design specifications require operation at 43 GHz with aperture efficiency of 50%. For the central 11 m of the antennae, operation at 115 GHz is called for, with an aperture efficiency of 30%. The design pointing precision is 12″ arc under average weather conditions.

The antenna optics will be Cassegrain. The feed horns will be mounted on a turret to bring any feed 'on-axis' for operation. Simultaneous operation will be possible at two frequencies with both linear polarizations. The feed horn design ensures good polarization performance.

5. Receivers

The dual-frequency, dual-polarization receiver systems will be housed in cryogenic packages. The following pairs will be provided, each operating from a single wide-band feed:

1.25–1.75 GHz	H and OH lines,
2.2–2.5 GHz	VLBI,
4.5–6.1 GHz	H_2CO line,
8.3–10.8 GHz	VLBI.

The following frequencies are planned for later implementation:

21.0–25.5	H_2O and NH_3 lines,
42.0–50.0 GHz	SiO and CS lines,
84.0–98.5 GHz	HCN, HCO^+, SiO, ... lines,
105.0–116.0 GHz	CO lines.

Low-noise field effect transistor (FET) amplifiers have been built and tested for 1.4 GHz (9 K amplifier temperature), 4.5 GHz (35 K), and 8.4 (40 K). Masers have been built for 22 and 43 GHz, but it is expected that FET performance at these frequencies will improve significantly before 1988.

Low-frequency receivers ($f < 1$ GHz) are also proposed. They will be FETs operating at room temperature.

The instantaneous noise bandwidth currently achievable with FET amplifiers is about 15%, which is significantly less than the bandwidth available from the feed horns (as listed above).

6. Digital Correlator

The correlator design is interferometer (or baseline) oriented. There will be four independent correlator subsystems:

(1) Compact array, line correlator.
(2) Compact array, continuum correlator.
(3) Long baseline array, line correlator.
(4) Long baseline array, continuum correlator.

The basic element of the correlator is a VLSI* silicon chip designed for the project. Each chip consists of 64 correlators arranged in an 8 × 8 matrix. Sixteen of these chips can be connected in a 32 × 32 *module*, which can accept 32 parallel inputs to lower the data rate from 320 to 10 Mbaud. Alternatively each module can provide sixteen frequency channels.

For each interferometer eight modules can be connected in a *block* which is capable of producing all the correlated products from this baseline. Each block handles two frequencies and four polarization products.

Contracts for the production of 5000 VLSI chips were placed in September, 1984.

7. Computing

(a) Software

The AIPS (Astronomical Image Processing System) software package used on the VLA has been selected for the data processing software for the Australia Telescope. The AIPS system has been used extensively for the configuration studies, and experience with it is currently being gained in the processing of images from the Molonglo Observatory synthesis telescope and from the VLA.

(b) Hardware

Computer hardware at Culgoora will consist of two VAX 11/750 machines for telescope control, calibration and initial data reduction. Image processing will take place at Epping using a VAX 11/780, two array processors and an image processor.

8. Cost

The capital cost of the Australian Telescope is $A 30.7 million (in March, 1983, dollars). It has been funded as an Australian Bicentennial Activity for completion in 1988.

How can we hope to build an array of telescopes so cheaply? Some factors in keeping the price down are:

(1) Only seven antennae are to be built. They add up to half the capital cost. The disadvantage of few antennae is that it takes a long time to synthesize a map.

(2) The antenna size has been restricted to 22 m. This keeps the cost down and increases the field of view. On the debit side, the flux sensitivity is reduced and confusion is increased.

* Very Large Scale Integrated circuit 6.5 mm square containing 46000 transistors.

(3) The cost of the new antennae has also been kept down by the design of very economic reflector panels. These have been used for resurfacing the Parkes 64 m telescope.

(4) The long baseline array uses existing 64 m diameter antennae at Parkes and Tidbinbilla, and will re-use an existing NASA antenna for the Hobart site.

(5) We have been very successful in designing a VLSI chip and a modular configuration to keep down the cost of the massive digital correlator.

(6) We have extensive experience in the design of feed horns and low-noise receivers. The FET amplifier designs are cheap, simple, stable, and reliable.

(7) The wide bandwidth (160 MHz) of the compact array and the correlator design provide: (a) bandwidth synthesis to fill in gaps in the U–V plane and (b) a wide-field mapping capability.

9. Performance

(a) COMPACT ARRAY – 6 km BASELINE

Frequency (GHz)	1.4	2.3	4.8	8.4	22	43
Resolution ($''$ arc)	7	4.2	2.0	1.1	0.5	0.2
Sensitivity (4 days)	15 μJy (for 160 MHz bandwidth)					
Simultaneous frequencies						
Field size ($'$ arc)	40	25	12	7	2.6	1.3

(b) LONG BASELINE ARRAY – CULGOORA TO TIDBINBILLA

Frequency (GHz)	2.3	(4.8)	8.4	22
Resolution ($''$ arc)	0.06	(0.04)	0.01	0.005
Sensitivity (1 day)	400 μJy (for 10 MHz bandwidth)			
Simultaneous frequencies				

The Australian Telescope will be operated by CSIRO as a national facility. Telescope time will be allocated strictly on merit.

RESULTS FROM NOBEYAMA RADIO OBSERVATORY (NRO) – A PROGRESS REPORT*

MASAKI MORIMOTO

Nobeyama Observatory Minamimaki, Nagano, Japan

(Received 12 July, 1985)

Abstract. The title telescope – in operation since 1982 – gives a surface accuracy of better than 0.2 mm (r.m.s.), with an aperture efficiency of about 25% at 115 GHz. A 5-element interferometer is at final adjustment stage. Observations of proto-stellar objects, extragalactic objects, and spectral lines are briefly described.

The 45 m telescope has been in operation since 1982. A surface accuracy of better than 0.2 mm (r.m.s.) has been achieved according to measurements using a laser ranging theodolite. An aperture efficiency of about 25% was obtained at 115 GHz, the frequency of the important spectral line of CO. In conjunction with the 16000 channel acousto-optical spectrometer (AOS) the telescope is the most powerful instrument for mm wave molecular line observations.

Though the telescope is still in the development stage, some telescope time is being opened to outside users.

A 5-element interferometer is now at the stage of final adjustment, and it is a pleasure to report here that we obtained the first radio picture from the instrument recently. Though the tests are being performed mainly at 1.3 cm, the phase stability and mechanical accuracy of the whole system measured from the test observations show that a satisfactory performance will be obtained, at least at 115 GHz, and very probably at 230 GHz, the frequency of the $J = 2$ to 0 transition of CO. The interferometer is equipped with a very large spectral correlator of a new type having 1024 complex spectral resolution elements for each interferometer pair, and it will also be a very powerful instrument for spectral mapping observations of compact molecular line sources.

In this report a brief description will be given on the scientific output from observations using the 45 m telescope. More details for each program are given in the contributed papers in the various sessions of this Regional Meeting.

* Paper presented at the IAU Third Asian-Pacific Regional Meeting, held in Kyoto, Japan, between 30 September–6 October, 1984.

Astrophysics and Space Science **118** (1986) 63–65.

1. Proto-Stellar Objects

We observed many proto-stellar objects showing bipolar outflow in molecular lines. Observation in the $J = 1$ to 0 line of CS (corresponding to a gas density of greater than 10^5 H_2 molecules/cc) show that most of observed protostars are surrounded by a rotating gas disk, similar to what is often quoted as the proto-solar system nebula. From this we feel that a dense gas disk surrounding a proto-stellar is a common phenomenon in star formation processes.

Seen in the $J = 1$ to 0 line of CO (corresponding to a less dense region of H_2 gas; density of $\sim 10^3$) the outflow of gas starts to be visible. The 15 arc sec beam of the 45 m telescope can clearly resolve the structure of gas flow near a protostar. The shape of the flow mates quite well with that of the less intense part of the CS emission – the cavity of the rotating disk of dense gas. This implies that the shape and perhaps the mechanism of the outflow of gas is governed by the existence of the disk.

The velocity of the outflow is also seen through 'blue' and 'red' shifts of the line. With the 45 m telescope the fine velocity structure is also revealed. In some cases there is also seen a gradient of radial velocity of the flow, indicating an acceleration of the flow near the protostar.

2. Extragalactic Objects

The nucleus of M82, the spiral arm region of M31 and similar sources are being observed, and interesting features are revealed. Extensive observations still await improvements of pointing accuracy, receiver noise and baseline development.

3. Spectral Line Survey

The 35–50 GHz band of Orion KL have been almost completely covered. For other sources and other bands, the survey is much less complete. Many newly detected lines were shown to be due to methyl alcohol and methyl formate.

A strong unidentified line at 45.379 GHz found in TMC-1 has peculiar properties. It is strong (2.5 K in TMC-1), not found in Ori KL, shows self-absorption in Sgr B2 and has no harmonically related counterpart in TMC-1. The line is also strong in other dark clouds, indicating that the line originates from some fundamental constituent of the dark cloud gas. Laboratory measurements and a search for the line in other types of object are being done as an effort towards identifying the line.

A new molecule, C_3O (tricarbon monoxide) was confirmed. This is the first carbon chain molecule with oxygen, and will provide an important clue on interstellar chemistry.

The 4_{-1}–3_0 E and 7_0–6_1 A lines of methyl alcohol were found to be strong maser sources in Sgr B2. Excitation study shows only collisional excitation and high abundance of methyl alcohol are necessary for maser action to occur. The 4_{-1}–3_0 line exhibits also spatially extended components of very large velocity widths. Such a property has not been found in other maser transitions.

4. Other Observing Programs

A 10 GHz Galactic continuum survey and VLBI tests at 22 GHz are also in progress, and interesting results are being provided.

SPACE ASTRONOMY IN JAPAN*

MINORU ODA

Institute of Space and Astronautical Science, Komaba, Tokyo, Japan

(Received 15 May, 1985)

Abstract. Japan's two X-ray astronomy satellites, HAKUCHO and TENMA, have been actively in operation leading to deeper understandings of the neutron star physics. Other satellite and balloon projects have been performed under international collaborators.

First, the development of space science is briefly summarized in Figure 1. Important observational events and major spacecrafts since Sputnik are tabulated for the disciplines of STP, astronomy and astrophysics, and planetary exploration. In astronomy and astrophysics a most remarkable development has been brought about by X-ray astronomy.

The history of X-ray astronomy since the first discovery of the X-ray source in 1962–1963 by sounding rocket observations are indicated in Figure 2. Since the first X-ray astronomy satellite UHURU, eight satellites have been dedicated to X-ray astronomy. It was 1979 when Japan's first X-ray astronomy satellite HAKUCHO (Cygnus or Swan in Japanese) was launched.

Japan's scientific satellites since the first launch in 1970 are tabulated in Figure 3. Among the fourteen satellites, two were for X-ray astronomy and one was for solar X-ray observation during the last solar maximum. During 1985 two Halley's comet missions are scheduled, and in 1987 a launch of the third X-ray astronomy satellite ASTRO-C is planned. Figure 4 indicates the specifications of these satellites. With these small satellites, HAKUCHO and TENMA (Pegasus in Japanese), we have focused our efforts mainly to the study of the bright compact X-ray sources.

Now we have firm evidence that the bright compact sources are close binary systems each of which consists of a neutron star (or perhaps a black hole) and its companion star: Observational studies of the X-ray sources hence provide powerful tools for the physics of compact stars. Among a number of findings, I would like to pick up a few topics of the physics of the neutron star. They are, first, the spinning magnetized neutron star which manifests itself as an X-ray pulsar and, second, the thermonuclear flash of the neutron star atmosphere which appears as an X-ray burst.

Secular variations of the pulsar period have been extensively studied for several X-ray pulsars. It has become clear that the variation is not a simple monotonic decrease of the pulsar period as originally supposed in 1970's as a consequence of spin-up of the

* Paper presented at the IAU Third Asian-Pacific Regional Meeting, held in Kyoto, Japan, between 30 September–6 October, 1984.

Astrophysics and Space Science **118** (1986) 67–70.

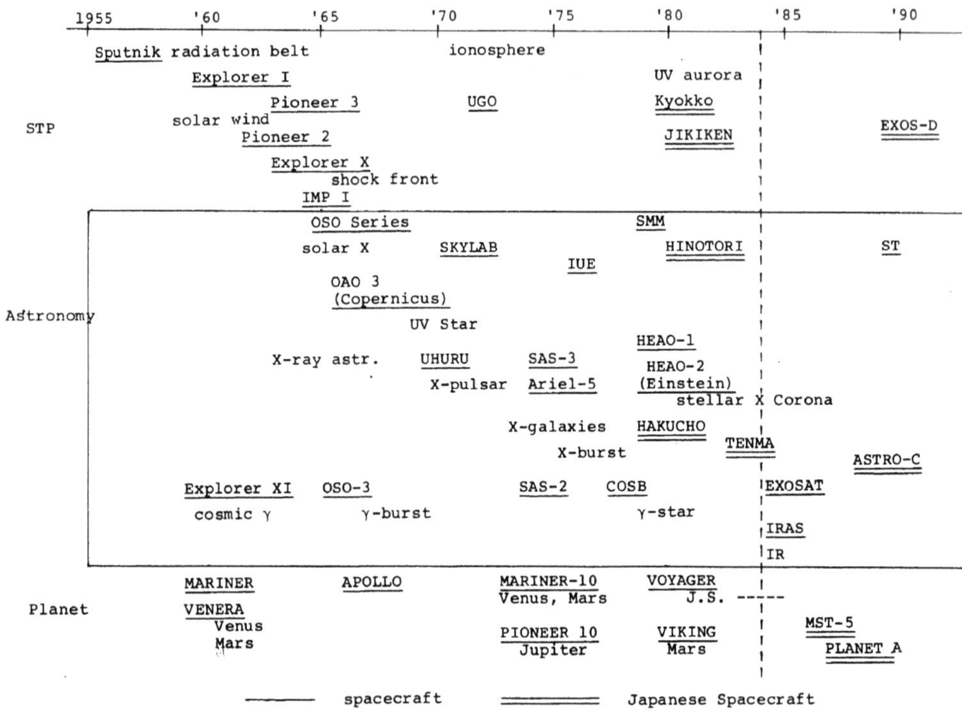
1955
'60
'65
'70
'75
'80
'85
'90
STP
Sputnik radiation belt
Explorer I
Pioneer 3
solar wind
Pioneer 2
Explorer X
shock front
IMP I
ionosphere
UGO
UV aurora
Kyokko
JIKIKEN
EXOS-D
Astronomy
OSO Series
solar X
OAO 3
(Copernicus)
UV Star
SKYLAB
IUE
SMM
HINOTORI
ST
X-ray astr.
UHURU
X-pulsar
SAS-3
Ariel-5
HEAO-1
HEAO-2
(Einstein)
stellar X Corona
X-galaxies
X-burst
HAKUCHO
TENMA
ASTRO-C
Explorer XI
cosmic γ
OSO-3
γ-burst
SAS-2
COSB
γ-star
EXOSAT
IRAS
IR
Planet
MARINER
VENERA
Venus
Mars
APOLLO
MARINER-10
Venus, Mars
PIONEER 10
Jupiter
VOYAGER
J.S.
VIKING
Mars
MST-5
PLANET A
spacecraft
Japanese Spacecraft

Fig. 1.

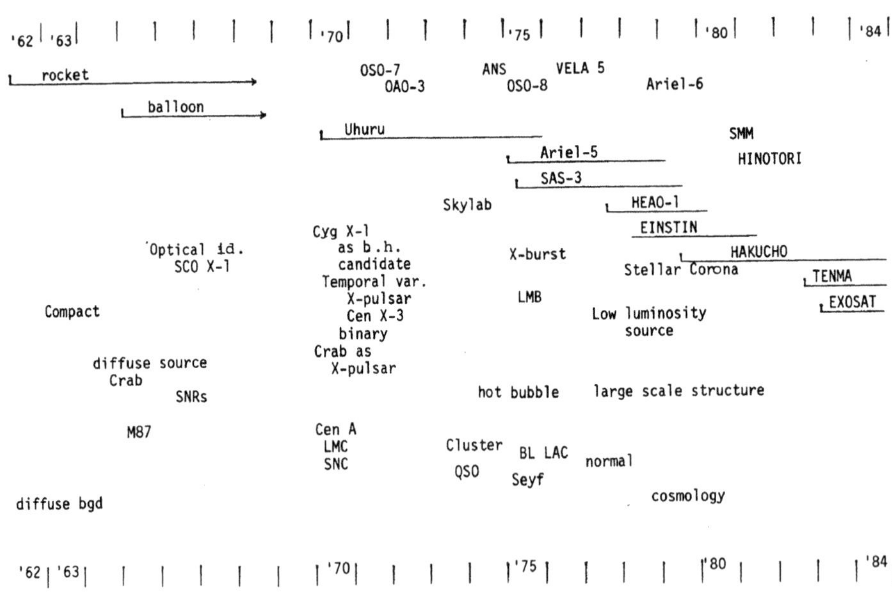
'62
'63
'70
'75
'80
'84
rocket
balloon
OSO-7
OAO-3
ANS
OSO-8
VELA 5
Ariel-6
Uhuru
Ariel-5
SAS-3
Skylab
HEAO-1
EINSTIN
HAKUCHO
SMM
HINOTORI
TENMA
EXOSAT
Optical id.
SCO X-1
Compact
Cyg X-1
as b.h.
candidate
Temporal var.
X-pulsar
Cen X-3
binary
Crab as
X-pulsar
X-burst
LMB
Stellar Corona
Low luminosity
source
diffuse source
Crab
SNRs
hot bubble
large scale structure
M87
Cen A
LMC
SNC
Cluster
QSO
BL LAC
Seyf
normal
cosmology
diffuse bgd

Fig. 2.

neutron star by the accretion flow, like a continuous whipping of a top. These long term variations and a short-term fluctuation of the period, as seen in the case of Vela X-1, are intriguing in the context of physics of the interaction between the neutron star and the accreting flow, and maybe physics of the interior of the neutron star.

As for the X-ray burst, several critical questions are raised in relation to an apparent excess of the peak luminosity over the Eddington limit, or the occasional succeeding of bursts with intervals which seem too short. Concentration of the estimated radius of sources, from a study of bursts from a number of bursters, at around 10 km confirms the source identification as a neutron star. If an absorption line at 4.5 keV of the

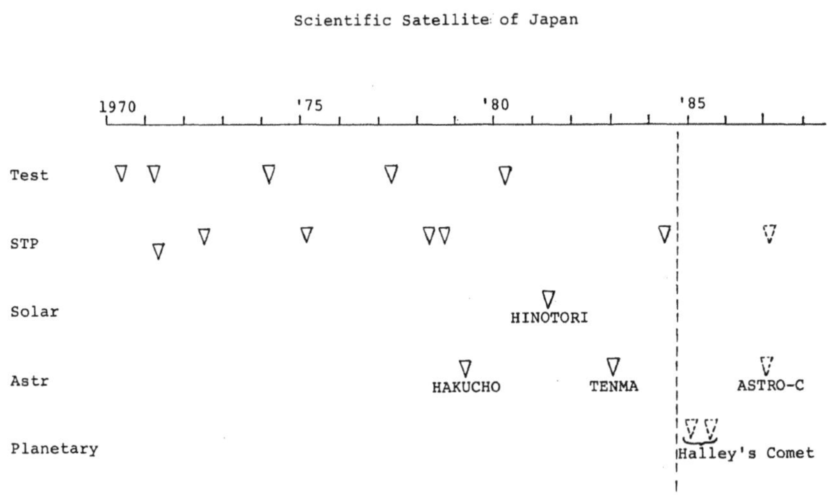

Fig. 3.

TABLE I

Space astronomy in Japan

Feb. 1979 'HAKUCHO' (Swan, Cygnus) X-ray astronomy	Temporal behavior bright galactic sources X-ray pulsar X-ray burster X-ray transient
Feb. 1981 'HINOTORI' Solar physics	Spatial structure: 5 arc sec Bragg spectroscopy: Fe-line
Feb. 1983 'TENMA' (Pegasus) X-ray astronomy	Spectroscopy bright galactic sources extragalactic sources
Scheduled Feb. 1987 ASTRO-C X-ray astronomy	 Collab. with U.K. U.S.

spectrum is granted to be redshifted from the 6.7 keV Fe line, the estimated gravitational field is consistent with what is theoretically expected at the neutron star surface.

Balloons provide a powerful means for astronomical observations. One of the key observations around 1970 which nailed down the precise location of Cygnus X-1 (most likely candidate of the black hole) was performed by balloons. A collaborative balloon project in the late 1970's by ISAS and UCSD resulted in hard X-ray images of the Crab nebula in the 10–60 keV range: a configuration of magnetized plasma, which consists of a cavity and a ring around the neutron star, is proposed.

Japan's third X-ray astronomy satellite is being produced: a part of the instrument will be contributed by our UK and US colleagues. Further international cooperations, concerning both hardware and software, are enthusiastically hoped for.

STARLAB: AN ULTRAVIOLET/OPTICAL SPACE TELESCOPE*

IAN R. TUOHY

Mount Stromlo and Siding Spring Observatories, Australian National University, Canberra, Australia

(Received 15 May, 1985)

Abstract. Mount Stromlo and Siding Spring Observatories are actively developing large format photon counting detectors suitable for operation in space over a broad wavelength range. It is intended that these detectors be used in the Scientific Instrument Package of the STARLAB telescope. A description is given of the STARLAB facility, with emphasis on the contribution being made by Australia.

1. Introduction

STARLAB is conceived as a one metre class, UV-optical space telescope intended to be placed in orbit for missions of 6–12 months duration. The principal capabilities of the proposed facility are high resolution imaging over a wide (30 arc min diameter) field of view, and spatially-resolved spectroscopy of extended objects in both long-slit and Echelle modes. The scientific rationale driving the STARLAB project is very great, and is discussed at length in the report of the Joint Science Working Group entitled 'STARLAB: An Australia–Canada–U.S.A. Orbiting Telescope'. Amongst the many astrophysical problems which would be addressed are: the early evolution of galaxies, clusters of galaxies, super-clusters, and quasars; calibration and extension of distance scales; ultraviolet surveys and stellar physics in other galaxies; spatial structure of supernova remnant shock-waves; and the physical conditions within extended plasmas associated with H II regions, planetary nebulae, supernova remnants, and galaxies.

Even though the faint-object sensitivity of STARLAB ($m_v \sim 25$) will be $\sim 10\times$ less than that of Space Telescope, STARLAB promises to be a far more efficient facility for conducting multicolour survey observations of large regions of sky (e.g., the Magellanic Clouds), and for long-slit spectroscopy of extended objects. STARLAB, therefore, both complements the capabilities of Space Telescope, while at the same time being an exceedingly powerful research tool in its own right.

The STARLAB project became a tripartite undertaking in 1980 between three countries, namely the U.S.A., Canada, and Australia. The original division of responsibility, as defined in a *Letter of Intent* between the three countries, is as follows: provision of the telescope and fine guidance system (Canada); provision of the scientific instrument package (Australia); and provision of launch, a space platform, and in-orbit operations (U.S.A.). Observing time is to be shared equally between the participating countries. This report presents an overview of the STARLAB facility and current status, with principal emphasis on Australia's contribution to the project.

* Paper presented at the IAU Third Asian-Pacific Regional Meeting, held in Kyoto, Japan, between 30 September–6 October, 1984.

Astrophysics and Space Science **118** (1986) 71–77.

2. The Starlab Facility

2.1. Telescope

An *f*/15 Ritchey–Chrétien telescope with a primary mirror diameter of 1 m is planned. The total unvignetted field of view will be 0.8°, with the central 0.5° reserved for science, and the outer annulus used for fine guidance purposes. The overall image quality of the complete telescope system (including fine guidance), is specified as 0.16 arc sec FWHM at 6330 Å, with 70% of the energy enclosed by a circle of diameter 0.3 arc sec. The bandpass of the optics will be 1150 to 12000 Å, but this is extensible down to ~900 Å by appropriate recoating of surfaces.

2.2. Scientific instrument package (SIP)

2.2.1. *Direct Imager and Spectrograph*

The STARLAB Scientific Instrument Package (SIP) is being designed to accommodate two major instruments, a Direct Imager and a Spectrograph. These are shown

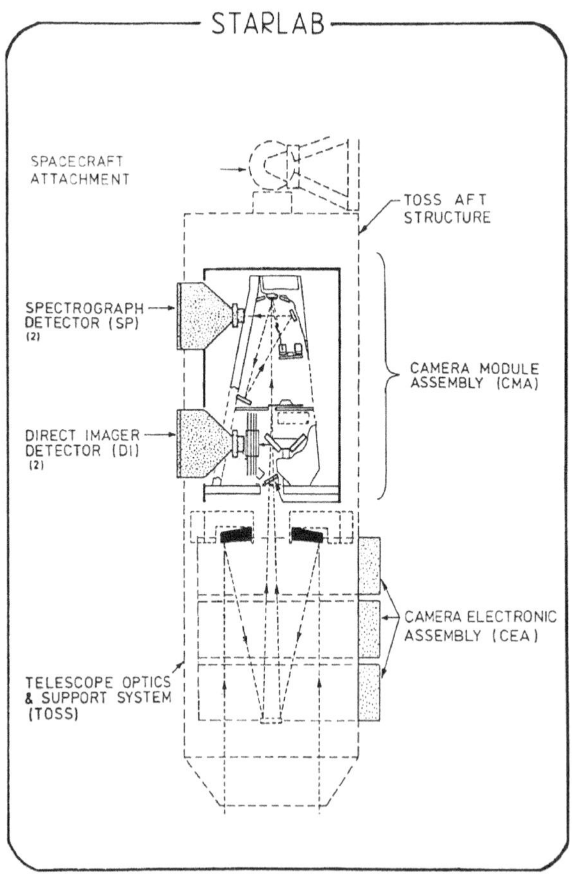

Fig. 1. Conceptual layout of the STARLAB telescope showing the Scientific Instrument Package located behind the primary mirror. The Fine Guidance Sensor components are also indicated.

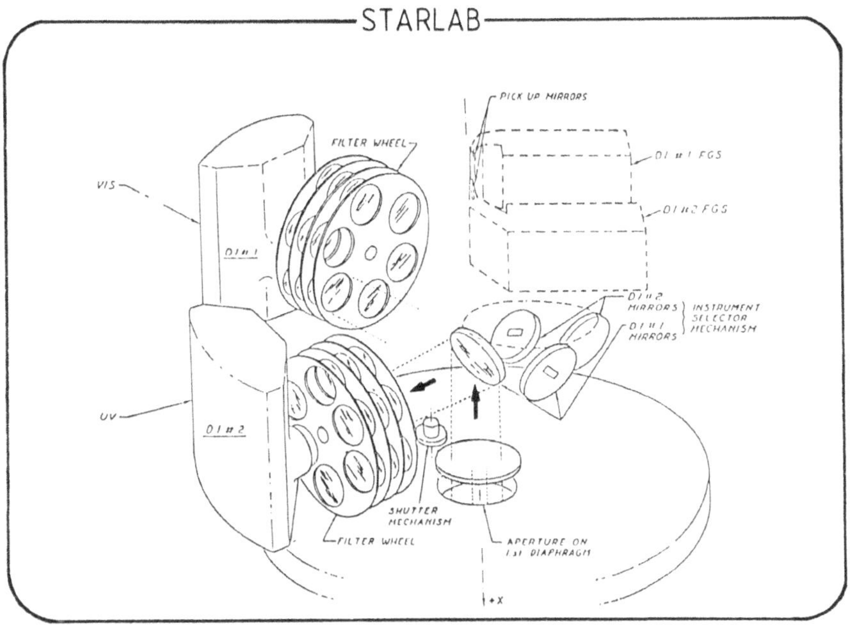

Fig. 2. Layout of the Direct Imager (DI), showing the UV and visible detectors together with filter wheels and a rotating instrument selector.

schematically in Figures 1–3. The Direct Imager comprises both an ultraviolet (1150–3000 Å) and a visible (3000–12000 Å) detector, either of which can be illuminated by rotating a turret in the path of the telescope beam. Individual filters from amongst a set of 20 for each detector can be selected.

The Spectrograph also includes separate detectors to encompass the ultraviolet and visible bands, again selectable by rotating mechanisms. A choice of four diffraction gratings is provided to efficiently cover the wavelength range from 1150 to 8000 Å. Three modes of operation are possible for the Spectrograph; a high resolution ($R \sim 10^5$) mode in which an Echelle grating is inserted into the beam to project multiple spectral orders (~ 30) onto the designated detector; a low resolution ($R \sim 5000$), long-slit (8 arc min) mode, where the Echelle grating is replaced by a plane mirror so that only a single order is projected onto the detector; and a direct imaging mode (for target acquisition), where both gratings are replaced by mirrors. The Spectrograph and Direct Imager are mutually exclusive in operation, except for a Serendipity Mode whereby the outer $\sim 50\%$ of the telescope beam can be fed to the Direct Imager while the centre of the beam illuminates the slit of the Spectrograph.

2.2.2. *Detectors*

The most critical element in both the Direct Imager and the Spectrograph, and, in fact, the key to Australia's involvement in STARLAB, is the detector system. All four detectors in the SIP are similar, and are based on the Photon Couting Arrays (PCA) which have been under vigorous development at Mount Stromlo Observatory since

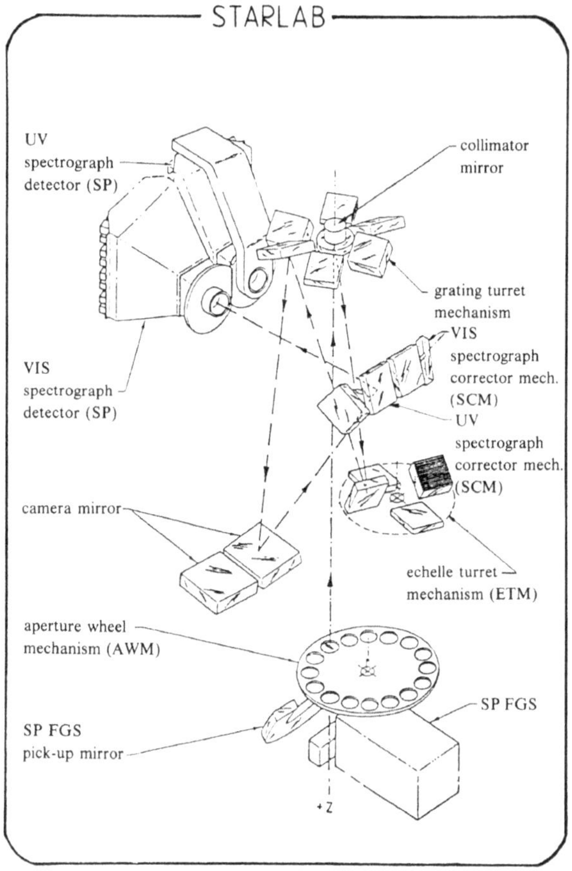

Fig. 3. Layout of the multi-mode spectrograph, showing the UV and visible detectors, and associated mechanised optics.

1976 (Stapinski *et al.*, 1981; Gorham *et al.*, 1982; Hobbs *et al.*, 1984; Roberts *et al.*, 1984). These very advanced low-light level detectors rely upon state-of-the-art technology utilising micro-channel plates (MCPs), coherent fiber optic image dissectors, and charge-coupled devices (CCDs).

The performance specifications of a STARLAB PCA are given in Table I, and the operational concept is illustrated in Figure 4. Briefly, a photon striking the photocathode, which is either bonded directly to the MCP or in close proximity focus, ejects a single photoelectron into one of the pores of the MCP. The ensuing electron gain of $\sim 10^6$ produces an electron avalanche which is proximity-focused at the end of the pore on to a phosphor screen. A flexible fiber optic coupler in intimate contact with the phosphor dissects the light into multiple sub-images (48 or 64) which are projected on to individual CCDs. Processing of video frames from each CCD involves subtraction of the previous frame to remove phosphor decay and fixed pattern components, and then thresholding to detect individual bursts of light corresponding to incident photons (i.e., at the photocathode). Since each detected CCD event has a very high signal-to-noise (~100 : 1) and covers multiple CCD photosites (~5 × 5), digital event-centering in real-time permits recovery of the original resolution of the MCP, defined by the pore spacing (~10 microns). The centroided events are entered into a high speed incrementing memory, which has a capacity of ~9000 × 9000 pixels of 16 bit depth. Inspection of the accumulating image can be undertaking while an exposure is in progress.

As emphasised by Roberts *et al.* (1984), the PCA concept being developed for STARLAB has tremendous flexibility. Tailoring of the photocathode/MCP front end is all that is required to produce a high resolution, large area, photon-counting detector suited for faint signal applications in energy bands ranging from X-rays through to the near infrared. Numerous ground-based versions of the PCA are in regular use at MSSSO, and have produced outstanding results in both direct imaging and spectroscopic modes (e.g., Zarro and Rodgers, 1983; Jenkins *et al.*, 1984).

2.3. Mission operations

It is proposed to launch STARLAB on the NASA Space Shuttle around 1991 for installation on a free-flying platform in low-earth orbit. A candidate platform, which is compatible with the requirements of STARLAB, is the Fairchild Leasecraft vehicle. Mission profile studies have shown that high observing efficiencies (~60%) can be achieved if a high slew rate between targets is possible. Typical STARLAB observations would involve exposures of 15–50 min, at the end of which the images would be transmitted to the operations centre via the TDRSS network. The intention is for STARLAB to be retrievable from orbit for upgrading and subsequent missions over a potential lifespan of 20 yr.

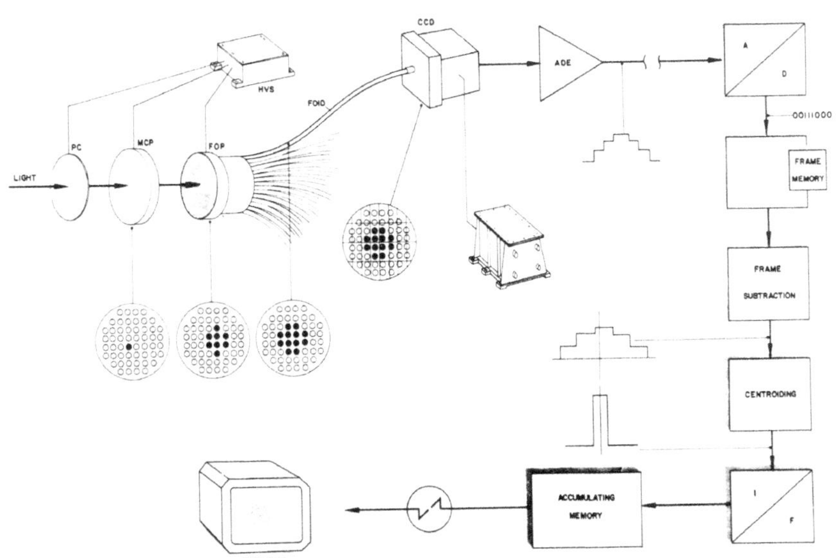

Fig. 4. Conceptual operation of a STARLAB-type Photon Counting Array, as discussed in the text (from Roberts *et al.*, 1984).

TABLE I

STARLAB detector performance[a]

Parameter	Baseline	Goal
Field of view	100 mm diameter	130 mm diameter
Resolution	< 15 microns FWHM	< 15 microns FWHM
Dark count rate	< 1E-4 Hz/pixel	< 5E-5 Hz/pixel
Maximum count rate	> 2 Hz/pixel	> 5 Hz/pixel
Dynamic range	> 2E4	> 5E4
Image stability	2 microns RMS	2 microns RMS
Photometric error	$< 1.1\sigma$ or 1%[b]	$< 1.1\sigma$ or 1%[b]
Visible QE (S20)	Peak > 15%	Peak > 20%
UV QE ($CsTe_2$)	Peak > 12%	Peak > 15%
Detective QE	> 80%	> 90%
QE uniformity	< 5% variation	< 5% variation
Overload damage	> 650 Hz/pixel	> 650 Hz/pixel

[a] Adapted from Roberts *et al.* (1984).
[b] Whichever is the greatest.

3. Status

Most of the emphasis in the STARLAB project so far has been directed towards the frontier technology area of the Scientific Instrument Package. A funding allocation of $4M by the Australian Government has enabled MSSSO, ANU and local industry to undertake a detailed phase B Feasibility Study. This study, completed in June 1984, has both verified and refined the conceptual definition of the SIP. A further allocation of $2M has been granted for the specific purpose of developing the detector system during the period 1984–1985. These detectors will be extensively tested on ground-based telescopes.

Regrettably, the overall progress of the STARLAB facility has received a major setback recently, following a decision by Canada that it is unable to allocate funds for the project. Negotiations are now under way with other potential partners.

4. COLUMBUS

A further UV space telescope project of potential interest to Australia is COLUMBUS. This project is a collaborative venture between NASA and ESA, and is currently in the definition phase. The principal objective of the mission will be to perform high-resolution spectroscopy ($R \sim 3 \times 10^4$) in the astrophysically important wavelength range of 900–1200 Å. This region of the electro-magnetic spectrum is of particular significance since it contains many resonance lines of the dominant ions and molecules found in astrophysical plasmas. Spectroscopy over the broader wavelength band of 100–2000 Å is also of considerable importance, and will be undertaken by COLUMBUS providing the 900–1200 Å performance is not compromised. Observations at the shorter wavelengths would necessitate grazing-incidence optics.

Australia has been invited by NASA and ESA to participate in the definition studies of the COLUMBUS project. Our potential contribution would be the provision of the photon-counting detectors for the spectrograph. Two different detector configurations are possible, depending on the spectrograph design. One format would require a highly rectangular array with a photocathode curved to match the circular focal surface of a Rowland Spectrograph. The other would require a square format to read-out the cross-dispersed orders of an Echelle Spectrograph. The performance specifications for either detector configuration are well within the capabilities of the detectors presently being developed under the STARLAB program.

5. Summary Remarks

There is a tremendous demand from Australian astronomers to be able to conduct observations from space, particularly in the ultraviolet regime. International collaboration is essential, however, and in this respect, we are able to offer the best large format photon-counting detectors available anywhere. Equally importantly, the MSSSO team is receiving enthusiastic support in this endeavour from the Australian Government and from local industry.

Acknowledgements

STARLAB is a multinational project involving many people from research institutions, government agencies, and private industry. Their contributions have been vital.

References

Gorham, R., Rodgers, A. W., and Stapinski, T.: 1982, *Proc. SPIE, Instrumentation in Astronomy IV* **331**, 490.
Hobbs, T., Carden, D., and Rodgers, A. W.: 1984, *Proc. SPIE, Instrumentation in Astronomy V* **445**, 402.
Jenkins, E. B., Rodgers, A. W., Harding, P., Morton, D. C., and York, D. G.: 1984, *Astrophys. J.* **281**, 585.
Roberts, E., Stapinski, T., Rodgers, A. W., and Carden, D.: 1984, *Proc. SPIE, State-of-the-Art Imaging Arrays and their Applications* **501**, 344.
Stapinski, T., Rodgers, A. W., and Ellis, M.: 1981, *Publ. Astron. Soc. Pacific* **93**, 242.
Zarro, D. and Rodgers, A. W.: 1983, *Astrophys. J. Suppl.* **53**, 815.

THE MT JOHN 1 METRE TELESCOPE PROJECT*

J. B. HEARNSHAW

Mt John Observatory, University of Canterbury, Christchurch, New Zealand

(Received 15 June, 1985)

Abstract. The Dall–Kirkham 1 m Cassegrain reflecting telescope to be installed at Mt. John University Observatory is described.

1. Introduction

The Physics Department at the University of Canterbury is currently building a 1 m reflector for installation at Mt John University Observatory in mid-1985. The Dall–Kirkham optical system was designed by, and is being produced with the assistance of, the Physics and Engineering Laboratory, Department of Scientific and Industrial Research, Lower Hutt. The telescope will operate at *f*/13.5 or *f*/8 cassegrain foci and be used mainly for high dispersion échelle spectroscopy, stellar photometry, and asteroid astrometry.

The Serrurier-truss tube and the single-pier asymmetric equatorial mounting are both conventional designs. Construction of the 8 m-diameter dome is also part of the project, and has been completed. A single five-phase Berger–Lahr stepping motor will drive the telescope in five different speeds in each coordinate. Our home-designed and built optical encoders will position the telescope to about 10 arc sec and be under microcomputer control.

The whole project is taking four years (1981–1985) and twelve man-years and is a good example of how a small but still scientifically significant telescope can be designed, constructed and funded within the limited resources of a university department. The cost of the telescope will be about one-fifth of a similar instrument obtained commercially.

2. Technical Details of the Telescope

(a) Optical system

Optical configuration: *f*/8 or *f*/13.5 Dall–Kirkham cassegrain.
Primary focal ratio: *f*/3.5.
Primary mirror: Zerodur, thickness 135 mm, weight 235 kg.
f/8 field of view: 1.7° with field correcting lenses.

* Paper presented at the IAU Third Asian-Pacific Regional Meeting, held in Kyoto, Japan, between 30 September–6 October, 1984.

Astrophysics and Space Science **118** (1986) 79–81.

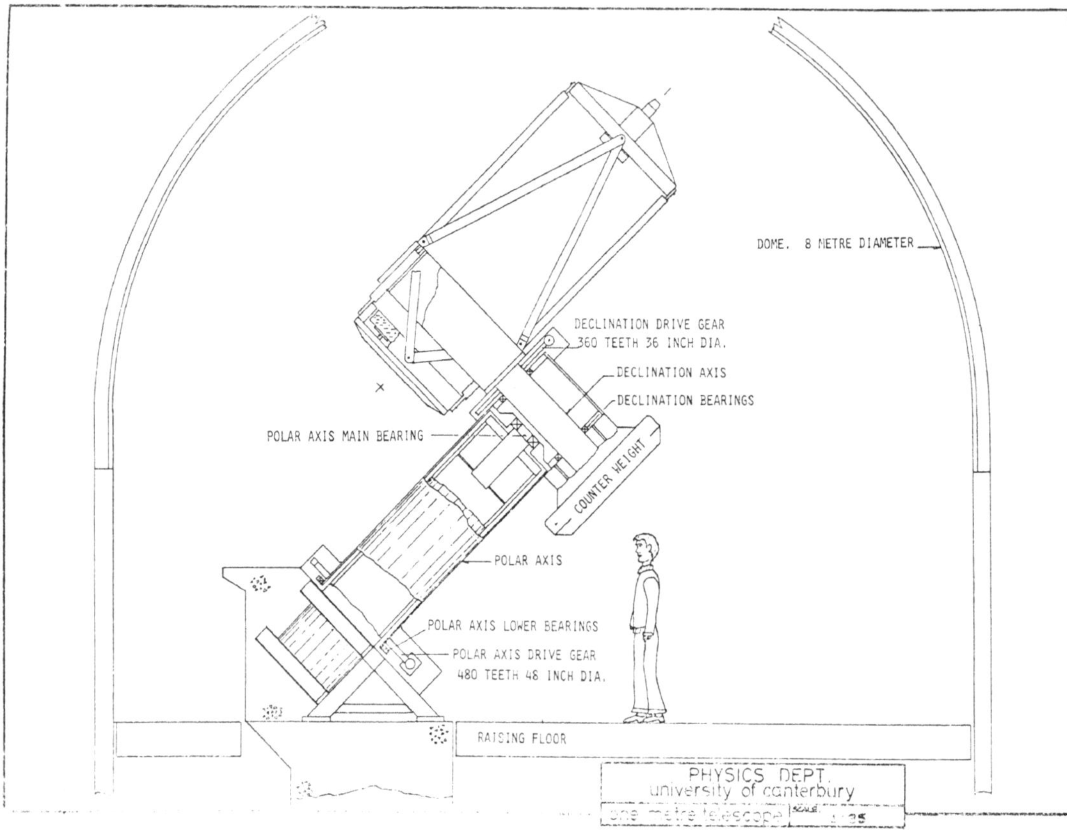

Fig. 1.

(b) Mechanical system and drive

Mounting: single-pier asymmetric equatorial.
Gears: wheels, nickel–bronze: R.A., 480 teeth, 1.2 m diameter; dec. 360 teeth, 0.9 m diameter; worms, stainless steel 97 mm diameter.
Motors: Berger–Lahr five-phase stepping motors; 1 step = $0''.178$.
Drive rates in R.A., dec: $1.6^\circ\ s^{-1}$, $10'\ s^{-1}$, $2.8'\ s^{-1}$, $15''\ s^{-1}$, $2''\ s^{-1}$.

(c) Electronics and control

Encoding: absolute encoding to 1° in dec. and to 3^m in R.A., incremental encoding to $10''$ in dec. and 1^s in R.A.
Microcomputer: Commodore 32K microcomputer will control telescope in R.A. and dec.

(d) Dome

Material: laminated wooden ribs, 11 mm-plywood cladding, aluminium skin.
Shutter: over-the-top opening.

(e) Main instrumentation and programmes

Echelle spectroscopy of stars at high dispersion (2 Å mm^{-1}).
Stellar spectroscopy at medium dispersion (60, 90 Å mm^{-1}).
Broad or intermediate band (*UBVRI*, *uvby*) photometry.
Asteroid and comet astrometry.

RADIO ASTRONOMY AT THE UNIVERSITY OF SYDNEY*

A. J. TURTLE
University of Sydney, NSW, Australia

(Received 15 May, 1985)

Abstract. The recent improvements to the radio telescopes at the Molonglo (MOST) and Fleurs (FST) Observatories are described along with some of the astronomical projects.

In 1960 W. N. Christiansen and B. Y. Mills moved from the CSIRO Division of Radiophysics to the University of Sydney departments of electrical engineering and physics, respectively. Mills then initiated the Molonglo Observatory near Canberra by building a cross-type radio telescope with EW and NS arms each 1.6 km long operating at 408 MHz (Mills *et al.*, 1963, Davies *et al.*, 1973). Christiansen had operated the grating cross (Christiansen and Warburton, 1955) at the CSIRO Fleurs Observatory near Sydney for solar observations. In 1963 this observatory was transferred to the University of Sydney and by 1973 the grating elements were incorporated into the 1415 MHz Fleurs Synthesis Telescope (FST) (Christiansen, 1973). Both these telescopes have undergone major changes in recent years.

The Molonglo cross ceased observing in 1978. The east and west arms (only) were then converted to operate from late 1981 at a frequency of 843 MHz in a tracking mode as the Molonglo Observatory Synthesis Telescope (MOST) (Mills, 1981). Each arm separately produces 64 simultaneous fan beams 22 arcsec apart. Though corresponding beams from each arm can be summed to form a total power instrument for objects such as pulsars, normally they are multiplied for use in synthesis (Durdin *et al.*, 1984). A real-time synthesis of 12 hr produces a beam of $43 \times 43 \csc \delta$ arc secs over a basic field of $23 \times 23 \csc \delta$ arc min. Often the field is increased to 46 or 70 arcmin by repeatedly setting the bank of 64 beams to 2 or 3 adjacent positions. For 12 hr observations *S* of $\delta = -30°$ the synthesised beam is very clean and further cleaning of the map may not be necessary. Sources of a few mJy are easily detected even on the larger fields. A variety of computing techniques exist for improving the map if required (Crawford, 1984). Circuits which detect significant transient increases in noise level will soon be attached to 32 total power beams to enable impulsive events with a time scale from 1 μs to 1 s to be recorded (Large *et al.*, 1984). The source of related extraterrestrial events which occur at several hour angles can be located to a few seconds of arc.

The FST now consists mainly of an EW array comprising the original thirty-two 5.6 m paraboloids plus four 13.7 m paraboloids covering an extended baseline of 1.585 km (Jones *et al.*, 1984). Two other 13.7 m paraboloids are available on a 0.60 km NS baseline. A new digital receiver is used to measure the correlations between the array

* Paper presented at the IAU Third Asian-Pacific Regional Meeting, held in Kyoto, Japan, between 30 September–6 October, 1984.

TABLE I

Telescope parameters

	MOST	FST	
Central frequency	843	1415	MHz
Effective bandwidth	3	6.4	MHz
H.A. coverage	±6.0	±4.0	h
Dec. range for this coverage	$-90° \rightarrow -30°$	$-75° \rightarrow -10°$	
Maximum Dec. coverage	$-90° \rightarrow +18\overset{\circ}{.}5$	$-85° \rightarrow +28°$	
Field size - basic	23 × 23 cosec δ	90 × 90	arc min
- time shared	70 × 70 cosec δ		
Effective synthesised beam	43 × 43 cosec δ	23 × 27 cosec δ	arc sec
System temperature	110	170	K
Measured r.m.s. background fluctuations on good maps	0.4	4.5	mJy

elements, and synthesis is subsequently performed on a VAX 11/780 in the usual way (Sault, 1984). The large antennae have a dual polarisation feed and polarisation measurements should soon be possible. The FST currently offers the highest angular resolution for far southern radio sources. Table I summarises the properties of the MOST and the FST.

Over 400 fields have so far been mapped with the MOST, covering a wide variety of objects. A survey of the Small Magellanic Cloud is complete and one of the Large Cloud will be finished by early 1985: these have already led to important results concerning the evolution of supernova remnants (Mills *et al.*, 1984). Galactic supernova remnants have also been mapped. A general survey along the Galactic plane will take some years to complete – results on the Galactic centre have been published (Mills and Drinkwater, 1984). Several fields have been observed with maximum sensitivity in order to obtain source counts. Other programs include nearby galaxies (Harnett, 1984), Abell clusters and structures of extended extra-galactic sources. The large collecting area of the MOST is also ideal for pulsar timing measurements. Observations with the FST include galaxies, supernova remnants, and the radio counterpart of Circinus X. Though the frequency separation from the MOST is insufficient for accurate spectral indices, this telescope is particularly useful where the higher resolution is required, for instance unresolved supernova remnants in the Large Magellanic Cloud and extragalactic jets. Elements of the FST have been used in an Australia-wide *VLBI* network.

Future plans for both telescopes include possible measures to achieve higher angular resolution.

Acknowledgements

Operation of the MOST and the FST is supported by the Australian Research Grants Scheme and by the University of Sydney.

References

Christiansen, W. N.: 1973, *Proc. Inst. Radio Elec. Engrs. Australian* **34**, 302.

Christiansen, W. N. and Warburton, J. A.: 1955, *Australian J. Phys.* **8**, 474.

Crawford, D. F.: 1984, in J. A. Roberts (ed.), *Indirect Imaging*, Cambridge University Press, Cambridge, p. 373.

Davies, I. M., Little, A. G., and Mills, B. Y.: 1973, *Australian J. Phys. Astrophys. Suppl.* **28**, 1.

Durdin, J. M., Large, M. I., and Little, A. G.: 1984, in J. A. Roberts (ed.), *Indirect Imaging*, Cambridge University Press, Cambridge, p. 75.

Harnett, J. I.: 1984, *Monthly Notices Roy. Astron. Soc.* **210**, 13.

Jones, I. G., Watkinson, A., Egau, P. C., Percival, T. M., Skellern, D. J., and Graves, G. R.: 1984, *Proc. Astron. Soc. Australia* **5**, 574.

Large, M. I., Vaughan, A. E., Durdin, J. M., and Little, A. G.: 1984, *Proc. Astron. Soc. Australia* **5**, 569.

Mills, B. Y.: 1981, *Proc. Astron. Soc. Australia* **4**, 402.

Mills, B. Y. and Drinkwater, M. J.: 1984, *J. Astrophys. Astron.* **5**, 43.

Mills, B. Y., Aitchison, R. E., Little, A. G., and McAdam, W. B.: 1963, *Proc. Inst. Radio Engrs. Australia* **24**, 156.

Mills, B. Y., Turtle, A. J., Little, A. G., and Durdin, J. M.: 1984, *Australian J. Phys.* **37**, 321.

Sault, R. J.: 1984, in J. A. Roberts (ed.), *Indirect Imaging*, Cambridge University Press, Cambridge, p. 367.

ASTRONOMICAL RESEARCH ACTIVITIES WITH THE 74 INCH TELESCOPE AT KOTTAMIA OBSERVATORY*

I. A. ISSA and A. I. GAMAL EL DIN

Helwan Institute of Astronomy and Geophysics, Cairo, Egypt

(Received 24 May, 1985)

Abstract. Two topics of research, namely extragalactic and variable star studies represent the main attitudes of astronomical work at the Helwan Observatory.

The 74 inch telescope at Kottamia (476 m, $2^h.12$, $29^\circ.93$ N) some 60 km to the east of Cairo is the main instrument. Some other auxiliary equipment (Cassegrain and echelle spectrographs, Westinghouse and Japanese cameras and photoelectric photometer) can be attached at both Cassegrain and Newtonian foci.

1. Introduction

It is not far from the truth when we say that astronomical researches in Egypt started on solid foundations as early as 1903, at which a time an observatory was erected at Helwan. It happened that an English amateur came to Aswan on an expedition to share in observations of the total solar eclipse. He became impressed by the clear atmospheric conditions prevailing in the Egyptian sky. As a result he presented the Egyptian government with a 30-inch reflector, which was then mounted at Helwan for its advantageous seeing conditions at that time.

After that time, the observatory at Helwan was equipped with a meridian circle and coelostat, which was considered as the base of a solar station.

Later on the Egyptian government contracted with Messrs Grubb Parsons (UK) to make a much bigger and more modern telescope. This became the 74-inch telescope at Kottamia, which then started its role in sharing with other international telescopes to remove the obscuration of the Universe. This was as early as 1963.

2. The 74 inch Telescope

The main mirror is parabolic in form, with 360 inch focal length. The outside diameter is 76 inch. It was made from Pilkington low-expansion glass. The surface of the primary and of the other secondary mirrors can be aluminised by equipment installed at Kottamia.

Observations can be done at three foci: namely, the Newtonianian, Cassegrain, and Coudé.

The form of the Cassegrain secondary mirror is hyperbolic, with a paraxial radius of 201 inch and made from fused quartz. The aperture is 19 inch.

* Paper presented at the IAU Third Asian-Pacific Regional Meeting, held in Kyoto, Japan, between 30 September–6 October, 1984.

Astrophysics and Space Science **118** (1986) 87–93.

The Newtonian secondary mirror is also made from fused quartz, with 20 inch aperture and a flat surface. It is provided with an optical glass correcting lens of 9 inch aperture. Later on, a camera was provided with a four element Newtonian focus field corrector lens. Other means for focusing and centering are present.

The aperture of the first and second Coudé mirrors is 14 and 10.5 inches, respectively. They are from fused quartz and flat in form.

The telescope is provided with two 6 inch finders each of 60 inch focal length. The finder eypieces are of the Kellner type and of 2 inch focal length.

Table I summarises the above data.

TABLE I

Mirror	Aperture	F ratio	Form	Material
Main	74″		Paraboloid	Pilkington low expansion glass
Newtonian	20″	4.85	Flat	Fused. quartz
Cassegrain	19″	18	Hyperbolic	Fused. quartz
1st Coudé	14″	28	Flat	Fused. quartz
2nd Coudé	10.5″	7.5	Flat	Fused. quartz
Finders	6″			

A central disk is provided to give the orientation and motions required.

3. Auxiliary Equipment

3.1. Cassegrain focus

At this focus a classical spectrograph with a focal ratio $F/18$ can be attached. The dispersing system consists of three similar prisms. The scale on the plate is ~6.5 arc sec per mm. Two cameras, a short one with $F/2.5$ and a long one with $F/6$, are used to record spectra. The later covers a wavelength range from 3600–8000 Å. The dispersion at Hγ for each camera is 100 Å mm^{-1} and 48 Å mm^{-1}, respectively. Comparison spectra can be provided by either iron arc or discharge tubes. Light passes through a condenser which focuses light onto the entrance slit.

3.2. Echelle spectrograph

The dispersion is provided by a Bausch and Lomb 315 lines mm^{-1} echelle grating. A flint prism is used to sort out the spectral orders. Accordingly, the separation between orders varies from 5 mm at 3500 Å to 2 mm at 6200 Å. An Olympus 35 mm reflex camera connected behind an EMI 3-stage image intensifier is used to record echellograms. The maximum photon gain (10^4–10^5) is attained at the blue end of the spectrum, but the sensitivity falls off rapidly at the red end. Three calibration lamps are used to provide comparison spectra through optical fibers. Table II indicates the spectral extent and linear dispersions for orders from 9 to 14.

TABLE II

Order No.	λ_c (Å)	Spec. range (Å)	Dispersion (Å mm^{-1})
9	–	27	1.12
10	5700	24	1.00
11	5175	22	0.91
12	4750	20	0.84
13	4390	18	0.76
14	4075	17	0.71

3.3. Photoelectric photometer

A two-channel photoelectric photometer can also be used at the Cassegrain focus. It consists mainly of an EMI photomultiplier of type 9683/35. The output is amplified and then recorded on a Honeywell recorder or a photon counter. Cooling by an inverse Peltier mechanism is provided to minimise, as far as possible, the dark noise

3.4. Westinghouse and Japanese cameras

The first consists mainly of a one stage image intensifier and can be attached to either Newtonian or Cassegrain focus. Plates of size 2 × 2 inches are used. It is simple in operation and of reasonable gain. A power of 15 kV is provided through a DC battery. It is used mainly for detecting variability in galaxies. It can also be operated at higher signal to noise after adding a cooling system to minimize dark current. The Japanese camera does not differ so much from the Westinghouse Camera except from the possibility of using plate holders and a 35 mm Nikon camera to record photographs.

4. Newtonian Focus

4.1. Newtonian camera

The focal ratio at this focus is 4.85. The scale of the plate equals 22.53 arc sec per mm. A flat Newtonian mirror of 50 cm aperture inclined at an angle of 45° is used to reflect light from the primary mirror. Plates of sizes 3.5 × 4.5 inches and 6.4 × 6.4 inches can be used. For the latter plates to be free from coma distortions a four-piece lens corrector can be attached. A flat field of approximately 1 sq. deg is then available.

4.2. Nebular spectrograph

An old prism spectrograph has been modified to become a nebular one by adding two small comparison prisms moving along the slit to provide spectra. The collimator (manufactured at the National Research Center in Cairo) was made to match the focal ratio of the telescope. It consists of two lenses, one from flint and the other from crown glass; to overcome colour aberration. The collimator has a 6 cm aperture and is of 28 cm focal length. To increase the efficiency of the spectrograph it is planned to provide it with an image intensifier.

5. Coudé Focus

At this focus a G51 Coudé spectrograph has been erected. A grating ruled with 600 lines per mm is used as the dispersing element. The spectrograph is provided with two cameras with focal ratios *F*/28 and focal length of 108 inch, and *F*/7.5 and focal length of 30 inch (Table III).

TABLE III

Optical data of the Coudé spectrograph

Grating	Blue wavelength	0.4 μ (1st order)
	Ruling frequency	600 lines mm^{-1}
	Ruled area (width)	104 mm
	(length)	101 mm
Collimator	Focal length	274.3 cm
Short camera	Relative aperture	*F*/7.5
	Focal length	76.2 cm
	Plate size	119 × 41
	Spectral range	3200–8000 Å
Long camera	Relative aperture	*F*/27
	Focal length	274.3 cm
	Plate size	240 × 50
	Spectral range	2900–8000 Å

The Coudé spectrograph is provided with guiding eyepieces. The comparison spectra can be obtained through either a discharge tube or arc electrodes.

Unfortunately, the Coudé spectrograph has not been used except for some laboratory experiments to evaluate the oscillator strengths of some alloys. This because of the great loss in light from the primary mirror to the plate.

6. Research Activities

Numerous fields of astronomy have been attacked by astronomers in the Helwan Observatory. Galactic clusters and interstellar matter, which represent the most powerful indicators of the spiral structure of our Galaxy, have been a field of work for a long time. *UBV* and *RGU* photographic programmes on the Newtonian focus were initiated by Prof. W. Becker. Distances, ages and extinctions were determined for more than 40 clusters.

A study of the diffuse interstellar bands represents a cooperative programme between the Helwan Observatory and the Mill Hill Observatory in London, under the supervision

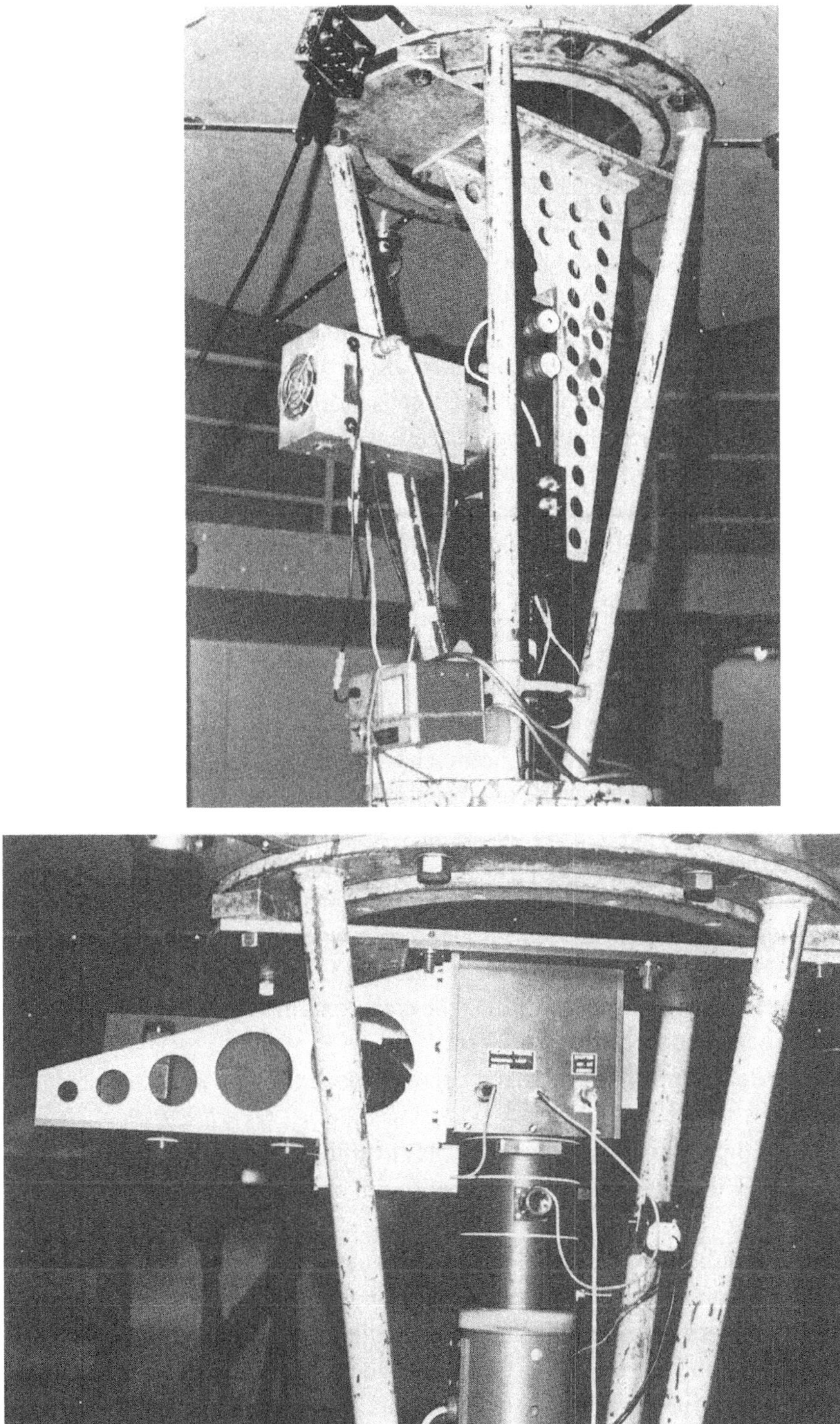

Fig. 1. (a) Photoelectric photometer. (b) Westinghouse camera for the 74 inch reflector at Kottamia.

of Dr D. McNally. Intercorrelations of these bands and correlations with other variables has resulted in:

(1) Equivalent widths of the diffuse bands are strongly intercorrelated.

(2) Equivalent widths of the diffuse bands are strongly correlated with interstellar reddening.

(3) Simple and partial correlation coefficients with colour excesses have supported the hypothesis that certain atoms or ions embedded in interstellar grains provide an origin for these bands. The possibility that more than a band can be produced by the same carrier has also been discussed.

Galactic dynamics has received a considerable attention by some workers. Encouragement in this field was given by Prof. R. v. d. R. Woolley from the Greenwich Observatory. Radial velocities of some thousand stars were measured. Motions perpendicular to the galactic plane under the effect of the gravitational field of the Galaxy provides the main subject of some researchers in Helwan.

Detailed studies of planets, Moon, and natural satellites depending on observations taken at different foci of the telescope has been also done and supervised by Prof. Z. Kopal.

The departure of many Egyptian astronomers to work in other Arab countries has caused remaining staff in the Helwan Observatory to confine themselves in two fields, extragalactic research and variable stars.

In the first field, detailed isophotometry of the galaxies NGC 128, NGC 3489, and NGC 4179 was the beginning. Isophotometry has been done using a Hilger microphotometer at Helwan. This work has been extended to many morphological types. Joyce-Loebl isodensitracers from Imperial College London and the University of Padova have been used for isophotometry. Plate calibration was done with the use of a tube sensitometer manufactured at Helwan. The light distribution functions obtained at Helwan has shown good agreement with other workers.

Such isophotometric data provided the material for a work about variation of the position angle of different isophotes along the major axis in 11 galaxies. Three important results have been deduced from this work:

(1) The main features of the position curves deduced in this work are similar to the rotation curves.

(2) Most of the position angle curves display a turnover, which defines the dimensions of the nuclei in these galaxies. Comparison with other means have shown a good agreement.

(3) If the standard deviation of the position angle is taken as a measure of the structure of the galaxy (which may be produced by rotation and/or any other process) the average standard deviation of all galaxies of the same type shows its smallest value in the lenticulars and the largest in the spirals and a medium value in the ellipticals.

Another subitem in extragalactic studies deals with estimating both the amount of dust and ionized gas in galaxies. The size distribution of dark clouds and of H II regions can be a good indicator for this purpose. In both cases the size distribution is best represented by an exponential relation. In all cases the dust has proved to be comparable to that in our Galaxy. The apparent size distributions can be considered as a tool to estimate distances of galaxies. The distances from both H II regions and dark cloud distributions can be given by

$$d = -\frac{\ln(n_1(\theta_1)/n_2(\theta_2))}{K(\theta_1 - \theta_2)},$$

K is a constant of 0.05 Mpc^{-1}.

Comparison of distances obtained from the above formulae and from other distance indicators have shown a good agreement. A later study of the distribution of dark clouds in NGC 4321 showed that a distance of about 28 Mpc can be derived. This allows homogeneity in estimating distances, which is not the case for other distance indicators. The use of the H II-regions as a distance indicator depending on the size distribution function can solve some of the difficulties arising in Scric's method.

One can also have some idea about gas to dust ratio from the distribution of both ingredients.

7. Activities in Variable Star Research

Researchers in this field are of different kinds. A subgroup deals with the variability in some contact systems like W UMa. Three colour photoelectric observations at the Cassegrain focus are taken to fulfill this study.

At the same focus using the photoelectric photometer another programme dealing with Be star variations is carried on. A spectrographic observational programme using the old Cassegrain spectrograph is carried on now to study the sequence of Be–B–Be variability.

Since 1981, a close astronomical cooperation has started with Japan, particularly with a sister 74 inch telescope at the Okayama Astrophysical Observatory. Modern equipment, including the Japanese camera with one-stage image intensifier, were supplied to the Kottamia Observatory by the Japan International Cooperation Agency (JICA), for which we are very grateful.

ASTRONOMY IN TADJIKISTAN*

P. B. BABADZHANOV

Astrophysical Institute, Dushanbe, U.S.S.R.

(Received 3 July, 1985)

Abstract. The astronomical investigations in the present Tadjikistan (USSR) are carried out by the Institute of Astrophysics of the Academy of Sciences of Tadjik SSR. The main scientific fields of investigation are meteors, comets, and variable stars. The Institute has two observational stations: the Gissar observatory with photographic and radar meteor patrols, a 70 cm telescope, a 40 cm Zeiss astrograph, etc., and the Sanglok observatory with the Ritchey–Chrétien 1 m telescope.

In 1932 an astronomical observatory was organized in Dushanbe. The main scientific fields of investigation in the observatory, namely, meteors, comets, and variable stars, were chosen with account of the geographical location and climatic conditions of Tadjikistan. The observatory had to investigate the astroclimate in different parts of Tadjikistan as well as to conduct a cultural-instructive work among the population.

In 1940 there were the following instruments at the observatory: an 8-inch refractor, a universal instrument, an astronomical clock, a 5-inch Zeiss apochromate, a 6.5-inch Steinheil refractor, a meteor patrol, two double astrographs, and one astrocamera. After the second World War the scientific activity of the Tadjik Astronomical Observatory was greatly extended. The old astronomical instruments were modernized and new ones obtained. The more modern methods of observations have been worked out. For example, the new meteor patrol system intended for photographic meteor observations, consisting of light-sensitive photocameras, has been constructed. The registration of meteors by radio methods has been started as well.

In 1958 the observatory was elevated into the Institute of Astrophysics of the Academy of Sciences of the Tadjik SSR. It consisted of three departments: the department of variables, of meteors, and of comets.

In 1963–1970 the Gissar Observatory of the Institute of Astrophysics was built at a distance of 14 km from Dushanbe. Its domes house a 70 cm telescope equipped with electron-optic, electrophotometric, and polarimetric receiving apparatus intended for observations of variables and comets; a 40 cm Zeiss astrograph, a 20 cm refractor, two meniscus telescopes, and a double astrograph. The observatory was also equipped with a large camera for photographing of artificial Earth satellites, comets and meteors.

The largest photographic meteor patrol system in the USSR is at Gissar Observatory. It consists of six aggregates including fourty cameras with large fields of view. Meteor investigations have been carried out by the use of an impulse radar, television, and electron-optic equipment.

* Paper presented at the IAU Third Asian-Pacific Regional Meeting, held in Kyoto, Japan, between 30 September–6 October, 1984.

Astrophysics and Space Science **118** (1986) 95–97.

In 1980, a mountain observational station of the Institute was built at the height of 2300 m (Sanglok Mt.). The first Ritchey–Chrétien 1 m telescope in the USSR was set up at the station.

1. Meteor Investigations

The Institute of Astrophysics is engaged in the study of the nature of meteoroids and related phenomena, accompanying the flight of these bodies in the Earth's atmosphere. In particular, atmospheric trajectories, meteor radiation, meteoroid orbits, the distribution of meteoric matter in the circumterrestrial space and, at last, the origin and evolution of meteor showers are studied.

Comprehensive meteor investigations over many years has given extensive information on radiants, velocities, masses, densities, and orbits of thousands of meteors. Physical parameters and the wind regime in the meteor zone of the Earth's atmosphere have been determined (heights 60–120 km). Observations of meteor trails and of ionospheric heterogeneities, carried out in Dushanbe, gave a possibility to study the variations of wind regime at altitudes of 80–120 km within the 11-yr cycle of solar activity.

The meteor contribution to the ionization of the upper atmosphere and the influx of meteoric matter to the Earth have been estimated. New meteor associations have been found as well. In 1964–1965, the author devised a method of 'instantaneous exposure'. This method allows us to obtain meteor photographs with exposure of 10^{-4} s. Hundreds of instantaneous photographs of meteors gave a possibility to study the process of disintegration of meteoroids in the atmosphere. Since 1976 this method has been used to obtain 'instantaneous' meteor spectra as well (Babadzhanov, 1983).

The expedition of the Tadjik Institute of Astrophysics, together with the Kharkov Polytechnical Institute, worked under the author's leadership for two years (1968–1970) in Equatorial Africa. Radiants and orbits of southern hemisphere meteors, and the wind regime of the upper atmosphere in equatorial regions have been determined. Radar observations of meteors and ionospheric heterogeneities extended our notions on the global circulation in the meteor zone.

2. Cometary Investigations

The institute had conducted investigations concerning all sections of cometary physics. In 1955, A. M. Bakharev, a scientist of the Tadjik Observatory, discovered a comet which became known as the Bakharev–McFarlan–Krinke comet. The expansion of the observational base and the training of staff allowed us to start spectral, photoelectric, and polarimetric cometary observations. Now we use a 40 cm astrograph, a 70 cm telescope and the 1 m telescope, at the mountain station of the Institute. In the fifties a mechanical theory of cometary forms was developed. We have also investigated a problem of interaction of corpuscular solar streams with cometary atmospheres. A theory of plasma cometary tails was developed. Non-stationary processes in comets have also been studied. It was shown that many such processes, for example, cometary outbursts, are associated with the solar activity.

Our scientists have also developed a theory of the disintegration of cometary nuclei. They have considered the nuclei to consist of two components with different values of heat of sublimation.

Disintegration of modeled cometary nuclei of H_2O, CO_2, NH_3, CH_4 ices, and the snow model of a cometary nucleus, have been studied at the institute. The heat regimes of such nuclei were estimated. Theoretical data were tested under laboratory conditions which simulated the cosmic ones. Cometary phenomena have been modeled in Dushanbe at the laboratory of experimental astrophysics since 1972. The process of disintegration of dust matrices, the velocities of ejection of matrix fragments, and so on were studied in the laboratory. Polarimetric and colorimetric investigations of comets have revealed negative polarization of the continuum radiation of comets at small phase angles (angl: Sun–comet–Earth).

Nonlinear growth of cometary brightness in the region of small phase angles has been found ('oppositional effect').

A search of the genetic interrelation among comets, asteroids, and meteoric matter is one of the important problems of minor bodies of the Solar system. For this reason asteroidal observations started in Dushanbe.

3. Sky Survey

Observations according to the 'Sky Survey' program have been carried out in Dushanbe for forty years. Now there are about 60000 negatives at the Institute. Various investigations of variables were done on the basis of this observational material. Some features of light curves, variability of period of variables of different types, and the oscillations in brightness of novae were studied. Observations on long-period variables were started in Dushanbe when the telescopes were equipped with photoelectric detectors. About one hundred novae have been discovered in T-associations. Carrying out the simultaneous polarimetric and photometric observations in different colours on some stars of the T–Taurus-type, scientists of the institute study physical conditions in the gas-dust shells of these stars.

The Tadjik astronomers include not only investigations of meteors, comets and variables. Theoretical investigations of the dynamical phenomenon of the collective gravitational interactions of stars in galaxies have been widely developed at the Institute.

References

Babadzhanov, P. B.: 1983, in C. I. Lagerkuist and H. Rickman (eds.), *Asteroids, Comets, Meteors*, Uppsala, pp. 439–444.

PRESENT AND FUTURE OF ASTRONOMICAL SCIENCE AT GADJAH MADA UNIVERSITY, CENTRAL JAVA, INDONESIA*

M. S. A. SASTROAMIDJOJO

Astronomy Department, Gadjah Mada University, Central Java, Indonesia

(Received 2 July, 1984))

Abstract. Reasons for the growth of interest in astrophysics and astronomy are discussed. The two main reasons are firstly the solar eclipse of June 1983 and secondly publicity given to the regular N.A.S.A. shuttle flights. However, radioastronomy will develop much faster than optical astronomy because the site requirements are much less stringent than for optical astronomy and the technical infrastructure is readily available.

Lately interest in astronomy and astrophysics has grown in Indonesia at an accelerated pace. Factors contributing to this somewhat surprising fact are the growing interest in shuttle-flights of the U.S.A., and reports of astronomical events like the total eclipse of June, 1983 and now the approaching comet Halley.

During the eclipse of June, 1983 the public demand for telescopes rose and as a result photoshops and department-stores started to offer all kinds of telescopes and photographic accessories. After the eclipse there was still astronomical optical gear on display and, therefore, still the opportunity for individuals or astronomical clubs, which have sprung up as a result of the 'eclipse-fever', to pursue their hobby. The daily papers frequently carry articles about astronomy or space events. Also, recently the U.S.A. offers to let an Indonesian payload specialist go on the June, 1986 shuttle flight to do research in outer space. In short, all the above added up to an heightened general interest in astronomy and astrophysics.

Activities in astronomy and astrophysics in the Gadjah Mada University are mainly centred in the Science Faculty; specifically in the Solar Energy Research Centre (SERC).

It might be argued that the application of solar energy is more connected with engineering than astronomy, and that astronomy, if needed at all, is only of marginal importance. However, solar energy is now using not only the visible part of the spectrum, but also radiation in the infrared and ultraviolet region, since biological conversion of solar energy into other forms of energy is becoming more and more important. With the growing need for more information came the realization that more fundamental knowledge was needed.

Solar physics cannot be done divorced from astrophysics. The growing need for solar information also in the field of radio communication, meteorology, and geophysics all added up to the realization that something had to be done about it.

* Paper presented at the IAU Third Asian-Pacific Regional Meeting, held in Kyoto, Japan, between 30 September–6 October, 1984.

Astrophysics and Space Science **118** (1986) 99–100.

Therefore, the SERC Gadjah Mada University is preparing a proposal to add astronomy and astrophysics to the existing physics courses. That in its turn would not be complete without facilities for doing research in astrophysics, and especially solar physics. To that purpose the SERC has been in the forefront of doing work in those fields.

Activities range from student experiments like identifying surface features of the Moon by making photographs with small telescopes to a spectroscopic analysis during the June, 1983 total Solar Eclipse. Work is in progress to make a spectrohelioscope – in fact, a monochromator for taking pictures of solar surface phenomena with different wavelengths, while students are building systems for radio telescopy in the 20–100 MHz region, using a chart recorder hooked up to the automatic gain control of a Yaesu FRG-7 receiver. In the field of instrumentation, the laboratory is trying to grind parabolic mirrors up to 35 cm diameter, and developing electronic tracking gear for coelostats and telescopes.

For the June, 1986 NASA shuttle flight, a proposal has been sent to the Indonesian Astronaut Selection Committee to include, as a payload, a Cassegrainian type of telescope for continuously recording of the movement of granulation. This piece of equipment was suggested by the Laboratory for Space Research in Utrecht, the Netherlands. A Hasselblad cinecamera capable of holding 100 exposures will be attached. If all goes well the system will be operated by an Indonesian payload specialist from this laboratory.

There is a lack of funds and expertise, and we have no accredited astronomers yet in four faculty. We have been looking for help and advice in other countries. Japanese astronomers, from the Tokyo Observatory and from the Kwasan Solar Observatory have been most helpful, as have been the people from the Space Laboratory, Utrecht in the Netherlands. In Indonesia, Lembang Observatory is playing a crucial role in our program.

The response of the students themselves has been most encouraging. Every foreign guest lecture on astronomy is attented by students in such numbers that often there is only standing room left. Discussions range from the Chandrasekhar limit to black holes and comets. Even if the questions are as yet at a rather shallow level, it should not be forgotten that the students of today will be the teachers of tomorrow, and, as such deserve all the attention we can give them.

It is hoped that foreign astronomer visiting Indonesia will be willing to give advice. Both institutional cooperation and cooperation on a person-to-person level are being explored. In science international cooperation is of the essence; this is even more true for astrophysics.

HIGH-ENERGY OBSERVATIONS OF SOLAR FLARES*

KATSUO TANAKA

Tokyo Astronomical Observatory, University of Tokyo, Mitaka, Japan

(Received 22 March, 1985)

Abstract. Extensive observations of solar flares made in high energy bands during the maximum of the present solar cycle are discussed with a special reference to the results from HINOTORI, and with attention to the relevant flare models. The hard X-ray (HXR) images from HINOTORI showed mostly coronal emission at 20–25 keV suggesting that the HXR is emitted from multiple coronal loops, consistent with the non-thermal electron beam model in a high density corona. The thermal HXR model seems to be inconsistent with some observations. Three types of flares which have been classified from the Hinotori results are described, along with newly discovered hot thermal component of 30–40 million K which contributes thermal HXR emission. A summary is given for the characteristics of the energy release in an impulsive burst; and an empirical model is described, which explains simultaneous energy releases in multiple loops and successive movements of the release site as suggested from the HXR morphology. The discovery of large blue-shifted hot plasma from the soft X-ray line spectrum leads to some quantitative arguments for the evaporating flare model. An electron-heated flare atmosphere appears to explain various observations consistently.

1. Introduction

A solar flare is an unparalleled phenomenon that can be studied in its details, among more generalized flare like processes in the Universe, which may come from slow energy storage and a subsequent explosive energy release. The energy storage, in a generalized flare, may occur in a region of high plasma β (convection zone, in the solar case) where energy is transferred from various differential motions; such as shear flow, convection, turbulence, rotation, and accretion; to the magnetic fields according to $\partial \mathbf{B}/\partial t = \nabla \times (\mathbf{v} \times \mathbf{B})$. An explosive release of this magnetic energy, or a flare, may occur when the fields are convected out to the low β region (corona) where magnetic fields with excess energy become unstable and dissipate by accelerated diffusion: $\partial B/\partial t = (c^2/4\pi\sigma)\nabla^2 B$. Sufficiently detailed observations of solar flares will clarify actual processes in this gross scenario, and may provide some solid basis for applications to other similar cosmic phenomena. Flare observations in the high energy bands may have an important role in revealing the basic nature of the explosive energy release, and thereby help us better to appreciate various proposed energy release mechanisms.

High-energy observations of flares from space began already in the previous solar cycle with NASA's OSO series, the Dutch TD1 and Soviet *Intercosmos* satellites. Most notable advances have been made by Skylab ATM (1973–1974), which provided high resolution images of the Sun in the bands of EUV and soft X-rays, and established the importance of loop structures in the magnetic morphology of flares. The *Skylab*

* Invited paper presented at the IAU Third Asian-Pacific Regional Meeting, held in Kyoto, Japan, between 30 September–6 October, 1984.

observations were made at 'thermal' bands, and towards the maximum of the present solar cycle, observations considered most important were imaging in 'non-thermal' bands, and observations with high time resolution covering the early phases of flares and coordinated over a sufficiently wide spectral range. During the present maximum (1979–1981) P78-1 (1979; Doschek *et al.*, 1979), Solar Maximum Mission (SMM, 1980; Chipman, 1981), TANSEI IV (1980; Tanaka, 1980), and HINOTORI 1981; (Tanaka, 1983) have been launched aiming at the above requirements. These observations provided new data on the location of primary energy release sites, on the origin and motions of very high temperature plasmas and on the impulsive nature of particle acceleration, and have advanced considerably our views of high energy aspects of solar flares.

SMM and HINOTORI succeeded, for the first time, in obtaining hard X-ray images in the effective energy range of 20–25 keV. HINOTORI, using a technique of a rotating

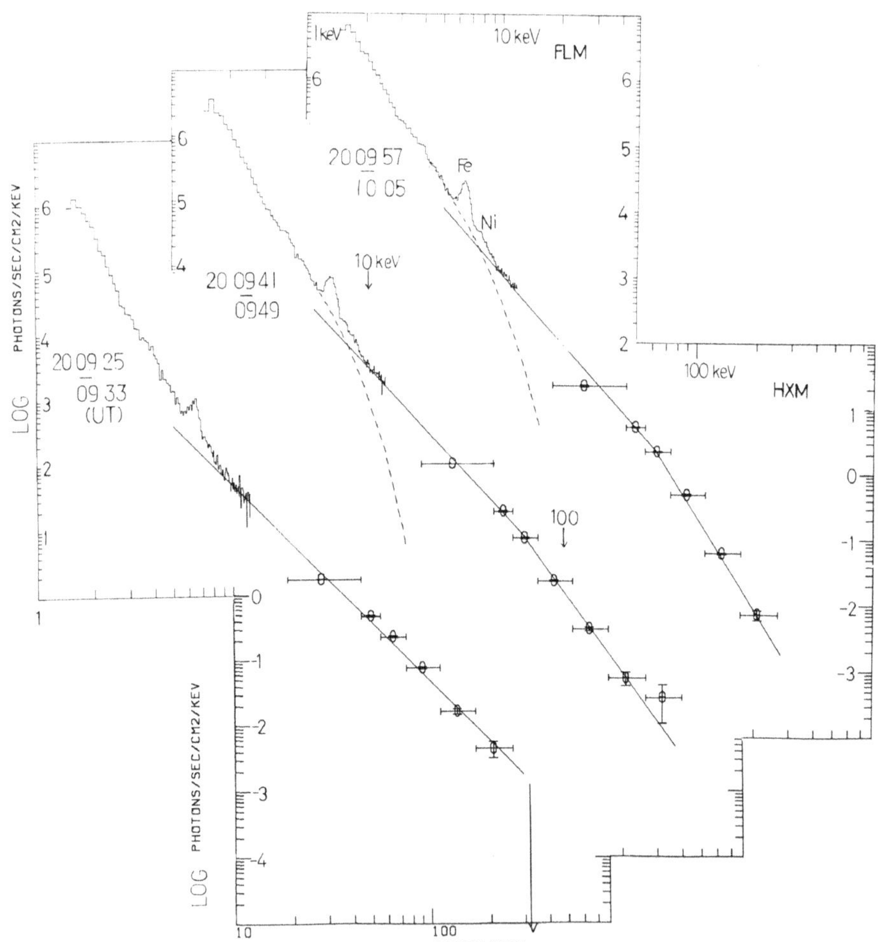

Fig. 1. Combined soft X-ray (1.5–12 keV) and hard X-ray (18–200 keV) spectra obtained by HINOTORI. A single power-law spectrum fits from below 10 to 200 keV in the early phase. Dashed lines denote thermal spectra expected from the line spectra. (From Tanaka *et al.*, 1984.)

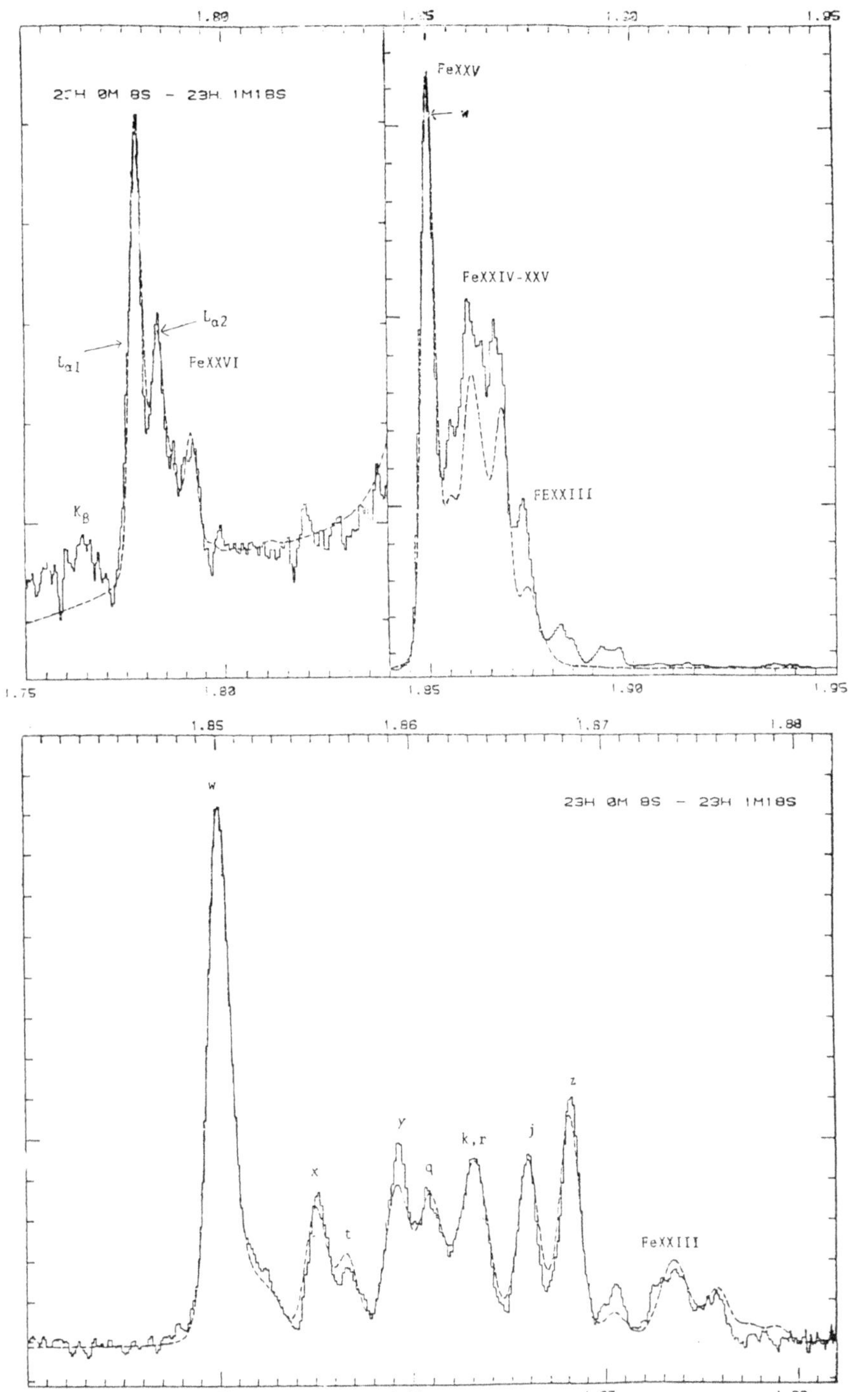

Fig. 2. Soft X-ray line spectrum around 6.7 keV obtained by quartz crystal spectrometers aboard HINOTORI. (a) Low resolution spectrum 1.75–1.95 Å. On the left, is the Fe XXVI spectrum; and on the right a complex from Fe XXV–XX. Dashed line shows a theoretical fit with electron temperature equal to 37 million K. (b) High resolution spectrum 1.84–1.88 Å at the same time as above. Dashed line shows a theoretical fit with electron temperature equal to 24 million K.

modulation collimator, obtained HXR images (15–40 keV) from about 150 flares with spatial resolution of 10″–15″, together with some soft X-ray images (5–12 keV). The SMM observed in six energy bands ranging from 3 to 30 keV with spatial resolution of about 8″, though the number of observed flares was much smaller. A wide spectral range, from soft X-ray to γ-ray was covered, in the case of HINOTORI, by five spectrometers with high time resolutions in the range of 10 ms to 10 s. The soft X-ray spectrum obtained from a gas scintillation proportional counter, enabled clear resolution of the soft X-ray continuum from the line emissions, as shown in Figure 1 (Inoue *et al.*, 1982). In this particular example combined spectra of soft and hard X-rays indicate a remarkable power-law distribution, extending from below 10 to 200 keV; although a thermal (exponential) spectrum generally prevails below 10 keV (Tanaka *et al.*, 1984). The prominent line emission at 6.7 keV is resolved, by two quartz crystal spectrometers, into many lines from the higher ionization stages of iron (Tanaka *et al.*, 1982). The emission consists of two line complexes at 1.85–1.94 Å (Fe XXV–II) and at 1.77–1.80 Å (Fe XXVI–XXV) as shown in Figure 2 (top). The spectrum of Fe XXVI – a H-like ion – diagnosing the hottest plasma regions, revealed an abundant new plasma with electron temperature higher than 30 million K. The second crystal further resolved the line complex at 1.84–1.89 Å into individual lines (Figure 2(b)), allowing detailed spectroscopic diagnosis of electron temperatures, non-thermal motions and the ionization equilibrium.

This review aims at discussing some essential points of the results; mainly from Hinotori and discussing relevant flare models. Some previous accounts of the HINOTORI results have been given by Tanaka (1983), Tsuneta (1983), Watanabe (1983), Takakura (1983), and Takakura *et al.* (1984).

2. The Hard X-Ray Morphology

The HXR (hard X-rays) are radiated when energetic electrons collide with ambient protons, while a more general energy loss occurs in collisions with ambient electrons. Thus the HXR morphology tells the location where energetic collisions occur. For comparisons with HXR imaging experiments there had been intensive discussions predicting the morphology. The non-thermal electron beam model (cf. Emslie, 1981) predicted that HXR should be mainly emitted from dense footpoints in a magnetic loop by a 'thick target' mechanism, while the thermal model, which attributed the HXR to thermal emission from a few hundreds million degree plasma, predicted that an HXR emitting region will expand rapidly from the top of a loop to the whole loop and footpoints (cf. Smith and Lilliquist, 1979).

Actual images made at 20–25 keV range showed a wide variety of morphologies, especially in the results from HINOTORI, which also provided an unexpected case of a stationary coronal source (Takakura *et al.*, 1983). The most common case observed from HINOTORI is a single source, extended through a widespread region of a flare, presumably covering the whole loops and footpoints. The source size is generally much larger than the instrumental resolution. The footpoint source is rather rare in the

HINOTORI results (about 10%). In one flare that occurred near the limb (20 July, 1981; Tsuneta *et al.*, 1983; Tanaka, 1983) a weak sub-source was found to be located at the footpoint, about 20000 km apart from the main source which extended to the corona. But no spread of emission was visible between the main source and the footpoint within the time resolution of 5 s, as would have been expected from the thermal model. A variety of flares have been classified by combining the HXR morphology, spectrum and time profile (Tanaka, 1983; Tanaka *et al.*, 1983; Tsuneta, 1983). Although physical differences in the classified types are not obvious yet, the following three should be described as separate types.

(a) IMPULSIVE BURST

This category is the most common type of flare. It shows a rapid HXR time variation of the order of a few seconds. Some results of HXR morphology from the SMM indicated a double source structure located at the footpoints in the initial phase, later evolving to a single source in between the double source (Hoyng *et al.*, 1981, 1983; Duijveman *et al.*, 1982). Hinotori showed mostly single source structure, and, in rare cases, double sources at the beginnings of flares (Takakura, 1983). A typical evolution of the single source structure is, as shown in Figure 3, that an extended source, covering the whole loop area between the footpoints (two Hα ribbons), shrinks gradually from the middle phase and tends to shift, presumably upwards, while a rapid flux variation continues. The apparent velocity of upward, and sometime lateral, motion (no downward motion is detected) is on the range of 10–150 km s^{-1}. It is likely that this motion manifests same change of the energy release site, in analogy with the rising motion of a loop prominence system. A small source size from the middle phase may provide another difficulty to the thermal model; because the confinement of a few hundred million K plasma near the loop top is unlikely, due to the effective thermal conduction (Takakura, 1984). The extended source in the early phase implies that energy releases occur simultaneously in multiple loops, as well as that HXR is emitted from these loops. In view of the high coronal density, of the order of 10^{11} cm^{-3}, as derived from soft X-ray and EUV observations (cf. Dere *et al.*, 1979), the coronal HXR emission would be compatible with a non-thermal electron beam model, taking into account a loop length larger than 10000 km and that the column density for stopping a 20 keV electron is equal to $10^{19.6}$ cm^{-2}. Only in a sufficiently early phase, in which the coronal density is low enough, will 20 keV electrons penetrate to the footpoint region showing HXR from the footpoints.

(b) THERMAL BURST

Among the family of impulsive flares there has been discovered a new type in which thermal emission is dominant in low energy HXR (Lin *et al.*, 1981; Tanaka *et al.*, 1982). From analyses of the Fe XXVI spectrum (Tanaka *et al.*, 1982; Tanaka, 1985) this new thermal component (hot thermal component) is found to show an electron temperature of some 30–40 million K, and a large emission measure of around 10^{49} cm^{-3}. It becomes conspicuous in the middle phase of a flare, and is manifested by a gradual

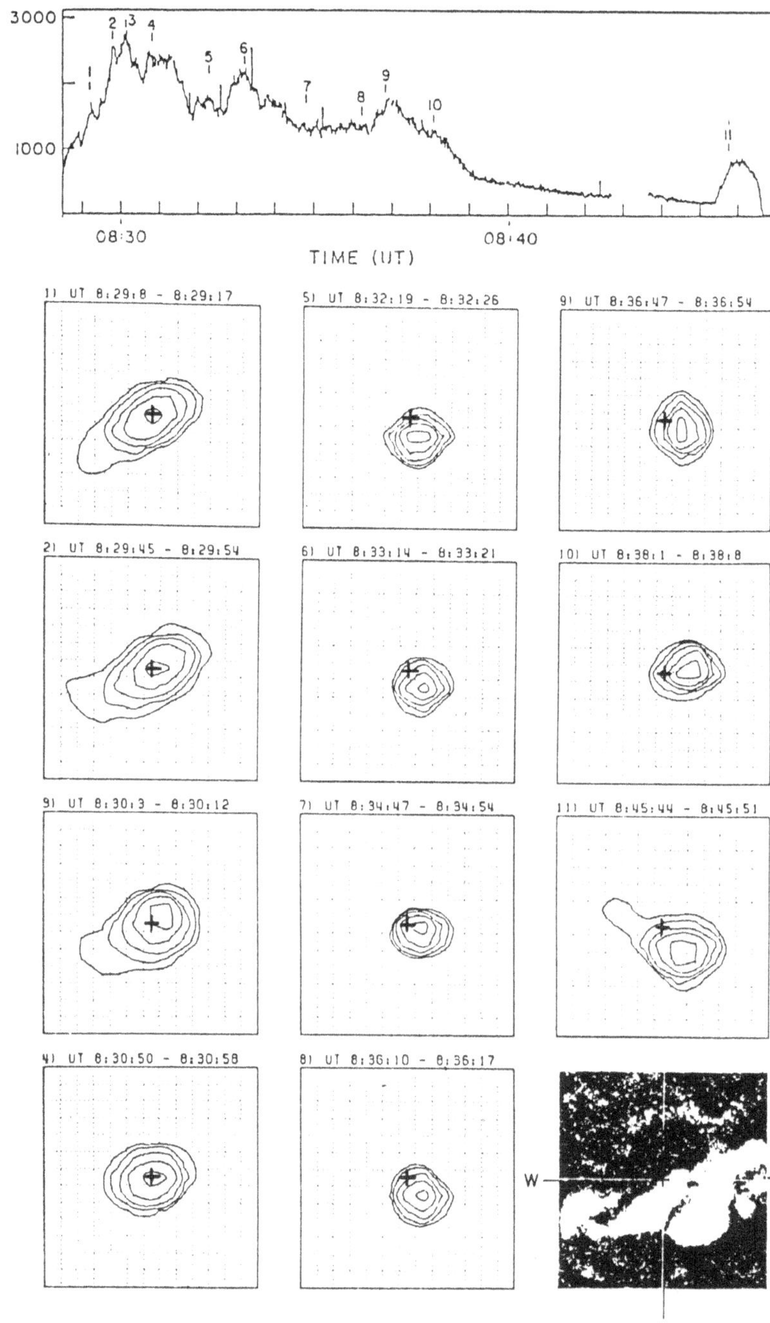

Fig. 3. HXR images (30–50 keV nominal energy range) in an impulsive burst of 21 August, 1981 from Hinotori. The top shows a X-ray time profile at the band of 18–46 keV. A Hα flare is shown at low right. One pixel in the map is 9.964 arc sec and the contour step is $\sqrt{3}$ times (from Takakura, 1983).

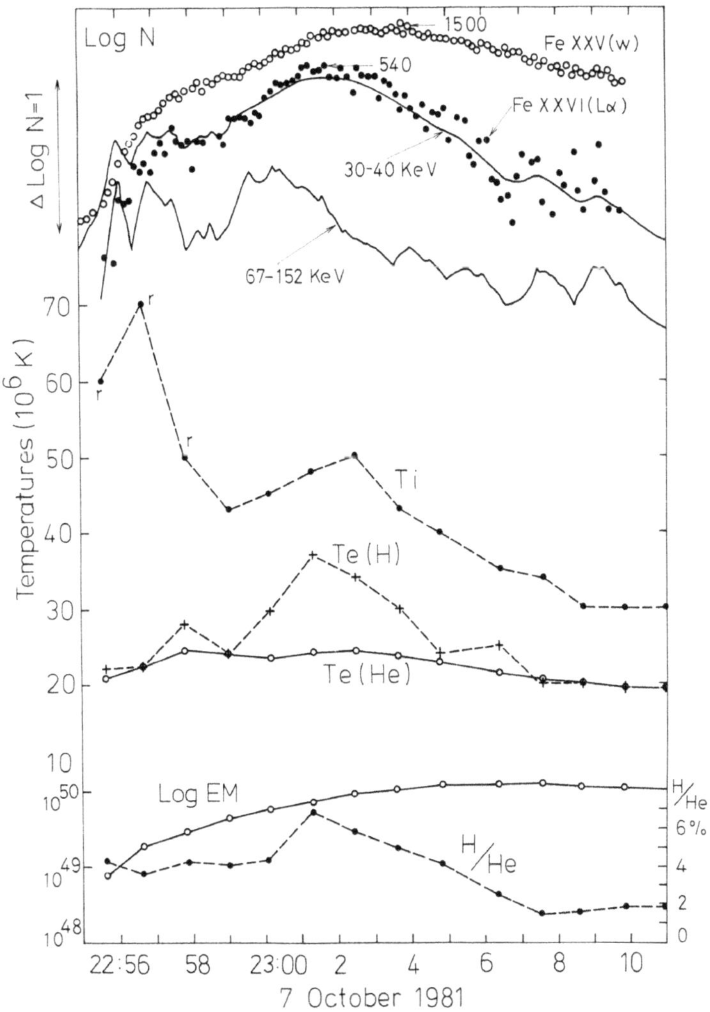

Fig. 4. Time profiles of 7 October, 1981 flare from HINOTORI. From top: fluxes of Fe XXV and Fe XXVI resonance lines, 30–40 keV and 67–152 keV. Ion temperature (Doppler broadening) of Fe XXV resonance line, electron temperatures: Te(H) and Re(He), respectively, derived from Fe XXVI and Fe XXV spectra, emission measure and abundance ratio of Fe XXVI (H-like) to Fe XXV (He-like) ions.

enhancement in the HXR below 40 keV. Figure 4 illustrates a typical variation of this type with time. While the electron temperature derived from the Fe XXV spectrum (Figure 2(b)) shows an almost flat distribution around 20 million K, the electron temperature derived from the Fe XXVI spectrum (Figure 2a) shows a secondary increase to 37 million K, coincident with a gradual HXR hump. The temperature increase starts together with the second series of impulsive bursts. Seven out of thirteen large flares observed from Hinotori showed hot thermal components hotter than 30 million K (Tanaka, 1985). Lin *et al.* (1981) detected one such flare from the high resolution HXR spectrum. The hot thermal component is emitted from the corona in close association with a bright Hα loop, which suggests a high density. Probably high coronal density would be a key factor to distinguish this type from the normal impulsive burst.

(c) GRADUAL HARD BURST

This HXR burst, which Hinotori often witnessed, is characterized by an extended duration (15–30 min) and broad peaks of duration 1 min or longer without sharp spikes. The HXR source is located high in the corona (above 15 000 km), extended and rather stationary (Tsuneta *et al.*, 1984). In this burst γ-rays show a delay of their peak (30 s–1 min) relative to the HXR peak, and spectral hardening is observed with time. This shows a contrast to the impulsive burst, in which the peaks of flux are simultaneous within a few seconds from HXR to γ-rays, and no spectral hardening is observed. These particular forms of behaviour in the gradual hard burst suggest trapping of high energy particles in the corona and continuous acceleration as well. Microwave morphology (Kawabata *et al.*, 1983) supports the coronal trapping, too. Differences in magnetic morphology seem essential to discriminate this class, since the large-scale magnetic fields plausible for this case tend to have large curvatures, which would result in the trapping of particles, due to the magnetic mirror effect.

3. The Energy Release in Impulsive Bursts

The nature of the energy release in impulsive bursts may be summarized as follows. First the acceleration takes place very rapidly for a very wide energy range. The SMM results indicated that from 40 keV to 65 MeV the flux rises in a few seconds and the peaks are simultaneous to within 2 s (SMM Workshop, 1984). The rapid fall of the γ-ray flux suggests that an energy loss of high energy particles takes place at high ambient density, of the order of 10^{13} cm^{-3}; i.e., in the chromosphere. Both electrons and protons are accelerated almost simultaneously, as demonstrated from the simultaneity between nuclear γ-ray lines, and HXR emissions. Very high resolution microwave mapping by VLA (Hoyng *et al.*, 1983) clearly shows brightenings in very small regions of a few arc sec at loop tops, suggesting that the acceleration actually takes place at the loop top. Impulsive UV and optical emissions in some flares show good time coincidences with the HXR spikes. From the observations of optical flashes at 3835 Å (Zirin and Tanaka, 1973), and UV lines of Si IV and O IV (Cheng *et al.*, 1981) it is found that brightening occurs sequentially at different footpoints corresponding to the individual HXR spikes. These observations suggest that the energy release site moves around different loops sequentially with a time-scale of a few seconds, and the produced electrons stream to the footpoints of individual loops, exciting the transition region or lower layers. The reported shortest variation in the HXR burst is 20 ms for the rise time, and 45 ms for the duration (Kiplinger *et al.*, 1983). This time-scale, much shorter than the electron transit time and collision decay time, would limit the shortest time-scale for the energy release, which must be explained by the theory.

From these results and the HXR morphology we may derive an energy release model as shown in Figure 5. In the initial phase the energy release may occur at widespread

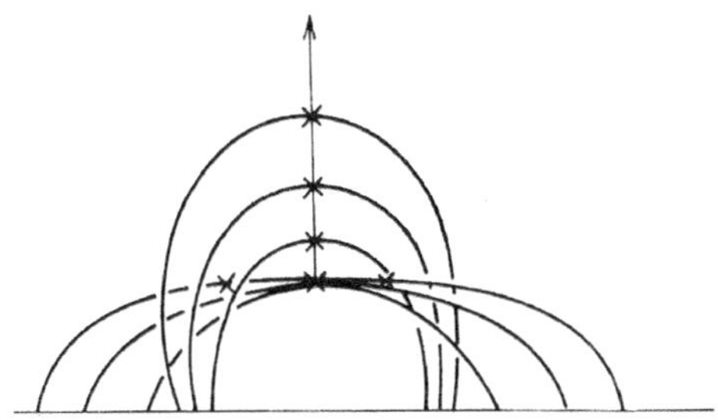

Fig. 5. An energy release model in the impulsive burst. Energy release site (shown by cross) moves around low-lying loops (shown by lines) and then shifts to higher loops.

low-lying loops, either simultaneously or in a fast sequence. In later phases the energy release site may move successively to higher or lateral loops as suggested from the HXR source motions. This empirical model may be further developed to a flare model, as illustrated in Figure 6. The pre-flare magnetic configuration would be an ensemble of sheared fields (Tanaka and Nakagawa, 1973), and a filament, which is located above the neutral line and merges with tops of many of the sheared fields. The sheared fields contain large currents, but there would not be a force-free field ($j \times B = 0$), since the Lorentz force can be balanced by the gravity of prominence matter, together with some magnetic forces arising from complex intertwinings between the sheared fields and filament fields as well. A flare may be triggered by a filament eruption, which separates the intertwined fields into rising filaments and many low-lying force-free fields, either through multiple reconnections and/or by slipping-out of the filament fields from the sheared fields. This separation will occur all at once for the low-lying sheared fields, with the extra-energy included in the sheared fields being released into individual (reconnected) force-free fields simultaneously. This realizes multiple energy releases in

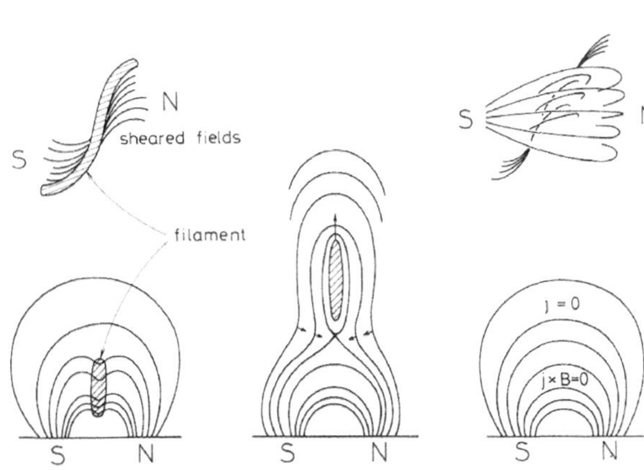

Fig. 6. Magnetic field configurations at pre-flare, flare, and post-flare states. Upper figures are views from above and lower figures are side views.

the widespread loops (force-free fields). The rising motion of erupted filaments will also stretch upper-lying fields, and cause further reconnections successively at higher levels (Hirayama, 1974). The energy will be released at the reconnected point or at the top of each loop formed by the reconnection. Post-flare loops appear successively from the lower to higher levels; explaining the apparent rising motions of the HXR source and Hα loop. The post-flare magnetic configuration will consist of force-free fields at the low levels, and potential fields at the higher levels.

4. Soft X-Ray Results and Electron-Heated Flare Atmospheres

High resolution soft X-ray spectra from highly ionized ions of iron and calcium provided a powerful tool for the diagnostics of flare-produced hot plasmas. One of the most important discoveries in the space-borne observations in this cycle would be the findings of large non-thermal broadening (100–250 km s^{-1}) and blue-shifted (400 km s^{-1}) components in the line spectra emitted at temperature of 10–20 million K (Doschek *et al.*, 1979; Antonucci *et al.*, 1982; Tanaka *et al.*, 1982). Large non-thermal broadening is found in all flares during the early phase, while the blue-shift is found in the early phase of disk flares only. In the largest class of soft X-ray flare (X12) the temporal behaviour of the blue-shifted component of the Fe XXV resonance line could be analyzed quantitatively (Tanaka and Zirin, 1985). It is found that the emission measure of the blue-shifted component is largest during the impulsive phase, showing best correlation with the HXR burst in 50–60 keV range, and that the emission measure of the stationary component of the hot cloud continues to rise, until the blue-shifted emission measure has declined. These relations suggest that rising plasma seen as the blue-shifted component supplies the coronal hot cloud of 20 million K; as well as that this rising motion occurs in association with the bombardment of 50–60 keV electrons, providing evidence for the evaporation hypothesis for the origin of the hot cloud (Hirayama, 1974). Further analyses have revealed (1) the time-integrated energy deposition by electrons above 50 keV can explain the total thermal energy (10^{31} ergs) in a 20 million K cloud; (2) the time-integrated mass input from the blue-shifted component roughly explains the total mass ($10^{15.6}$ g) of a stationary cloud; and (3) soft X-ray flux increases in proportion to the time-integral of the HXR flux.

From the above empirical results we may reach a model of an electron-heated flare atmosphere (Figure 7) for an energy-released loop. A stream of 50 keV electrons will penetrate to the chromosphere, and heat it instantaneously up to 20 million K, where upon the heated plasma will expand up to the corona forming a coronal hot cloud. Heating by the energetic electrons can be very efficient, as revealed from numerical simulations (cf. MacNeice *et al.*, 1984), and may be consistent with the observed rapid heating (Tanaka, 1980), although it is not obvious how to make this agree with a finding of very small polarization in the Fe line spectrum (Akita *et al.*, 1983). Lower energy electrons will be stopped before reaching the footpoints and contribute to the HXR emission from the corona at the 20 keV range. These electrons will also heat the coronal plasma present before the rising plasma arrives. The coronal energy deposition by these

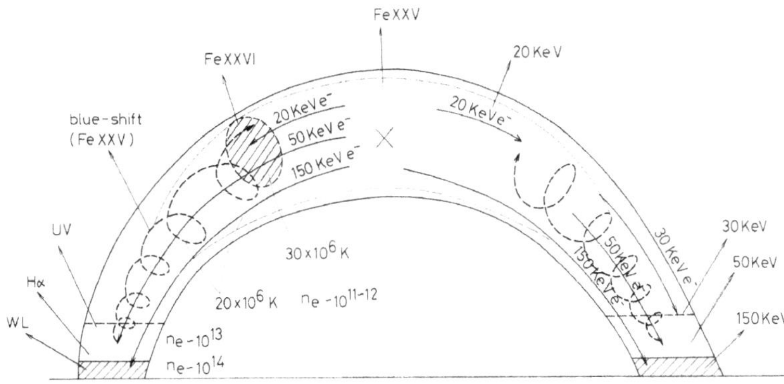

Fig. 7. A loop model of an electron-heated flare atmosphere. Cross indicates the energy release site. Dashed lines represent evaporating plasmas from the chromosphere. Electron streams are shown by arrows inside the loop while various radiations are shown by arrows outside the loop.

electrons is about one or two orders of magnitude larger than the energy deposited by electrons above 50 keV or the soft X-ray source thermal energy. However, the thermal energy of *in situ* coronal plasma will not become large, because of its low density, and most of the coronal thermal energy will come from the evaporating dense plasma. Presumably, initial large energy deposition in the corona may be immediately balanced by the downward conduction flow, which is eventually radiated in the UV, EUV, and optical regions, and possibly by the mass motion (coronal transient) which may arise associated with the large coronal energy deposition. The energy radiated in the UV region appears to be much larger than the thermal energy in the soft X-ray region (Canfield *et al.*, 1980). On the other hand, the finding of a non-thermal power-law spectrum below 10 keV (Figure 1) raises a serious problem to the flare energy budget as enormous energy must be deposited from electrons below 20 keV if this spectrum suggests the existence of such electrons (Tanaka *et al.*, 1984). Actually, the non-thermal behaviour of HXR below 10 keV explains satisfactorily the observed transient iron Kα emission as due to fluorescence (Tanaka *et al.*, 1984). The energy budget is certainly an important problem to be clarified in the future.

The newly discovered hot thermal component originates in the corona. Its large emission measure and appearance from the middle phase, together with a good correlation with the enhanced phase of the impulsive burst, may suggest that its origin is due to re-heating of a massive amount of evaporating plasma by collisions with low energy electrons. Such collisions may be realized when acceleration continues in a loop for a sufficient time, and also when acceleration occurs in neighbouring fields, to which the rising plasma drifts. At the present stage of analysis the relation between the hot thermal and the normal hot plasma is still not clear, thus leaving a possibility that the hot thermal component is produced, independently of the evaporation, and simultaneously with the accelerations, by direct conversion (Joule heating) from the magnetic field (Tsuneta, 1985).

The evidence of particle bombardment to the lower, dense atmosphere may be given, as noted before, by the short decay time of γ-ray emission, and also by impulsive UV and optical flashes. White light flares were previously attributed to the bombardment of high energy protons to the photosphere. However, the protons' total energy seems to be not sufficient. In the largest flare observed from Hinotori the white light emission showed best temporal correlation with the HXR burst in the 100–150 keV range, while correlation with nuclear γ-line at 4–6 MeV was not so good (Tanaka and Zirin, 1985). The total energy of the white light emission (10^{30} ergs) is consistent with energy deposition by electrons above 100 keV, thus suggesting this energy of electron as a possible exciter of the white light flare. Direct evidence of bombardment by 100 keV electrons has been provided from stereoscopic observations using two spacecrafts: ISEE3 and PVO (Kane *et al.*, 1982), which showed that 100 keV HXR originates from a lower level than 2500 km, or from the chromosphere.

In conclusion, particle accelerations and subsequent energy transfers seem to explain the heatings of various atmospheric levels and the formation of hot coronal clouds as well. For further clarification of this model we will need HXR images in higher energy ranges, and with higher spatial resolution, together with more comprehensive data for spectroscopic diagnostics.

References

Akita, K., Tanaka, K., and Watanabe, T.: 1983, *Solar Phys.* **86**, 101.

Antonucci, E., Gabriel, A. H., Acton, L. W., Culhane, J. L., Doyle, J. G., Leibacher, J. W., Machado, M. E., Orwig, L. E., and Rapley, C. G.: 1982, *Solar Phys.* **78**, 107.

Canfield, R. C., Cheng, C.-C., Dere, K. P., Dulk, G. A., Mclean, D. J., Robinson, R. D., Schmahl, E. J., and Schoolman, S. A.: 1980, in P. A. Sturrock (ed.), *Solar Flares*, Colorado Associated University Press, p. 451.

Cheng, C.-C., Tandberg-Hanssen, E., Bruner, E. C., Orwig, L., Frost, K. J., Kenny, P. J., Woodgate, B. E., and Shine, R. A.: 1981, *Astrophys. J.* **248**, L39.

Chipman, E. G.: 1981, *Astrophys. J.* **244**, L113.

Dere, K. P., Mason, H. E., Widing, K. G., and Bhatia, A. K.: 1979, *Astrophys. J. Suppl.* **40**, 341.

Doschek, G. A., Kreplin, R. W., and Feldman, U.: 1979, *Astrophys. J.* **233**, L157.

Duijveman, A., Hoyng, P., and Machado, M. F.: 1982, *Solar Phys.* **81**, 137.

Emslie, A. G.: 1981, *Astrophys. J.* **245**, 711.

Hirayama, T.: 1974, *Solar Phys.* **34**, 323.

Hoyng, P., Duijveman, A., Machado, M. E., Rust, D. M., Svestka, Z., Boelee, A., de Jager, C., Frost, K. J., Lafleur, H., Simnett, G. M., van Beet, H. F. and Woodgate, B. E.: 1981, *Astrophys. J.* **246**, L155.

Hoyng, P., Marsh, K. A., Zirin, H., and Dennis, B. R.: 1983, *Astrophys. J.* **268**, 865.

Inoue, H., Koyama, K., Mae, T., Matsuoka, M., Ohashi, T., Tanaka, Y., and Waki, I.: 1982, *Nuclear Instr. Method.* **196**, 69.

Kane, S. R., Fenimore, E. E., Klebesadel, R. N., Laros, J. G.: 1982, *Astrophys. J.* **254**, L53.

Kawabata, K., Ogawa, H., and Suzuki, I.: 1983, *Solar Phys.* **86**, 247.

Kiplinger, A. L., Dennis, B. R., Emslie, A. G., Frost, K. J., and Orwig, L. E.: 1983, *Astrophys. J.* **265**, L99.

Lin, R. P., Schwartz, R. A., Pelling, R. M., and Hurley, K. C.: 1981, *Astrophys. J.* **251**, L109.

MacNeice, P., McWhirter, R. W. P., Spicer, D. S., and Burgess, A.: 1984, *Solar Phys.* **90**, 357.

Smith, D. F. and Lilliequist, C. G.: 1979, *Astrophys. J.* **232**, 582.

Takakura, T.: 1983, *Proc. Flare Workshop at Kumming,* China (in press).

Takakura, T.: 1984, *Solar Phys.* **91**, 311.

Takakura, T., Tanaka, K., and Hiei, E.: 1984, *Adv. Space Res.* **4**, 143.

Takakura, T., Tsuneta, S., Ohki, K., Nitta, N., Makishima, K., Murakami, T., Ogawara, Y., Oda, M., and Miyamoto, S.: 1983, *Astrophys. J.* **270**, L83.

Tanaka, K.: 1980, in F. Moriyama and J. C. Henoux (eds.), *Proc. Japan–France Seminar on Solar Physics*, JSPS, Tokyo, p. 219.

Tanaka, K.: 1983, in P. B. Byrne and M. Rodono (eds.), 'Activity in Red-Dwarf Stars', *IAU Colloq.* **71**, 307.
Tanaka, K.: 1985, *Publ. Astr. Soc. Japan* **38** (in press).
Tanaka, K. and Nakagawa, Y.: 1973, *Solar Phys.* **33**, 187.
Tanaka, K. and Zirin, H.: 1985, *Astrophys. J.* (in press).
Tanaka, K., Nitta, N., Akita, K., and Watanabe, T.: 1983, *Solar Phys.* **86**, 91.
Tanaka, K., Watanabe, T., Nishi, K., and Akita, K.: 1982, *Astrophys. J.* **254**, L59.
Tanaka, K., Watanabe, T., and Nitta, N.: 1984, *Astrophys. J.* **282**, 793.
Tanaka, Y.: 1983, *Solar Phys.* **86**, 3.
Tsuneta, S.: 1983, in J. C. Pecker and Y. Uchida (eds.), *Proc. Japan–France Seminar on Active Phenomena in the Outer Atmosphere of the Sun and Stars*, C.N.R.S. and Observatory of Paris, Paris, p. 243.
Tsuneta, S.: 1985, *Astrophys. J.* **290**, 353.
Tsuneta, S., Takakura, T., Nitta, N., Ohki, K., Makishima, K., Murakami, T., Oda, M., and Ogawara, Y.: 1983, *Solar Phys.* **86**, 313.
Tsuneta, S., Takakura, T., Nitta, N., Ohki, K., Tanaka, K., Makishima, K., Murakami, T., Oda, M., Ogawara, Y., and Kondo, I.: 1984, *Astrophys. J.* **280**, 887.
Watanabe, T.: 1984, in J. C. Pecker and Y. Uchida (eds.), *Proc. Japan–France Seminar on Active Phenomena in the Outer Atmosphere of the Sun and Stars*, C.N.R.S. and Observatory of Paris, Paris, p. 230.
Zirin, H. and Tanaka, K.: 1973, *Solar Phys.* **32**, 173.

DYNAMICAL PROCESSES IN THE SOLAR CORONA AND INTERPLANETARY SPACE (INVITED REVIEW)*

R. T. STEWART

Division of Radiophysics, CSIRO, Sydney, Australia

(Received 22 March, 1985)

Abstract. Observations of the solar corona collected over the past decade are discussed from the point of view of short-term (< 1 day) and long-term (> 1 year) effects. Various phenomena are described, including coronal mass ejections, shock waves, and magnetic sector structures. It is argued that emerging magnetic flux is probably the prime cause of these phenomena, although the details of the interaction processes in the corona are not fully understood.

Short-term changes are caused by the sudden release of energy in the lower corona or chromosphere. The prime cause is thought to be due to the build-up of highly sheared magnetic flux. Theoretical work of late has concentrated on attempting to explain the time-scales of flare events (< 10 s) in terms of magnetic reconnection. Other work has concentrated on attempting to explain observed features of coronal mass ejections which last for periods of several hours. Long-term changes last for several years and are characterized by the slow evolution of coronal structures, especially magnetic sectors, which extend into interplanetary space out and beyond the Earth. Recent observations place new restraints on the solar dynamo which is thought to be responsible for the emerging magnetic flux involved in these long-term changes.

1. Introduction

There appears to be a growing amount of observational evidence, supported by semi-quantitative analytical and numerical models, which indicates the importance of magnetic field reconnection in coronal dynamical processes such as solar flares and coronal mass transients. I intend to review some of this evidence and then briefly discuss magnetic reconnection in global coronal structures.

Theoretical models of solar flares have been reviewed recently by Priest (1981, 1983). The observational evidence has been supplied by excellent ground-based facilities and by recent spacecraft experiments aboard the Skylab (1973), Solar Maximum Mission (1980, 1984), and Hinotori (1981–1983) satellites.

High-resolution Hα flare patrol observations at Big Bear Observatory and elsewhere have highlighted the importance of emerging magnetic flux near active regions for the production of solar flares. Soft X-ray and EUV spatial observations from Skylab showed that the basic element in a simple flare was a magnetic loop, whilst in a more complex 'two-ribbon' flare it was an arcade of loops. Recent SMM and Hinotori hard X-ray images have confirmed that the bremsstrahlung emission comes mainly from the feet of a loop (where the density is highest), except in some extended bursts, when the whole loop appears to light up (Figure 1; Kosugi and Tsuneta, 1983).

* Paper presented at the IAU Third Asian-Pacific Regional Meeting, held in Kyoto, Japan, between 30 September–6 October, 1984.

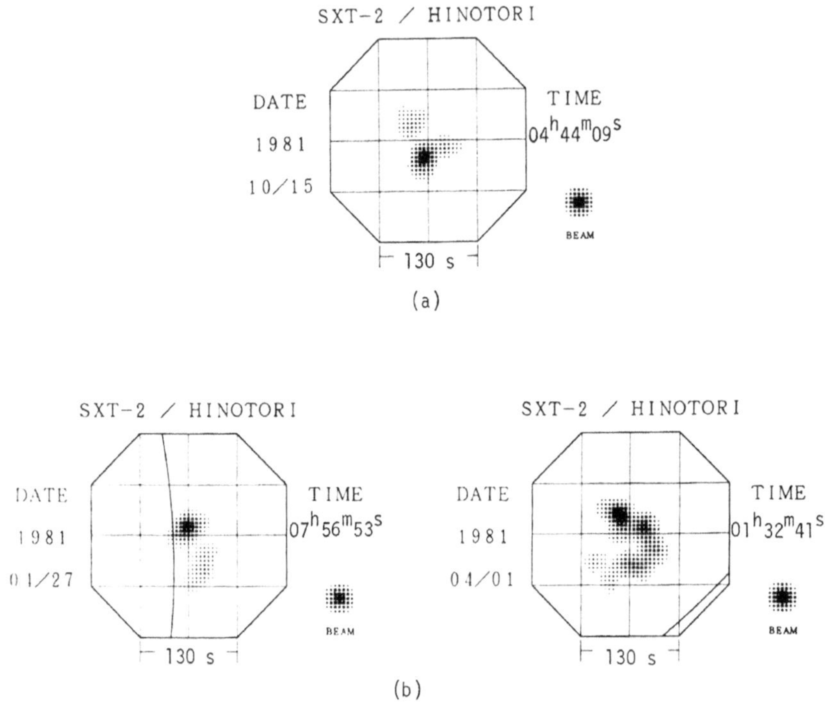

Fig. 1. (a) A snapshot map of the multiple impulsive hard X-ray burst of 15 October, 1981 showing three distinct source regions. Top is to the north and right is to the west. (b) The extended outburst of 27 April, 1981 (left) and 1 April, 1981 (right). Note the arch-like structure. (After Kosuga and Tsuneta, 1983.)

VLA observations at microwaves have shown that the loop sources are moderately circularly polarized, as expected for gyro-synchrotron emission from mildly relativistic electrons. The disposition of the polarized radio sources suggest that the magnetic topology changes during the flare and raises the possibility of an X-type neutral point forming between two diametrically opposed magnetic loop systems at the time of the impulsive phase of a flare. One of these loop systems could be a contender for emerging flux (dashed loops of Figure 2).

Two well-observed impulsive flares during the SMM emission (Figure 3) show similar quasi-periodic time profiles in microwaves, hard X-rays, and γ-rays (Nakajima *et al.*, 1983). The microwave plots show two sub-pulses, the first appearing along with the hard X-ray spike and the second along with the γ-ray spike. The two sub-pulses are separated by a time interval 1–2 s in the first event, and ~ 5 s in the second event. These observations indicate that electrons are accelerated up to energies ~ 350 keV in the first pulse, while protons are accelerated to energies ~ 50 MeV in the second pulse. Hence, if a *second-phase acceleration* process occurs in solar flares, as mooted previously to explain solar cosmic-ray events, then this additional acceleration occurs extremely rapidly within ≤ 1 s of the first acceleration. Such an observation removes the need for second-stage acceleration of fast particles by a coronal shock wave, which has been a long-standing unresolved question in solar physics. It also places severe constraints on

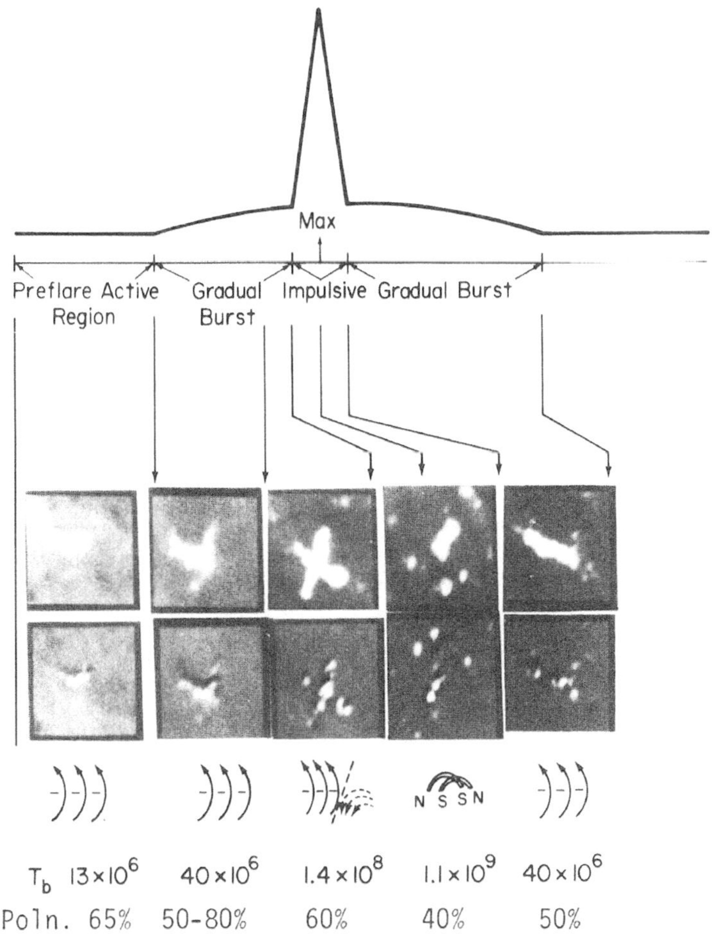

Fig. 2. VLA snapshot maps for the 14 May, 1980 radio burst at 6 cm. Central row shows total flux I and bottom row circular polarization V. Source brightness temperature T_b and degree of circular polarization are also given. (After Kundu, 1983.)

the impulsive acceleration mechanism, requiring the responsible instability to act in a time period ≤ 1 s and to be quasi-periodic with a time-scale ~ 10 s.

Finally, SMM soft X-ray observations have provided confirmation that chromospheric 'evaporation' occurs during the impulsive phase of the flare. The emission lines show strong Doppler broadening with upward motions of a few hundred kilometres per second (Antonucci *et al.*, 1982). Spatial observations (de Jager, 1983) show that the initial flare heating occurs at the top of the loop (at B of Figure 4(c)) followed by a gradually increasing X-ray emission in one leg of the loop (near A of Figure 4(c)), which becomes the brightest kernel of emission during the impulsive phase (Figure 4(a)). When this fades an extended tongue of weak X-ray emission (Figures 4(a), (b)) can be seen expanding outwards from the small loop towards the larger set of nested loops (D and F of Figure 4(c)). De Jager (1983) has suggested that the gradual component of this soft

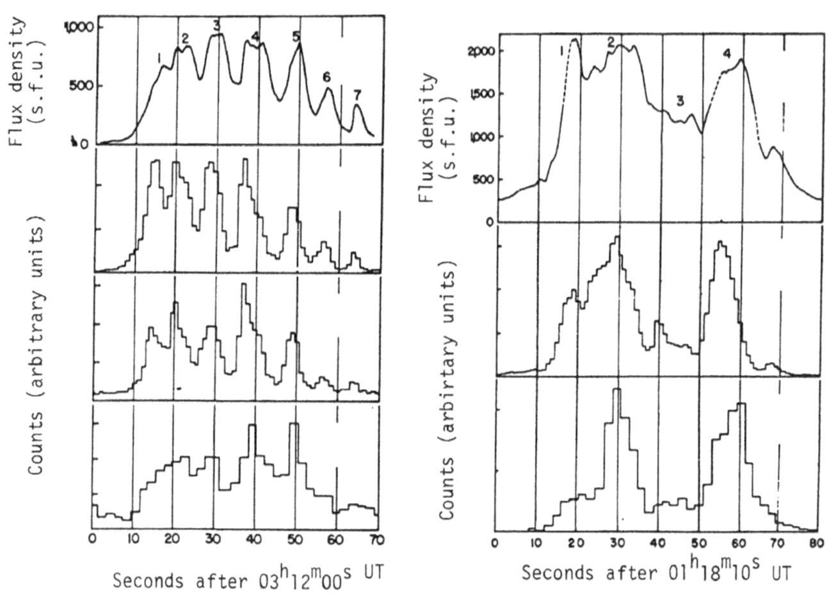

Fig. 3. (a) Time profiles of microwave (a, 17 GHz), hard X-ray (b, 40–140 keV), (c, 300–350 keV), and prompt γ-ray line (d, 4.1–6.4 MeV) emissions for the 7 June, 1980 solar flare. Time constants for SMM X-ray spectrometer were ~ 1 s (b), ~ 1 s (c), and ~ 2 s (d). Time constant for the 17 GHz Nobeyama polarimeter was 0.3 s (a). (b) Time profiles of microwave (a, 17 GHz), hard X-ray (b, 40–140 keV), and γ-ray line (c, 4.1–6.4 MeV) emissions for the 21 June, 1980 solar flare. Time constants of 0.3 s (a), ~ 1 s (b), and ~ 2 s (c). (After Nakajima *et al.*, 1983.)

X-ray flare (the extended tongue of Figure 4) is due to convectively rising hot material ($T \approx 10^7$ K) caused by chromospheric evaporation.

Such a picture is consistent with coronal loop flare models (see Priest, 1981, 1983), in which a sudden heat source ($T \approx 10^8$ K) at the top of the loop causes a conduction front and shock wave to propagate along the loop down to the chromosphere, which is then heated and expands upwards (evaporation). Beams of fast electrons streaming down the loop legs from the impulsive heat source produce microwave gyro-synchrotron emission and hard X-ray bremsstrahlung. The cause of the heat source is undetermined.

2. Simple Loop Flare Model

Heyvaerts *et al.* (1977) suggest that simple-loop flares occur in three stages as the emerging magnetic flux interacts with the overlying field (Figure 5). During the pre-flare heating stage steady reconnection occurs in the current sheet (shaded area of Figure 4a) formed when the new flux pushes against the old. Plasma heating occurs between the four shocks (dashed lines) emerging from the neutral point (Petschek, 1964; Sato and Hayashi, 1979). The current sheet grows in height into a region of decreasing density and increasing temperature until a critical height is reached where the current density exceeds the level for the onset of plasma micro-instabilities (turbulence). At this

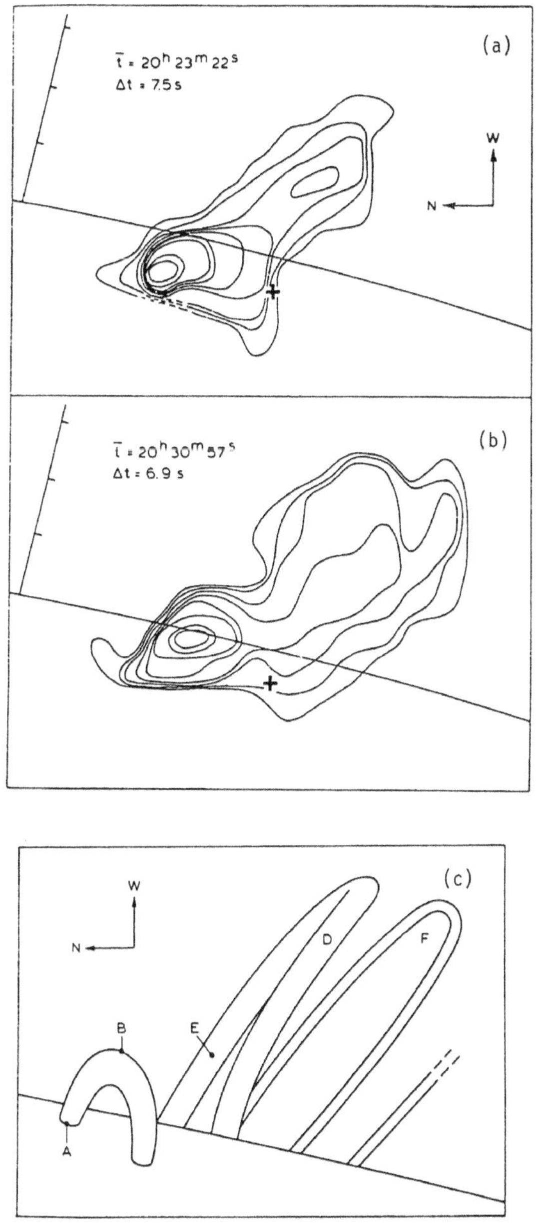

Fig. 4. (1) and (b) Soft X-ray (3.5–8.0 keV) maps obtained by the HXIS instrument aboard SMM during the phase of maximum brightness of the kernel (a) and at the end of the impulsive phase (b) of a solar flare. The curved line is the solar limb. The dashes denote height intervals of 10^4 km. (c) Pre-flare loop configuration derved from C IV observations from the UVSP instrument aboard SMM. (After de Jager, 1983.)

stage the impulsive phase is triggered (Figure 5(b)) and particle acceleration results from the strong electric fields generated within the current sheet. (Magnetic energy is converted at the current sheet into various forms of mechanical energy.) Radio and

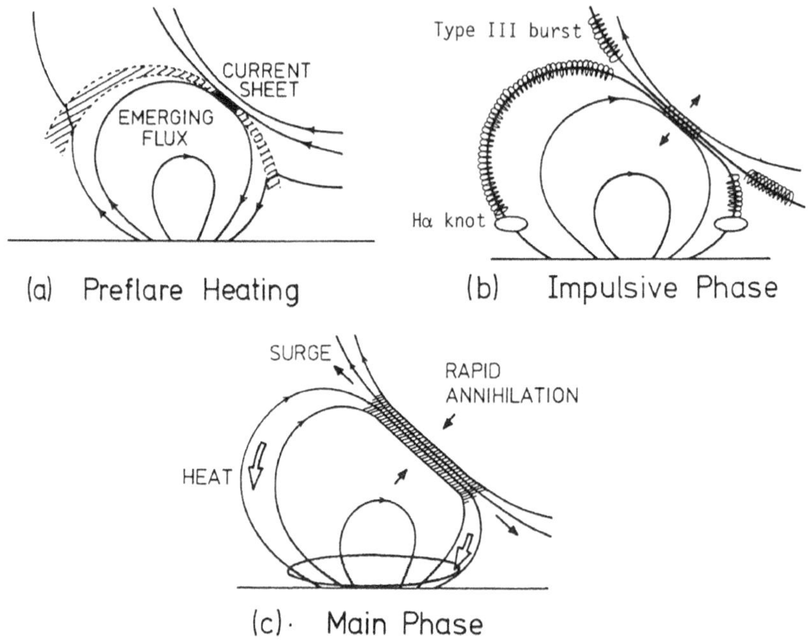

Fig. 5. The emerging flux mechanism for a simple-loop flare. (a) *Pre-flare phase.* The emerging flux reconnects with the overlying fields. (b) *Impulsive phase.* The onset of turbulence in the current sheet (when it has reached a critical height) causes a rapid expansion. The resulting electric field accelerates particles, which then escape along field lines and produce an impulsive mirowave burst, hard X-rays and type III radio bursts. (c) *Flash and main phases.* The current sheets reach a new steady state, with reconnection based on a turbulent resistivity. It is much larger than before and both heat and particles are conducted down to the lower chromosphere to produce the Hα flare (after Priest, 1981).

X-ray bursts result. Priest (1981) estimates that the condition for turbulence is reached when $T_c > 4 \times 10^5$ K, where T_c is the plasma temperature inside the current sheet. Once turbulence starts the plasma heats up rapidly (impulsive phase) until a new thermal equilibrium state is reached based on turbulent resistivity (main phase). Priest (1981) estimates that if $B_0 = 100$ G and $L = 10^4$ km, where B_0 is the magnetic induction outside the current sheet and L_0 is its length, then the amount of energy released is $w = 10^{29}$ erg, which is characteristic of a small solar flare. During the main phase (Figure 5(c)) the current sheet grows and widens and plasma may be ejected from the sheet as a surge.

3. Two-Ribbon Flare

For a large two-ribbon flare the production of a small turbulent region by the above process stimulates the release of stored magnetic energy from a much larger overlying force-free field (non-potential, sheared field). Hα observations suggest that this can occur when new mangetic flux emerges near or directly beneath an active-region filament (Zirin, 1974). The speed of the flux emergence may be similar to that of the emerging arch filament system, i.e., ~ 1 to 10 km s^{-1} (Bruzek, 1969). The free energy contained in the force-free fields above a large complex active region is sufficient (~ 10^{32} erg) to account for the energy released in a two-ribbon flare (Tanaka and Nakagawa, 1973). The fast release of this energy excess may be made possible by the presence of a large (~ 10^4 km) turbulent current sheet between the emerging flux and the overlying arch filament fields. This could provide a local increase in diffusivity to allow fast reconnection to occur (Priest, 1981).

As shown by Anzer and Pneuman (1982), once reconnection occurs then field lines are pushed together (Figure 6) and new magnetic flux is added beneath the filament which creates a magnetic pressure which becomes the driving force for the coronal mass transient material trapped in the closed magnetic fields above the filament. The X-type neutral point (Figure 6) continues to rise, creating a system of nested loops with the hottest ones above cooler ones, as shown by soft X-ray and Hα observations of post-flare loops. Hα flare ribbons form at the footpoints of the arcade of magnetic loops, presumably owing to the streaming of particles or the propagation of conduction fronts and shocks down the legs of the loops away from the neutral point. In three dimensions the internal magnetic field of the coronal mass transient might resemble nested helices rather than closed loops as in Figure 6. For example, in a sheared magnetic structure

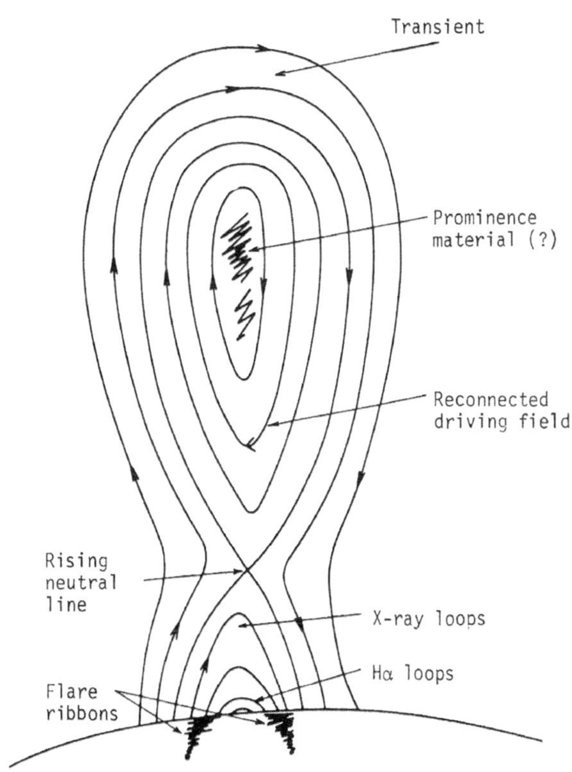

Fig. 6. Schematic of the magnetic configuration (end-on view) of a coronal transient showing locations of the rising neutral line due to reconnection, flare ribbons and loops, and the reconnected driving field (after Anzer and Pneuman, 1982).

the leg of one loop can reconnect with the opposite leg of an adjacent loop thus producing a helical field. Examination of high-resolution white-light coronal mass transient pictures sometimes suggests that the internal magnetic structure is of this kind (Anzer and Pneuman, 1982). These helical, magnetic fields affect the picth-angle distribution of the fast electrons accelerated during the reconnection process so as to produce the metre-wavelength moving type IV burst by either gyro-synchrotron or plasma emission. Other fast electrons probably stream down the post-flare loops to produce extended hard X-ray and microwave bursts. Finally, if the mass ejection is fast enough, a shock wave exciting a metre-wavelength type II burst accompanies the coronal mass ejection out into interplanetary space. The total kinetic and potential energy in the coronal mass outflow is estimated to be $\sim 10^{32}$ erg, similar to the energy release in the impulsive phase of a large flare. Upon arrival at the Earth the shock wave and mass outflow produce geomagnetic disturbances.

4. Evolution of Global Magnetic Fields

The global coronal magnetic field consists of two components: (a) large-scale open field regions (coronal holes) near the poles, and sometimes at other latitudes; (b) small-scale closed field structures interconnecting oppositely-polarized sunspot and plage fields associated with new (emerging) and old magnetic flux.

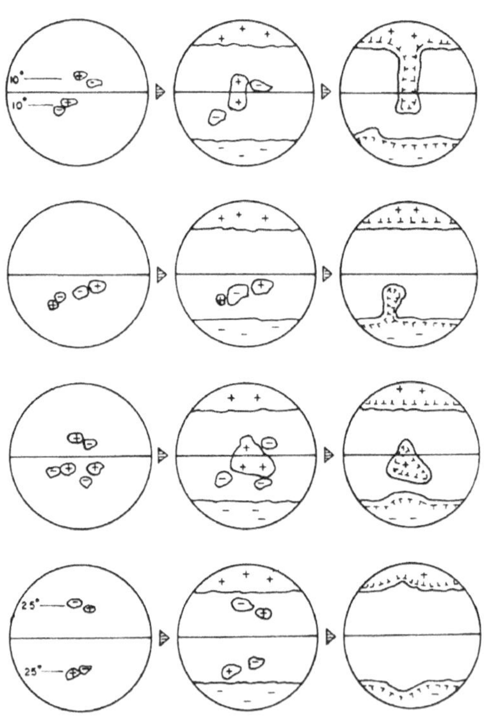

Fig. 7. Schematic examples of magnetic cell formation; + and − signs refer to polarity and ⊥ signs denote coronal hole (after Bohlin and Sheeley, 1978).

4.1. Coronal holes and polar field reversal

Coronal holes are probably formed from old active region fields as illustrated in Figure 7. These diagrams are based on observations taken during the Skylab mission (Bohlin and Sheeley, 1978). Note for example how a large coronal hole forms out of following polarity flux from two active regions which then merge with the north polar coronal hole (top row of Figure 7). In the bottom example we see how the coronal holes near the poles

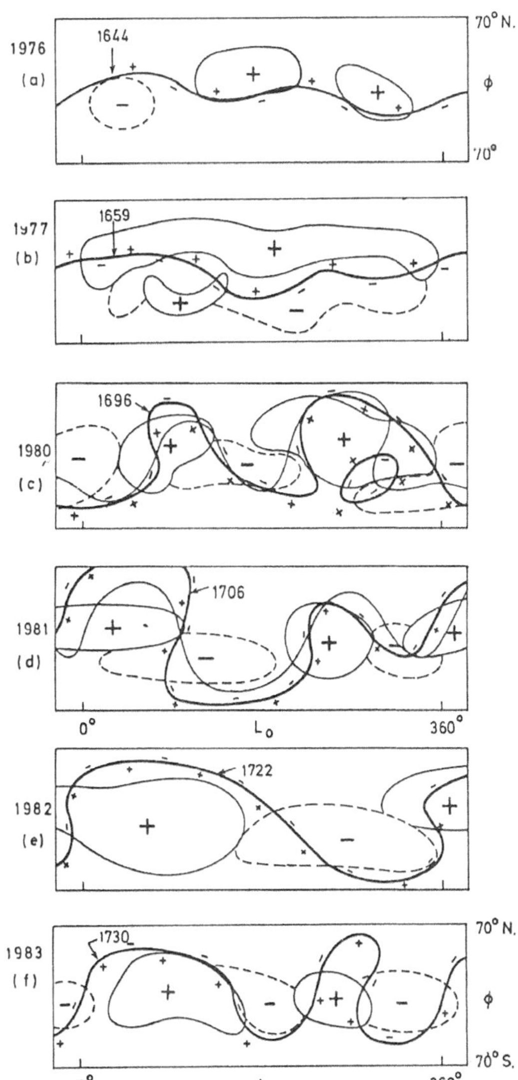

Fig. 8. Yearly, 160 MHz noise storm sector structure (full and dashed outlines denoting + and − magnetic polarity) compared with computed heliospheric current sheets, heavy and light wavy curves, heavy giving monthly value for a specified Carrington rotation, light giving yearly averages. ϕ denotes solar latitude, L_0 heliographic longitude (after Stewart, 1984; computed heliospheric current sheet data from Hoeksema *et al.*, 1983).

are reduced by the arrival of following polarity flux. This process when repeated many times during the rising part of the solar cycle eventually leads to the reversal of the polar fields near solar maximum.

The polar fields take approximately one year to reverse sign. This effect can be simply explained in terms of the observed meridional drift of sunspot magnetic flux towards the poles at a rate ~ 10 km s^{-1} (Topka *et al.*, 1982). (There is a slight excess of follower sunspot polarity drifting towards the poles.) Here, for simplicity, second-order diffusive effects due to supergranular convective motions are ignored. For a more quantitative discussion of these effects (see De Vore *et al.*, 1984). At the beginning of the new solar cycle the poles contain old-cycle follower flux. New-cycle flux first appears at latitudes $\sim 40°$ and drifts slowly towards the poles, arriving at $\sim 80°$ near the edge of the polar crown some four years later. This new flux is oppositely polarized to the old polar field and over the next one-year period reconnects with the polar field and annihilates it. New-cycle follower flux continues to arrive and replenish the polar fields. Hence, within a period of approximately a year or so near solar maximum the polar field reverses sign and then retains this sign until the next solar maximum.

4.2. Sector structures and noise storms

Metre-wavelength noise storms seem to be closely associated with emerging flux near active regions. For example, noise forms first emit at 160 MHz about one day after an increase in the slowly varying microwave component (Kai and Sekiguchi, 1973) and after about the same delay from the emergence of ephemeral active regions visible as growing loops of EUV or soft X-ray emission. The rate of growth of these loop systems, ~ 1 to 10 km s^{-1}, is of the correct magnitude to explain the one-day delay before the onset of noise storm activity. The latter probably arises from the micro-stabilities generated when new emerging flux interacts with old sunspot fields. The sense of circular polarization of the noise storm agrees with the dominant sunspot polarity in the active region (Stewart, 1984; Payne-Scott and Little, 1951).

However, these noise-storm associated sunspots do not obey the hemisphere polarity rule of Hale (1913). Rather they obey a sector polarity rule which matches the sector structure observed in the interplanetary magnetic field (Stewart, 1984). Consequently, noise-storm polarities map out large-scale magnetic cells in the low corona which correspond with the interplanetary magnetic field. This result is illustrated in Figure 8, where the heavy line represents the boundary between oppositely directed fields in interplanetary space – commonly referred to as the heliospheric current sheet.

This new result from radio noise storm observations suggests that the large-scale coronal magnetic cells have lifetimes $t_c \sim 1$ yr and that such cells are closely associated with emerging flux in active regions: the cells are formed and maintained by the appearance of sunspots at preferred solar latitudes and longitudes; furthermore, there is a net flux emerging from the active region which has the sign of the dominant sunspot. How can we explain the lifetimes of these magnetic structures? Firstly consider the observed latitudinal drifts of sunspots. Tuominen and Kyröläinen (1982) find that sunspots within $\pm 20°$ latitudes drift towards the equator with a maximum drift rate

$V_d \approx 3$ m s^{-1}. A similar, if somewhat greater, drift towards the poles occurs for sunspots in the 20° to 30° range. If we require sunspots to drift on an average, say, ± 10° before a large coronal magnetic cell is replaced by magnetic convection, then the lifetime of the cell would be $t_c \approx 500$ days. This is somewhat long. A more reasonable figure would be obtained if the drift rate were $v_d \approx 6$ m s^{-1} closer to that observed for meridional flows towards the poles. Secondly, we note that if global convective flows are mainly responsible for replacing the large-scale magnetic cells by magnetic reconnection of new flux with old flux then a systematic longitudinal drift is also required. Large-scale convective motions beneath the solar surface have been postulated by Yoshimura (1971, 1983) to be necessary for the maintenance of solar differential rotation by dynamo action, and as a possible explanation for the large-scale ordering of solar active regions. Whether such global convective cells occur and emerge from below the solar surface is still debatable, but the recent radio result presented above is probably further evidence for such an effect.

References

Antonucci, E., Gabriel, A. H., Acton, L. W., Culhane, J. L., Doyle, J. G., Leibacher, J. W., Machado, M. E., Orwig, L. E., and Rapley, C. G.: 1982, *Solar Phys.* **78**, 107.
Anzer, U. and Pneuman, G. W.: 1982, *Solar Phys.* **79**, 129.
Bohlin, J. D. and Sheeley, N. R.: 1978, *Solar Phys.* **56**, 125.
Bruzek, A.: 1969, *Solar Phys.* **8**, 29.
de Jager, C.: 1983, *Solar Phys.* **86**, 21.
De Vore, C. R., Sheeley, N. R., and Boris, J. P.: 1984, *Solar Phys.* **92**, 1.
Hale, G. E.: 1913, *Astrophys. J.* **38**, 27.
Heyvaerts, J., Priest, E. R., and Rust, D. M.: 1977, *Astrophys. J.* **216**, 123.
Hoeksema, J. T., Wilcox, J. M., and Scherrer, P. H.: 1983, *J. Geophys. Res.* **88**, 9910.
Kai, K. and Sekiguchi, H.: 1973, *Proc. Astron. Soc. Australian* **2**, 217.
Kosugi, T. and Tsuneta, S.: 1983, *Solar Phys.* **86**, 333.
Kundu, M. R.: 1983, *Solar Phys.* **86**, 205.
Nakajima, N., Kosugi, T., and Kai, K.: 1983, *Nature* **305**, 292.
Payne-Scott, Ruby and Little, A. G.: 1951, *Australian J. Sci. Res.* **A4**, 508.
Petschek, H. E.: 1964, *The Physics of Solar Flares*, AAS-NASA Sump., NASA Sp. 50, 425.
Priest, E. R. (ed.): 1981, *Solar Flare Magnetohydrodynamics*, Gordon and Breach, Vol. 1, p. 139.
Priest, E. R.: 1983, *Solar Phys.* **86**, 33.
Sato, T. and Hayashi, T.: 1979, *Phys. Fluids* **22**, 1189.
Stewart, R. T.: 1985, *Solar Phys.* **96**, 381.
Tanaka, K. and Nakagawa, Y.: 1973, *Solar Phys.* **33**, 187.
Topka, K., Moore, R., LaBonte, B. J., and Howard, R.: 1982, *Solar Phys.* **79**, 231.
Tuominen, J. and Kyröläinen, J.: 1982, *Solar Phys.* **79**, 161.
Yoshimura, H.: 1971, *Solar Phys.* **18**, 417.
Yoshimura, H.: 1983, *Astrophys. J. Suppl. Ser.* **52**, 363.
Zirin, H.: 1974, *Vistas Astron.* **16**, 1.

MAGNETODYNAMIC PHENOMENA IN THE SOLAR AND STELLAR OUTER ATMOSPHERES*

YUTAKA UCHIDA

Tokyo Astronomical Observatory, University of Tokyo, Mitaka, Tokyo, Japan

(Received 22 March, 1985)

Abstract. Magnetic effects causing anomalous heating and drastic flarings in the atmospheres of various types of stars are discussed. The best studied examples of these magnetic effects with spatially resolved observations are those in the case of the Sun, but no simple application of the solar knowledge to the stellar cases is allowed, since there are generally very large differences in the physical conditions between the Sun and other types of stars. We examine possible effects of the magnetic field in the respective situations of several types of stars which show X-ray and radio emissions indicating the presence of such activities, and it is concluded that the magnetic field may be playing important roles in producing anomalous heating and drastic flarings also in those stars having very different physical conditions, in ways seemingly very different from, but intrinsically closely related to, those in the case of the Sun.

1. Introduction

The Sun is, so-to-speak, a showcase of miniature prototypes of cosmic magnetodynamic phenomena. Amplification and regeneration of the primordial magnetic field take place in the Sun, and the processed magnetic field emerges to the surface. The high temperature outer atmosphere permeated by the extension of the photospheric magnetic field behaves as a typical medium of astrophysical magnetohydrodynamics exposed to a disturbance coming up from the convection zone below. Motions in the high β ($\equiv p_g/p_m$) region in the sub-photospheric convection zone disturb the magnetic field, and the disturbance coming out along the extension of the magnetic field into the low β part in the chromosphere and the corona, in turn, governs the situation there and causes the heating and drastic flarings which occur from time to time. The presence of the magnetic field serves as a means for transmitting disturbances, confining heated plasmas, controlling equilibria and flows, and producing drastic effects by releasing mass and energy through sudden transitions of the stressed magnetic configurations to lower energy states through instabilities. If the motion happens to be highly time-dependent, it can also produce prominent electrodynamic effects through electric charge separation in the plasma, due to the different responses of electronic and ionic components in it; and may lead to non-thermal effects through the acceleration of particles to high energies.

Recent development in the observations in optical, UV, X-ray, and radio information channels has revealed evidence indicating that phenomena similar to those observed on the Sun are also occurring in various types of stars. In the present paper we discuss the

* Invited review paper presented at the IAU Third Asian-Pacific Regional Meeting, held in Kyoto, Japan, between 30 September–6 October, 1984.

Astrophysics and Space Science **118** (1986) 127–148.

magnetodynamic activities in the outer atmospheres of various types of stars with the solar examples as guides into this new field.

2. Magnetodynamic Phenomena in the Solar Atmosphere

In order to provide some insight into the magnetodynamic phenomena taking place in the outer atmospheres of other stars, we first briefly see examples of the magnetodynamic phenomena in the solar atmosphere. These are the best studied examples of astronomical magnetodynamic phenomena observed with high spatial resolution. These include, among others, the anomalous heating of the rarefied outer atmosphere (= corona); which turned out to be magnetic field related phenomenon as shown by the Skylab observations. The corona was found to consist of aggregates of looped structures, coinciding with the magnetic field patterns there. Also the drastic release of energy and mass in the solar flares is known to be governed by the magnetic field, which plays the major roles in holding the plasma in a high-energy state and in releasing it to lower-energy states drastically through instabilities.

2.1. Anomalous heating of rarefied outer atmosphere by magnetic effects

The importance of the magnetic field in the steady anomalous heating of the solar outer atmosphere was first made clear by the soft X-ray and EUV observations from Skylab (Vaiana *et al.*, 1973; Sheeley *et al.*, 1974). The hot and dense parts of the corona are centered above active regions and consist of aggregates of fine bright loops, and some of these bright loops connect separate active regions, even across the equator. Also the fianter general corona is made up of fine loops and threads. The shapes of the loops were shown to coincide roughly with the coronal magnetic field pattern calculated from the photospheric distribution of the magnetic field (Figure 1; Sakurai and Uchida, 1977).

Prior to the Skylab observations, coronal heating was believed to be due to shock dissipation of acoustic waves, produced in turbulent convection in the sub-photospheric layers (Biermann, 1948; Schwarzschild, 1948; Schatzman, 1949; De Jager and Kuperus, 1961; Osterbrock, 1961; Uchida, 1963; Ulmschneider, 1967; Kopp, 1968), but the close relation of the corona with the magnetic field pattern, as discovered by Skylab, strongly suggested that the magnetic field plays essential roles in heating, as well as in confining, the corona. It was difficult for the acoustic heating mechanism to reproduce the strong relation of the loop corona with the magnetic field pattern. The presence of coronal holes (Krieger *et al.*, 1973) just above the normal photosphere where turbulent convection obviously exists was also difficult to explain in the scheme of the acoustic heating theory.

The possibility of magnetic-mode wave heating of the corona was originally proposed by Alfvén (1947), but forgotten after several criticisms behind the then-successful acoustic wave theory. The difficulties with magnetic-mode waves (Alfvén-mode and fast-mode of magnetohydrodynamics) were the heavy damping in propagating upward

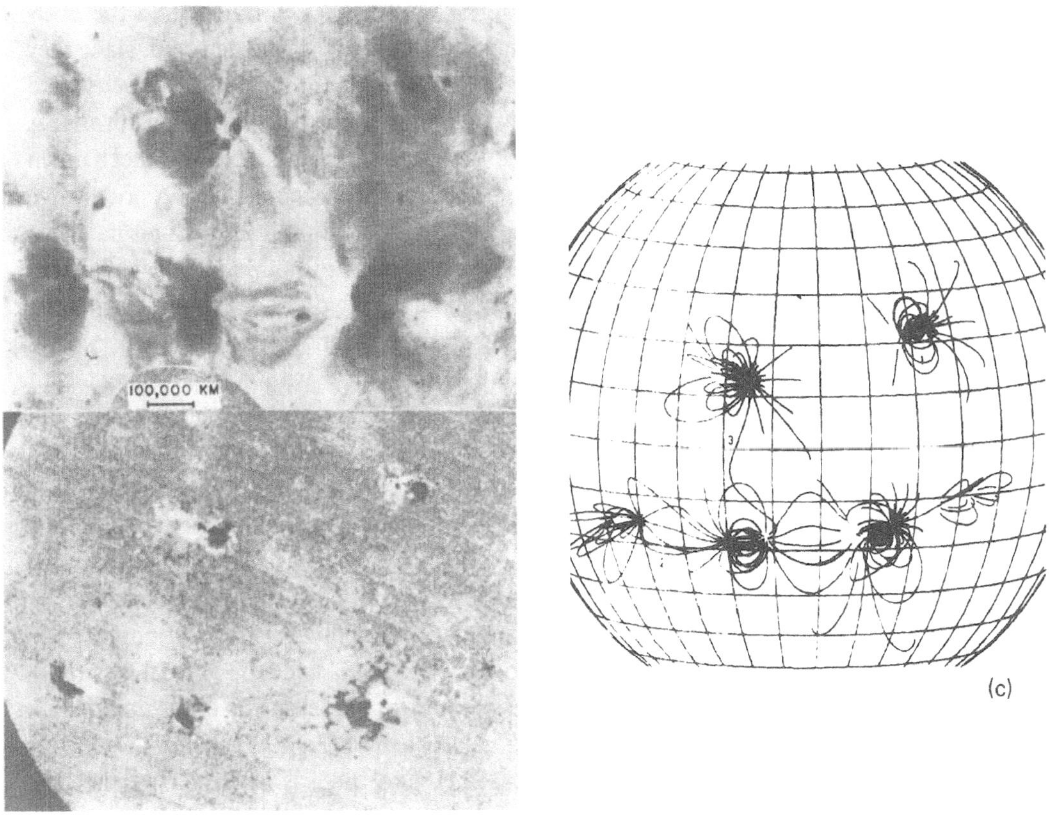

Fig. 1. (a) Skylab EUV picture in Fe XV (Sheeley *et al.*, 1974). (b) Magnetogram of the same day (Kitt Peak). (c) Coronal magnetic field calculated from the photospheric distribution of magnetic field (Sakurai and Uchida, 1977).

through the normal photosphere, after being generated in the convection zone, and, more essentially, the fact that there was no mechanism effective for the thermalization of these waves in the coronal condition (Piddington, 1956; Osterbrock, 1961). The importance of the magnetic-mode wave heating, however, was noted again after the *Skylab* observation, and a revival of magnetic-mode wave heating was proposed by Piddington (1973) and Uchida and Kaburaki (1974). The latter authors pointed out the 'transparent window' character of the photospheric layers in the magnetic regions for magnetic-mode waves. They also suggested that these waves coming out to the corona along field lines would be dissipated if mode coupling to the decayable compressional mode waves occurs effectively as the amplitude of the magnetic-mode waves grow large in the corona. Also Wentzel (1974) discussed a wave mode coupling in which the long wavelength Alfvén waves would decay into shorter wavelength ones which suffer more from dissipation. These wave mode coupling processes, however, take place in a nonlinear regime, in which the wave amplitude becomes large; but it was claimed that observations showed no particularly large Doppler motions in the hot corona above active regions (Beckers and Schneeberger, 1977). An alternative possibility was suggested by Ionson

(1977), in which the energy carrier was a surface Alfvén wave, rather than the body Alfvén wave. These would propagate into the corona without being detected, since they propagate along the surface of loops having a discontinuity in the Alfvén velocity, and dissipate energy at the point of resonance with the ion gyrofrequency in the corona. A third possibility, discussed more recently, is the phase mixing dissipation mechanism proposed by Heyvaerts and Priest (1983) and Sakurai and Granik (1984). A body wave in an Alfvén mode, with an oscillating velocity $\mathbf{v}$, propagating in a medium having a gradient in the magnetic field strength $|\mathbf{B}|$, perpendicular to both $\mathbf{v}$ and $\mathbf{B}$, will attain a large phase discrepancy with waves in a neighbouring region after propagating a long enough distance, due to the difference in the propagation velocity. An initially homogeneous oscillation grows in an indefinitely thin sheet with surface current, and the Joule and viscous dissipations become effective to heat the plasma. In applying the mechanism to the heating of a coronal loop, the loop should be long enough in terms of the wavelength; or the power supplied in the short wavelength range of the wave spectrum should be rather large in order for this mechanism to be effective.

Another category of mechanism of loop heating is the current dissipation mechanism (Rosner *et al.*, 1978; Hinata, 1979; Vlahos, 1979). In this picture, one considers the setting up of an anomalous resistivity in a steady current flowing along a loop, due to microscopic plasma instabilities (Spicer and Brown, 1980), since the normal resistivity at the coronal temperature is too low to do the job. As for the setting up of the anomalous resistivity, it is pointed out that too large a value of the current is needed to start up current-driven instabilities (Kuperus *et al.*, 1981), and there is an argument that the tearing mode instability will occur before anomalous resistivity sets in (Spicer, 1976).

Uchida (1984b) noted that although the current dissipation mechanisms have dealt with the decay of an already built-up current constantly flowing along the loop, there must also be a transiency for the wave to fill up the loop. Sturrock and Uchida (1981) discussed the intermediate statistical building up process of the current, and Uchida and Shibata (1984b) noted that the incidence of packets of magnetic twist into a loop can provide mass and energy to the loops by nonlinear effects. They proposed this for the interpretation of loop flares, but a weak version of this may operate also in coronal loops.

2.2. Drastic release of magnetic energy–flares and related phenomena

Another solar phenomenon intrinsically related to the magnetodynamic effects is the solar flare phenomenon. We do not have enough space for the detailed description of a solar flare-complex here, and refer the readers to, e.g., Švestka (1976), Kane *et al.* (1983) and references therein, and concentrate ourselves to its magnetodynamical significance.

The solar flare was previously recognized as a sudden heating up of the chromosphere as observed in Hα, and considered to be a chromospheric phenomenon. Skylab X-ray and EUV observations, however, have indicated that the Hα flares are, so-to-speak, mere footprints of the thus far invisible main body of a phenomenon occurring in the corona. For example, a typical two-ribbon flare in Hα was found to have a hot arcade-like structure connecting the two-ribbons above them (Švestka, 1976). These loops were obviously related to magnetic field lines. Smaller flares with two bright patches were two bright footpoints of a loop.

It was revealed by radio and X-ray observations that there are two distinct phases in a flare (De Jager, 1969). The phase near the start of a flare, called the impulsive phase, is characterized by a short rise time of 10–10^2 s. It is related to the creation of nonthermal particles (electrons producing type-III radio bursts and microwave and hard X-ray impulsive bursts; Kane, 1969). The relatively quieter phase thereafter, called the thermal phase, is characterized by the appearance and persistence of high temperature plasmas in loops or arcades (Švestka, 1976). Various dynamical effects are produced around the starting time of a flare, in the form of shocks, like metric radio type-II burst sources (Smerd, 1970; Uchida, 1974), or Moreton waves (Moreton, 1964; Uchida *et al.*, 1973), or in the form of mass ejections such as flare-related surges (Bruzek, 1974) and sprays (Stewart *et al.*, 1974), though their actual appearance may show some delay.

The fact that flares occur in active regions with the shape of a magnetic arcade, or loops, across the magnetic polarity reversal line shows their relation to the magnetic field. A comparison of the total amount of the flare energy, of the order of 10^{32} ergs, with the estimated amount of all kinds of energies available in the site of the flare suggests that the source of the energy may be magnetic. It is, however, found that there is no drastic change in the active region photospheric magnetic field at the moment of a flare itself, although there are changes, like a long term build up of the field strength in the region before the flare (Rust, 1972). This leads us to the view that the magnetic energy to be liberated is that part due to the local current. Proposed model situations of the magnetic energy storage may be classified into the neutral sheet models (Sweet, 1969; Parker, 1964; Sturrock, 1968) and the current loop models (Alfvén and Carlqvist, 1967; Carlqvist, 1979; Spicer, 1976). The mechanism proposed to explain the fast release of the stored magnetic energy in the former case is magnetic reconnection enhanced by anomalous resistivity (Friedman and Hamberger, 1968), with an external drive (Sato and Hayashi, 1979). Those in the latter picture of the current loops may be classified into the models in which microscopic instabilities cause anomalous resistivity and thermalize the current in the loop (Carlqvist, 1979; Spicer and Brown, 1980; Rosner *et al.*, 1978), models in which differentially twisted loops dissipate their current through the tearing mode instability (Spicer, 1976), and models in which the energy of the current loop decreases through a structural deformation into a double helix by the macroscopic kink instability (Sakurai, 1976). The hot and dense mass which did not exist in the loop prior to the flaring, but is observed in X-rays, is considered to come up as an expansion of the chromospheric material heated either by bombarding high energy particles, or by heat conducted down from the heat source above (Syrovatsky and Shmeleva, 1972; Kostyuk, 1976; Nagai, 1980). A point thus far unnoticed in the loop model is the transiency in which a 'truncated' current (e.g., coaxially closed current system in the form of a packet of magnetic twist) invades into a loop, either open or closed, through, e.g., magnetic reconnection with another fully twisted loop, or through some other

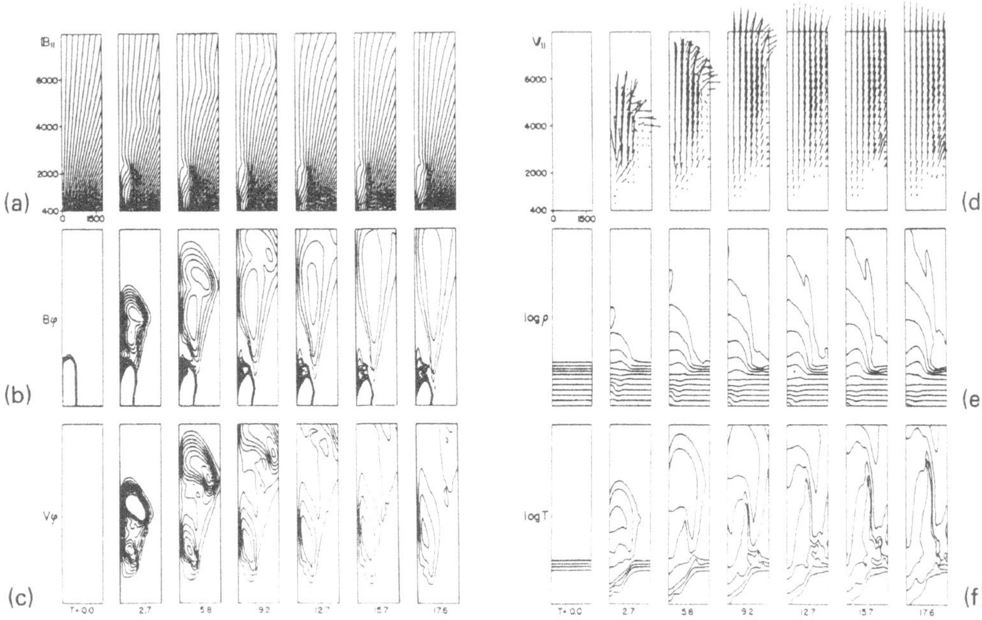

Fig. 2. Acceleration of jets by magnetic twist relaxing into low β regions. (a) Magnetic field lines in $r - z$ plane. (b) Contours of B_φ.(c) Contours of v_φ. (d) Velocity vectors in $r - z$ plane. (e) Contours of density. (f) Contours of temperature. Left edge in each figure is the axis of symmetry, and numbers on abscissa and ordinate are in km. It is shown that a high velocity jet is formed as a packet of B_φ relaxes into low β regions (Uchida and Shibata, 1984b).

process. The mass and energy are driven into the loop by the sweeping pinch effect, as well as by the $\mathbf{j} \times \mathbf{B}$ force in the incident packet of magnetic twist, which thus explains the mass ejection from the flaring regions (Bruzek, 1974; Brueckner, 1983) in the case of an open flux tube (Uchida and Shibata, 1984b); or the loop heating in the case of the closed loops (Tsuneta *et al.*, 1983; Uchida and Shibata, 1984b) (cf. Figure 2).

Although no ultimate solution of the problem has been reached as yet, we may say that various flare-related phenomena will be systematically explained as due to the sudden release of the energy accumulated in the magnetic stress in a short Alfvén transit time-scale by one or other instability inherent to the magnetic field configuration under consideration. The energy will be converted into the kinetic energy of shock and mass ejections, and into heat, either directly by anomalous Joule dissipation, or by shocks in the mass motions. The presence of the magnetic field also facilitates the acceleration of particles in events like shocks and collisions of shocks by providing the electric field (e.g., electric double-layers) for the initial acceleration of particles. The latter is also expected to operate in a statistical way in the process which causes anomalous resistivity.

The effects of the magnetic field may thus be summarized as follows (Uchida, 1984b): (a) Magnetic field may provide a low-loss channel (Osterbrock, 1961; Uchida and Kaburaki, 1974) for a dynamical energy supply (e.g., in magnetic mode waves) to the rarefied part of the atmosphere. The rarefied atmosphere, which is a poor radiator and has a small heat capacity, can be heated to an X-ray emitting temperature if a large enough amount of energy is dissipated into heat and deposited there. (b) Magnetic field can sustain plasma in a situation which would be out of mechanical equilibrium *without magnetic field*. For example, it can sustain the plasma in a highly stressed magnetic field, or in a high gravitational potential state. It can also contain heated plasma which would escape away from the star without it. Magnetic field can modify gaseous flow to a pattern different from the one without it, and this would affect the velocity, density, etc., of the flow (e.g., Bernoulli flow with the de Laval point; Parker, 1963; Holzer and Axford, 1970). (c) Magnetic field can insulate the heated plasma from the surrounding low temperature part, because of the reduction of the heat conduction perpendicular to the field lines. (d) Finally and most conspicuously, a magnetic field can make drastic transitions from quite a high energy state to another lower energy state through a variety of instabilities. The sudden reconfiguration of the system can release the potential energy into kinetic energy drastically in a short Alfvén transit time-scale, $\tau_A \equiv L/V_A$, and into heat by shocks, or by Joule dissipation with anomalously enhanced resistivity. If magnetic reconnection is involved, there can be transfer of mass and momentum, as well as of the magnetic twist, to the newly connected part of the flux tubes, which may be either open or closed. If the injection is into an open flux tube, jet-like ejection may occur, and if it is into a closed loop the resulting phenomenon may be loop flares (Uchida and Shibata, 1984b).

3. Magnetodynamic Effects in Other Stars

The classical acoustic heating picture of the solar corona was transplanted to the stellar case before the paradigm change in the solar coronal research took place from acoustic to magnetic heating (Kuperus, 1969; De Loore, 1970; Landini and Monsignori-Fossi, 1975; Hearn, 1975; Mullan, 1976). The prediction made there was that a corona should be expected mainly in F- and G-type dwarfs which have surface convection vigorous enough to produce turbulence, as well as strong enough gravity to hold back the heated plasma to the stellar surface. Observations by HEAO-1, EINSTEIN, COPERNICUS, IUE, and VLA, etc., however, revealed in contradiction to this expectation that stars of almost every type on the HR-diagram, ranging from late-type dwarfs to early-type stars, from very young stars emerged from nebulosities to much evolved late-type giants, exibit strong X-ray emissions (Vaiana *et al.*, 1981). The new results were striking because: (i) late-type giants without sufficiently strong gravity have X-ray emitting regions, (ii) early-type stars without surface convection layers show X-ray emissions, and (iii) the luminosity of X-ray emission from stars having the same spectral type and luminosity class have a scatter as large as 10^3. Observational fact (i) suggests strongly the possible role of the magnetic confinement of the heated plasma, (ii) indicates that the heating may not be due to the acoustic heating and some other mechanisms including the magnetic-field-related mechanism may be in operation, and (iii) indicates that there may exist some hidden parameter(s) controlling the heating, and again the mechanism is not likely to be acoustic heating. This is because in the classical theory of stellar structure, the mass, the chemical composition, and the age of the star, completely specify the state of the star, including its convection; and the acoustic heating theory, which is a quite deterministic theory, would then uniquely specify the special range of stars having coronae on the HR-diagram, as mentioned above. The hidden parameter may be the rotation of the star, which was out of the scope of the classical theory, and again may be related to the magnetic field.

There are thus strong reasons to think that the magnetic field may play important roles also in stellar atmospheres as in the solar atmosphere (Uchida, 1984b; Montmerle *et al.*, 1983). No simple transplantation of the solar picture to other stars, however, is allowed, since there are very large differences in the physical conditions in other type stars. The gravity, gas and radiation pressures and temperature at their surface, the states of the surface convection zone and of rotation and, therefore, of the magnetic field, in other type stars, all differ markedly from those of the Sun. Also, the effect of duplicity, or the influence of a companion star in close binary systems, can be another important factor which does not exist in the case of the Sun. What is to be sought for is, therefore, not the simple similarity of the apparent phenomenology, but of the physics underlying these phenomena.

We see, in the following, some examples of these, in the cases of late-type stars, early-type stars, close binaries (we confine ourselves to binary systems composed of normal stars in this paper), and young stars, by considering that the presence of observed X-ray activities can be an indication of magnetic activities.

3.1. Magnetodynamic Effects in Late-Type Stars

The Sun is the best studied example of an isolated, slowly rotating late-type dwarf star, but there are also a number of late-type dwarfs whose quiet and bursting X-ray and radio emissions have been detected. Typical ones are dMe flare stars like UV Cet and YY Gem, emitting $L_x \sim 10^{27-29}$ erg s^{-1}. The flux ratio of X-ray emission to visual, amounts to $10^{-3 \sim -1}$ which is very large partly due to the dark optical luminosity of the star itself. Also flares have been observed in X-rays, and analysed in detail. For example, in the case of Prox Cen (Haisch *et al.*, 1982) L_x amounts to $10^{28.3}$ erg s^{-1} ($E \sim 3 \times 10^{31}$ ergs), and the time variations in the temperature and emission measure mimic very closely those of solar 'disparition brusque' type flares.

Among late-type dwarfs, also spotted stars (e.g., EQ Peg, CR Dra, BY Dra) as well as the rapid rotators, π^1 UMa, are X-ray emitters. In view of the fact that the X-ray luminosity is correlated with the rotational velocity $L_x \propto (v \sin i)^2$; Pallavicini *et al.*, 1981) and that physical conditions are similar to those of the Sun, it is highly likely that the X-ray emitting regions in the late-type dwarfs are similar to those in the magnetic corona of the Sun. In particular, the coronae of these stars probably have looped structures closely related to the magnetic field, as well as the flares in them do. The magnetic field in turn is produced by a dynamo mechanism, in which the cooperative action of rotation and a non-axisymmetric motion like convection is essential. Furthermore, the magnetic field should not be a mere potential field, but must be agitated, e.g., by the convective motion, in order to produce any activity, including the heating. In this context, it is interesting to see that the chromospheric activity by Mg II h and k lines

(Ayres and Linsky, 1980) is correlated to the period of rotation derived from Wilson's Ca II H, K intensity variation (Wilson, 1976), and this activity is nicely a unique function of a parameter related to the Rossby number (Noyes *et al.*, 1984; Mangeney and Praderie, 1984).

Giants and supergiants, of F, G, and K-type also have X-ray emissions. For example, an F0 type supergiant Canopus has $L_x \sim 10^{32}$ erg s^{-1}, and $f_x/f_v \sim 10^{-6 \sim -7}$. Since $r\mathcal{R}T/GM > 10$ for giants, it is difficult for the gravity alone to hold the X-ray emitting high temperature plasma back to the stellar surface; and the picture will be a hybrid model, in which magnetically confined hot loops are imbedded in an outflowing warm wind (Linsky, 1983; Dupree and Hartmann, 1980) which escapes from the magnetic field-free part of the stellar surface with much lower temperature. The heating process of the hot loops may be an analogue of the solar active region loops as in late-type dwarfs.

A problem arises about the dynamo process for giants or supergiants, however. The dynamo action is expected to become weak, since the star spins down as it expands due to the evolution into the giant stage. The chromospheric activity for giants seems to be maintained, however (Linsky, 1980). Uchida and Bappu (1982) proposed one possible solution for this difficulty by considering the following: in the evolution, the core of the star contracts and spins up, while the envelope expands and spins down, and the convection zone invades deeper into the envelope from the surface as the envelope expands. An internal dynamo layer permeated with the convection and possessing a strong shear in rotational velocity will appear around the rapidly spinning core. The corresponding surface active region with magnetic flux lifted by convective or buoyant transport will be diffuse and large in size, reflecting the large depth of the convection zone.

3.2. Magnetodynamic effects in early-type stars

Early-type stars (O and B types) are also found to be strong X-ray emitters. Their X-ray emission amounts to $L_x \sim 10^{31 \sim 34}$ erg s^{-1} and fulfills a nice linear relation with L_{bol} (Pallavicini *et al.*, 1981).

In the case of early-type stars, there is a strong mass outflow (Morton, 1967) which is believed to be due to the strong pressure of their photospheric radiation exerting on resonance lines (Lucy and Solomon, 1970). It is, therefore, possible that some instability inherent to this driving mechanism may disturb the flow and generate turbulence and shocks, and thus convert the energy of the mean flow into heat (Lucy and White, 1980; Owocki and Rybicki, 1984).

Underhill, however, claims that there must be mechanisms other than radiative, and the magnetic field may play important roles also in the atmospheres of early-type stars (Underhill, 1980, 1983; Underhill and Doazan, 1982). Her argument is based on (i) the presence of nonthermal radio emissions and optical emission lines in some of the early-type stars (Underhill, 1984), (ii) evidence of anisotropic mass ejection from early-type stars (Underhill and Fahey, 1984), and (iii) the fact that v_∞ is not a unique function of L_{bol} and, therefore, there may be some other hidden factors modifying the outflow.

These are not supposed to exist in the spherically-symmetric outer atmosphere of a star in a purely radiative regime.

From the theoretical point of view, there are a number of ways in which the magnetic field, if present, may interact with the outflow. It may contribute either (i) in converting a certain part of the flow energy into heat, or (ii) in modifying the momentum of the flow by providing excess heating to boost up the flow (Hearn, 1979), or (iii) in providing excess momentum by waves (Hartmann and MacGregor, 1980), or (iv) in modifying the flow by nozzling (Holzer and Axford, 1970; Cannon and Thomas, 1977), or (v) in transmitting angular momentum to the flow (MacGregor and Friend, 1984), or (vi) in converting the rotational energy to that of the flows in the polar directions (Draine, 1983; Uchida and Shibata, 1984b, c, 1985). We refer the reader to, e.g., *Mass Loss and Evolution of O-type Stars* (1979) concerning radiatively driven flow and related mechanisms proposed for the heating along those lines, and concentrate ourselves here to the magnetic effects in the early-type star atmospheres.

The situation with the magnetic field in early-type stars, however, seems to be very different from that of late-type stars. For example, early-type stars do not have surface convection zones and, therefore, no dynamo mechanism like that proposed for the solar case (= self-regenerative, oscillating dynamo) can operate. We have to start by asking even whether there is a magnetic field in early-type stars at all. Unfortunately, it is not possible to directly answer this question based upon observations yet due to the difficulty in measuring the Zeeman-effect hindered by the large width of the spectral lines in early-type stars. Absence of the appropriate observations, however, *does not* mean that there is no magnetic field. We guess theoretically that considerable amount of magnetic flux is certainly brought into early-type stars from the nebulosity. Observations that A-type stars, which do not have surface convection zones either, have strong magnetic field, supports our guess.

The presence of the regenerative dynamo process should indeed be a must for maintaining the magnetic field in less massive Main-Sequence stars like the Sun, having an exceedingly long life. In early-type stars whose life is only of the order of 10^7 yr, which is short compared with the magnetic decay time $\tau_D = 4\pi\sigma R^2/c^2$, the regenerative dynamo process is not necessarily indispensable in maintaining their magnetic fields. Furthermore, some of the functions of the dynamo mechanism are in operation even without the effect of the convection. For example, strengthening of the B_φ-field by the differential rotation in r may take place, and the B_φ-component would increase continuously if the dissipative effect is weak. The differential rotation in r may be maintained if the magnetic field extending out of the star exerts a braking effect on the outer shells of the rotating star (Figure 3). In such a case, a toroidal magnetic field of considerable strength may exist in the star, and some parts of a flux tube may float up to the surface in the low latitude zones. There is a poloidal current system induced inside the star corresponding to the toroidal field component. The rotational drag may be superseded by the effect of the outflow in the outer atmosphere, and field lines are pulled outward at the equatorial part, which may cause an equatorial sheet current to flow in the azimuthal direction.

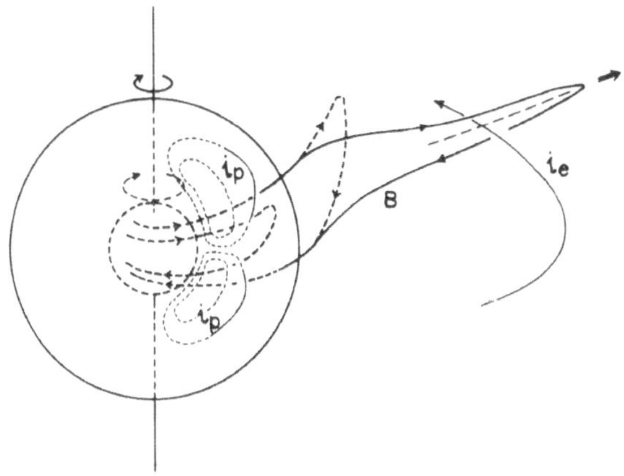

Fig. 3. Schematic presentation of the system of magnetic field and current in a rotating early-type star with a strong wind. The magnetic field outside the star which would otherwise be dragged by rotation is stretched outwards by the strong wind, and an equatorial sheet current is formed (Uchida, 1984b).

Heating of the outer atmosphere to an X-ray emitting temperature may occur if the current in the rarefied part is destroyed in certain efficient ways. This is equivalent to releasing some free part of the magnetic energy, and the distortion in the magnetic field pattern associated with deviation from the potential field will disappear. This process can occur in the possible situations of the outer atmosphere of early-type stars described above as either a steady or a non-steady process (Galeev, 1982; Uchida *et al.*, 1986, in preparation). The former can take the form of the dissipation with a slippage of the plasma through the magnetic field at the tip of the stretched magnetic field where the radius of curvature becomes small (top figure of Figure 4). The field line which would

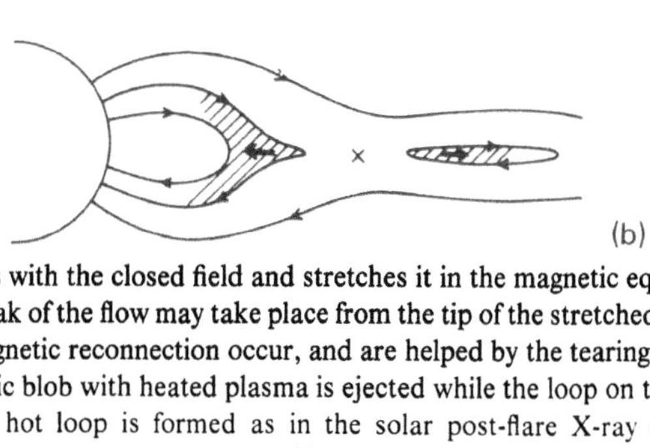

Fig. 4. Outflow interacts with the closed field and stretches it in the magnetic equatorial plane. (a) In the steady state, a diffusive leak of the flow may take place from the tip of the stretched loop with weak heating. (b) If the outflow and magnetic reconnection occur, and are helped by the tearing mode instabilities at the squeezed point, a magnetic blob with heated plasma is ejected while the loop on the stellar side shrinks as a sling-shot, and a very hot loop is formed as in the solar post-flare X-ray emitting loops (Uchida, 1984b).

otherwise be pulled out indefinitely in $\sigma = \infty$ plasma, recedes to the coronal location by slippage if the resistivity is finite. The region is heated due to the Joule dissipation in the slippage.

If the process takes place nonsteadily, a blob of warm or hot plasma may be produced by magnetic reconnection and carried away by the outflow. The part of the flux tube on the stellar side shrinks at the Alfvén velocity, like a rubber belt in a sling-shot, and will be heated up due both to the resistive heating and to compression and shocks, as in the solar post-flare loops (Sturrock, 1968; Kopp and Pneuman, 1976).

Although we refrained from going into the details of non-magnetic heating mechanisms, most of which rely on instabilities of the flow driven by the strong radiation pressure from the early-type star photosphere, we should mention that there is a possible difficulty in setting up the radiatively driven wind, as pointed out by Marlborough and Roy (1970). Their point was that if the outward force by radiation dominates over gravity from the beginning, no critical point in the Parker-flow (Parker, 1963) would appear in the flow,and thus a subsonic flow should stay subsonic all the way, contradicting the observed velocity of 1000–2000 km s^{-1}. This point was argued further by several authors (e.g., Nerney and Suess, 1975). It should be mentioned here that the difficulty may be avoided if the initial acceleration takes place in a region in which $dT/dr > 0$ (Rogerson and Lamers, 1975) and this may be possible if the initial heating is somehow realized by the magnetic effect as in other type stars. If this is the case, the magnetic effects would be very essential in helping the radiation pressure to produce the major effect in the early-type star atmosphere, namely, the radiatively driven flow.

Other magnetic effects which may modify the outflow include the nozzling effect (Holzer and Axford, 1970; Cannon and Thomas, 1977). This is the early-type star analogue of the origin of fast and slow streams in the solar wind (Kopp and Holzer, 1977), and may be expected if the magnetic field is strong enough to control the flow pattern of the initial part of the flow. The polar fast stream seen in the Sun (Svargaad *et al.*, 1975) may thus be expected also in early-type stars. Alfvén wave driving (Hartmann and MacGregor, 1980) may be ineffective, due to the lack of the surface convection which buffets the footpoints of the magnetic field to produce the Alfvén waves. Instead, a flow in the polar direction is expected by the interaction of rotation and the magnetic field (Draine, 1983; Uchida and Shibata, 1984b, c, 1985). The mechanism is essentially due to the pressure of the torsional Alfvén wave, which is created continuously by rotation instead of convection. This again may provide the initial acceleration by which the fluid can receive stronger radiation of the continuum from the photosphere owing to the Doppler shift out of the absorption line center.

Finally, the flow steadily slipping out of the stretched magnetic field, or the magnetic blob produced in the reconnection (Figure 4), may add warm or hot components to the outflow, and modify the thermal state of the stellar surroundings. These may contribute to the fluctuation of X-ray emission from the extended source around early-type stars which is considered to primarily due to the collision of the high-speed outflow with the surrounding cloud.

3.3. Magnetodynamic effects in close binary systems

It is well-known that close binaries form a category of strong X-ray emitters, among others, especially when they have collapsed components (X-ray binaries with a neutron star, or AM Her-type with a white dwarf). The availability of the mass supply either by the overflowing Roche lobe of the low-mass primary, or by the mass-loss flow from early-type massive primary, allows the mechanism of mass accretion to operate. Here, without going into the well-studied problems of close binaries having a mass accreting collapsed star in the system (X-ray binaries; e.g., Rappaport and Joss, 1981), or those having a nondegenerate star, but with heavy mass exchange (W UMa type; cf. Dupree, 1982), we confine ourselves to the case of RS CVn type binaries which are detached systems consisting of normal stars, without apparent mass exchange.

RS CVn stars (such as UX Ari, RS CVn, σ CrB, AR Lac, HR 1099) are strong X-ray emitters (Walter *et al.*, 1980). The steady component of X-rays from UX Ari, for example, is claimed to consist of two thermal components, one with $T \sim (3\text{–}7) \times 10^6$ K and $EM \sim 10^{53}$ cm^{-3}, and the other with $T \sim (1.5\text{–}10) \times 10^7$ K and $EM \sim 10^{54}$ cm^{-3} (Swank and White, 1980). The extent of the sources was examined (Walter *et al.*, 1980) and the latter component had the binary scale, while the former had a more normal extent as a stellar corona. The rough configuration of the X-ray emitting regions was deduced by using eclipses in the case of AR Lac (Walter *et al.*, 1983), and it was found in this case that the corona existed on the preceding side of both component stars with respect to the direction of rotation. Flares in RS CVn's are seen to occur both in X-rays (HR 1099, SAO 15338, HD 8359; Walter *et al.*, 1980) and in radio (HR 1099, HR 5110, UX Ari; Gibson *et al.*, 1978; Feldman *et al.*, 1978). Since there is no apparent mass exchange, a softer version of the mechanism proposed for X-ray binaries with degenerate components cannot be applied to the present case, and some magnetic mechanisms are suggested.

RS CVn binaries have features favorable for the magnetic interpretation of their activities. These include the so-called photometric wave (= PW) which drifts across the light curve to the direction of smaller phase once in a cycle with a period of $8 \sim 10$ yr (Rodono, 1981). The variation of the depth of the dip with phase in the cycle, and also the colour change accompanying PW, suggest that this darkening may actually be due to something like a sunspot. Thus the gigantic star-spot model was proposed by Hall (1972), and the possibility that it is a cumulative effect of many smaller spots was discussed by Eaton and Hall (1979). In these models, the drift in phase is attributed to the increasingly faster rotation of the lower latitude zones into which the spot migrates down with the phase in the cycle.

Since the rate of the drift – one rotation on the stellar surface in $8 \sim 10$ yr – is actually very slow, the rate of differential rotation $\Delta\Omega/\Omega$ estimated with this is two orders of magnitude smaller than that scaled from the Sun by using Durney and Robinson's (1982) scaling law. Although it is generally thought that the tidal effect caused by the companion tends to bring the rotation of the star into a synchronized rotation (Zahn, 1977), this seems to be too perfect an effect.

Is the migrating gigantic spot picture correct? Although it is a fascinating picture, does it have a good enough ground? In this connection, it is worth noting that spots, in the case of the Sun, do not themselves migrate to lower latitude, although the gigantic starspot hypothesis claims an analogy to the solar case. It is the latitudinal zone in which the spots appear that migrates equatorward. Furthermore, the sign of $d\varphi_{min}/dt$ should be negative in the solar analogy, but in some of the RS CVn systems it is not negative. For example, in HK Lac the sign is positive, and $d\varphi_{min}/dt$ even changes sign in the cases of SS Boo and V711 Tau, indicating that the migration picture, if applicable, introduces a rather irregular behaviour of spots.

The very slow drift, either eastward or westward, or even changing direction, reminds us of a solar feature called the 'active longitude belt' (Bumba and Howard, 1969); and Uchida and Sakurai (1983) proposed that the PW may correspond to an active longitude belt ($\equiv$ ALB) rather than to a gigantic starspot. The ALB is a longitudinal zone on the Sun in which many spot-pairs appear, drift across, and disappear before the preceding edge is reached, and it was actually seen either to stand still, or to drift as a whole either eastward or westward with a slow drift velocity relative to the solar surface (Gaizauskas *et al.*, 1982) of magnitude similar to that of the PW on RS CVn's.

The introduction of the notion of the ALB gives us a much more activated situation, while a single gigantic starspot, which scarcely moves in the co-rotating frame, would be quite static in nature. Now, spot pairs are born in, drift across, and disappear from this belt, and thus the magnetic field configuration of the systen invariably changes, with magnetic reconnections taking place from time to time. Furthermore, the rather stringent

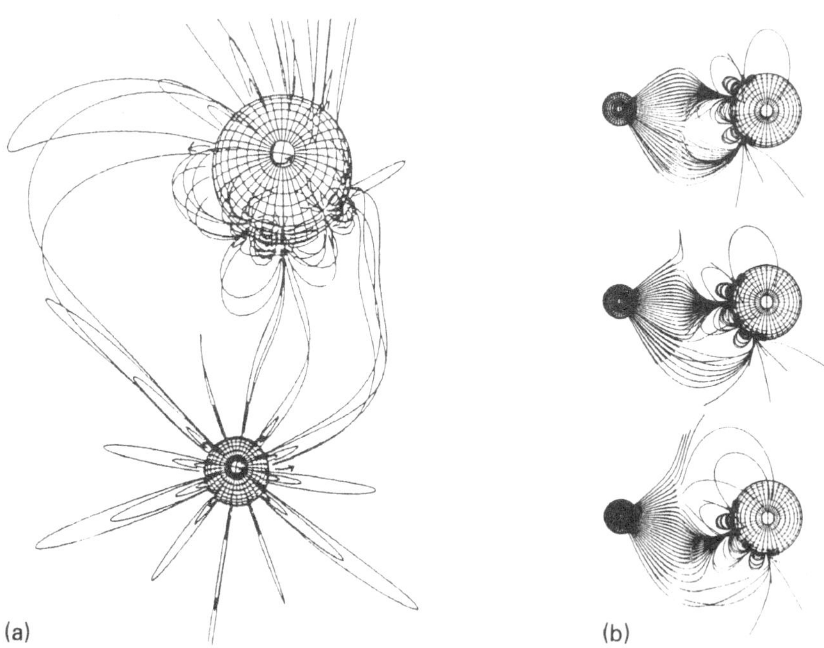

Fig. 5. (a) Model magnetic field calculated for RS CVn in the ALB picture. (b) Top views of magnetic field connections and reconnections as spot-pairs drift in the ALB (Uchida and Sakurai, 1983, 1984).

restriction imposed on the magnitude of the differential rotation by the small drift rate of PW is now relaxed, and the stars can have considerable differential rotation, though with synchronized latitudes at medium latitudes in north and south hemispheres. Uchida and Sakurai (1983, 1984) discussed the possible magnetic field configurations and reconnection in them by taking RS CVn itself as an example (Figure 5). They showed that there can be reconnections of the field mediated by the field of the companion star, and reconnections may transfer the hot and dense plasma in an emerging spot-to-spot connection into the newly formed spot-to-companion star connection on the foreside of the K star, for example, by the process of the relaxation of magnetic twist from highly twisted loops into less twisted weaker field loops (Figure 2, Uchida and Shibata, 1984b) to form hot loops of the binary size.

3.4. Magnetodynamic effects in young stars

Another stellar category of strong emitters of X-rays is the very young stars. Some of the T Tau stars are known to be strong X-ray emitters emitting $L_x \sim 10^{30 \sim 31}$ erg s^{-1}. The steady component of X-rays is known to fluctuate (Montmerle *et al.*, 1983), and enhanced variability is seen in T Tau, SU Aur, and GW Ori, and flaring, which is considerably different from that of dMe flare stars, is detected, e.g., in DG Tau (Feigelson and De Campli, 1981). T Tau stars are also known to show widely separated optical spectral lines (Hartmann, 1982; Mundt and Giampapa, 1982) which may be interpreted as due to coexisting in- and outflows (review by Bertout, 1984).

More recently, young T Tau stars, or their progenitor IR stars, have been shown to be related to the bipolar flows observed in CO molecular lines at millimetric radio wavelengths (Snell *et al.*, 1980; Bally and Lada, 1983; Cohen, 1982), and to the high velocity small-scale optical jets which lie also in the general directions of bipolar flows and sometimes connect the central T Tau star with Herbig–Haro objects (Mundt and Fried, 1983). It has been found in millimetric CS molecular lines that an accretion disk is rotating around the central star in a plane perpendicular to the bipolar flow (Kaifu *et al.*, 1984).

The activity in these young stars was already known from the variability of the optical lower Balmer lines, which show a corresponding and sometimes controversial behaviour of the spectral line profiles. Stellar wind models (Kuhi, 1964) as well as accretion models (Lynden-Bell and Pringle, 1974; Ulrich, 1976; Bertout, 1979) have been discussed to interpret these. The origin of the outflow was also attributed to some unknown process of energy liberation of a still unsettled magnetic configuration in the star after its formation (Gershberg, 1982); or to a vigorous generation of Alfvén waves, whose dynamic pressure may drive the outflow (Hartmann and McGregor, 1980). In order to explain the coexistence of the X-ray emitting region and the outflow of cool material emitting Hα, a hybrid model in which active region loops and coronal holes coexist, as in the Sun, has also been discussed (Imhoff and Giampapa, 1980).

The magnetic effects in the stars of this category, however, can be quite different from that of the Sun. Uchida (1984a), Uchida and Shibata (1984c, 1985) recently proposed a model to explain activity on such a star, and its optical bipolar jets, as well as the large

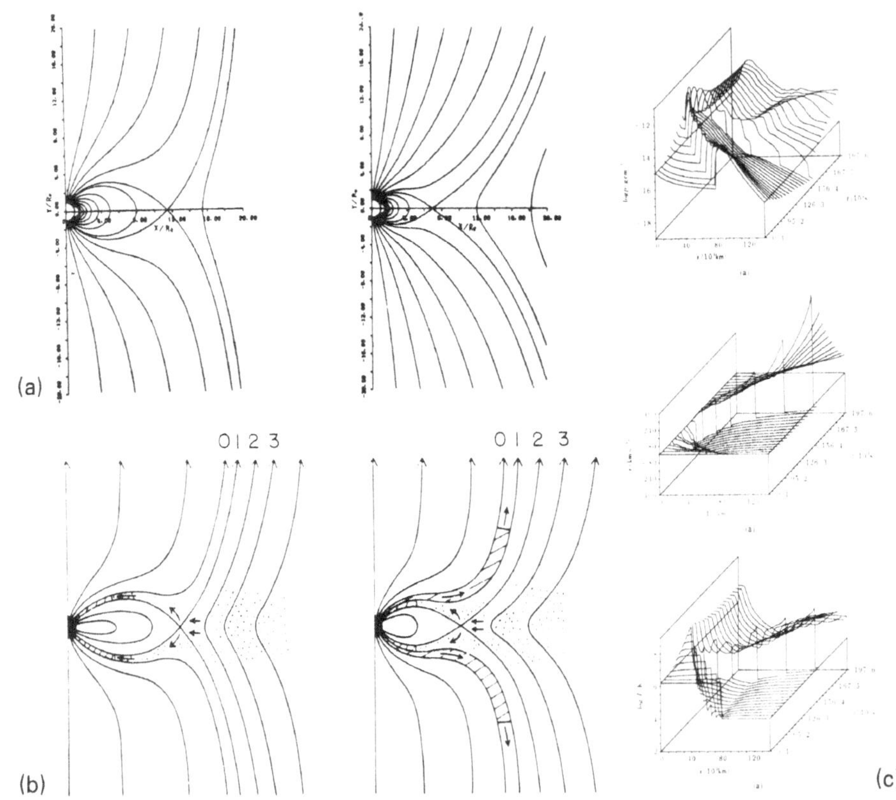

Fig. 6. (a) Calculated magnetic field structure around a star loaded by accreted nebular mass (Uchida and Low, 1982). Magnetic fields of the star and the accreted mass are assumed to have a common origin in the large-scale nebula. (b) Schematic figure explaining the behaviour of the mass leaking by magnetic reconnection occurring at the neutral ring (Uchida and Shibata, 1984a). (c) From top to bottom: density, velocity, and temperature as functions of t and s (coordinate along the curved flux tube in which the flow takes place) (Uchida and Shibata, 1984a).

scale CO-bipolar flows in a consistent way, based on a global magnetic picture. They first pointed out that the magnetic field configuration around a young stellar object formed in a magnetized nebula has a magnetically neutral ring in the equatorial plane around the star (Figure 6; Uchida and Low, 1982), and the accreted magnetized material will be supported by the magnetic structure having this ring of magnetically neutral points at its base. It is noted that the magnetic field configuration may be considered as a global version of the magnetic configuration supporting a solar prominence in a planar geometry.

The nebular mass accreted to the star after this stage is buffered by this magnetic structure, and falls onto the stellar surface by being transferred to the stellar field through magnetic reconnections taking place at the neutral ring. For a star having a polar magnetic field strength of the order of 10^3 G (Nakano and Umebayashi, 1982) the distance of the neutral ring from the star is estimated to be several stellar radii, and the potential energy of the accumulated material is $\sim 6 \times 10^{14}$ erg g^{-1}. The free fall time is of the order of 10^4 s, the terminal velocity in the free fall is roughly 300 km s^{-1}, and the temperature attained by the conversion of this kinetic energy is 3×10^6 K.

The problem is whether the rate of mass release $\dot{M}$ by the magnetic reconnection from the reservoir is large enough. A simulation of an MHD flow with magnetic reconnection by Sato and Hayashi (1979) shows that the flow is very like that of Petschek's (1964) earlier suggestion, in which the velocity of the inflow is $\sim 0.1\ V_A$. By using this, $\dot{M}$ is estimated to be 10^{18} g s^{-1}, or $2 \times 10^{-8} M_\odot$ yr^{-1}, and this corresponds to the luminosity of 6×10^{32} erg s^{-1} (Uchida and Shibata, 1984a, c, 1985). These numbers suggest that this buffered accretion may explain the observed inflow and the production of the X-ray emitting region closer to the stellar surface than Hα-emitting region, which may be the inner part of the disk (Walter and Kuhi, 1981). The outflow is explained by Uchida and Shibata (1984a) as mass driven out by the recoiled shock which is created in the crash of the infalling mass at the stellar surface and strengthens itself in propagating upward along the tail of the infalling mass (Figure 6(c)). The paths of in- and outflows are separated by the reconnection process, and in- and outflows can exist separately without being degraded into turbulence (Figure 6(b)).

On the other hand, cold bipolar flows whose scale is of the order of a pc, and coming from the general region of the new-born star, have been given an explanation also by assuming strong stellar wind, or blast waves from the central star. The high collimation of the flows is a difficult point besides the energy source problem, and mechanisms

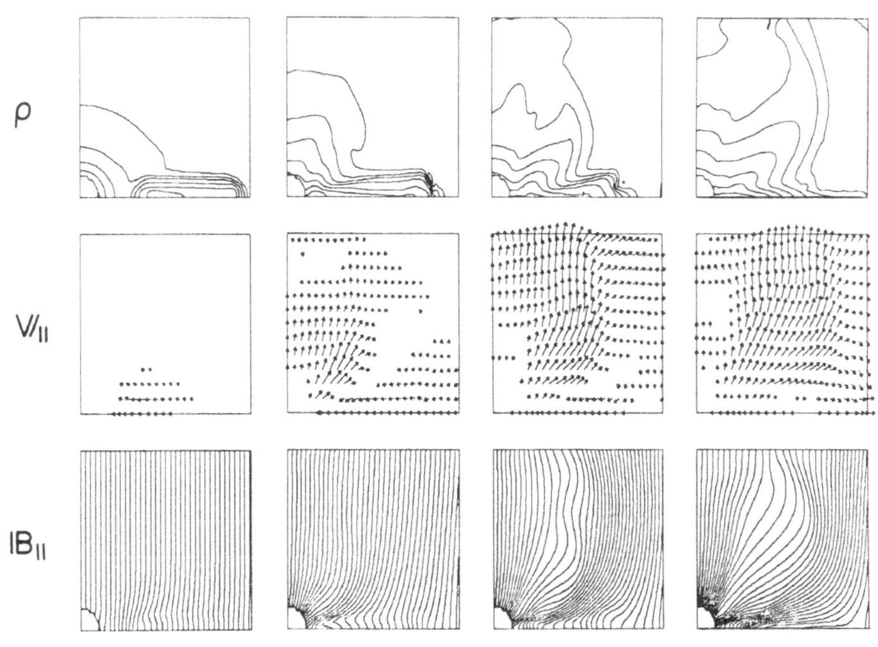

Fig. 7. Formation of a large-scale cold bipolar flow. *Top to bottom*: contour of ρ, poloidal velocity $\mathbf{v} = (v_r, v_z)$, and poloidal magnetic field $\mathbf{B} = (B_r, B_z)$. It is seen that a flow (helical when combined with v_φ) is caused by the relaxing B_φ-field created by the rotation of the disk (Uchida and Shibata, 1984c, 1985).

consider polar funnels in the disk (Blandford and Rees, 1974; Fukue, 1982), opened up by the strong initially isotropic outflow from the central star (Koenigl, 1982) have been discussed. An alternative mechanism discussed was to deal with stellar wind blockaded in the directions of the disk (Sakashita *et al.*, 1984; Okuda and Ikeuchi, 1984). An entirely different point of view was discussed by Draine (1983) and independently by Uchida and Shibata (1984b, 1985) by introducing the effect of a twisted magnetic field. Draine discussed semi-quantitatively the effect of the twisted magnetic field of the star itself, while Uchida and Shibata considered the magnetic twist due to the rotation of the contracting accretion disk. It was shown that, as the magnetic twist created by the rotation of the contracting disk relaxes into the low β regions above and below the disk, the gas is accelerated (Uchida and Shibata, 1984b) along helical paths toward $\pm z$-directions (Figure 7; Uchida and Shibata, 1985). The velocity of the outflow is of the order of the Alfvén velocity ($10 \sim 100$ km s^{-1} in the case of the environment of new-born stars), which is keeping with the observed velocity, and the hollow cylindrical shape of the jet can also be seen in typical cases of stellar bipolar flows, such as L1551, GL490, or NGC 2071. The large mass, as well as the large rate of mass loss, $\sim 10^{-4} M_0$ yr^{-1}, which is difficult to explain otherwise, can be explained in this model in a natural way (see details in Uchida and Shibata, 1984c, 1985; Shibata and Uchida, 1984, 1985).

4. Conclusions

Magnetic effects in the outer atmospheres of stars showing activities in X-rays have been discussed by generalizing the discussion of the magnetic effects known from solar activities. As a result, it may be concluded that, although there are ample differences in the physical conditions in stars of different mass, different state of evolution, and so on, the magnetic field-related phenomena, considerably different in appearance but physically similar to their solar counterparts, may be taking place. For example, the build up of current and its dissipation, expected from the change of relative location of the magnetic flux sources (e.g., spots in the active longitude belt and stellar global field) in RS CVn binaries, as depicted in Section 3.3, may correspond to those occurring due to the relative motion of sunspots in the solar active regions. The relationship of the suspension and the dumping of the mass of an accretion disk to the stellar surface through magnetic reconnection at the neutral point *to* the X-ray emission in the case of young stars as discussed in Section 3.4, may correspond to the relationship of the suspension and release of the solar dark filament material *to* Hα and thermal flares on the Sun. The picture of acceleration of the large scale cold bipolar flows from an accretion disk, as discussed in Section 3.4, has a similarity to the proposed process of acceleration of surges on the solar surface in our picture. The heating from slippage and magnetic reconnection with the production of an escaping hot blob in the radiatively driven wind from early-type stars may have that process of the opening up of closed arches and production of post-flare hot loops on the Sun as its solar counterpart. These physical similarities, not necessarily the similarities of appearance, are of course not a matter of surprise. If the set of the dimensionless coefficients of terms in the non-dimensionalized basic equations is the same in two events, the phenomena taking place under the corresponding boundary conditions should be similar between these two events, even though they belong to different astronomical hierarchies, just as in the case of two flows in the laboratory having a different scale but the Reynolds number in common.

It may thus be expected that the field of research of the magnetodynamical phenomena in the stellar atmospheres will see a fruitful development in the coming years by starting from the proper extension of the knowledge of the phenomena taking place on the Sun.

Acknowledgements

The author thanks Drs J. Linsky, F. Walter, D. M. Gibson, R. W. Noyes, A. Underhill, N. Kaifu, Y. Sofue, T. Simon, for helpful discussions, and Mrs H. Suzuki, Messrs N. Shibuya, and Y. Shiomi for their assistance in preparing the manuscript.

References

Alfvén, H.: 1947, *Monthly Notices Roy. Astron. Soc.* **107**, 211.
Alfvén, H. and Carlqvist, P.: 1967, *Solar Phys.* **1**, 220.
Ayres, T. R. and Linsky, J. L.: 1980, *Astrophys. J.* **241**, 279.
Bally, J. and Lada, C. J.: 1983, *Astrophys. J.* **265**, 824.
Beckers, J. M. and Schneeberger, T. J.: 1977, *Astrophys. J.* **215**, 356.
Bertout, C.: 1979, *Astron. Astrophys.* **80**, 138.
Bertout, C.: 1984, *Rep. Progress Phys.* **47**, 111.
Biermann, L.: 1948, *Z. Astrophys.* **25**, 161.
Blandford, R. D. and Rees, M. J.: 1974, *Monthly Notices Roy. Astron. Soc.* **169**, 395.
Brueckner, G.: 1983, *Solar Phys.* **86**, 259.
Bruzek, A.: 1974, in G. Newkirk (ed.), *Coronal Disturbances*, D. Reidel Publ. Co., Dordrecht, Holland, p. 323.
Bumba, V. and Howard, R.: 1969, *Solar Phys.* **7**, 28.
Carlqvist, P.: 1979, *Solar Phys.* **63**, 353.
Cannon, C. J. and Thomas, R. N.: 1977, *Astrophys. J.* **210**, 910.
Cohen, M.: 1982, *Publ. Astron. Soc. Pacific* **94**, 266.
De Jager, C.: 1969, in C. de Jager and Z. Švestka (eds.), *Solar Flares and Space Research*, D. Reidel Publ. Co., Dordrecht, Holland, p. 1
De Jager, C. and Kuperus, M.: 1961, *Bull. Astron. Inst. Neth.* **16**, 71.
De Loore, C.: 1970, *Astrophys. Space Sci.* **6**, 60.
Draine, B. T.: 1983, *Astrophys. J.* **270**, 519.
Dupree, A. K.: 1982, in M. Giampapa and L. Golub (eds.), *Cool Stars, Stellar Systems, and the Sun* (2), Vol. II, Center for Astrophysics, Cambridge, p. 3.
Dupree, A. K. and Hartmann, L.: 1980, in D. F. Gray and J. L. Linsky (eds.), *Stellar Turbulence*, IAU Colloq. No. 51, Springer, Berlin, p. 279.
Durney, B. R. and Robinson, R. D.: 1982, *Astrophys. J.* **253**, 297.
Eaton, J. A. and Hall, D. S.: 1979, *Astrophys. J.* **227**, 907.
Feigelson, E. D. and De Campli, W. N.: 1981, *Astrophys. J.* **243**, L89.
Feldman, P. A., Taylor, A. R., Gregory, P. C., Seaquist, F. R., Balonek, T. J., and Cohen, N. L.: 1978, *Astron. J.* **83**, 1471.
Friedman, M. and Hamberger, S. M.: 1968, *Astrophys. J.* **152**, 667.
Friend, D.: 1984, in A. B. Underhill (ed.), *The Origin of Non-Radiative Energy/Momentum in Hot Stars*, Proc. NASA Workshop, NASA Printing Office, p. 142.
Fukue, J.: 1982, *Publ. Astron. Soc. Japan* **34**, 163.

Gaizauskas, V., Harvey, K. L., Harvey, J., and Zwaan, C.: 1982, Herzberg Inst. Preprint.
Galeev, A. A.: 1982, in A. Nishida (ed.), *Magnetospheric Plasma Physics*, D. Reidel Publ. Co., Dordrecht, Holland, p. 143.
Gershberg, R. E.: 1982, *Astron. Nachr.* **303**, 251.
Gibson, D. M., Hicks, P. D., and Owen, F. N.: 1978, *Astron. J.* **83**, 1945.
Haisch, B. M., Linsky, J. L., Borman, P. L., Stencel, R. E., Antiochos, S. K., Golub, L., and Vaiana, G. S.: 1982, JILA Preprint.
Hall, D. S.: 1972, *Publ. Astron. Soc. Pacific* **84**, 323.
Hartmann, L.: 1982, *Astrophys. J. Suppl.* **48**, 109.
Hartmann, L. and MacGregor, K. B.: 1980, *Astrophys. J.* **242**, 260.
Hearn, A. G.: 1975, *Astron. Astrophys.* **40**, 355.
Hearn, A. G.: 1979, in P. S. Conti and C. de Loore (eds.), *Proc. Mass Loss and Evolution of O-type Stars*, D. Reidel Publ. Co., Dordrecht, Holland, p. 169.
Heyvaerts, J. and Priest, E. R.: 1983, *Astron. Astrophys.* **111**, 104.
Hinata, S.: 1979, *Astrophys. J.* **232**, 915.
Holzer, T. E. and Axford, W. I.: 1970, *Ann. Rev. Astron. Astrophys.* **8**, 31.
Imhoff, C. L. and Giampapa, M. S.: 1980, *Astrophys. J.* **239**, L115.
Ionson, J. A.: 1977, PhD. Thesis, University of Maryland.
Kaifu, N., Suzuki, S., Hasegawa, T., Morimoto, M., Inatani, J., Nagane, K., Miyazawa, K., Chikada, Y., Kanzawa, T., and Akabane, K.: 1984, *Astron. Astrophys.* **134**, 7.
Kane, S. R.: 1969, *Astrophys. J.* **157**, L139.
Kane, S. R., Uchida, Y., Tanaka, K., and Hudson, H. S.: 1983, *Solar Phys. Special Issue* **86**, Nos. 1/2.
Koenigl, A.: 1982, *Astrophys. J.* **261**, 115.
Kopp, R. A.: 1968, Ph.D. Thesis, Harvard University.
Kopp, R. A. and Holzer, T. E.: 1977, *Solar Phys.* **49**, 43.
Kopp, R. A. and Pneuman, G. W.: 1976, *Solar Phys.* **50**, 85.
Kostyuk, N. D.: 1976, *Soviet Astron.* **19**, 458.
Krieger, A. S., Timothy, A. F., and Roelof, E. C.: 1973, *Solar Phys.* **29**, 505.
Kuhi, L. V.: 1964, *Astrophys. J.* **140**, 1409.
Kuperus, M.: 1969, *Space Sci. Rev.* **9**, 713.
Kuperus, M., Ionson, J. A., and Spicer, D. S.: 1981, *Ann. Rev. Astron. Astrophys.* **19**, 7.
Landini, M. and Monsignori-Fossi, B. C.: 1975, *Astron. Astrophys.* **42**, 413.
Linsky, J. L.: 1980, *Ann. Rev. Astron. Astrophys.* **18**, 439.
Linsky, J. L.: 1983, in J. O. Stenflo (ed.), *Solar and Stellar Magnetic Fields*, D. Reidel Publ. Co., Dordrecht, Holland, p. 313.
Lucy, L. B. and Solomon, P. M.: 1970, *Astrophys. J.* **159**, 879.
Lucy, L. B. and White, R. L.: 1980, *Astrophys. J.* **241**, 300.
Lynden-Bell, D. and Pringle, J. E.: 1974, *Monthly Notices Roy. Astron. Soc.* **168**, 603.
Mangeney, A. and Praderie, F.: 1984, *Astron. Astrophys.* **130**, 143.
Marlborough, J. M. and Roy, J.-R.: 1970, *Astrophys. J.* **160**, 221.
Montmerle, T., Koch-Miramond, L., Falgarone, E., and Grindley, J. E.: 1983, *Astrophys. J.* **269**, 182.
Moreton, G. E.: 1964, *Astrophys. J.* **69**, 145.
Morton, D. C.: 1967, *Astrophys. J.* **150**, 535.
Mullan, D. J.: 1976, *Astrophys. J.* **209**, 171.
Mundt, R. and Fried, J. W.: 1983, *Astrophys. J.* **274**, L83.
Mundt. R. and Giampapa, M. S.: 1982, *Astrophys. J.* **25**, 156.
Nagai, F.: 1980, *Solar Phys.* **68**, 351.
Nakano, T. and Umebayashi, T.: 1982, in *Proc. ISAS Lunar and Planetary Symp.*, ISAS, Tokyo, p. 223.
Nerney, S. F. and Suess, S. T.: 1975, *Astrophys. J.* **196**, 837.
Noyes, R. W., Hartmann, L. W., Baliunus, S. L., Duncan, D. K., and Vaughan, A. H.: 1984, *Astrophys. J.* **279**, 763.
Okuda, T. and Ikeuchi, S.: 1985, *Publ. Astron. Soc. Japan* (submitted).
Osterbrock, D. E.: 1961, *Astrophys. J.* **134**, 347.
Owocki, S. P. and Rybicki, G. B.: 1984, *Astrophys. J.* **284**, 337.
Pallavicini, R., Golub, L., Rosner, R., Vaiana, G. S., Ayres, T. R., and Linsky, J. L.: 1981, *Astrophys. J.* **248**, 279.

Parker, E. N.: 1963, *Interplanetary Dynamical Processes*, Interscience, New York.
Parker, E. N.: 1964, in W. N. Hess (ed.), *Proc. NASA Symp. Physics of Solar Flares*, NASA Printing Office, p. 417.
Petschek, H. E.: 1964, in W. N. Hess (ed.), *NASA Symp. Physics of Solar Flares*, NASA Printing Office, p. 425.
Piddington, J. H.: 1956, *Monthly Notices Roy. Astron. Soc.* **116**, 314.
Piddington, J. H.: 1973, *Solar Phys.* **33**, 363.
Rappaport, S. and Joss, P.: 1981, in R. Giacconi (ed.), *X-ray Astronomy with the Einstein Satellite*, D. Reidel Publ. Co., Dordrecht, Holland, p. 123.
Rodono, M.: 1981, in E. B. Carling and Z. Kopal (eds.), *Photometric and Spectroscopic Binary Systems*, D. Reidel Publ. Co., Dordrecht, Holland, p. 285.
Rogerson, J. B. and Lamers, H.: 1975, *Nature* **256**, 190.
Rosner, R., Golub, L., Coppi, B., and Vaiana, G. S.: 1978, *Astrophys. J.* **222**, 317.
Rust, D. M.: 1972, *Solar Phys.* **25**, 141.
Sakashita, S., Hanami, H., Umemura, M.: 1985, *Astrophys. Space Sci.* **111**, 213.
Sakurai, T.: 1975, *Publ. Astron. Soc. Japan* **28**, 177.
Sakurai, T. and Granik, A.: 1984, *Astrophys. J.* **277**, 897.
Sakurai, T. and Uchida, Y.: 1977, *Solar Phys.* **52**, 397.
Sato, T. and Hayashi, T.: 1979, *Phys. Fluids* **22**, 1189.
Schatzman, E.: 1949, *Ann. Astrophys.* **12**, 203.
Schwarzschild, M.: 1948, *Astrophys. J.* **107**, 1.
Sheeley, N. R., Bohlin, J. D., Brueckner, G. E., Purcell, J. D., Scherrer, V., and Tousey, R.: 1974, *Solar Phys.* **40**, 103.
Shibata, K. and Uchida, Y.: 1984, in these proceedings.
Shibata, K. and Uchida, Y.: 1985, *Publ. Astron. Soc. Japan* **37**, 31.
Smerd, S. F.: 1970, *Proc. Astron. Soc. Australia* **1**, 305.
Snell, R. L., Loren, R. B., and Plambeck, R. L.: 1980, *Astrophys. J.* **239**, L17.
Spicer, D. S.: 1976, NRL Report No. 8036.
Spicer, D. S. and Brown, J. C.: 1980, in S. Jordan (ed.), *Sun as a Star*, NASA Printing Office, p. 413.
Stewart, R. T., McCabe, M. K., Koomen, M. J., Hansen, R. T., and Dulk, G. A.: 1974, *Solar Phys.* **36**, 203.
Sturrock, P. A.: 1968, in K. O. Kiepenheuer (ed.), 'Structure and Development of Solar Active Regions', *Proc. IAU Symp.* **35**, 471.
Sturrock, P. A. and Uchida, Y.: 1981, *Astrophys. J.* **246**, 331.
Svargaad, L., Wilcox, J. M., Scherer, P. H., and Howard, R.: 1975, *Solar Phys.* **45**, 83.
Švestka, Z.: 1976, *Solar Flares*, D. Reidel Publ. Co., Dordrecht, Holland.
Swank, J. H. and White, N. E.: 1980, in A. K. Dupree (ed.), *Cool Stars, Stellar Systems, and the Sun*, Center for Astrophysics, Cambridge, p. 47.
Sweet, P. A.: 1969, *Ann. Rev. Astron. Astrophys.* **7**, 149.
Syrovatsky, S. I. and Shmeleva, O. P.: 1972, *Astron. Zh.* **49**, 334.
Tsuneta, S., Takakura, T., Nitta, N., Ohki, K., Makishima, K., Murakami, T., Oda, M., and Ogawara, Y.: 1983, *Solar Phys.* **86**, 313.
Uchida, Y.: 1963, *Publ. Astron. Soc. Japan* **15**, 376.
Uchida, Y.: 1974, *Solar Phys.* **39**, 431.
Uchida, Y.: 1984a, in P. B. Byrne and M. Rodono (eds.), *Activity of Red Dwarf Stars*, p. 625.
Uchida, Y.: 1984, in A. B. Underhill (ed.), *Proc. NASA Workshop "The Origin of Non-Radiative Energy/Momentum in Hot Stars'*, NASA Printing Office.
Uchida, Y., Altschuler, M. D., and Newkirk, G.: 1973, *Solar Phys.* **28**, 495.
Uchida, Y. and Bappu, M. K. V.: 1982, *Astrophys. Astron.* **3**, 277.
Uchida, Y. and Low, B. C.: 1982, *Astrophys. Astron.* **2**, 405.
Uchida, Y. and Kaburaki, O.: 1974, *Solar Phys.* **35**, 451.
Uchida, Y. and Sakurai, T.: 1983, in P. B. Byrne and M. Rodono (eds.), *Activity of Red Dwarf Stars*, p. 629.
Uchida, Y. and Sakurai, T.: 1984, in M. Kundu (ed.), *Unstable Current Systems Plasma Instabilities in Astrophysics*, D. Reidel Publ. Co., Dordrecht, Holland, p. 281.
Uchida, Y. and Shibata, K.: 1984a, *Publ. Astron. Soc. Japan* **36**, 105.
Uchida, Y. and Shibata, K.: 1984b, in M. Kundu (ed.), *Unstable Current Systems and Plasma Instabilities in Astrophysics*, D. Reidel Publ. Co., Dordrecht, Holland, p. 287.

Uchida, Y. and Shibata, K.: 1984c, in A. B. Underhill (ed.), *Proc. NASA Workshop 'The Origin of Non-Radiative Energy/Momentum in Hot Stars'*, NASA Printing Office, 169.
Uchida, Y. and Shibata, K.: 1985, *Publ. Astron. Soc. Japan* **37**, 515.
Ulmschneider, P.: 1967, *Z. Astrophys.* **67**, 193.
Ulrich, R. K.: 1976, *Astrophys. J.* **210**, 377.
Underhill, A. B.: 1980, *Astrophys. J.* **240**, L153.
Underhill, A. B.: 1983, *Astrophys. J.* **265**, 933.
Underhill, A. B.: 1984, *Astrophys. J.* **276**, 583.
Underhill, A. B. and Doazan, V.: 1982, in A. B. Underhill (ed.), *B Stars with and without Emission Lines*, NASA SP-456.
Underhill, A. B. and Fahey, R. P.: 1984, *Astrophys. J.* **280**, 712.
Vaiana, G. S., Krieger, A. S., and Timothy, A. F.: 1973, *Solar Phys.* **32**, 81.
Vaiana, G. S., Cassinelli, J. P., Fabbiano, Gr., Giacconi, R., Golub, L.: 1981, *Astrophys. J.* **245**, 163.
Vlahos, L.: 1979, Ph.D. Thesis, University of Maryland.
Walter, F. M. and Kuhi, L. V.: 1981, *Astrophys. J.* **250**, 254.
Walter, F. M., Cash, W., Charles, P. A., Bowyer, C. S.: 1980, *Astrophys. J.* **236**, 212.
Walter, F. M., Gibson, D. M., and Basri, G. S.: 1983, *Astrophys. J.* **267**, 665.
Wentzel, D. G.: 1974, *Solar Phys.* **39**, 129.
Wilson, O. C.: 1976, *Astrophys. J.* **205**, 823.
Zahn, J. P.: 1977, *Astron. Astrophys.* **57**, 383.

HIGH RESOLUTION OBSERVATION OF Hα SOLAR FLARES AND TEMPORAL RELATION BETWEEN Hα AND X-RAY, MICROWAVE EMISSION*

H. KUROKAWA, T. KITAHARA, Y. NAKAI, Y. FUNAKOSHI, and K. ICHIMOTO

Kwasan and Hida Observatories, University of Kyoto, Kamitakara, Gifu, Japan

(Received 22 March, 1985)

Abstract. We studied the evolutional characteristics of fine structures in Hα flare emitting regions and their relation to X-ray and microwave emissions for selected events observed with the 60 cm Domeless Solar Telescope at Hida Observatory, University of Kyoto. The principal conclusions of this investigation are: (1) Hα kernel consists of some finer bright points or Hα *flare points* whose individual size is less than 1 arc sec. (2) Impulsive brightnenings of Hα flare points occurred simultaneously with the spikes of the hard X-ray and microwave bursts within the time resolution of our Hα observations which varied from 1 to 10 s. (3) It is concluded that fast electron beams must be the principal mechanism of heating Hα flares during the impulsive phase of a flare.

1. Introduction

Optical observations of solar flares can give us detailed information on the morphological features of flare development with the advantage of their much higher spatial resolution than at other wavelengths. There have been, however, only a few published optical observations with high temporal and spatial resolution which can correctly follow the rapidly developing features of Hα bright points in the impulsive phase of flares (Zirin and Tanaka, 1973; Kämpfer and Magun, 1983).

From 1981, using the 60 cm Domeless Solar Telescope at Hida Observatory, University of Kyoto, we obtained many Hα monochromatic images and Hα spectra of solar flares with high temporal and spatial resolution. Ichimoto and Kurokawa (1984) studied the evolutional change of Hα line profiles of the flares and found that the red-asymmetric profile of the Hα line is due to the downward motion of the compressed chromospheric flare region associated with the impulsive heating of an energetic electron beam or thermal conduction.

In this paper we demonstrate how closely correlated the successive and abrupt brightenings of Hα flare fine structures are with the spikes of hard X-ray and microwave bursts at the impulsive phase of a flare.

2. Fine Structures in Hα Flare Kernel

Vorpahl (1972) has summarized properties of Hα flare kernels. According to him, the effective kernel diameter ranged from 3000 to 6000 km for flares of importance 1, or sub-flares.

* Paper presented at the IAU Third Asian-Pacific Regional Meeting, held in Kyoto, Japan, between 30 September–6 October, 1984. Contributions from the Kwasan and Hida Observatories, University of Kyoto, No. 265.

Astrophysics and Space Science **118** (1986) 149–152.

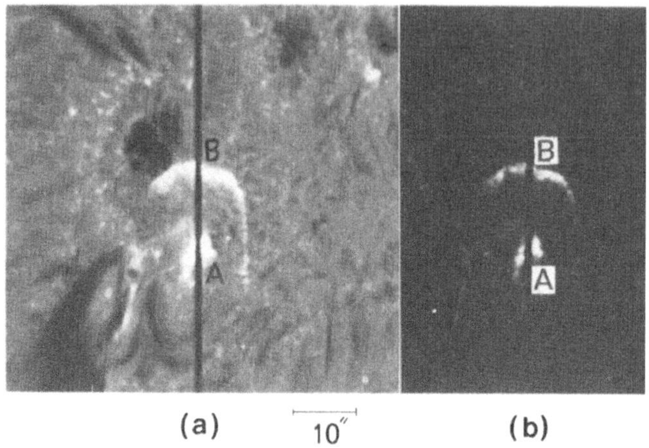

Fig. 1. Flare points in Hα flare kernels. (a) 6 August, 1982 flare at 23 : 36 UT in Hα − 0.8 Å. (b) Dark print of the same picture as (a). Flare kernels *A* and *B* consist of many Hα flare points.

With higher spatial resolution we found an Hα flare kernel consisting of several small bright points whose characteristic size is about 1 arc sec. And the successive brightenings of these fine structures are closely connected with the spikes of hard X-ray and microwave bursts at the impulsive phases of the 7 September, 1981, the 23 March, 1982 and the 12 June, 1982 flares (Kurokawa and Kanno, 1982; Kurokawa *et al.*, 1985).

In Figure 1 we show another example of the fine structure of an Hα flare region observed in the light of Hα − 0.8 Å, where the width of the spectrograph slit (vertical black line) corresponds to 1 arc sec. In Figure 1(b) we can find two Hα flare kernels *A* and *B* consisting of many small bright points, which will be called as Hα *flare points* hereafter. The characteristic size of these flare points are estimated to range from 0.5 to 0.8 arc sec, and a little larger than the moustaches and filigree-like fine bright points also seen in this picture. The 3835 Å flash found by Zirin and Tanaka (1973) may correspond to the brightest of the Hα flare points.

3. Temporal Relation Between Hα, X-Ray, and Microwave Emissions

We selected five events observed with the highest spatial resolution and measured the intensity variations of Hα flare points in them with a microphotometer. The light curves of these Hα flare points were compared with the time profiles of corresponding X-ray and microwave flux. The time differences between Hα intensity and hard X-ray or microwave flux peaks were estimated and given in the last line of Table I, together with the earlier results of other authors. Hard X-ray data were supplied by HINOTORI and ISEE-3 satellites, and 17 GHz data by Nobeyama Solar Radio Observatory. The plus sign of the time difference means that the Hα peak is delayed from the corresponding hard X-ray or microwave peak. The sign $\pm t$ is used when the peaks of Hα and hard X-ray or microwave emissions coincide with each other to within t s. In Table I our data show that a smaller time difference between Hα and hard X-ray or microwave emissions

TABLE I

Time differences between the peak of Hα, hard X-ray, and microwave emissions at the impulsive phases of flares

Reference	This study	This study	This study	This study	This study	Vorpahl (1972)	Zirin and Tanaka (1973)	Zirin (1978)	Kämpfer and Schöchlin (1982)
Date and time of flare	7 Spt., 1981 00:54 UT	12 Oct., 1981 05:27 UT	23 Mar., 1982 05:18 UT	27 Mar., 1982 22:11 UT	12 June, 1982 06:24 UT	1 Aug., 1968	2 Aug., 1972	(Statistical study of 41 flares)	17 Dec., 1980
Flare importance	1N	3B	1N	sB	2B	1N	1B		2B
Wavelength (Å)	Hα ± 0.0	Hα – 1.0	Hα – 1.0	Hα – 0.5	Hα – 2.4	Hα – 0.5	3835		Hα ± 0.0
Hard X-ray and microwave data	HINOTORI 40–67 keV	ISEE-3 70 keV	HINOTORI 40–67 keV	17 GHz	17 GHz	19.2 to 32 keV	53 to 75 keV		8.4 GHz
Temporal resolution of optical observation	10 s	7 s	1 s	3 s	5 s	20 s	5 s		1.6 s
Time difference between Hα and hard X-ray microwave peaks	< ± 10 s	< ± 5 s	< ± 1 s	< ± 1.5 s	< ± 5 s	+ 22s	< ± 3 s	< ± 5 s	+ 8 s

was obtained with better temporal resolution of Hα observation. Especially the temporal resolution for the 23 March, 1982 flare is better than any which have ever been published. And it is remarkable to find that there is a very good temporal coincidence (within 1 s) between the Hα and hard X-ray impulsive peaks for this flare (Kurokawa *et al.*, 1985).

4. Conclusion

The principal observational results described in the preceding sections may be summarized as follows: (1) The Hα flare kernel consists of some finer bright points or Hα *flare points* whose characteristic size is less than 1 arc sec. (2) The progressive brightenings of these Hα flare points are closely correlated with the spikes of hard X-ray and microwave flux at the impulsive phase of a flare. The temporal coincidence between Hα and hard X-ray spikes for the 23 March, 1982 flare was especially good, or within 1 s.

These results strongly suggest that the impulsive Hα flare brightenings occur simultaneously with the initial explosive energy release in the flare loops. The Hα flare points must be located at the footpoints of flare loops (Kurokawa, 1983), while the location of the initial explosive energy deposition is considered to be around the tops of the loops (Craig, 1981). The speed of a heat conduction front moving down along the loop is too slow to account for such a nearly instantaneous energy transfer from the top to the footpoint of a flare loop.

Then we can conclude that a fast electron beam must be the principal mechanism of heating the Hα flare points during the impulsive phase of a flare.

Acknowledgements

The authors wish to thank Mr Matsui for preparing the photographic prints. This work was supported in part by the Scientific Research Fund of the Ministry of Education, Science and Culture under Grant No. 59420003 (I. Kawaguchi in 1984 and 1985).

References

Craig, I. J. D.: 1981, in E. R. Priest (ed.), *Solar Flare Magnetohydrodynamics*, Gordon and Breach Science Publishers, New York, p. 277.
Ichimoto, K. and Kurokawa, H.: 1984, *Solar Phys.* **93**, 105.
Kämpfer, N. and Magun, A.: 1983, *Astrophys. J.* **274**, 910.
Kämpfer, N. and Schöchlin, W.: 1982, *Solar Phys.* **78**, 215.
Kurokawa, H.: 1983, *Solar Phys.* **86**, 195.
Kurokawa, H., Takakura, T., and Ohki, K.: 1985, *Publ. Astron. Soc. Japan* (submitted).
Kurokawa, H. and Kanno, M.: 1982, *Proceedings of the HINOTORI Symposium on Solar Flares, Tokyo*, p. 199.
Vorpahl, J. A.: 1972, *Solar Phys.* **26**, 397.
Zirin, H.: 1978, *Solar Phys.* **58**, 95.
Zirin, H. and Tanaka, K.: 1973, *Solar Phys.* **32**, 173.

THREE-DIMENSIONAL PROPERTIES OF INTERPLANETARY DISTURBANCES IN 1978–1981*

TAKASHI WATANABE and TAKAKIYO KAKINUMA

Research Institute of Atmospherics, Nagoya University, Toyokawa, Japan

(Received 22 March, 1985)

Abstract. This paper presents the average three-dimensional configuration of solar flare- or disappearing filament-associated interplanetary disturbances on the basis of IPS (interplanetary scintillation) and spacecraft observations in 1978–1981. The angular distribution of the propagation speed at 1 AU is largely isotropic over the range of 110° in solar longitude centered at the normal of the solar source. In the latitudinal direction, the characteristic angular extent is about 60°. Thus the three-dimensional shape of an interplanetary disturbance can be approximated by a half of an ellipsoid having an axial ratio of about 1.8.

1. IPS Observations of Interplanetary Disturbances

The IPS (interplanetary scintillation) technique has been proved to be a useful means to study the large-scale, three-dimensional configurations of interplanetary disturbances, because solar wind data in the interplanetary region away from the ecliptic plane can be obtained by this technique (e.g., Kakinuma and Watanabe, 1976; Houminer, 1977; Watanabe and Kakinuma, 1984b). We can identify the post-shock plasma of the interplanetary disturbance by a transient increase in the solar wind flow speed accompanied with enhancement in the level of scintillation. Since several observations (e.g., Tappin *et al.*, 1983) showed that the radial extent of the turbulent post-shock plasma is about 0.2 AU, the post-shock plasma will be observed for about 17 hr (the propagation speed of 500 km s^{-1} is assumed). Three-station IPS observations, which provided us with the solar wind flow speed, have been carried out by two IPS groups at La Jolla (UCSD), U.S.A. and at Toyokawa, Japan. Since the maximum time difference between the observations of a radio source is 17 hr, the majority of interplanetary disturbances having thick and turbulent post-shock plasmas will be detected by one of or both of the IPS groups, although some very high-speed disturbances will escape detection. The observed flow speed is the weighted average of the transverse components of the flow vectors in the principal scattering zone along the line-of-sight to a radio source. The typical width of the zone is 0.5 AU. According to a numerical simulation of the IPS observation of a spherical disturbance, the observed flow speed is close to that of the high-density post-shock plasma (Watanabe and Kakinuma, 1984b). Thus we can observe density enhancements situated in the post-shock region by the IPS technique. The approximate propagation speed of the disturbance (shock speed plus

* Paper presented at the IAU Third Asian-Pacific Regional Meeting, held in Kyoto, Japan, between 30 September–6 October, 1984.

Astrophysics and Space Science **118** (1986) 153–157.

pre-existing flow speed) is 1.2 times as high as the observed post-shock flow speed (Watanabe and Kakinuma, 1984a).

The local propagation speed obtained at each observation point (peak of the scattering weighting function) in the interplanetary space is normalized to that at a specific heliocentric distance (1 AU). We assume that the dynamical evolution of an interplanetary disturbance can be expressed by a power-law relationship $V_s \propto R^{-\alpha}$. The deceleration coefficient α is estimated from the local shock speed near the Earth, which is determined using the solar wind data obtained by space probes and the traveling time of the disturbance between the Sun and 1 AU heliocentric distance. In this paper, we used the IPS data obtained at UCSD (Solar-Geophysical Data, U.S. Department of Commerce, Boulder, Colorado, U.S.A.) and those obtained at Toyokawa (Solar Wind Speed from IPS Measurements).

TABLE I

Solar sources of the interplanetary disturbance

Date	Source	Position
23 Aug., 1978	DSF	N 15 E 03
18 Feb., 1979	F	N 16 E 13
20 Feb., 1979	F	N 16 W 15
22 Apr., 1979	DSF	S 20 E 10
4 July 1979	F	N 14 E 35
27 Nov., 1979	DSF	N 05 W 03
21 May, 1980	F	S 13 W 15
23 July 1980	F	S 20 E 20
8 Apr., 1981	F	N 11 W 03
8 May, 1981	F	N 09 E 38
10 May, 1981	F	N 07 E 31
16 May, 1981	F	N 10 E 14

DSF: disappearing solar filament.
F: solar flare.

2. Three-Dimensional Shape of the Interplanetary Disturbance

The probable configuration of the interplanetary disturbance is determined through an averaging procedure of the 'directivities' of many disturbances. The directivity of each disturbance is determined in such a way that the estimated propagation speeds of several portions of the disturbance, at 1 AU heliocentric distance, are normalized with respect to the observed maximum propagation speed of the disturbance. As compared with the simple average of propagation speeds employed in our earlier analysis (Watanabe and Kakinuma, 1984b), the directivity is free from the ambiguity in the multiplying factor of 1.2 in estimation of the local propagation speed, and the scattering of the data points is small. To obtain the directivity, we select 12 of the 17 examples which were analysed by Watanabe and Kakinuma (1984b). We exclude the disturbance whose solar source

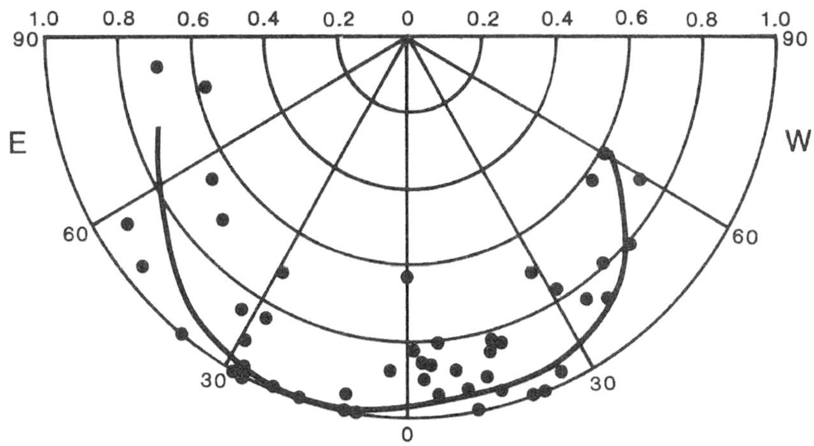

Fig. 1. The most probable directivity of propagation speed of the interplanetary disturbance at 1 AU in the longitudinal direction (see text). The longitude is measured from the flare normal. The dots represent the data points.

was near the solar limb because only the propagation speeds on the eastern or the western side were observed. The solar sources of the solar-interplanetary events which are analysed in this paper are listed in Table I. The data of solar flares and disappearing solar filaments are taken from the Solar-Geophysical Data. The solar wind data obtained by space probes (Ogilvie *et al.*, 1982; Sanahuja *et al.*, 1983; Sheeley *et al.*, 1983; King, 1983; Zastenker *et al.*, 1984) are also used in the present analysis.

The most probable directivity in the longitudinal direction is drawn in Figure 1 as a function of longitude measured from the flare normal (or the normal of the center of a disappearing filament). We use the data points whose latitudes measured from the flare normal are less than 20° to minimize the effect of anisotropic nature in the latitudinal direction. The curve for the latitudinal direction is shown in Figure 2. The latitude is measured from the flare normal. In this case, we use the data points whose longitudes are less than 50° to minize the effect of anisotropic expansion in the longitudinal direction (Watanabe and Kakinuma, 1984b). Since the distribution of data points is not uniform, we determine the most probable directivity assuming north–south symmetry.

3. Summary and Discussion

The directivity in the longitudinal direction at 1 AU heliocentric distance (Figure 1) is largely isotropic over the range of 110°. This result is consistent with other estimates, depending on the magnitude of sudden commencements of geomagnetic storms (Hirshberg, 1968), and shock surface observations by space probes (e.g., Chao and Lepping, 1974). The large-scale geometry of the diagram given in Figure 1 is well represented by a model of a shock shape proposed by Smart *et al.* (1984). There exists a tendency that the maximum propagation speed appears in the region to the east of the flare normal, as proposed by several authors (e.g., Sakurai, 1973; Pudovkin *et al.*,

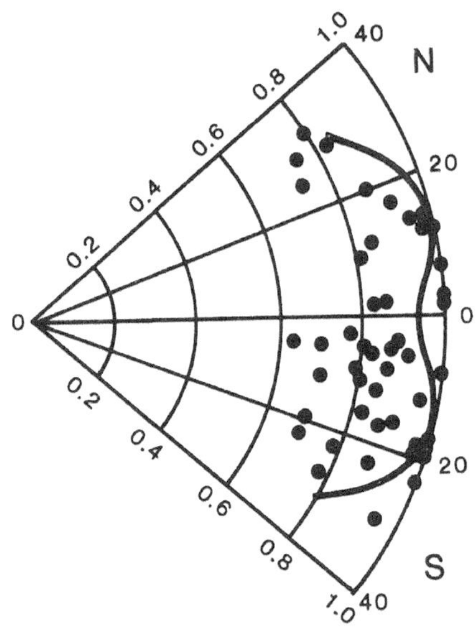

Fig. 2. The most probable directivity of the interplanetary disturbance at 1 AU in the latitudinal direction. The north–south symmetry is assumed. The latitude is measured from the flare normal.

1979). For the latitudinal direction (Figure 2), the propagation speed drops when the latitude of the observation point measured from the flare normal approaches $\pm 30°$. Thus the characteristic angular extent of the interplanetary disturbance in the latitudinal direction is $60°$. If we neglect a dip seen around the latitude of flare normal, the thre-dimensional shape of the disturbance can be represented by a half of an ellipsoid with the ratio of longitudinal and latitudinal axes equal to 1.8. This axial ratio is consistent with that obtained by Ivanov (1973).

The spatial evolution of an interplanetary disturbance can be traced over the range from the solar corona to 1 AU heliocentric distance when we combine white-light observations of coronal mass ejections (CMEs) and IPS observations. The characteristic latitudinal angular extent of an interplanetary disturbance at 1 AU is comparable to those of high-speed ($> 500 \text{ km s}^{-1}$), loop-like, curved-front, and complex CMEs (Howard *et al.*, 1984). Michels *et al.* (1984) also obtained similar conclusions; interplanetary disturbances are usually observed when the plasma probe in the solar wind lies within $15°$ of the solid angle subtended by the CME at 10 solar radii. As a speculation, lateral expansion of the disturbance in the latitudinal direction is suppressed by coronal and solar wind magnetic structures, e.g., coronal streamers and their extensions, and the quasi-ellipsoidal configuration of the interplanetary disturbance forms as a result.

The latitudinal directivity (Figure 2) shows a local minimum at the latitude of the solar source of the disturbance. Although such a fine structure should be confirmed by a statistical analysis of many examples, the local minimum may be produced in the course of propagation within a nonuniform solar wind plasma. The flow speed of the pre-existing solar wind will be low in the radial directions of bright coronal regions (Watanabe, 1975). Thus low propagation speed of the interplanetary disturbance (shock speed plus pre-existing flow speed) is expected around the latitudes of high solar activity.

References

Chao, J. K. and Lepping, R. P.: 1974, *J. Geophys. Res.* **79**, 1799.
Hirshberg, J.: 1968, *Planetary Space Sci.* **16**, 309.
Houminer, Z.: 1977, in M. A. Shea, D. F. Smart, and S. T. Wu (eds.), *Study of Travelling Interplanetary Phenomena*, D. Reidel Publ. Co., Dordrecht, Holland, p. 119.
Howard, R. A., Sheeley, N. R., Jr., Koomen, M. J., and Michels, D. J.: 1984, *Adv. Space Res.* **4**, No. 7, 307.
Ivanov, K. G.: 1973, *Soviet Astron.* **17**, 94.
Kakinuma, T. and Watanabe, T.: 1976, *Space Sci. Rev.* **19**, 611.
King, J. H.: 1983, *Interplanetary Medium Data Book Supplement 2, 1978–1982*, NSSDC/WDC-A-R&S.
Michels, D. J., Howard, R. A., Sheeley, N. R., Jr., and Koomen, M. J.: 1984, in M. A. Shea, D. F. Smart, and S. M. P. McKenna-Lawlor (eds.), *STIP Symp. on Solar/Interplanetary Intervals*, Book Crafters Inc., Chelsea, MI, U.S.A., p. 319.
Ogilvie, K. W., Coplan, M. A., and Zwickl, R. D.: 1982, *J. Geophys. Res.* **87**, 736.
Pudovkin, M. I., Zaitseva, S. A., and Benevolenska, E. E.: 1979, *J. Geophys. Res.* **84**, 6649.
Sakurai, K.: 1974, *Indian J. Radio Space Phys.* **3**, 289.
Sanahuja, B., Domingo, V., Wenzel, K.-P., Jocelyn, J. A., and Keppler, E.: 1983, *Solar Phys.* **84**, 321.
Sheeley, N. R., Jr., Howard, R. A., Koomen, M. J., Michels, D. J., Schwenn, R., Muhlhauser, K. H., and Rosenbauer, H.: 1983, in M. Neugebauer (ed.), *Solar Wind Five*, NASA Conference Publication 2280, p. 693.
Smart, D. F., Shea, M. A., Barron, W. R., and Dryer, M.: 1984, in M. A. Shea, D. F. Smart, and S. M. P. McKenna-Lawlor (eds.), *STIP Symposium on Solar-Interplanetary Intervals*, Book Crafters Inc., Chelsea, MI, 48118, U.S.A., p. 139.
Tappin, S. J., Hewish, A., and Gapper, G. R.: 1983, *Planetary Space Sci.* **31**, 1171.
Watanabe, T.: 1975, *Publ. Astron. Soc. Japan* **27**, 385.
Watanabe, T. and Kakinuma, T.: 1984a, *Proc. Res. Inst. Atmospherics, Nagoya Univ.* **31**, 11.
Watanabe, T. and Kakinuma, T.: 1984b, *Adv. Space Res.* **4**, No. 7, 331.
Zastenker, G. N., Borodkova, N. L., Vaisberg, O. L., Omel'chenko, A. N., Yermolaev, Yu. I., and Balabanov, V. M.: 1983, preprint.

ON THE SPÖRER MINIMUM*

YAOTIAO JIANG

Department of Astronomy, Nanjing University, Nanjing, China

and

ZHENTAO XU

Purple Mountain Observatory, Nanjing, China

(Received 7 October, 1985)

Abstract. In this paper we have examined the real behaviour of solar activity during the period AD 1400–1600. The results are as follows: (1) the distributions of the 20 naked-eye sunspot records are inhomogeneous. There are 2 sightings in the 15th century and 18 sightings in the 16th century; (2) the distributions of auroral records are similar to sunspot. There are 33 records in the 15th century and 315 records in the 16th century; (3) the climatic fluctuations in China shows that the period AD 1430–1520 was cold while the period AD 1520–1620 was warm. These facts clearly demonstrate that the Spörer Minimum, if it extended from AD 1460 to 1550, could be a specious results and it, if its extent was AD 1400–1510, is a real feature of solar variability in that time.

Few people studied the behaviour of solar activity in the period AD 1400–1600. Clark and Stephenson (1978), by means of the Chinese ancient sunspot records, showed that this period could have been a period of prolonged inactivity. However, Wang (1979), by means of Chinese ancient auroral records, pointed out that AD 1400–1500 could

TABLE I

Naked-eye sunspot records

No.	Julian date	Description
1	1402, 15 Nov.	Black spot
2	1511, May–June	Black gas
3	1518, Feb.	Ripples of Sun's light
4	1520, 9 Mar.	Black gas
5	1529, Jan./Mar.	Black shadows
6	1556, 17 Apr.	As large as a hen's egg
7	1562	Black spot
8	1566, Feb.	As large as a hen's egg
9	1567, Spring	Black spot
10	1569, Summer	Black light
11	1590, May	Black gas
12	1593, 3 Jan.	Two spots
13	1597, 15 June	Like a black cake

* Paper presented at the IAU Third Asian-Pacific Regional Meeting, held in Kyoto, Japan, between 30 September–6 October, 1984.

Astrophysics and Space Science **118** (1986) 159–162.

have been a very active period, and AD 1500–1600 a very inactive one. It is interesting that Eddy (1976a), by means of the ^{14}C deviation, concluded that AD 1460–1550, that is the Spörer Minimum, was a period of prolonged minimum of solar activity. These contradictory results stimulate us to search for the real behaviour of solar activity in the period AD 1400–1600.

First of all, we have examined the distributions of ancient sunspot records of China. We have scrutinized the Chinese official histories, and the local topographies, and have found 13 naked-eye sunspot records in the period AD 1400–1600. Table I gives the Julian dates and descriptions of these records. It shows that there was one sighting in the 15th century and twelve sightings in the 16th century. A similar result existed in the Western naked-eye sunspot records (Wittmann, 1978). In his sunspot catalogue there were six sightings in the 16th century, but only one sighting in the 15th century. Combining both Chinese and Western sunspot we find that the distributions of sunspot activity in the two centuries are very different: two sightings and eighteen sightings during the 15th and 16th century, respectively. This could mean that the solar activity between 15th and 16th centuries was completely different: the former was inactive but the latter was active.

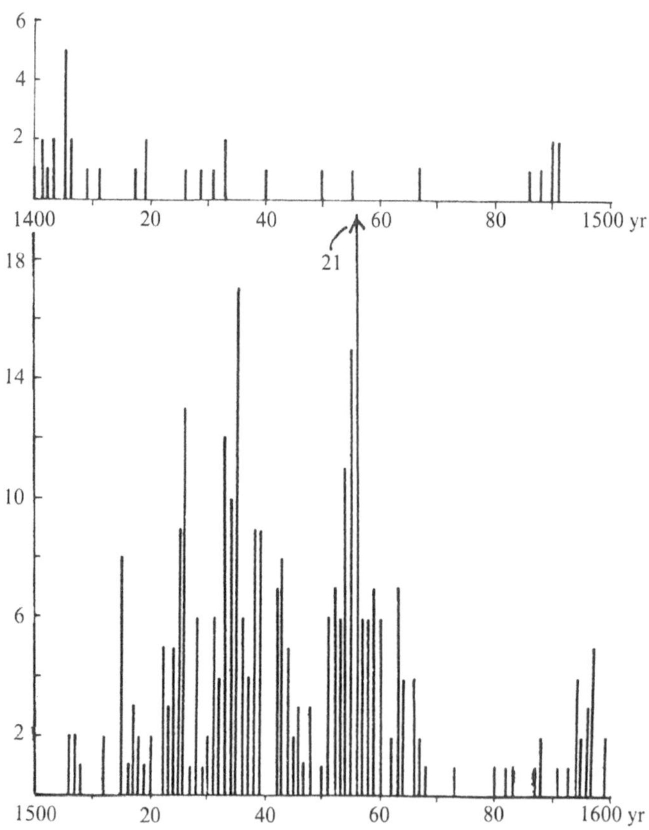

Fig. 1. Frequency distribution of oriental auroral records.

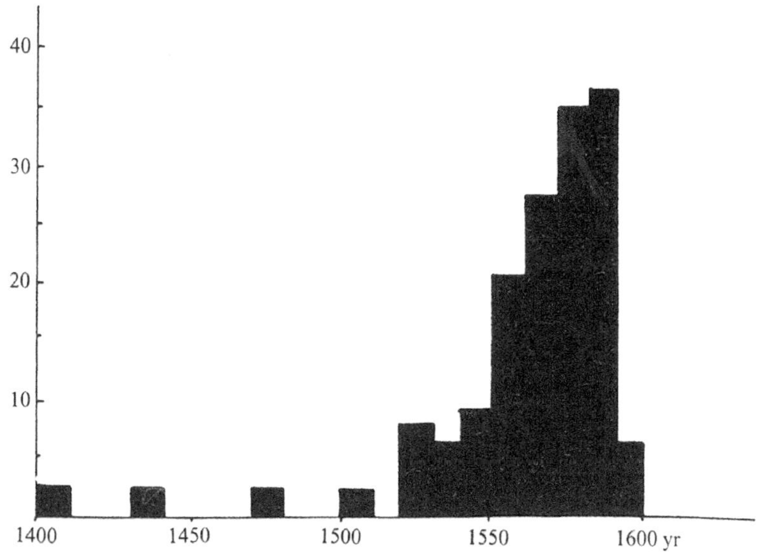

Fig. 2. Distribution of European auroral records.

Secondly, we have examined the historical records of aurorae. From the recent list of oriental auroral records (Dai and Chen, 1980), we can see that there were 33 records in the 15th century, but 315 records in the 16th century. The distribution of frequency is shown in Figure 1. A similar distribution occurred in the European auroral records. Figure 2 (Siscoe, 1980) shows that the number of auroral days in the 15th century was very much less than that in the 16th century. It is well known that auroral activity is a good indicator of solar activity. So the above result implies that solar activity of the 15th century is very much weaker that that of the 16th century.

According to Zhu Ke-zhen (1973), climatic fluctuations are also an indirect display of solar activity, a cold climate corresponding to low solar activity and vice versa. On the basis of tree ring analyses (Zhuo *et al.*, 1978), the climatic fluctuation of the last 900 years in China has shown that the period AD 1430–1520 was a cold period, while the period AD 1520–1620 was a warm period. Figure 3 gives the measurement results

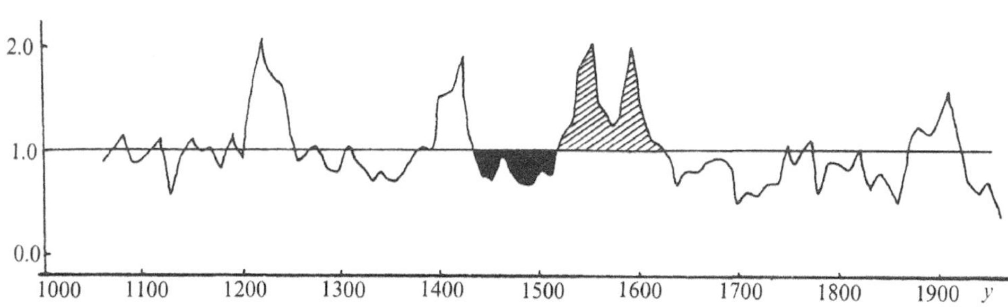

Fig. 3. Variation of tree ring index.

of this analysis. The difference between both periods is very apparent. The shaded region means a cold period while the hatched region means a warm period. If we suppose that the effect of solar activity on the terrestrial climate may have about a 20 year lag, it is safe to say that solar activity of the 15th century is clearly weaker than that of the 16th century.

It is very interesting that the above three statistical results accord well with each other. This is unlikely to represent some occasional coincidence. Therefore, it should be a real reflection of solar activity at that time. However, the originally proposed Spörer Minimum was just situated between the 15th and 16th centuries. This appears to be contradictory with the above mentioned facts. The existence of the Spörer Minimum may, therefore, appear to be in doubt. Fortunately, Eddy (1976b), by accurate measurement, determined the period AD 1400–1510 as the real extent of the Spörer Minimum. Evidently, the redetermined Spörer Minimum is quite consistent with the above statistical results. It shows the ^{14}C deviation, already known as an indirect indicator of solar activity, may also describe the fundamental behaviour of solar variability. So we have good reason to say that the Spörer Minimum is a real feature of solar variability in the 15th century. On the other hand, it is very interesting that the reality of the Maunder Minimum is still a debatable problem (Xu and Jiang, 1982; Eddy, 1983). As to why the solar activity was so weak during the Spörer Minimum, we need further research.

References

Clark, D. H. and Stephenson, F. R.: 1978, *Quant. J. Roy. Astron. Soc.* **19**, 387.
Dai, N. Z. and Chen, M. D.: 1980, *Collected Works of the History of Science and Technology*, No. 6.
Eddy, J.: 1976a, *Science* **192**, 1189.
Eddy, J.: 1976b, *Physics of Solar Planetary Environment.*
Eddy, J.: 1983, *Solar Phys.* **89**, 195.
Siscoe, G. L.: 1980, *Rev. Geophys. Space Phys.* **18**, 647.
Wang, J. C. H.: 1979, *Bull. Am. Astron. Soc.* **11**, 610.
Wittmann, A.: 1978, *Astron. Astrophys.* **66**, 93.
Xu, Z. T. and Jiang, Y. T.: 1982, *CAA* **6**, 84.
Zhu, K. Z.: 1973, *Scientia Sinica* **16**(2), 226.
Zhuo, Z. D. *et al.*: 1978, *J. Lanzhou Univ.*, No. 2.

SOLAR VECTOR MAGNETOGRAMS OF THE OKAYAMA ASTROPHYSICAL OBSERVATORY*

MITSUGU MAKITA, SHIGEO HAMANA, KEIZO NISHI, MINORU SHIMIZU

Tokyo Astronomical Observatory, University of Tokyo, Mitaka, Tokyo, Japan

TAKASHI SAKURAI

Department of Astronomy, University of Tokyo, Bunkyo-ku, Tokyo, Japan

and

KIYOTO SHIBASAKI

Research Institute of Atmospherics, Nagoya University, Honohara, Toyokawa, Japan

(Received 22 March, 1985)

Abstract. The title instrument is mounted on the 65 cm solar Coudé telescope at the Okayama observatory. Observation is usually of the Fe I 5250 Å line. The data obtained are briefly described.

The solar vector magnetograph of the Okayama Astrophysical Observatory is mounted on the 65 cm solar Coudé telescope as shown in Figure 1. The polarimeter is put at the image focus and in front of the 10 m Littrow spectrograph. Two compensators, each composed of a plane parallel glass plate and a Babinet compensator, cancel the instrumental polarization of, at most, several percent produced by the two tilt mirrors in the imaging optical system. A 40 Hz rotating wave plate and a Wollaston prism analyse the incident polarization. Two spectra, modulated by the rotating wave plate and separated by the Wollaston prism, are formed at the exit of the spectrograph. Six photomultipliers, as illustrated in Figure 2, pick up the signals at the wings of the spectral line and the neighbouring continuum. Each signal may be described as;

$$I_{\text{obs}} = \frac{1}{2}\left[I + \frac{1 + \cos\delta}{2} Q + \frac{1 - \cos\delta}{2} (Q \cos 4\Omega t + U \sin 4\Omega t) + \right.$$

$$\left. + V \sin\delta \sin 2\Omega t + R \cos 2\Omega t \right],$$

where (I, Q, U, V) are the Stokes parameters, δ and Ω are the retardation and the angular frequency of the rotating wave plate, respectively, and R is the reflection coefficient of

* Paper presented at the IAU Third Asian-Pacific Regional Meeting, held in Kyoto, Japan, between 30 September–6 October, 1984.

Astrophysics and Space Science **118** (1986) 163–167.

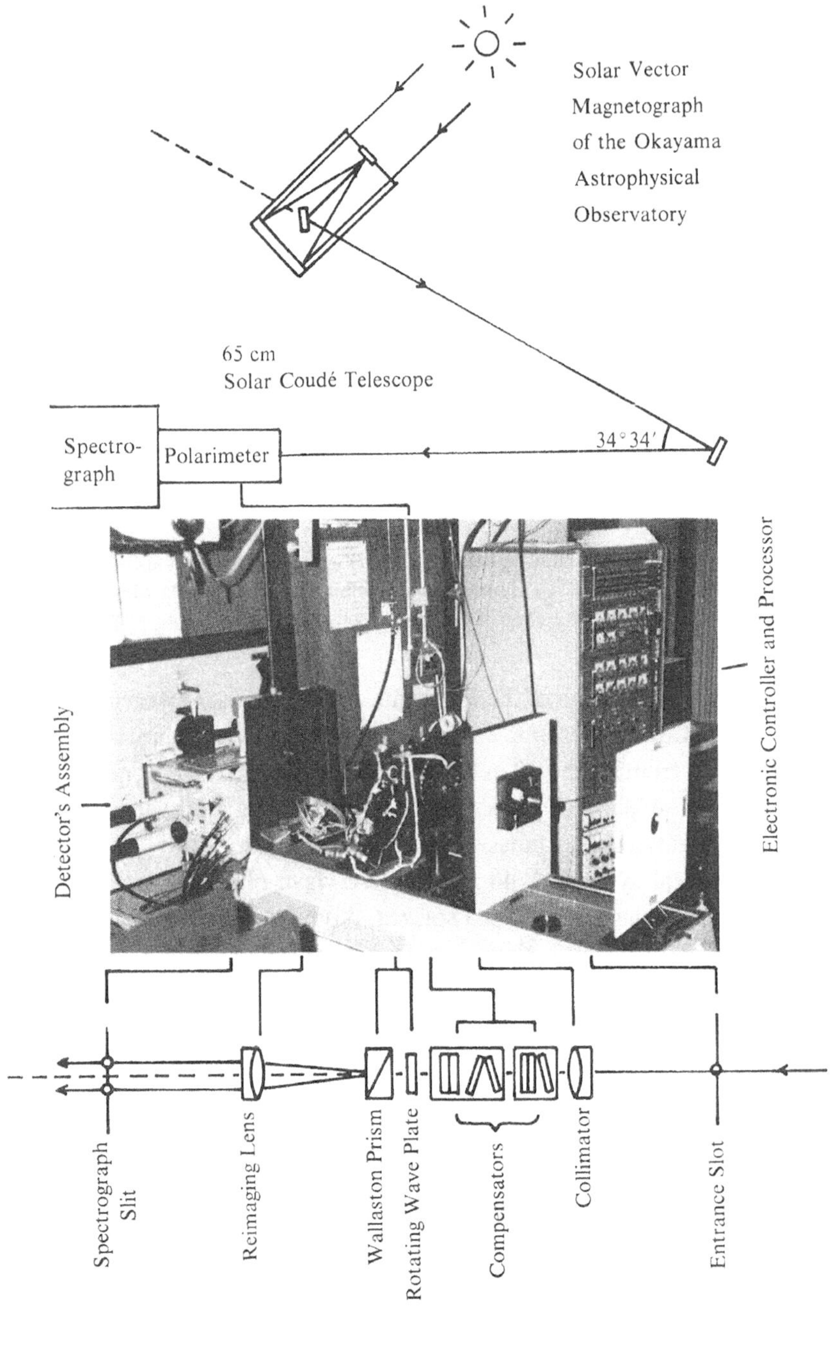

Fig. 1.

the rotating wave plate. A line shifter in front of the spectral line makes the wing intensities equal, and gives the velocity signal. The electronic processor combines the photocurrents, makes the frequency and phase analysis, and gives the Stokes parameters

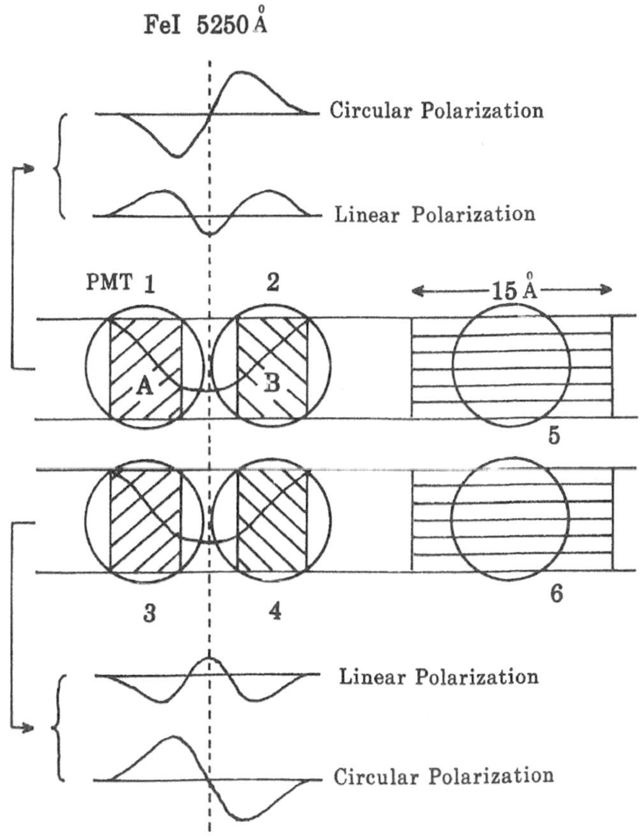

Fig. 2. Arrangement of the photomultipliers (PMT) and polarization signals of the spectral line.

of the line and the continuum. The Stokes parameters, as well as the velocity signal, are sent to the mini-computer, MELCOM 70/60, digitized and stored in the magnetic disk.

The observation is usually made with the Zeeman sensitive line, Fe I 5250 Å, for which $\delta = \frac{7}{4}\lambda$; and with the neighbouring continuum from 5282 to 5297 Å. A photoelectric guider in the side telescope shifts stepwise in right ascension or declination under the control of the mini-computer, and makes the telescope direct to the point to be observed. A typical spatial resolution and step width are $12'' \times 0\overset{''}{.}6$ and $10''$, respectively. In covering the area of (50×45) points, it takes roughly 80 min with the integration time of 1 s. The measuring accuracy of the polarization is limited by the photon shot noise, and can be as good as 10^{-3} and 10^{-4} for the line and continuum observations, respectively.

The data obtained are displayed as shown in Figure 3. That shows the active region which produced a white light flare on 25 April, 1984. From top (1) to bottom (6):

(1) The location of the scanned area on the Sun.

(2) The intensities of the line wing (green) and the continuum (pink). Sunspots observed at the line wing usually look larger. The isophots are drawn at every 10%.

(3) The line polarizations. Blue and red contours are the longitudinal magnetic field with north and south polarities, respectively. Green bars are the direction of the

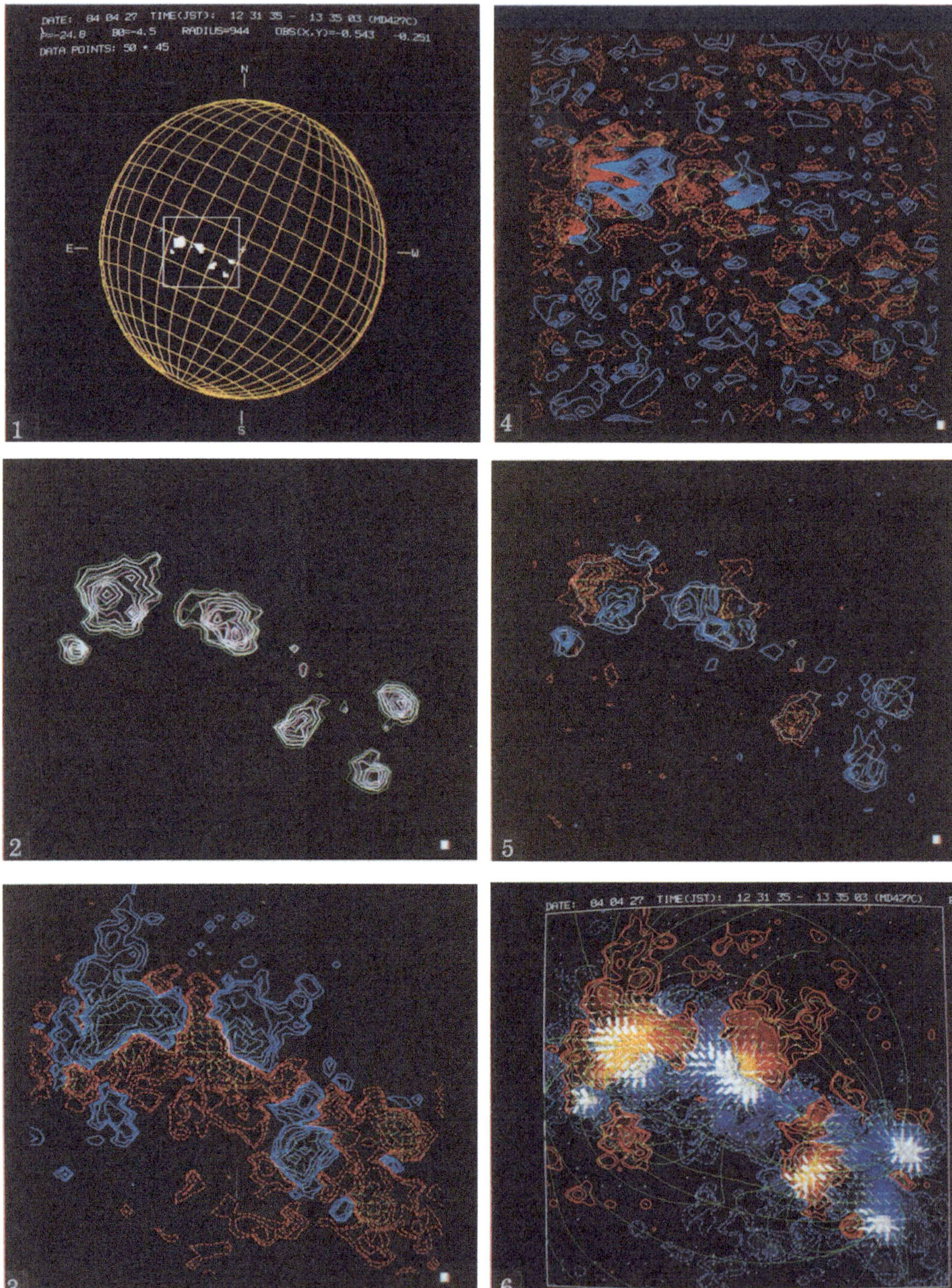
DATE: 84 04 27 TIME(JST): 12 31 35 - 13 35 03 (MD427C)
P=-24.8 B0=-4.5 RADIUS=944 OBS(X,Y)=-0.543 -0.251
DATA POINTS: 50 * 45
N
E
W
S
1
2
3
4
5
DATE: 84 04 27 TIME(JST): 12 31 35 - 13 35 03 (MD427C)
6

Fig. 3.

transverse magnetic field and the deeper green shows the stronger field. The translation of the polarization degree to the magnetic field strength has not been done.

(4) The radial velocity. The five minutes oscillation is dominant outside sunspots. Blue and red contours are shifts toward and away, respectively, and drawn every 110 m s^{-1}.

(5) The continuum polarizations. The representation is the same as (3). The polarizations correspond to the net Zeeman effect of the spectral lines.

(6) The magnetic lines of force (green) calculated by assuming the potential field. Magnetic polarities are red for N and blue for S here.

All the data except the last can be displayed on the colour display tube within 15 min after finishing the observation. The calculation of the magnetic lines of force takes, at longest, several hours, and this depends on the complexity of the measured region.

References

Makita, M., Hamana, S., and Nishi, K.: 1985, in M. J. Hagyard (ed.), *Measurements of Solar Vector Magnetic Fields*, NASA CP 2374, p. 173.

Makita, M., Hamana, S., Kawakami, H., and Nishi, K.: 1982, *Ann. Tokyo Astron. Obs.* **19**, 24.

Makita, M., Hamana, S., Nishi, K., Shimizu, M., Koyano, H., Sakurai, T., and Komatsu, H.: *Publ. Astron. Soc. Japan* **37** (in press).

Makita, M., Nishi, K., Shimizu, M., Hamana, S., Sakurai, T., Grigoryev, V. M., Kuklin, G. V., and Selivanov, V. L.: 1985, in M. J. Hagyard (ed.), *Measurements of Solar Vector Magnetic Fields*, NASA CP 2374, p. 399.

Sakurai, T.: 1982, *Solar Phys.* **76**, 301.

Sakurai, T. and Makita, M.: 1984, *Vector Magnetograms of Solar Active Regions*, December 1982–December 1983, Okayama Astrophys. Obs., Tokyo Astron. Obs.

SOLAR FIVE-MINUTE OSCILLATIONS OF LOW, INTERMEDIATE, AND HIGH DEGREE*

H. M. ANTIA, S. M. CHITRE, and D. NARASHIMA
Tata Institute of Fundamental Research, Bombay, India

(Received 22 March, 1985)

Abstract. The overstability of acoustic modes trapped inside the Sun is studied with mechanical and thermal effects of turbulence included in an approximate manner through the eddy diffusivities. Many of the acoustic modes are found to be overstable with the most rapidly growing modes occupying a region centred around 3.3 mHz and spread over a wide range of length-scales. The numerical results turn out to be in reasonable accord with the observed power-spectrum of the five-minute oscillations of arbitrary degree. It is demonstrated that these oscillations are most likely to be driven by a simultaneous operation of the κ-mechanism and the convective Cowling mechanism, the dominant contribution to the generation of self-excited acoustic waves arising from the turbulent diffusion.

1. Introduction

The discovery of solar five-minute oscillations by Leighton *et al.* (1962) has provided a very valuable tool to probe the interior of the Sun. Later, Deubner (1975) resolved the spatial and temporal structure of these oscillations, thus initiating the science of helioseismology. A detailed power-spectrum of the five-minute oscillations of high degree (spherical harmonic degree $l \gtrsim 150$) was provided by the work of Deubner *et al.* (1979), while the low-degree oscillations ($l \leq 3$) were detected by Claverie *et al.* (1981) and Grec *et al.* (1983) using integrated sunlight. The gap between observations of high degree and those of low degree was bridged by the recent observations of solar five-minute oscillations of intermediate degree ($1 \leq l \lesssim 150$) by Duvall and Harvey (1983).

The important question of the excitation mechanism responsible for these oscillations was addressed by Ando and Osaki (1975) and Ulrich and Rhodes (1977). These authors investigated the stability of non-radial oscillations of high degree for a realistic solar envelope model by including the full effects of radiative exchange, but neglecting the interaction between turbulent convection and oscillations. This situation was remedied by Goldreich and Keeley (1977) by incorporating the influence of turbulent convection on the stability of acoustic modes which were shown to be stabilized by the presence of turbulent viscosity.

* Paper presented at the IAU Third Asian-Pacific Regional Meeting, held in Kyoto, Japan, between 30 September–6 October, 1984.

Astrophysics and Space Science **118** (1986) 169–172.

2. Frequencies of Five-Minute Oscillations and the Excitation Mechanism

It is well known that, except for the top few tens of kilometers, a major fraction of the total flux in the solar convection zone is transported by convection. Furthermore, the turbulent conductivity is much larger than the radiative conductivity for the most part of the convection zone. Turbulence is, therefore, expected to play a major role in modulating the heat flux and the oscillatory motion. It should be recognized that both radiative and turbulent diffusion mechanisms have their origin in the strongly super-adiabatic sub-surface layer, but the efficiency of the turbulent diffusion mechanism turns out to be significantly larger (cf. Unno, 1976). This motivated Antia *et al.* (1982) to undertake a study of the overstability of acoustic modes in the solar envelope by approximately including the mechanical and thermal effects of turbulence by means of eddy transport coefficients. We investigate the stability of acoustic modes, of arbitrary degree, trapped inside the Sun by adopting for the present Sun the following characteristics: $X = 0.744$, $Z = 0.018$, $T_c = 1.52 \times 10^7$ K, $\rho_c = 145$ g cm^{-3}, $X_c = 0.360$, $L = z + 486$ km. Here L is the mixing length and z the distance measured downwards from the level $\tau = 1$. The solar model was constructed using a non-local prescription, and also including the contribution from the turbulent pressure (cf. Antia *et al.*, 1983).

The results of the stability calculations are displayed in Figures 1 and 2. It is gratifying to note that the theoretically computed frequencies are in reasonable agreement with the observations of five-minute oscillations by Deubner *et al.* (1979) for high degree acoustic modes and those by Duvall and Harvey (1983) for intermediate and low degree modes. For the intermediate and low values of l, the departure of theoretically calculated

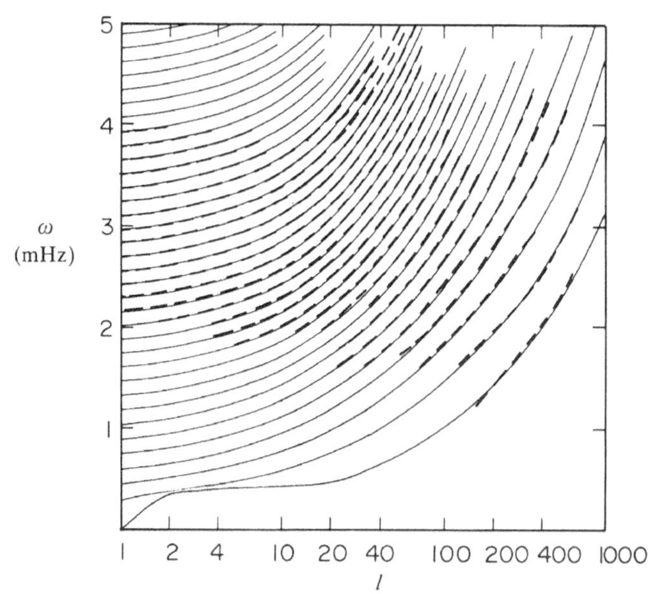

Fig. 1. The frequencies ω as a function of l for the first 35 acoustic modes ($f - p34$). The continuous lines represent the theoretically calculated frequencies, while the dashed lines observed frequencies.

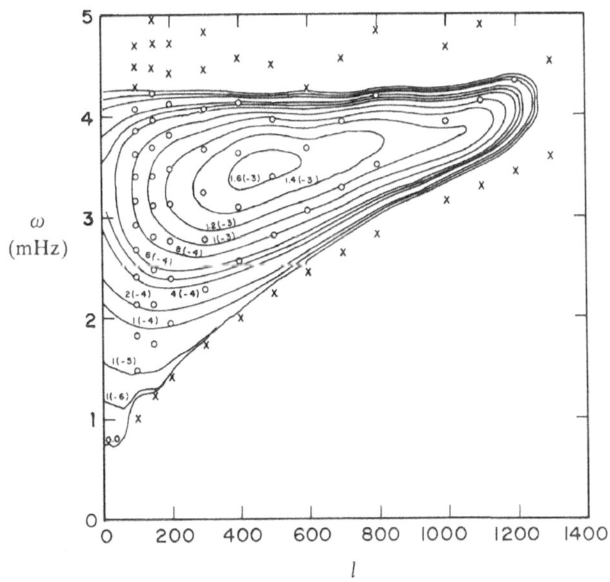

Fig. 2. The contours of equal stability coefficient η are shown in the $(\omega - l)$ plot where the crosses ($\times$) denote stable modes and open circles (O) unstable modes. The outermost contour refers to the marginally stable case of $\eta = 0$.

frequencies is no more than 0.7% of the observed frequencies and in most cases the agreement is to within 15 μHz. However, for the high degree modes the observed frequencies were read from the power-spectrum plots, which can contribute errors comparable with the discrepancy between the observed and theoretical values, but nevertheless the agreement seems to be good.

The contours of constant stability coefficient η ($\equiv$ growth rate/frequency) of a given acoustic mode are shown in Figure 2 in a frequency (ω)-spherical harmonic degree (l) plot. The outermost contour corresponds to the marginally stable case ($\eta = 0$), within which all the modes are unstable, while the modes outside this region are stable. We make the plausible assumption that only those modes with significant growth rates will have substantial observed power. It is interesting to note that the region in the figure where $\eta \gtrsim 10^{-4}$ approximately coincides with the region where a substantial amount of power has been observed. In particular, the high-frequency cut-off yielded by our analysis around 4.3 mHz, more or less independent of l, is in rough agreement with the observations of Duvall and Harvey (1983). Furthermore, for low l, the lower harmonics are either stable or have an extremely small growth rate, which is consistent with the low observed power in these harmonics. Our analysis is generally in agreement with the earlier results of Ando and Osaki (1975), save for a significant difference in that we get closed contours of the stability coefficient η, with a distinct and pronounced peak (having $\eta = 1.66 \times 10^{-3}$) around $l = 475$ and $\omega = 3.5$ mHz, while Ando and Osaki have open contours with η increasing with l. This is clearly the influence of turbulent viscosity included in our work which becomes effective at short wavelength and depresses the growth rates at high l.

3. Conclusions

The stability of acoustic modes trapped inside the Sun is investigated in the framework of a linearized theory, by incorporating in an approximate manner, the mechanical and thermal effects of turbulence on the mass flow through turbulent transport coefficients. Several of these acoustic modes are found to be overstable, with the most rapidly growing modes occupying a region centred around 3.3 mHz with a wide range of horizontal length-scales. The theoretical frequencies turn out to be in reasonable accord with the observed power-spectrum of the five-minute oscillations of low, intermediate, and high degree. The numerical results reveal that the acoustic modes which are stable for the viscous case when only the κ-mechanism is present, are significantly destabilized when turbulent diffusion is included. Likewise, in the inviscid situation the degree of instability of acoustic modes is stronger by a substantial amount when the full effects of radiative and turbulent diffusion mechanism are present over the case when only the κ-mechanism operates. We, therefore, conclude that the five-minute oscillations are driven by a combination of the κ-mechanism and the convective Cowling mechanism, the dominant contribution to the generation of self-excited acoustic waves arising from the diffusion mechanism.

References

Ando, H. and Osaki, Y.: 1975, *Publ. Astron. Soc. Japan* **27**, 581.
Antia, H. M., Chitre, S. M., and Narashima, D.: 1982, *Solar Phys.* **77**, 303.
Antia, H. M., Chitre, S. M., and Narashima, D.: 1983, *Monthly Notices Roy. Astron. Soc.* **204**, 865.
Claverie, A., Isaak, G. R., McLeod, C. P., van der Raay, H. B., and Roca Cortes, T.: 1981, *Nature* **293**, 443.
Deubner, F.-L.: 1975, *Astron. Astrophys.* **44**, 371.
Deubner, F.-L., Ulrich, R. K., and Rhodes, E. J., Jr.: 1979, *Astron. Astrophys.* **72**, 177.
Duvall, T. L., Jr. and Harvey, J. W.: 1983, *Nature* **302**, 24.
Goldreich, P. and Keeley, D. A.: 1977, *Astrophys. J.* **211**, 934.
Grec, G., Fossat, E., and Pomerantz, M.: 1983, *Solar Phys.* **82**, 55.
Leighton, R. B., Noyes, R. W., and Simon, G. W.: 1962, *Astrophys. J.* **135**, 474.
Ulrich, R. K. and Rhodes, E. J., Jr.: 1977, *Astrophys. J.* **218**, 521.
Unno, W.: 1976, in E. A. Spiegel and J. P. Zahn (eds.), 'Problems of Stellar Convection', *IAU Colloq.* **38**, 315.

REFERENCE MODELS OF SUNSPOT CHROMOSPHERES*

H. S. YUN
Seoul National University, Korea

and

H. A. BEEBE
New Mexico State University, Las Cruces, N.M., U.S.A.

(Received 22 March, 1985)

Abstract. Recent observational and theoretical studies of the structure of umbral chromospheres have led us to propose new models of the region. We combine space and ground based determinations of the density-temperature structure of the transition region with existing Ca II H, K, and IR triplet spectral data to establish relatively high density, 'gradient' models of the sunspot chromosphere.

Avrett in 1981 presented the *Sunspot Sunspot Model* which can be characterized as a physically extended, low density, plateau model. However, in a similar study, the authors (Beebe *et al.*, 1982) pointed out that the relatively high transition region pressures derived from space observations indeed are required to reproduce high resolution Ca II spectral features. Studies of the umbral chromosphere as a resonant cavity for slow-mode magneto-acoustic waves producing umbral oscillations (Zhugzhda *et al.*, 1984) also lead to atmospheres of relatively small physical thickness, thus higher densities.

Observations of several line core emission features made at Sacramento Peak Observatory are compared with calculated profiles to determine a range of conditions for chromospheric models. High resolution spectra of Ca II H, K, λ8542, and λ8498 were made over a sunspot umbra (SPO 5007) with the Sacramento Peak Observatory's echelle spectrograph at the Vacuum Tower telescope on 17 December, 1979 (see Yun *et al.*, 1980). The spot was single, round, and about 10 arc sec in diameter. Line profiles were calculated by solving equations of transfer and statistical equilibrium, using a complete linearization method and assuming complete redistribution of photons in the line.

Electron densities in the zone above the line formation region are based upon Solar Maximum Mission results reported by Gurman *et al.* (1982), rocket UV spectra reported by Dere *et al.* (1982), and a number of other studies. Electron densities representative of those investigations correspond to a column mass m_0 of 2×10^{-5} g cm^{-2} at the base of the transition zone. This constraint sets the pressure-height scale in the upper chromosphere and, therefore, with the temperature model, fixes the overall physical extent of the chromosphere.

Recent studies by Zhugzhda *et al.* (1984) describe the umbral chromosphere as a resonant cavity for slow-mode waves which produce umbral oscillations. The resulting

* Paper presented at the IAU Third Asian-Pacific Regional Meeting, held in Kyoto, Japan, between 30 September–6 October, 1984.

Astrophysics and Space Science **118** (1986) 173–175.

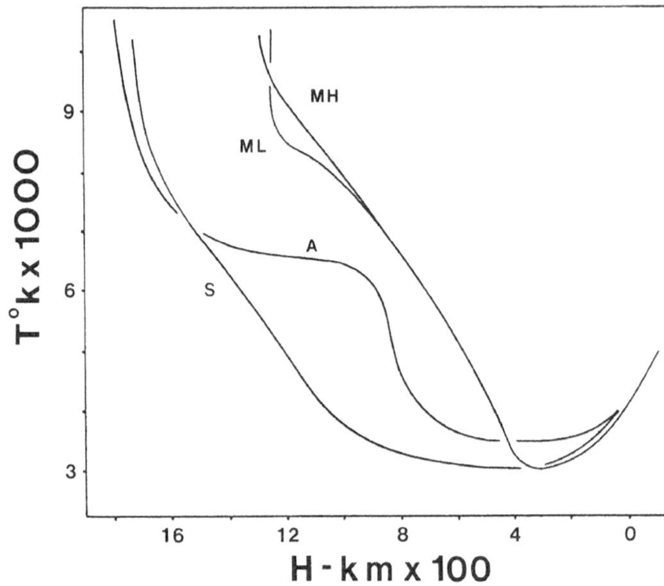

Fig. 1. Temperature variation with height in umbral chromospheres according to Aurett (A) (1981), Staude (S) (1982), and this study, ML and MH.

limits on physical thickness work in conjunction with the column mass m_0 to further specify the chromospheric model.

Figure 1 shows several representative physical height vs temperature models of umbral chromospheres. Staude's (1981) model is itself an extension of a model by Teplitskaya *et al.* (1978). The extension was based on electron density determinations from transition region observations similar to those used in this study. Staude does not show, however, calculated spectral lines.

Avrett's (1981) sunspot model is based on a model by Lites and Skumanich (1982) which in turn was determined by OSO-8 Mg and Ca line profiles. The model assumes an overlying column mass of $m_0 = 3.5 \times 10^{-6}$ g cm^{-2} and produces extremely spiked H2 and K2 features with a K3/H3 intensity ratio of about 1.5, surprisingly at odds with the ratio of 1.1 determined by Lites and Skumanich.

The models of the present study ML and MH represent cool and hot components of the umbral chromosphere, producing the extremes in brightness, of the observed H and K lines. The models are characterized by a transition zone column mass of $m_0 = 2 \times 10^{-5}$ g cm^{-2} and steep chromospheric temperature gradients relative to the others shown in Figure 1. We find that Ca II H, K, and IR triplet line-core intensity profiles can be predicted with accuracy, yielding both the narrowness of the emission features and K3/H3 intensity ratios consistent with observations. Thus Ca II H, K, and infrared triplet lines, transition region electron densities, and umbral oscillation periods, all observables, may be combined to establish well-defined chromospheric models. Detailed model listings will be sent upon request.

Acknowledgements

This work was partially supported by NSF Grant INT-8215502 and KOSEF grant K53 under the US-ROK Cooperative Science Programs. We also wish to thank the staff at SPO, the NMSU Computer Center, and graduate students W. E. Baggett, A. K. Dobson-Hockey, and B. C. Koo.

References

Avrett, E. H.: 1981, L. E. Cram and J. H. Thomas (eds.), *The Physics of Sunspots, Proceedings*, Sac. Peak Obs., p. 235.

Beebe, H. A., Baggett, W. E., and Yun, H. S.: 1982, *Solar Phys.* **79**, 31.

Dere, J. P., Bartoe, J. D. F., and Brueckner, G. E.: 1982, *Astrophys. J.* **259**, 366.

Gurman, J. B., Leibacher, J. W., Shine, R. A., Woodgate, B. E., and Henze, W.: 1982, *Astrophys. J.* **253**, 939.

Lites, B. W. and Skumanich, A.: 1982, *Astrophys. J. Suppl.* **49**, 293.

Staude, J.: 1981, *Astron. Astrophys.* **100**, 284.

Teplitskaya, R. B., Grigoryeva, S. A., and Skochilov, V. G.: 1978, *Solar Phys.* **56**, 293.

Yun, H. S., Beebe, H. A., and Baggett, W.: 1980, *J. Korean Astron. Soc.* **13**, 1.

Zhugzhda, Y. D., Staude, J., and Locans, V.: 1984, *Solar Phys.* **91**, 219.

RESONANT EXCITATION OF THE SOLAR g-MODES THROUGH COUPLING OF 5-MIN OSCILLATIONS*

HIROYASU ANDO

Tokyo Astronomical Observatory, University of Tokyo, Mitaka, Tokyo, Japan

(Received 22 March, 1985)

Abstract. It is proposed that the interacting solar g-modes are driven by the beat forcing of g-modes by two 5-min oscillations, excited by a κ-mechanism. The observational implications of the proposed mechanism are discussed.

1. Introduction

The discovery of solar g-modes other than the 160 min oscillation (Severny *et al.*, 1976; Brookes *et al.*, 1976) was reported by several groups (Delache and Scherrer, 1983; van den Raay *et al.*, 1983; Severny *et al.*, 1984). Some peaks in the observed power spectra were attempted to be identified as g-modes with low degree (l = 1–3) and high radial order (n = 6–30) (Delache and Scherrer, 1983; Severny *et al.*, 1984; Gabriel, 1984), though a satisfactory diagnosis is not obtained yet. These results give rise to some questions: Why are no other g-modes observed? Why has the 160 min oscillation so large an amplitude if it is a g-mode? Dziembowski (1983) suggested that these facts could be accounted for by three-mode resonant interactions of g-modes and Guenther and Demarque (1984) extended this idea to find a coincidence between peaks in the resonant count diagram and the observationally identified g-modes. However, the excitation mechanism of interacting g-modes is unknown. We here propose the beat forcing of g-modes by two 5-min oscillations which are excited by κ-mechanism (Ando and Osaki, 1975) and/or by turbulent convection (Goldreich and Keeley, 1977).

2. Beat Forcing

The coupling between two p-modes and a g-mode may well be regarded as a result of wave, mean-flow interactions (Andrews and McIntyre, 1978; McIntyre, 1980), since the time-scales for both motions are much different. The wave equations of a forced g-mode (mean-flow) by two acoustic modes (waves) can be obtained by averaging over a short time-scale of a 'wave' as

$$\rho \frac{\partial u_i^{(3)}}{\partial t} - \mathscr{L}_{ik} u_k^{(3)} = \rho f_i \,, \tag{1}$$

* Paper presented at the IAU Third Asian-Pacific Regional Meeting, held in Kyoto, Japan, between 30 September–6 October, 1984.

Astrophysics and Space Science **118** (1986) 177–181.

where $u_i^{(3)}$ and $\mathscr{L}_{ik}$ are the velocity vector of a forced g-mode, and the spatial differential operator describing stellar linear oscillations (Unno *et al.*, 1979), respectively. The beat forcing f_i is given by

$$f_i = -\frac{1}{\rho}\frac{\partial}{\partial x_k}(\rho u_k u_i) - \frac{\partial P_{(2)}}{\partial x_k}, \tag{2}$$

where the first term comes from Reynolds stresses, and $P_{(2)}$ is the second-order fluctuation of pressure, and it comes from the continuity and entropy equations. An overbar is taken to be the average over a short time-scale of p-modes. The rapidly varying term corresponding to frequency $\omega_2 + \omega_1$ is cancelled out in Equation (1), and thus frequency ω_3 of a forced g-mode is given by $\omega_3 = \omega_2 - \omega_1$.

We can easily evaluate the growing time-scale T of a forced g-mode using the energy relation derived from Equation (1): i.e.,

$$T \sim \frac{KE}{W}, \tag{3}$$

where KE is kinetic energy of a forced g-mode and W is the work done by the beat forcing of two p-modes: i.e.,

$$KE = \frac{1}{2}\int \rho u_i^{(3)} u_i^{(3)}\,\mathrm{d}V, \tag{4}$$

and

$$W = \int \rho u_i^{(3)} f_i\,\mathrm{d}V = -u_i^{(3)}\frac{\partial}{\partial x_k}(\rho u_k u_i) - \rho\nabla_a\left(\frac{\delta P}{P}\right)^{(3)} T\,\delta S_{(2)} - \\ -\frac{P'^{(3)}}{\rho}\frac{\partial}{\partial x_k}(\rho' u_k), \tag{5}$$

where ∇_a is $(\mathrm{d}\ln T/\mathrm{d}\ln P)_{ad}$, and $\delta P^{(3)}$ and $P'^{(3)}$ are Lagrangian and Eulerian fluctuations of pressure of a g-mode, respectively, and $\delta S_{(2)}$ is the second-order entropy fluctuation of two p-modes. The integral W includes a well-known integral, with angular coordinates, appearing in the angular momentum vector rule in quantum mechanics. It is not vanishing if, and only if,

$$l_1 + l_2 + l_3 = \text{even}, \qquad m_2 - m_1 = m_3. \tag{6}$$

We consider here the resonance coupling of p-modes with $l = 1$–5. It is consistent with observations to assume that the surface radial velocity amplitudes ($\sim 10\ \mathrm{cm\ s^{-1}}$) for a forced g-mode and two p-modes are comparable (Delache and Scherrer, 1983; Severny *et al.*, 1984; Grec and Fossat, 1983; Claverie *et al.*, 1981). We have found that the fluctuation of mass flux is the main driving source of g-modes and is dominant in the convective envelope, since a g-mode has an acoustic nature here. According to our computations for a solar model (Ando and Osaki, 1975), the time-scale T ranges from

10^2 to 10^4 days for beat frequencies 67–200 μHz, while the damping time of a *g*-mode due to radiation loss is long enough (10^5–10^6 yr). Therefore, it may well grow up to the observed amplitude. A remaining problem is whether energy gained by a *g*-mode can be be normally supplied by the energy source of *p*-modes, since the back reaction of *g*-mode coupling may work on *p*-modes. The magnitude of the coupling energy per one cycle of a *p*-mode is $\sim 10^{26}$ ergs from our calculation, while the excitation energy of *p*-modes by turbulent convection becomes $\sim 10^{20}$ ergs (Goldreich and Keeley, 1977). This implies that the proposed coupling is still unsatisfactory. However, this discrepancy should not be considered seriously, since turbulent convection is not enough (0.1 cm s^{-1}) (Goldreich and Keeley, 1977) to explain the observed amplitude. If any preferred excitation mechanism is proposed, the above discrepancy would be removed.

3. Observational Implications

In what follows, we discuss the observational aspects expected from the beat forcing. The beat frequency ω, for our case (low-degree and high-order *p*-modes), is roughly given according to the WKBJ approximation (Vandakurov, 1967; Tassoul, 1980) as

$$\omega = \omega_0(2\Delta n + \Delta l)\,, \tag{7}$$

where Δn and Δl are the differences of radial order n and degree l, respectively, and ω_0 is estimated as 67 μHz (Shibahashi *et al.*, 1983). When combined with the selection rule (6) and the relation (7), it is obvious that the beat frequencies are approximately given by $(67 \times j)$ μHz ($j = 1, 2, \ldots$) and also that *g*-modes with even j should have even degree and those with odd j should have odd degree.

In practice, frequency differences of *p*-modes deviate a little from the relation (7), and are dispersive around the values given by (7). In Figure 1, we give a comparison of the frequency differences of the observed *p*-modes (Grec and Fossat, 1983; Claverie *et al.*, 1981; Scherrer *et al.*, 1983; Woodard and Hudson, 1983) with the observed frequencies of *g*-modes (Delache and Scherrer, 1983; Severny *et al.*, 1984), together with the theoretical frequency differences of *p*-modes (Shibahashi *et al.*, 1983). They are evaluated with a condition of $\Delta l = 1, 2$ for the same radial order. The original pairs of degree l_1 and l_2 for each Δl are also given in the parentheses. From this figure, the coincidence near 65 μHz between observed peaks and results of beat forcing of *g*-modes is found to be fairly good. On the other hand, no strong concentration near 104 μHz (160 min) is found, though the frequency difference corresponding to 160 min oscillation is seen for the degree differences (2, 4) and (3, 5). The large power of the 160 min oscillation in the observed power spectra still needs further explanation.

The second consequence may put a restraint on the mode identification of the observed *g*-modes (in particular degree l), if the beat forcing of *g*-modes is favourable. This aspect can give a clear clue to distinguish our beat forcing by *p*-modes from the three-mode resonant coupling of *g*-modes. We need a further accumulation of qualified observations, and a general improvement of the solar model.

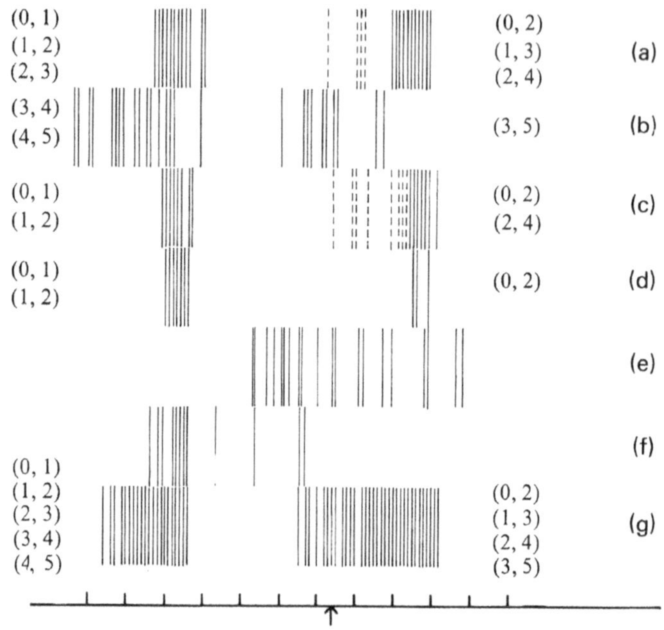

Fig. 1. Comparison of the observed frequency differences of *p*-modes. (a) Grec *et al.* (1983); (b) Scherrer *et al.* (1983); (c) Claverie *et al.* (1981); (d) Woodard and Hudson (1983) with the observed frequencies of *g*-modes; (e) Severny *et al.* (1984); (f) Delache and Scherrer (1983). The dashed vertical lines in (a) and (c) show the frequency differences between $l = 2$ and 4, using the observed *p*-modes with $l = 4$ by Scherrer *et al.* (1983). The theoretical frequency differences are also shown in the bottom. (g) Shibahashi *et al.* (1983) for the *p*-modes with $l = 0$–5. These differences are calculated with a condition of $\Delta l = 1, 2$ for the same radial order. The degrees l_1 and l_2 of the two *p*-modes for each Δl are given in the parentheses. The abscissae give frequency in units of μHz, and an arrow indicates the position of 160 min.

Acknowledgements

The author would like to express his sincere gratitude to Dr M. E. McIntyre for his stimulative discussions, in particular, his suggestion of the importance of mass flux and entropy fluctuations apart from Reynolds stresses in the beat forcing. He is grateful to Drs M. E. McIntyre and R. Kimura for their elegant beaker experiment, showing the excitation of gravity waves by sound waves, which underlies this work. He also thanks Prof. Y. Osaki and Dr H. Shibahashi for their kind discussions. The computational part of this work was carried out on the FACOM 380R at the Tokyo Astronomical Observatory. This work was supported in part by a Grant-in-Aid for Scientific Research (59540137), Ministry of Education, Science, and Culture.

References

Ando, H. and Osaki, Y.: 1975, *Publ. Astron. Soc. Japan* **27**, 581.
Andrews, D. G. and McIntyre, M. E.: 1978, *J. Atmos. Sci.* **35**, 175.
Brookes, J. R., Isaak, G. R., and van der Raay, H. B.: 1976, *Nature* **259**, 92.

Claverie, A., Isaak, G. R., McLeod, C. P., and van der Raay, H. B.: 1981, *Nature* **293**, 443.
Delache, P. and Scherrer, P. H.: 1983, *Nature* **306**, 651.
Dziembowski, W.: 1983, *Solar Phys.* **82**, 259.
Gabriel, M.: 1984, *Astron. Astrophys.* **134**, 387.
Goldreich, P. and Keeley, D. A.: 1977, *Astrophys. J.* **212**, 243.
Grec, G. and Fossat, E.: 1983, *Solar Phys.* **82**, 55.
Guenther, D. B. and Demarque, P.: 1984, *Astrophys. J.* **277**, L17.
McIntyre, M. E.: 1980, *Pure Appl. Geophys.* **118**, 152.
Scherrer, P. H., Wilcox, J. M., Christensen-Dalsgaard, J., and Gough, D. O.: 1983, *Solar Phys.* **82**, 75.
Severny, A. B., Kotov, V. A., and Tsap, T. T.: 1976, *Nature* **259**, 87.
Severny, A. B., Kotov, V. A., and Tsap, T. T.: 1984, *Nature* **307**, 247.
Shibahashi, H., Noels, A., and Gabriel, M.: 1983, *Astron. Astrophys.* **123**, 283.
Tassoul, M.: 1980, *Astrophys. J. Suppl.* **43**, 469.
Unno, W., Osaki, Y., Ando, H., and Shibahashi, H.: 1979, *Nonradial Oscillations of Stars*, University of Tokyo Press, Tokyo.
Vandakurov, Y. V.: 1967, *Astron. Zh.* **44**, 786,
Van der Raay, H. B., Claverie, A., Isaak, G., McLeod, J. M., Roca Cortes, T., Palle, G., and Delache, P.: 1983, *Proc. Oscillations as a Probe of the Sun's Interior*, University of Catania, Catania, p. 25.
Woodard, M. and Hudson, H. S.: 1983, *Nature* **305**, 589.

PLASMA AND RADIATION PROCESSES (INVITED REVIEW)*

R. G. HEWITT and D. B. MELROSE
School of Physics, University of Sydney, Sydney N.S.W., Australia

(Received 13 June, 1985)

Abstract. The physical processes leading to metre-wave and microwave emissions from the solar corona are reviewed and the possibility of observing analogous phenomena on other stars is discussed. Particular emphasis is placed on three alternative processes for the emission from flare stars: cyclotron maser emission, second harmonic plasma emission, and gyro-synchrotron emission.

1. Introduction

There has been a rapid expansion of interest in the radio emissions from stars in recent years, due in large part to the suitability of facilities like the VLA for observing these phenomena (e.g., the reviews by Hjellming and Gibson, 1980; and Gibson, 1983). The emission mechanisms which operate in stars are likely to have solar counterparts, and the interpretation of stellar radio emission has indeed been based almost entirely on the interpretation of solar phenomena. The specific mechanisms which are known to operate in the solar corona are bremsstrahlung, fundamental and second harmonic plasma emission, gyro-synchrotron emission, and electron-cyclotron maser emission. Bremsstrahlung is the basic thermal emission process for a plasma but, although it is important in some stellar emissions (e.g., from stellar winds), it seems unnecessary to discuss it in detail here. In this review we concentrate on those mechanisms which seem most favourable for stellar emissions – i.e., electron-cyclotron maser emission, second harmonic plasma emission, and gyro-synchrotron emission.

Radio observations of the Sun have been made since the 1940's (e.g., the reviews by Wild *et al.*, 1963; Kundu, 1965; and, more recently, by Kundu and Gergely (eds.), 1980 and McLean and Labrum (eds.), 1985). The characteristics of solar radio emission change markedly as the wavelength varies from the microwave range through decimetre wavelengths to metre and longer wavelengths. In the microwave range the dominant emission mechanisms are gyro-synchrotron emission in bursts, and thermal emission in the 'slowly-varying' component from localized hot spots in the corona. A phenomenon thought to be of particular relevance to some stellar radio flares is that of 'spike bursts' which occur at about 1 GHz and which are thought to be due to electron-cyclotron maser emission. Figure 1 shows observations of 'spike bursts' by Slottje (1978). There is a rich variety of other types of emission at decimetric wavelengths (e.g., the review by Kuijpers, 1980). Studies of the decimetre wave phenomena have concentrated largely on fine structures, which could not be detected from a stellar source. The

* Paper presented at the IAU Third Asian-Pacific Regional Meeting, held in Kyoto, Japan, between 30 September–6 October, 1984.

Astrophysics and Space Science **118** (1986) 183–193.

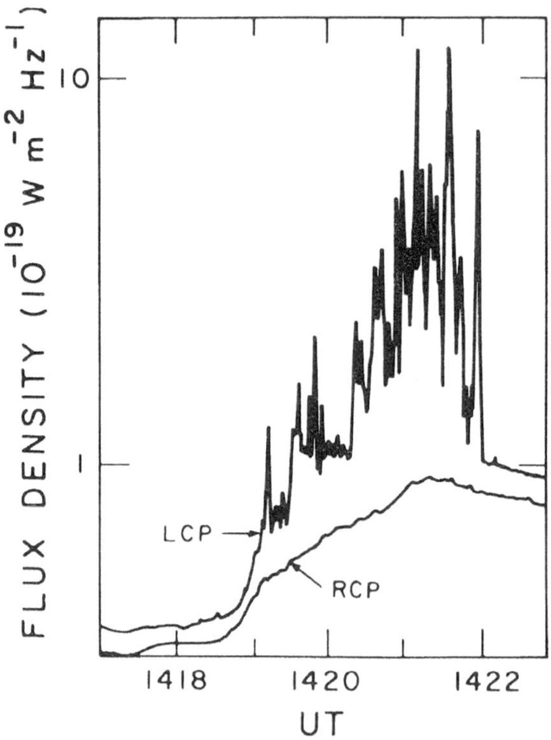

Fig. 1. Solar microwave 'spike bursts' as observed by Slottje (1978).

bursts at metre and longer wavelengths were classified by Wild (Wild and McCready, 1950) as types I, II, and III. Type-III bursts are due to electron streams which can propagate through the corona and the interplanetary medium, and have been studied out to beyond the orbit of Jupiter. Type-II bursts are due to flare-initiated shock waves in the corona. Both of these involve plasma emission at the fundamental and/or the second harmonic. Type-I emission is thought to arise from particles confined in a flux tube near an active region. The emission is at the fundamental of the plasma frequency; its generation is not adequately understood.

The radio emission from stars has been studied most extensively at frequencies $\gtrsim 1$ GHz; stellar flares, when observed, can be up to 10^4 more intense than solar flares at similar wavelengths (Lovell, 1969; Spangler and Moffet†, 1976). Clearly gyro-synchrotron emission and electron-cyclotron maser emission are reasonable candidates for emission mechanisms since they are associated with radiation at $\gtrsim 1$ GHz from the Sun. Second harmonic plasma emission also needs to be considered because it is sometimes observed at these relatively high frequencies. However, fundamental plasma emission is never observed above a relatively low frequency, say $\gtrsim 200$ MHz. The likely reason for this is the strong free–free absorption of fundamental plasma emission in relatively dense plasmas. Because of this frequency restriction we will not consider the process further.

In Section 2 we discuss the interpretation of two important parameters for solar and stellar radio emission: the brightness temperature and the degree of polarization. In Sections 3 and 4 we review the 'coherent' radiation mechanisms, electron-cyclotron maser emission and second harmonic plasma emission, respectively. Gyro-synchrotron emission is discussed briefly in Section 5.

2. Brightness Temperature and Polarization

Solar, and presumably stellar, radio emissions are due to non-relativistic or mildly relativistic electrons. An estimate of the brightness temperature of the source can provide important information on whether the emission mechanism is incoherent or coherent. Brightness temperatures less than about 10^{10} K can be explained in terms of gyro-synchrotron emission, which is an incoherent mechanism, and brightness temperatures much greater than 10^{10} K require a coherent mechanism such as plasma emission or electron-cyclotron maser emission. Similarly the degree of polarization can be useful in distinguishing between mechanisms, with second harmonic plasma emission being weakly polarized and electron-cyclotron maser emission being highly polarized.

The brightness temperature $T_B(\nu)$ is defined in terms of the specific intensity $I(\nu)$ by

$$I(\nu) = \frac{2\nu^2}{c^2} k_{\rm B} T_B(\nu)\,, \qquad (1)$$

where $k_{\rm B}$ is Boltzmann's constant. In a non-thermal source $T_B(\nu)$ is a function of frequency ν.

In the solar corona, and stellar coronae and winds, the temperature is typically of order 10^6 K. Thermal emission from such a source would have $T_B(\nu) \lesssim 10^6$ K, or perhaps somewhat higher if it were dominated by hot spots.

For an incoherent emission mechanism such as gyro-synchrotron emission, $T_B(\nu)$ is limited by the effects of self-absorption, as in the thermal case. If emission at frequency ν is due to electrons with a mean energy $\langle E \rangle$ then self-absorption limits $T_B(\nu)$ to

$$T_B(\nu) \lesssim \langle E \rangle / k_{\rm B}\,. \qquad (2)$$

An energy of 1 eV corresponds to a temperature of 1.16×10^4 K, and hence mildly relativistic electrons ($\langle E \rangle \lesssim 1$ MeV say) cannot produce emission with $T_B(\nu) \gtrsim 10^{10}$ K.

Coherent mechanisms are implied by brightness temperatures $\gg 10^{10}$ K and indirectly by rapid large-scale intensity variations. Such variations indicate small sources and, hence, relatively high brightness temperatures. Thus, for example, the time-scale ($\lesssim 10$ ms) for the rise and decay of solar 'spike bursts' indicates source sizes as small as 30 km to 3000 km and brightness temperatures $\gtrsim 10^{15}$ K (Slottje, 1978).

In the coronae of stars radio emission splits into two natural modes which propagate independently of each other – i.e., the modes rapidly get out of phase. The modes are referred to as the o mode and the x mode and for almost all purposes they may be regarded as oppositely circularly polarized. The x mode has the same handedness as a spiralling electron and, hence, is strongly favoured by electron-cyclotron maser emission. Gyro-synchrotron emission also favours the x mode but nowhere near as strongly. Although fundamental plasma emission can favour the o mode very strongly, second harmonic plasma emission is intrinsically weakly polarized.

The degree of circular polarization r_c may be defined in terms of the intensities $I_o(\nu)$ and $I_x(\nu)$ in the two modes by

$$r_c = \frac{I_x(\nu) - I_o(\nu)}{I_x(\nu) + I_o(\nu)} . \tag{3}$$

Values of $|r_c| \cong 1$ from a stellar source are indicative of electron-cyclotron maser emission.

Note that large Faraday rotation should preclude any detection of linear polarization. From time to time linear polarization has been claimed in solar and stellar sources, but experiments which eliminate any possibility of instrumental polarization (e.g., Grognard and McLean, 1973) have led to negative results for the detection of linear polarization.

3. Electron-Cyclotron Maser Emission

Twiss (1958) first showed that electrons with a distribution f which is an increasing function of the momentum $p_\perp$ perpendicular to the magnetic field lines can lead to negative absorption just below the harmonics $s = 1, 2, \ldots,$ of the nonrelativistic gyrofrequency $\Omega_e = eB/m$. Twiss restricted his discussion to the case of zero-parallel momentum, $p_\parallel = 0$. The full resonance condition for a wave with angular frequently ω and wavevector $\mathbf{k}$, and an electron with velocity $\mathbf{v}$ is

$$\omega - k_\parallel v_\parallel = s\Omega_e \left(1 - \frac{v_\parallel^2}{c^2} - \frac{v_\perp^2}{c^2}\right)^{1/2} , \tag{4}$$

and it is obvious that $\omega < s\Omega_e$ for $v_\parallel = 0$. There is an extensive literature on electron-cyclotron instabilities but it was not until quite recently that the general case, with $v_\parallel \neq 0$, $k_\parallel \neq 0$ and with the relativistic effect retained on the right-hand side, was explored in detail. This general case turns out to be very favourable, allowing electron-cyclotron maser emission to occur under quite mild conditions.

One important requirement is that the plasma frequency ω_p be less than or comparable to Ω_e, and this condition is not satisfied in most plasmas. It is, however, satisfied above the auroral zones of the Earth, in the magnetospheres of Jupiter and Saturn and in some magnetic flux tubes in the lower solar corona. All these regions are sources of specific radio emissions whose only plausible interpretations involve electron-cyclotron maser emission. These are the auroral kilometric radiation (AKR), Jovian decametric radiation (DAM), Saturnian kilometric radiation, and solar 'spike bursts', respectively. The interpretations have been discussed by Wu and Lee (1979), Lee *et al.* (1980), Holman *et al.* (1980), Hewitt *et al.* (1981), Melrose *et al.* (1982, 1984), Omidi and Gurnett (1982), Melrose and Dulk (1982a, b) amongst others. The emissions are all very bright, e.g., $\gtrsim 10^{17}$ K for DAM and $\gtrsim 10^{15}$ K for solar 'spike bursts' and, with the exception of AKR, are essentially 100% circularly polarized. The most intense AKR bursts are $\cong 100\%$ x-mode but weaker bursts often have an o-mode component which tends to be observed only when ω_p/Ω_e is relatively close to unity.

Two important points were not recognized prior to the work of Wu and Lee (1979). First, the emission can occur above $\omega = s\Omega_e$ for $k_\parallel v_\parallel > 0$. This is important since it allows fundamental ($s = 1$) emission to occur above the cutoff frequency for the x mode. Second, the instability can be driven by electrons with $\partial f/\partial v_\perp > 0$ in only a localized region of momentum space and such a distribution is automatically produced during driven precipitation.

These points can be appreciated with the aid of a particularly useful concept, that of a resonance ellipse. Equation (4) implies that resonance is possible only if $\omega^2 < s^2\Omega_e^2 + c^2k_\parallel^2$ and that, if this inequality is satisfied, the resonant electrons are those which lie on an ellipse in $v_\parallel - v_\perp$ space. The position and shape of the ellipse are completely determined by s, ω, Ω_e, and $k_\parallel$ (which in turn is determined by the propagation angle θ and the refractive index of the plasma). Depending on the distribution $f(v_\parallel, v_\perp)$ of electrons in velocity space, the wave either extracts energy from the resonant electrons and grows, or loses energy to them and is damped. In the former case, the intensity of the wave at any point increases exponentially with time, i.e., $I \propto \exp(2\Gamma_s t)$, until the maser saturates. The dominant contribution to the growth rate Γ_s for the sth harmonic is given by an expression of the form

$$\Gamma_s \cong \int d^3\mathbf{v} A_s(\mathbf{v}, \mathbf{k}) \frac{\partial f}{\partial v_\perp} \delta(\omega - k_\parallel v_\parallel - s\Omega_e/\gamma), \tag{5}$$

with $A_s > 0$ and $\gamma = (1 - v_\parallel^2/c^2 - v_\perp^2/c^2)^{-1/2}$. Because of the delta function this corresponds to an integration around the resonance ellipse. Growth can occur if the resonance ellipse passes through a source of free energy where $\partial f/\partial v_\perp > 0$. This condition on $\partial f/\partial v_\perp$ is analogous to the requirement of population inversion for the operation of a laser.

The source of free energy for all the maser emissions discussed earlier is believed to be a single-sided loss-cone distribution. This type of distribution is formed when electrons are accelerated downwards along converging magnetic field lines near the surface of a star or planet. Electrons with large pitch angles are reflected by the inhomogeneous magnetic field while those with small pitch angles precipitate. There are no electrons with small pitch angles in the upgoing part of the distribution, i.e., there is an upgoing loss cone. Figure 2 shows a contour map of a Maxwellian distribution with such a loss cone. The figure also shows a resonance ellipse for a typical growing wave passing through the region where $\partial f/\partial v_\perp > 0$.

The parameters of the resonance ellipse depend strongly on ω and θ. Small changes in either of these quantities can make $\omega^2 > s^2\Omega_e^2 + c^2k_\parallel^2$ so that resonance is impossible, or can alter the size and position of the ellipse sufficiently for damping rather than growth to occur. The sth harmonic radiation from the loss cone driven maser has ω just

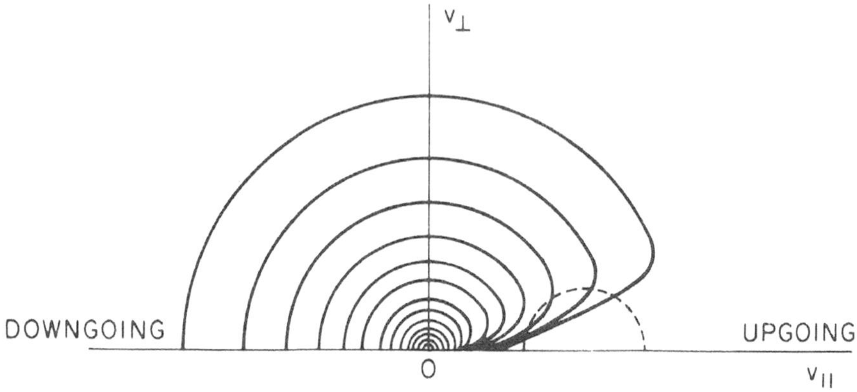

Fig. 2. Contour map in $v_\parallel - v_\perp$ space of a Maxwellian distribution of energetic electrons with a loss cone in the upward direction. The superimposed dashed curve is a resonance ellipse for a growing wave.

above $s\Omega_e$ and a narrow frequency bandwidth, typically $\cong 0.001\ \omega$ to $0.01\ \omega$; it is beamed upwards into a hollow cone with cone half angle $\cong 70°$ to $80°$, cone width $\lesssim 10°$ and axis aligned with the magnetic field (Hewitt *et al.*, 1982).

In models for the solar corona, the electrons driving the maser typically have energies of 10–100 keV. A colder relatively dense background plasma is also present in the source region and this has a significant effect on the relative growth rates of the various modes because it changes the refractive index and, hence, $k_\parallel$. For energetic distributions of the type shown in Figure 2, the growth rates for the fundamental ($s = 1$) x-mode are appreciably larger than those for the other modes when $\omega_p/\Omega_e \lesssim 0.3$. At higher densities the $s = 1$-x mode is suppressed, the $s = 1$ o-mode becomes dominant when $0.3 \lesssim \omega_p/\Omega_e \lesssim 1.0$, the $s = 2$ x-mode when $1.0 \lesssim \omega_p/\Omega_e \lesssim 1.4$. Since a single mode dominates in each range of ω_p/Ω_e the emissions from a localized, reasonably homogeneous source would be highly circularly polarized.

Melrose and Dulk (1984) estimated the brightness temperature of the maser to be

$$\frac{k_B T_B}{mc^2} \cong \eta \frac{c}{r_0 \omega} \frac{\Gamma}{\Omega_e}, \tag{6}$$

where r_0 is the classical radius of the electron and η is the fraction of the energy density of the driving electrons converted to maser radiation. The maximum growth rates for the fundamental x mode for energetic electron number densities $\cong 10^{13}\ \mathrm{m}^{-3}$ and temperatures $\cong 10$ keV at frequencies $\cong 1$ GHz are typically $\cong 10^{-3}\ \Omega_e$ (Hewitt *et al.*, 1982); modest values of η, $\cong 10^{-3}$ say, lead to brightness temperatures $\cong 10^{17}$ K.

The above estimates suggest that electron cyclotron maser emission can account satisfactorily for the high brightness temperatures and high degrees of circular polarization seen in solar 'spike bursts' and in some components of stellar emissions. Figure 3, for example, shows observations by Brown and Crane (1978) of flare emissions from the RS CVn-star V 711 Tau at two frequencies, 2.7 and 8.1 GHz. The rapidly varying left hand circularly polarized component at 2.7 GHz is probably due to this mechanism.

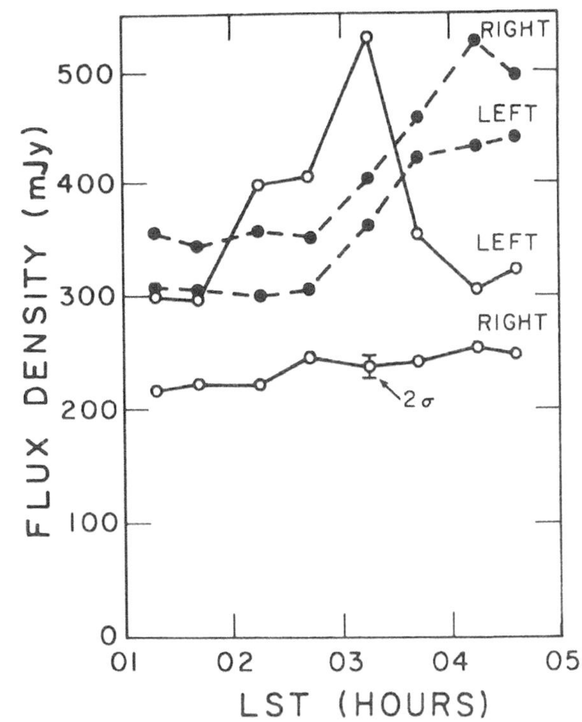

Fig. 3. Circularly polarized flare emissions from the RS CVn star V 711 Tau at two frequencies, 2.7 GHz (hollow circles) and 8.1 GHz (solid circles) (after Brown and Crane, 1978).

The absence of similar activity at 8.1 GHz is consistent with the narrow bandwidth of a maser.

There is, however, a difficulty with solar and stellar applications of this mechanism. There is an average decrease in magnetic field strength with increasing distance from the surface of the Sun or star, and the escaping radiation must, therefore, pass through layers where its frequency is an integer multiple of the local gyrofrequency and where gyromagnetic absorption by thermal electrons occurs. This absorption should prevent the fundamental modes from escaping and would allow the second harmonic modes to escape only under rather special conditions. It is by no means clear how this difficulty can be overcome, because growth at $s \geq 3$ is certainly too weak to account for observed emissions.

4. Second Harmonic Plasma Emission

Plasma emission was first treated quantitatively by Ginzburg and Zheleznyakov (1958). It involves emission at either the fundamental or the second harmonic of the plasma frequency ω_p. Here we concentrate on the latter. The magnetic field plays only a minor role in determining the gross properties of the emissions and can be neglected for most purposes. Plasma emission cannot be treated as simply as the cyclotron maser mechanism since it proceeds in at least two stages, each of which introduces complications. In the first stage resonant interactions between energetic electrons and Langmuir waves generate high levels of Langmuir turbulence. The waves are longitudinal oscillations of the plasma electrons relative to the ions and propagate at frequencies just above ω_p. The most favourable sources of free energy for the interaction are either an energetic electron stream or a double-sided loss cone distribution with a 'gap', i.e., a region with $\partial f/\partial|\mathbf{v}| > 0$ at $|\mathbf{v}| \neq 0$. The second stage involves a nonlinear interaction between two Langmuir (L) waves with wavevectors $\mathbf{k}_1$ and $\mathbf{k}_2$, respectively, and the plasma to create an electromagnetic (t) wave with wavevector $\mathbf{k}$ and frequency just above $2\omega_p$. A necessary condition for this coalescence process ($L + L \rightarrow t$) to occur is that $\mathbf{k}_1 + \mathbf{k}_2 = \mathbf{k}$. Since $|\mathbf{k}| \ll |\mathbf{k}_1|, |\mathbf{k}_2|$; the Langmuir waves must have $|\mathbf{k}_1| \cong |\mathbf{k}_2|$ and propagate in essentially opposite directions: i.e., $\mathbf{k}_2 \cong -\mathbf{k}_1$.

An electron stream generates Langmuir waves in a filled cone in the direction of the stream so that for this free-energy source, an intermediate stage is required to produce Langmuir waves travelling in the reverse direction. This greatly complicates the treatment of the mechanism, making it a three-stage one, and effectively precluding a quantitative treatment in which one can have confidence.

Type-III solar radio bursts are generated by electron streams through plasma emission and it is reasonable to consider the possibility of observing analogous emissions from other stars. There are two classes of type-III events: storm-associated and flare-associated. The storm events start at frequencies $\lesssim 100$ MHz and drift to lower frequencies; they are, therefore, not relevant to current observations of most flare stars. Although flare type-III bursts can start at frequencies $\gtrsim 500$ MHz the highest values of T_B occur at frequencies lower than this. Moreover, the duration of these bursts at the highest frequency is short ($\cong 0.1$ s). Thus type-III bursts at frequencies $\gtrsim 500$ MHz could not be observed on other stars with current techniques.

The one plasma emission mechanism we discuss here involves generation of Langmuir waves by a loss cone distribution. Although this is known to be a possible mechanism for solar radio bursts (e.g., Zaitsev and Stepanov, 1983; and references given therein), it is not nearly as familiar as type-III-like plasma emission. It seems more favourable than the latter for scaling up to the higher frequencies and longer time-scales associated with flare star emissions.

A double-sided loss cone distribution with a gap can be formed when a group of accelerated electrons is injected near the top of a magnetic arch connecting two points on the surface of a star and containing a relatively dense thermal plasma. The energetic electrons with small pitch angles precipitate at the feet of the arch producing loss cones in both forward and backward directions. Over a longer time-scale the energetic electrons are scattered by the thermal plasma into the loss cone from which they are rapidly precipitated. A gap results because the scattering is more probable for the slower electrons. This type of distribution can develop only after the injection ceases and can last for only a few collision times (Hewitt and Melrose, 1985).

The loss cone distribution gives positive contributions to the growth rates for Langmuir waves propagating nearly perpendicular to the symmetry axis with phase speeds $\cong \omega_p/k$ which are less than the speed for which the energetic distribution function is a maximum. The thermal plasma gives negative contributions which prevent growth when the phase speed is too small. The net growth rates are quite large if the loss cone is large – i.e., loss cone angles $\gtrsim 45°$ – if the pitch angle distribution falls off rapidly within the loss cone and if the ratio of typical speeds for the energetic and thermal electrons is greater than about 10. Growth is possible for longer periods if the initial injected distribution is relatively hard, i.e., if its distribution function does not fall off rapidly with increasing electron energy. An attractive feature of this source of free energy is that it produces pairs of Langmuir waves satisfying $\mathbf{k}_2 \cong -\mathbf{k}_1$ directly and no intermediate stage is necessary. The coalescence process produces electromagnetic waves with weak x-mode polarization.

The brightness temperature of plasma emission is limited by the balance between the coalescence process $L + L \to t$ and the inverse decay process $t \to L + L$. In the present case, balance occurs when the effective temperatures of the two wave distributions are equal. These effective temperatures $T_{\text{eff}}(\mathbf{k})$ are functions of $\mathbf{k}$ with

$$W = \int \frac{\mathrm{d}^3\mathbf{k}}{(2\pi)^3}\, k_{\mathrm{B}} T_{\text{eff}}(\mathbf{k}) \,, \tag{7}$$

representing the energy density in the respective waves. For transverse waves T_{eff} is just the brightness temperature T_B, which is a constant along a ray.

We may estimate the maximum T_B as follows. The observed T_B must be no greater than T_{eff} for the transverse waves at the source, and this can be no greater than T_{eff} for the Langmuir waves producing the transverse waves. If W_{free} is the free-energy density available to drive the instability generating the Langmuir waves, and Δk, $\Delta\Omega$ are, respectively, the ranges of wavenumbers and solid angles to which the Langmuir waves are confined, then

$$T_{\text{eff}} < \frac{(2\pi)^3 W_{\text{free}}}{k_{\mathrm{B}} k^2 \Delta k \Delta\Omega} \,. \tag{8}$$

This can be rewritten in the form

$$T_{\text{eff}} < \eta \,\frac{mv^2}{2k_{\mathrm{B}}}\, \frac{8nv^3}{\nu^3}\, \frac{n_1}{n} \left(\frac{\Delta k}{k}\, \Delta\Omega\right)^{-1} , \tag{9}$$

where $W_{\text{free}} = \eta^{1/2} n_1 mv^2$, n_1 and n are the number densities of the energetic and thermal electrons, respectively, $v = \omega_p/k$ is the resonant speed and $\nu = 2\omega_p/2\pi$ is the frequency of observation. For 10–100 keV energetic electrons, $n \cong 10^{15}\ \mathrm{m}^{-3}$ and $\nu \cong 1$ GHz we have $mv^2/2k_{\mathrm{B}} \cong 10^8$–$10^9$ K and $8nv^2/\nu^3 \cong 10^{11}$. Except for quite large values of n_1/n (e.g., $\gtrsim 10^{-4}$) these numbers suggest that it is unlikely that this mechanism could produce radiation with a very high brightness temperature. The observed maximum $\cong 10^{13}$ K for plasma emission from the solar corona is unlikely to be exceeded to any significant extent in other sources.

5. Gyro-synchrotron Emission

Electrons spiralling in a magnetic field radiate at frequencies ω, given by (4), with positive integral values of s. Simple analytic treatments for this emission have long been known in the nonrelativistic or 'cyclotron' limit and in the ultrarelativistic or 'synchrotron' limit. In the intermediate energy range from about 10 keV to 1 MeV, neither of these approximate treatments is valid, and this is just the range of relevance to gyro-synchrotron emission. Considerable progress has been made over the past few years in treating this intermediate range. One useful approach is to calculate various quantities exactly and use curve-fitting techniques to derive relatively simple functional forms for the results. This approach has been pursued by Dulk and Marsh (1982) whose formulas have already been used extensively in semi-quantitative applications. The other approach is analytic. Petrosian (1981) showed that one could derive quite accurate results for gyro-synchrotron emission from a distribution of electrons by carrying out the integrals over pitch angle and energy by the method of steepest descents. This method has been extended and generalized by Petrosian and McTiernan (1983) and by Robinson and Melrose (1984). Although this approach leads to particularly useful simplified formulas for Maxwellian distributions, it is more convenient to use Dulk and Marsh's (1982) numerically-derived formulas for power-law distributions.

In the gyro-synchrotron regime, radiation emitted at harmonics $\omega \cong s\Omega_e$ with $10 \lesssim s \lesssim 100$ merges to form a continuum. Dulk and Marsh (1982) found that, for a power-law distribution function proportional to $E^{-\delta}$ with $E > 10$ keV and $2 \lesssim \delta \lesssim 7$, the brightness temperature was bounded above by

$$T_B \lesssim 2.2 \times 10^9 \times 10^{-0.31\delta} (\sin\theta)^{-0.36-0.06\delta} (\omega/\Omega_e)^{0.50+0.085\delta}, \tag{10}$$

where $\theta \gtrsim 20°$. This implies that brightness temperatures are typically $\lesssim 10^{10}$ K, with the largest values occurring for small values of δ and Ω_e – i.e., for relatively flat distributions and weak magnetic fields. Dulk and Marsh also gave formulas similar to (10) for the emission and absorption coefficients, the degree of circular polarization, and for the frequency where the brightness temperature is largest. The radiation pattern has a broad maximum perpendicular to the magnetic field (i.e., $\theta = 90°$) where the degree of circular polarization is typically $<40\%$. There are no values of δ, Ω_e, and θ for which $T_B(\nu)$ and $|r_c|$ are both large. Gyro-synchrotron emission is the best candidate for the non-flaring microwave background emissions from the Sun and other stars.

References

Brown, R. L. and Crane, P. C.: 1978, *Astron. J.* **83**, 1504.
Dulk, G. A. and Marsh, K. A.: 1982, *Astrophys. J.* **259**, 350.
Gibson, D. M.: 1983, in M. Rodono and P. Byrne (eds.), 'Activity in Red Dwarf Stars', *Proc. IAU Colloq.* **71**, 273.
Ginzburg, V. L. and Zheleznyakov, V. V.: 1958, *Soviet Astron. A.J.* **2**, 653.
Grognard, R. J.-M. and McLean, D. J.: 1973, *Solar Phys.* **29**, 149.
Hewitt, R. G., Melrose, D. B., and Rönnmark, K. G.: 1981, *Proc. Astron. Soc. Australia* **4**, 221.
Hewitt, R. G. and Melrose, D. B.: 1985, *Solar Phys.* **96**, 157.

Hewitt, R. G., Melrose, D. B., and Rönnmark, K. G.: 1982, *Australian J. Phys.* **35**, 447.
Hjellming, R. M. and Gibson, D. M.: 1980, in M. R. Kundu and T. E. Gergely (eds.), 'Radio Physics of the Sun', *IAU Symp.* **86**, 209.
Holman, G. D., Eichler, D., and Kundu, M. R.: 1980, in M. R. Kundu and T. E. Gergely (eds.), 'Radio Physics of the Sun', *IAU Symp.* **86**, 457.
Kuijpers, J.: 1980, in M. R. Kundu and T. E. Gergely (eds.), 'Radio Physics of the Sun', *IAU Symp.* **86**, 341.
Kundu, M. R.: 1965, *Solar Radio Astronomy*, Interscience, New York.
Kundu, M. R. and Gergely, T. E. (eds.): 1980, 'Radio Physics of the Sun', *IAU Symp.* **86**.
Lee, L. C., Kan, J. R., and Wu, C. S.: 1980, *Planetary Space Sci.* **28**, 703.
Lovell, B.: 1969, *Nature* **222**, 1126.
McLean, D. J. and Labrum, N. R. (eds.): 1985, in *Solar Radiophysics*, Cambridge University Press, Cambridge.
Melrose, D. B., Rönnmark, K. G., and Hewitt, R. G.: 1982, *J. Geophys. Res.* **87**, 5140.
Melrose, D. B. and Dulk, G. A.: 1982a, *Astrophys. J.* **259**, 844.
Melrose, D. B. and Dulk, G. A.: 1982b, *Astrophys. J.* **259**, L41.
Melrose, D. B. and Dulk, G. A.: 1984, *Astrophys. J.* **282**, 308.
Melrose, D. B., Hewitt, R. G., and Dulk, G. A.: 1984, *J. Geophys. Res.* **89**, 897.
Omidi, N. and Gurnett, D. A.: 1982, *J. Geophys. Res.* **87**, 2377.
Petrosian, V.: 1981, *Astrophys. J.* **251**, 727.
Petrosian, V. and McTiernan, J. M.: 1983, *Phys. Fluids* **26**, 3023.
Robinson, P. A. and Melrose, D. B.: 1984, *Australian J. Phys.* **37**, 675.
Slottje, C.: 1978, *Nature* **275**, 520.
Spangler, S. R. and Moffett, T. J.: 1976, *Astrophys. J.* **203**, 497.
Twiss, R. Q.: 1958, *Australian J. Phys.* **11**, 564.
Wild, J. P. and McCready, L. L.: 1950, *Australian J. Sci. Res.* **A3**, 387.
Wild, J. P., Smerd, S. F., and Weiss, A. A.: 1963: *Ann. Rev. Astron. Astrophys.* **1**, 291.
Wu, C. S. and Lee, L. C.: 1979, *Astrophys. J.* **230**, 621.
Zaitsev, V. V. and Stepanov, A. V.: 1983, *Solar Phys.* **88**, 297.

OSCILLATIONS AND PULSATIONS IN THE SUN AND STARS (INVITED REVIEW)*

YOJI OSAKI and HIROMOTO SHIBAHASHI
Department of Astronomy, Faculty of Science, University of Tokyo, Tokyo, Japan

(Received 13 June, 1985)

Abstract. The discovery of the solar global oscillations, and their identification to many radial and nonradial eigenmodes of the Sun, have opened a new important field of research called 'solar seismology', in which we may probe the internal structure of the Sun by using its oscillations. In recent years, pulsations and oscillation-related phenomena have also been discovered in many stars which were hitherto regarded as non-pulsating stars. They include white dwarfs (ZZ Ceti-stars), Ap-stars, early-type stars with slow and rapid rotation (53 Per and ζ Oph-stars, respectively). Developments in high spectral and high time-resolution observations will be expected to reveal variability in many other stars as well, in near future.

In this review, the general properties of stellar eigenmode oscillations are first discussed. Recent observational and theoretical develepments in the solar seismology and on pulsations and oscillations of newly discovered variable stars will then be reviewed.

1. Introduction

The theory of stellar pulsations was originally developed in order to explain the pulsations of classical variables such as the Cepheid and RR Lyrae stars. Although there remain a few problems to be solved, we have now a good understanding on the basic nature of pulsations in these stars.

However, we have recently witnessed a new development in observations and theories of stellar pulsations. This new development comes basically from the discovery of pulsations and oscillation-related phenomena in many stars, which were hitherto regarded as non-pulsating stars. They include our Sun itself, white dwarfs (their variables are called ZZ Ceti-stars), Ap-stars (rapidly oscillating Ap-stars), and early-type stars with slow and rapid rotation (53 Per-stars and ζ Oph-stars, respectively). The most important characteristics of pulsations in these newly discovered variables are that their pulsations are usually the so-called nonradial pulsations, i.e., more general types of stellar oscillations than a simple radial pulsation; and that their pulsations are multi-periodic (i.e., many stellar eigenmode oscillations are involved). This new development has opened an important field of research called 'solar and stellar seismology', in which one can probe the internal structures of the Sun and stars by using their oscillations.

Along with these observational developments, much progress has also been made in the theoretical side of the solar and stellar nonradial oscillations. Now we have a good understanding on the seismic properties of various stellar models.

* Paper presented at the IAU Third Asian-Pacific Regional Meeting, held in Kyoto, Japan, between 30 September–6 October, 1984.

Astrophysics and Space Science **118** (1986) 195–207.

In this short review, the general properties of stellar eigenmode oscillations are first presented. Recent observational and theoretical developments in solar seismology, and on pulsations and oscillations of some newly discovered variable stars will then be reviewed briefly. For more complete reviews on this subject, other review articles such as those by Cox (1980) and by Unno *et al.* (1979) should be consulted.

2. Eigenmodes of Stars

Any persistent oscillations of a star may be considered as a superposition of normal modes of the star. There exist two kinds of oscillations: radial oscillations and nonradial oscillations. The radial oscillation (or radial pulsation) is a simple type of oscillation in which a star oscillates around its equilibrium figure, by executing expansion and contraction, but always keeping its spherical shape. Nonradial oscillations are a more general type of oscillation in which a star oscillates in such a way as to deviate from its spherical shape. The radial oscillation may be regarded as one of the special cases of nonradial oscillations with the spherical-harmonic index $l = 0$.

The normal modes of nonradial oscillations may be written for perturbations of physical variables (e.g., for the pressure p) as

$$p = p_0 + p' , \tag{1}$$

$$p' = p'_{nl}(r)\, Y_l^m(\theta, \phi)\, e^{i\omega t} , \tag{2}$$

where the subscript 0 and the symbol '′' stand for the equilibrium and the Eulerian perturbation, respectively, $Y_l^m(\theta, \phi)$ stands for spherical harmonics, and ω is the angular frequency of oscillation. A normal mode of nonradial oscillation is specified by three integers (n, l, m). The quantum number l represents the number of zonal lines by which the stellar surface is divided to oscillate in a different phase, while the quantum number m indicates the number of nodal lines in the longitude. The quantum number n represents the radial ordinal number, which can take both positive and negative values. The absolute value of this quantum number corresponds to the number of nodal surfaces in the radial direction in simple stellar models, but this correspondance breaks down in complicated stellar models, such as those of evolved stars. For a more detailed discussion on this point, see Osaki (1975) and Unno *et al.* (1979).

The eigenfrequency ω and the radial dependence of the eigenfunction $p'_{nl}(r)$ are described by the two quantum numbers n and l of the mode, but are independent of m for the case of a spherical star in its unperturbed state (i.e., in a non-rotating, non-magnetic star). For a given l, the quantum number m takes integer values from $-l$ to l, and thus the nonradial modes are, in this case, $(2l + 1)$-fold degenerate. This degeneracy is lifted if rotation of the star is taken into account. That is, for a slowly rotating star, we have

$$\omega_{nlm} = \omega^0_{nl} - m(1 - C_{nl})\Omega , \tag{3}$$

where Ω denotes the angular frequency of rotation of the star, C_{nl} is a numerical constant (usually $C_{nl} < 1$) which depends on the particular mode of oscillation and the stellar model under consideration. This 'line splitting' corresponds to that of the Zeeman-levels in atomic spectra. The rotational splitting of nonradial modes serves as a major tool for us to probe the internal rotation law of the Sun and stars.

The existence of two extra indices (l, m), describing the horizontal dependence of eigenfunctions, gives nonradial oscillations a complicated appearance as compared with the radial oscillation, but it also gives them an extra richness in the eigenvalue spectrum. Besides that, two different kinds of restoring forces (i.e., the pressure force and the buoyancy force) operate in nonradial oscillations, and there exist, therefore, two different kinds of modes: pressure (acoustic) modes and gravity modes. The pressure modes (p-modes) form a sequence of increasing eigenfrequency with the order of modes specified by the number of radial nodes (i.e., the quantum number n), that is,

$$\omega_{p_1} < \omega_{p_2} < \omega_{p_3} < \cdots \to \infty\ ;$$

while the gravity modes (g-modes) form a sequence a decreasing frequency, i.e.,

$$\omega_{g_1} > \omega_{g_2} > \omega_{g_3} > \cdots \to 0\ .$$

The positive and negative sequences of the quantum number n correspond to these two sequences.

3. Solar Oscillations

The oscillatory motion called the 'five-minute oscillation' was first discovered by Leighton *et al.* in 1960. Some fifteen years later, Deubner (1975) succeeded in resolving observed oscillations into discrete modes in the so-called diagnostic diagram, and a comparison between observation and theoretical prediction has established that solar five-minute oscillations are global nonradial p-modes of the Sun with high spherical-harmonic degree l (l = 200–1000). Since the late 1970s, observations using integrated sunlight over the whole disk have been performed to study both the radial velocity variation, and the total irradiance variation, by several groups (Birmingham, Crimea, Nice, Stanford, and SMM satellite). These studies have revealed the existence of other p-mode oscillations with low degree l ($l = 0 \sim 5$). Furthermore, p-mode oscillations with intermediate degree l ($l = 1 \sim 200$) have recently been detected; and the Sun must now be regarded as one of the pulsating stars, pulsating in many p-modes with $l = 0 \sim 1000$.

There are a number of good reviews on the solar oscillations and solar seismology, but a most recent one is that given by Deubner and Gough (1984).

3.1. High l p-modes

As is well known, Deubner (1975) was the first to resolve the observed power of the five-minute oscillations into several discrete ridges in the (k, ω)-diagram, where k and ω are the horizontal wavenumber ($k = l/R_\odot$) and the angular frequency of oscillations, respectively. The general pattern of observed ridges agreed fairly well with those of theoretical prediction (Ulrich, 1970; Ando and Osaki, 1975) for nonradial p-modes of the Sun with l = 200–1000 and overtones of the radial order $n = 1 \sim 10$. These p-modes are acoustic eigenmodes which are trapped in the solar convection zone, and their frequencies must reflect the structure of the latter. It is found from detailed comparison of observed frequencies with theoretical ones that a rather deep convection zone (the depth of the convection zone $d \gtrsim 0.3\ R_\odot$ implying the ratio of the mixing length to the pressure scale height in the mixing length theory $l/H_p \gtrsim 2$) is required in order to have a good agreement between observation and theory (Ulrich and Rhodes, 1977; Berthomieu *et al.*, 1980).

An attempt was also made to measure the sub-photospheric rotational velocity from the slight difference in positions of ridges between eastward and westward propagating waves by Deubner *et al.* (1979), but their results are still inconclusive as judged from recent observations. Observations of some daily variation in the positions of ridges were reported by Hill *et al.* (1983), who interpreted them as a result of the passage of giant cells.

3.2. Low l p-modes

The Birmingham group (Claverie *et al.*, 1979) discovered many discrete peaks in the power spectrum for the Doppler shift measurements of the integrated sunlight over the whole disk, with periods of about five minutes and amplitudes less than 1 ms^{-1}. Further analysis (Claverie *et al.*, 1980, 1981, 1982), and an independent observation performed at the South Pole by the Nice group (Grec *et al.*, 1980, 1983), have shown that these oscillations consist of many discrete peaks in the power spectrum with a nearly equal spacing of $\sim 68\ \mu$Hz in the frequency range of $2.5 \sim 5$ mHz. Such a well-defined structure in the power spectrum of oscillations has also been confirmed in the measurements of the total irradiance variation of the Sun from the SMM satellite (Woodard and Hudson, 1983a, b).

Since whole-disk measurements are sensitive only to modes with low l ($l \lesssim 3$), the observed oscillations are interpreted as high overtones of p-modes with low l. According to the asymptotic theory (Tassoul, 1980), the frequencies ν_{nl} of p-modes with $n \gg l \sim 1$ are given by

$$\nu_{nl} \simeq (n + l/2 + \varepsilon)\nu_0 , \tag{4}$$

where

$$\nu_0 \equiv \left[2 \int_0^R c^{-1}\, \mathrm{d}r \right]^{-1} ; \tag{5}$$

and c denotes the sound velocity and ε is a small correction due to the boundary effects. Frequencies of p-modes with even and odd l will then alternate with a separation of $\nu_0/2$, which corresponds to the observed spacing of 68 μHz. Long observational runs by those three groups now yield much higher resolution in the power spectrum than those given by the asymptotic formula, and individual p-modes with l and n have been resolved for each sequence of $l = 0$, 1, 2, and 3. Doppler shift measurements, using the modified magnetograph at Stanford University (Scherrer *et al.*, 1983), have a different sensitivity to the harmonic degree l, and they have detected global p-modes with $l = 3, 4,$ and 5.

The p-modes are eigenmodes of acoustic waves and, therefore, their frequencies reflect the sound velocity in the solar and stellar interior and the temperature structure $T = T(r)$. Since p-modes with low degree l penetrate deeply into the solar radiative core, they can be used as a probe to investigate deeper parts of the solar interior. Various group have compared observed frequencies of low l p-modes with theoretical frequencies of various solar models (Christensen-Dalsgaard and Gough, 1980, 1981; Scuflaire *et al.*, 1981; Shibahashi and Osaki, 1981; Ulrich and Rhodes, 1983). It is found that the standard solar model yields a fairly good fit with observations, but there remains a small discrepancy between observed frequencies and theoretical ones in the sense that theoretical frequencies are slightly lower than observed ones. Several attempts were made to reduce such discrepancies by experimenting with the equation of state and convection (Gabriel *et al.*, 1982; Noels *et al.*, 1984; Scuflaire *et al.*, 1982; Shibahashi *et al.*, 1983; Ulrich and Rhodes, 1984), but a slight discrepancy as much as $5 \sim 10$ μHz has not been removed yet. It is noted here that an extreme care is now required in constructing equilibrium models of the Sun, and in calculating their eigenfrequencies in order to get sufficiently high accuracy to compare with observations, as the observational error of frequencies is as small as 1 μHz for solar p-mode oscillations with frequency of ~ 3 mHz.

3.3. Intermediate Degree l p-Modes

Duvall and Harvey (1983) performed Doppler-shift observations on the solar image divided into zonal and sectoral components, and by analysing these data they succeeded in identifying intermediate l p-modes near the five-minute range. Their detection of intermediate l p-modes now fills the gap that had existed between low l and high l p-modes, and the Sun is now found to be pulsating in many p-modes with $l = 0 \sim 1000$.

Comparison of the observed frequencies with the theoretical ones confirms the systematic discrepancy known before in low and high l-modes: the theoretical frequencies are slightly lower for $l \lesssim 30$ but higher for $l \gtrsim 50$ than observations.

The rotational frequency splitting was also detected for p-modes with $l = 1 \sim 100$ by Duvall and Harvey (1984). Average frequency shifts due to rotation (i.e., $\Delta\nu \simeq (\nu_{m=-l} - \nu_{m=l})/2l$) where found to range from 0.423 μHz for $l = 6$ to 0.660 μHz for $l = 1$.

3.4. g-Modes

The observational detection of solar oscillations other than those of the five-minute range was first announced by the Crimean group (Severny *et al.*, 1976) and then by the Birmingham group (Brookes *et al.*, 1976). The period reported by them was 160 minutes, which is very close to one ninth of a day. Because of this, there has been controversy as to whether or not the 160 min oscillation is of solar origin. From the theoretical point, the 160 min oscillation must be one of the high overtone g-modes with low l, if it is of solar origin, and one of difficulties is why only a single mode is excited among densely populated spectrum of high overtone g-modes.

Hill *et al.* (1976) reported the detection of solar oscillations with periods near 50 min in their measurement of the solar diameter variation. However, the interpretation of its solar origin still remains inconclusive.

The detection of low l g-modes has recently been reported by Delache and Scherrer (1983) based on the Stanford data, by Isaak *et al.* (1984) based on the Birmingham data, and by Fröhlich and Delache (1984) based on the SMM data. In the asymptotic theory of nonradial oscillations, periods P_{nl} of high overtone g_n-modes with low l are given by

$$P_{n,l} \simeq (n + l/2 + \varepsilon) P_0 \sqrt{l(l+1)}\,; \tag{6}$$

with

$$P_0 = 2\pi^2 \left[\int_0^{r_*} \frac{N}{r}\, dr \right]^{-1}, \tag{7}$$

where r_* is the radius at the bottom of the convective envelope and N is the Brunt–Väisälä frequency. On the basis of this formula, these groups searched for peaks having uniformly-spaced periods in the power spectrum, and found them. The values of P_0 thus obtained were $P_0 \simeq 38.6$ min from the Stanford data, $P_0 \simeq 41.2$ min for the Birmingham data, and $P_0 \simeq 38.6$ min for the SMM data. If the identification of g-modes in the Sun is real, it will give important information about the stratification in the radiative core of the Sun.

The theoretical values of P_0, estimated for the standard models by various people, lie in the range 32 ~ 34 min. Berthomieu *et al.* (1984) have proposed a non-standard model with turbulent diffusion mixing which can resolve this discrepancy. However, Gabriel (1984) has pointed out another possibility of mode identification which gives the observed value of $P_0 \simeq 32$–34 min.

3.5. Seismological Approach

The most important and unique aspect of solar oscillations is the possibility of a seismological approach, by which one can probe the solar deep interior by using oscillations. This has been realized only by precise measurements of frequencies for many eigenmodes of the Sun. We have already mentioned it, for instance, concerning the depth of the convection zone.

Most works so far made in this respect are, however, essentially those of model fitting, in which observed frequencies are compared with theoretical ones calculated with solar models of varying parameters (such as the helium abundance Y, the mixing length ratio l/H_p, and so on), and parameters giving the best fit are then determined. This approach is called the forward problem. Another approach, called the inverse problem, would be, in principle, more useful as a seismological treatment. In the latter approach, functional forms of certain physical quantities such as the temperature distributions $T = T(r)$ or the rotational angular velocity distribution $\Omega = \Omega(r)$ are determined directly by solving integral equations for eigenfrequencies.

Based on the rotational splitting $\Delta\nu$ of prograde $(m = -l)$ and retrograde $(m = +l)$

sectoral modes with $l = 1 \sim 100$ obtained by Duvall and Harvey (1984), Duvall *et al.* (1984) have determined the internal rotation rate of the Sun by solving the appropriate integral equation (i.e., the inversion problem for the rotation law $\Omega = \Omega(r)$). It has been found that much of the solar interior rotates slightly less rapidly than the surface, while the innermost part rotates more rapidly.

4. Stellar Oscillations

4.1. Solar-type oscillations in stars

The successful detection of solar global oscillations by using integrated light has stimulated attempts to search for solar-type oscillations of stars in general. Since the amplitudes of such oscillations are expected to be very small, long periods of continuous observational data are required to build up sufficiently high signal-to-noise ratio. Two groups (Fossat *et al.*, 1984; Noyes *et al.*, 1984) have so far reported the detection of such oscillations in stars α Centauri and ε Eridani, respectively. Since the signal-to-noise ratios of the data in both cases are still marginal, their results need to be confirmed. However, their attempts give us a hope to develop stellar seismology in the near future.

4.1.1. α Centauri A

This star is a star of spectral type G2V, very similar to the Sun, with a mass of 1.1 $M_\odot$. By using their sodium cell spectrophotometer and the ESO 3.6 m telescope, Fossat *et al.* (1984) obtained good quality data on the radial velocity variation in α Cen over 20 hr. They first calculated the power spectrum of the data, and then searched for a regular pattern of equi-distant peaks on it. To do this, they performed Fourier analysis of the power spectrum itself, and then found such a pattern, which implied solar-type oscillations around frequencies $\nu \sim 3$ mHz (again a five-minute oscillation!) with an equi-distance of 81.3 μHz. This means that ν_0 of Equation (5) would be $\nu_0 \simeq 160$ μHz for α Cen, which may be compared with $\nu_0 \simeq 135$ μHz for the Sun.

4.1.2. ε Eridani

Noyes *et al.* (1984) detected variability in the intensities of the Ca H and K lines of ε Eri (a young K2V star with strong chromospheric emission), in the course of nightly monitoring for the Mt. Wilson H–K program. Power spectrum analysis for each of two six-hour data sets revealed spikes located near $\nu \sim 1.7$ μHz with an equi-distance of about 86 μHz, indicating *p*-mode oscillations in this star with $\nu_0 = 172$ μHz for the spacing of successive orders of a fixed l. The amplitude $\Delta I/I$ of this variation is found to be 6×10^{-3}, which is some two orders of magnitude greater than the solar case.

4.2. Rapidly oscillating Ap-stars

Kurtz (1982) discovered rapid light variability in several cool Ap-stars by high-speed photometric observations, and named these stars the rapidly oscillating Ap-stars. Ten stars have so far been recognized as members (Kurtz, 1984; and references given therein). Their periods are in the range of 6 to 14 min, which are much shorter than those of δ Sct variables (typically 2 hr), although these two groups of pulsating stars overlap each other on the H–R diagram. The amplitudes of light variation are very low ($\Delta m \lesssim 10$ m mag.) and are modulated with the same period and phase as the magnetic strength variation.

The very high frequencies of observed oscillations are most likely to be explained by high overtone p-mode pulsations. Kurtz (1982) has proposed a model called the oblique pulsator model, which can explain the observed properties of these stars fairly well. In this model, the observed oscillations are interpreted as very high overtone p-modes of zonal ($m = 0$) nonradial oscillations of low degree ($l = 1 \sim 2$), whose symmetry axes are aligned to the magnetic axis, but oblique to the rotation axis of the star. As the star rotates, the amplitudes of oscillations are modulated in parallel with the magnetic strength.

Detailed power spectrum analysis has revealed that some of these stars are pulsating in several modes with uniformly-spaced frequencies (Kurtz and Seeman, 1983) in a similar way to the solar oscillations. The observed frequency spacing (e.g., 33 μHz for HD 24712) is consistent with the theoretical values of $\nu_0/2$ for stellar models with $M \simeq 2M_\odot$ in the Main-Sequence stage (Shibahashi and Saio, 1985), which indicates that observed oscillations may be an alternation of even and odd l p-modes. As for radial order n of p-modes, the combination of the asymptotic theory and observed frequencies yields $n \simeq 40$.

The discovery of the rapidly oscillating Ap-stars has raised many questions: Why do these stars pulsate in such high overtones? Why are the pulsations aligned with the magnetic axis instead of the rotation axis? What is the relationship between the pulsation and the anomalous chemical abundances? What is the relationship between these stars and the δ Sct variables?

Two explanations have so far been proposed for the excitation mechanism of the rapid oscillations in Ap-stars: the κ-mechanism (Dolez and Gough, 1982) and the magnetic overstability (Shibahashi, 1983; Cox, 1984). In the former model, the diffusion process of chemical elements is supposed to occur along the magnetic axis, resulting in chemical stratification, which is not spherically symmetric but axisymmetric with respect to the magnetic axis. Overstability due to the κ-mechanism, if it works, is likely to induce oblique pulsation, because the κ-mechanism is a sensitive to the chemical stratification. In the latter model, the ordinary convective instability is supposed to be suppressed by the presence of magnetic fields, but it manifests itself as overstable convection, a cause of magnetic overstability of the oscillatory motion. This magnetic overstability can naturally explain the coincidence of the pulsation axis with the magnetic axis, and it seems to explain the rather short periods of excited oscillations. However, much work is necessary before definitive answers are given to these questions.

4.3. Pulsating White Dwarfs

Light variability in white dwarfs was first discovered in 1968 for a DA white dwarf called HL Tau-76 by Landolt (1968). Since then, about a dozen of DA white dwarfs have been found to be pulsating variables. McGraw and Robinson (1976) have shown that these luminosity-variable white dwarfs make up a homogeneous class of pulsating variables, called ZZ Ceti-stars, with the following characteristics: (1) all are (apparently single) DA white dwarfs; (2) the colour of the variable white dwarfs is confined to a narrow range $0.16 \leq (B - V) \leq 0.20$ which corresponds to an effective-temperature range of about $10000 \sim 13000$ K; (3) the light variations are multi-periodic, with periods in the range of 200–1000 s.

Since the fundamental period of radial pulsation is about 10 s for white dwarfs, these oscillations can not be explained by radial pulsations. However, nonradial g-mode oscillations have rather long periods in white dwarfs, and two periods of 213 s and 274 s in ZZ Ceti itself may naturally be explained by nonradial g_1 and g_2 modes of $l = 1$ or $l = 2$ (Osaki and Hansen, 1973; Brickhill, 1975). However, the periods of 700–1000 s observed in large-amplitude variables are too long even for nonradial g-modes, unless they are high overtone g_n-modes with $n \gtrsim 5$.

The observed range of effective temperature in these variables coincides with that region on the H–R diagram where a low-luminosity extension of the Cepheid instability strip would cross the white dwarf sequence. This suggests that the pulsations of ZZ Ceti-stars may be driven by the so-called κ-mechanism of the hydrogen- and helium-ionization zones, which is thought to be responsible for the Cepheid pulsation. Recent theoretical calculations of non-adiabatic nonradial pulsations (Dziembowski and Koester, 1981; Dolez and Vauclair, 1981; Winget *et al.*, 1982a) confirm that white dwarf stars are indeed unstable against nonradial g-modes in the range appropriate to ZZ Ceti-stars due to the κ-mechanism of the hydrogen- and helium-ionization zones. These calculations show that some nonradial g-modes could be 'trapped' in the compositionally stratified layers (Winget *et al.*, 1981) which are formed due to the gravitational settling of chemical elements in the envelopes of white dwarf stars, and that these 'trapped' modes could be preferentially excited among densely populated high-overtone g_n-modes: a possible explanation of the rather long periods of 700–1000 s observed in large-amplitude ZZ Ceti-stars. This indicates another good example of the stellar seismology, in that the pulsations of ZZ Ceti-stars could be used to probe the envelope structure of white dwarfs.

Period analyses of light curves of ZZ Ceti-stars often reveal fine structures in the power spectra, such as two closely spaced peaks (e.g., in ZZ Ceti itself), or several distinct peaks equally spaced in frequency (e.g., in HL Tau-76). They are usually interpreted as the rotational m-mode splitting of nonradial g-mode oscillations. The rotational periods of white dwarf stars thus estimated are found to be from several hours to a day. Such a relatively slow rotation of white dwarfs is consistent with other observational evidence (such as the narrowness of certain spectral line cores).

Pulsations in white dwarfs other than the classical ZZ Ceti stars have recently been discovered – for DB white dwarfs (Winget *et al.*, 1982b), and for some hot white dwarfs, called PG1159–035 variables (McGraw *et al.*, 1979; Grauer and Bond, 1984). Now there exist, therefore, three distinct classes of pulsating white dwarfs.

4.4. Early-type variables

The β Cephei-stars have been known for a long while as pulsating variables of early spectral type. A new class of early-type pulsating stars called line-profile variable B stars have recently been discovered, owing to the development of high precision spectroscopy with the use of solid-state detectors. The latter class of variables manifest their variability most conspicuously in spectral line-profiles, with little accompanying change in luminosity or radial velocity (line centroid). Two sub-classes of line-profile variable B stars are recognized: the one is that of sharp-lined variables, called 53 Per-stars (Smith, 1977), and the other is that of the rapidly rotating B or Be stars, called ζ Oph-stars (Vogt and Penrod, 1983).

Variability in the line-profiles of some B stars had been noticed before, but its full significance as a distinct class of pulsating variables was first recognized by Smith (1977). He has shown that many of the sharp-lined B stars exhibit variation in line-profiles in such a way that their width and asymmetry change in time, and that their variation can be explained by low-l nonradial pulsations of traveling wave type (with $l = 2$ or 3). 53 Per itself shows photometric variations with two closedly spaced periods of near 2 days (Buta and Smith, 1979), implying that 53 Per-stars may belong to the more general class of variable stars called 'slow variables' (Burki, 1983). The 53 Per-stars are sharp-lined O8 to B5 stars near the Main Sequence, surrounding the classical β Cephei variables in the H–R diagram, and a possibility is entertained that the β Cephei-stars may represent the most easily observed 'tip of the iceberg' of a much wider class of variables. However, the exact relation between β Cephei and 53 Per-stars is still unknown, largely because the very excitation mechanisms of pulsations in these two classes of variables have not been established as yet.

Another interesting development in recent years has been the discovery of a different type of line-profile variability, in that quasi-absorption/emission 'bumps' are traveling across the rotationally broadened line-profiles, either from blue to red or from red to blue, in certain rapidly rotating B and Be stars (e.g., ζ Oph by Walker *et al.*, 1979; μ Cen by Baade, 1984). Stars showing this type of line-profile variability are called ζ Oph-stars. This phenomenon is most naturally explained by nonradial traveling-wave modes of intermediate l with $l = 6 \sim 10$ and $m = \pm l$ (Vogt and Penrod, 1983). It was thought before that high l ($l \gtrsim 5$) nonradial modes could not be observed in stars because of the cancellation effects over the visible disk, and that line-profile variability would be difficult to be detected in rapidly rotating stars. Such arguments were found to be something of a prejudice, and it has turned out that rapid rotation rather helps to increase the visibility of intermediate l nonradial oscillations in line-profiles, as it helps to resolve the stellar disk through rotational Doppler shift. Thus, observational manifestation of low l nonradial modes in the 53 Per-stars and of high l nonradial modes in the ζ Oph-stars may simply reflect different visibility of different modes depending on stellar rotational velocities. There exists a good possibility that these two classes of stars could be basically similar.

An intrigueing suggestion has been recently made that the Be phenomenon may be caused by nonradial pulsations (Vogt and Penrod, 1983). In this scenario, nonradial pulsations are first built up in B stars, resulting in the deposit of angular momentum in the stellar atmosphere. If the amplitudes of nonradial oscillations become large enough, mass ejection occurs near the equatorial regions, which in turn damps nonradial pulsations. This kind of episode then repeats, explaining an alternation of B and Be phenomenon in a single star. This suggestion is one of the most promising so far proposed concerning the mechanisms for formation of the circumstellar shell in Be stars.

5. Concluding Remarks

The discovery of very small amplitude pulsations, especially those found from observations of the Sun as a star, has raised a question concerning the concept of pulsating and non-pulsating stars. We now expect that oscillations, at least with very small amplitudes, will be ubiquitous among stars in general. However, since the amplitudes are expected to be small in most cases, a high precision in photometry and spectroscopy is required to detect their variability. One of the most important characteristics of newly discovered variables, discussed here, is that their oscillations are multi-periodic, and thus information included in their oscillations (i.e., in the power spectrum) are rich as compared with singly-periodic variables, such as most of the classical pulsating stars. Long strings of uninterrupted observational data are, however, necessary to secure high resolution in the power spectrum. The best way to achieve these requirements will be to go to space, and, in fact, some such projects are now under study. Ground-based observations by internationally linked observatory networks will also be desirable.

On the theoretical side, the effects of rotation and magnetic fields on stellar oscillations must be clarified, as they play essential roles in some newly discovered variables, such as the Be and Ap-stars. Effects of nonlinear coupling among various modes should also be investigated, as large-amplitude ZZ Ceti-stars and some early-type variables seem to show mode-switching phenomena. It is quite likely that some sort of mode coupling may occur among the various modes of solar oscillations, as they have such a rich spectrum. The excitation mechanisms of pulsations for the newly-discovered variables remain to be investigated, except for ZZ Ceti-stars. Also, the inversion technique to extract information about solar and stellar internal structure from observational data should be improved.

Finally, we conclude this review by stating that developments in observations and theories will make the 'solar and stellar seismology' one of the most important fields of research in astrophysics of the Sun and stars in the near futures.

References

Ando, H. and Osaki, Y.: 1975, *Publ. Astron. Soc. Japan* **27**, 581.
Baade, D.: 1984, *Astron. Astrophys.* **135**, 101.
Berthomieu, G., Cooper, A. J., Gough, D. O., Osaki, Y., Provost, J., and Rocca, A.: 1980, in H. A. Hill and W. A. Dziembowski (eds.), *Nonradial and Nonlinear Stellar Pulsation*, Springer-Verlag, Berlin, p. 307.
Berthomieu, G., Provost, J., and Schatzman, E.: 1984, *Mem. Soc. Astron. Ital.* **55**, 107.
Brickhill, A. J.: 1975, *Monthly Notices Roy. Astron. Soc.* **170**, 405.
Brookes, J. R., Isaak, G. R., and van der Raay, H. B.: 1976, *Nature* **259**, 92.

Burki, G.: 1983, *Astron. Astrophys.* **121**, 211.
Buta, R. J. and Smith, M. A.: 1979, *Astrophys. J.* **232**, 213.
Christensen-Dalsgaard, J. and Gough, D. O.: 1980, *Nature* **288**, 544.
Christensen-Dalsgaard, J. and Gough, D. O.: 1981, *Astron. Astrophys.* **104**, 173.
Claverie, A., Isaak, G. R., McLeod, C. P., van der Raay, H. B., and Roca Cortes, T.: 1979, *Nature* **282**, 591.
Claverie, A., Isaak, G. R., McLeod, C. P., van der Raay, H. B., and Roca Cortes, T.: 1980, *Astron. Astrophys.* **91**, L9.
Claverie, A., Isaak, G. R., McLeod, C. P., van der Raay, H. B., and Roca Cortes, T.: 1981, *Nature* **293**, 443.
Claverie, A., Isaak, G. R., McLeod, C. P., van der Raay, H. B., and Roca Cortes, T.: 1982, *Solar Phys.* **74**, 51.
Cox, J. P.: 1980, *Theory of Stellar Pulsation*, Princeton University Press, Princeton, N.J.
Cox, J. P.: 1984, *Astrophys. J.* **289**, 220.
Delache, P. and Scherrer, P. H.: 1983, *Nature* **306**, 651.
Deubner, F.-L.: 1975, *Astron. Astrophys.* **44**, 371.
Deubner, F.-L. and Gough, D. O.: 1984, *Ann. Rev. Astron. Astrophys.* **22**, 593.
Deubner, F.-L., Ulrich, R. K., and Rhodes, E. J., Jr.: 1979, *Astron. Astrophys.* **72**, 177.
Dolez, N. and Gough, D. O.: 1982, in J. P. Cox and C. J. Hansen (eds.), *Pulsations in Classical and Cataclysmic Variable Stars*, JILA, University of Colorado, Boulder, Colorado, p. 248.
Dolez, N. and Vauclair, G.: 1981, *Astron. Astrophys.* **102**, 375.
Duvall, T. L., Jr. and Harvey, J. W.: 1983, *Nature* **302**, 24.
Duvall, T. L., Jr. and Harvey, J. W.: 1984, *Nature* **310**, 19.
Duvall, T. L., Jr., Dziembowski, W. A., Goode, P. R., Gough, D. O., Harvey, J. W., and Leibadher, J. W.: 1984, *Nature* **310**, 22.
Dziembowski, W. and Koester, D.: 1981, *Astron. Astrophys.* **97**, 16.
Fossat, E., Grec, G., Gelly, B., and Decanini, Y.: 1984, *The Messenger* **36**, 20.
Fröhlich, C. and Delache, P.: 1984, *Mem. Soc. Astron. Ital.* **55**, 99.
Gabriel, M.: 1984, *Astron. Astrophys.* **134**, 387.
Gabriel, M., Scuflaire, R., and Noels, A.: 1982, *Astron. Astrophys.* **110**, 50.
Grauer, A. D. and Bond, H. E.: 1984, *Astrophys. J.* **277**, 211.
Grec, G., Fossat, E., and Pomerantz, M.: 1980, *Nature* **288**, 541.
Grec, G., Fossat, E., and Pomerantz, M.: 1983, *Solar Phys.* **82**, 55.
Hill, H. A., Stebbins, R. T., and Brown, T. M.: 1976, in J. H. Sanders and A. H. Wapstra (eds.), *Atomic Masses and Fundamental Constants*, Plenum Press, New York, p. 622.
Hill, F., Toomre, J., and November, L. J.: 1983, *Solar Phys.* **82**, 411.
Isaak, G. R., van der Raay, H. B., Pale, P. L., Roca Cortes, T., and Delache, P.: 1984, *Mem. Soc. Astron. Ital.* **55**, 91.
Kurtz, D. W.: 1982, *Monthly Notices Roy. Astron. Soc.* **200**, 807.
Kurtz, D. W.: 1984, *Monthly Notices Roy. Astron. Soc.* **209**, 841.
Kurtz, D. W. and Seeman, J.: 1983, *Monthly Notices Roy. Astron. Soc.* **205**, 11.
Landolt, A. U.: 1968, *Astrophys. J.* **153**, 151.
Leighton, R. B., Noyes, R. W., and Simon, G. W.: 1962, *Astrophys. J.* **135**, 474.
McGraw, J. T. and Robinson, E. L.: 1976, *Astrophys. J.* **205**, L155.
McGraw, J. T., Starrfield, S. G., Liebert, J., and Green, R.: 1979, in H. M. Van Horn and V. Weidemann (eds.), 'White Dwarfs and Variable Degenerate Stars', *IAU Colloq.* **53**, 377.
Noels, A., Scuflaire, R., and Gabriel, M.: 1984, *Astron. Astrophys.* **130**, 389.
Noyes, R. W., Baliunas, S. L., Belserene, E., Duncan, D. K., Horne, J., and Windrow, L.: 1984, *Astrophys. J.* **285**, L23.
Osaki, Y.: 1975, *Publ. Astron. Soc. Japan* **27**, 237.
Osaki, Y. and Hansen, C. J.: 1973, *Astrophys. J.* **185**, 277.
Scherrer, P. H., Wilcox, J. M., Christensen-Dalsgaard, J., and Gough, D. O.: 1983, *Solar Phys.* **82**, 75.
Scuflaire, R., Gabriel, M., and Noels, A.: 1981, *Astron. Astrophys.* **99**, 39.
Scuflaire, R., Gabriel, M., and Noels, A.: 1982, *Astron. Astrophys.* **113**, 219.
Severny, A. B., Kotov, V. A., and Tsap, T. T.: 1976, *Nature* **259**, 87.
Shibahashi, H.: 1983, *Astrophys. J.* **275**, L9.
Shibahashi, H. and Osaki, Y.: 1981, *Publ. Astron. Soc. Japan* **33**, 713.
Shibahashi, H. and Saio, H.: 1985, *Publ. Astron. Soc. Japan* **37**, 245.

Shibahashi, H., Noels, A., and Gabriel, M.: 1983, *Astron. Astrophys.* **123**, 283.
Smith, M. A.: 1977, *Astrophys. J.* **215**, 574.
Tassoul, M.: 1980, *Astrophys. J. Suppl.* **43**, 469.
Ulrich, R. K.: 1970, *Astrophys. J.* **162**, 993.
Ulrich, R. K. and Rhodes, E. J., Jr.: 1977, *Astrophys. J.* **218**, 521.
Ulrich, R. K. and Rhodes, E. J., Jr.: 1983, *Astrophys. J.* **265**, 551.
Ulrich, R. K. and Rhodes, E. J., Jr.: 1984, in A. Noels and M. Gabriel (eds.), *Theoretical Problems in Stellar Stability and Oscillations*, Université de Liège, Liège, p. 250.
Unno, W., Osaki, Y., Ando, H., and Shibahashi, H.: 1979, *Nonradial Oscillations of Stars*, University of Tokyo Press, Tokyo.
Vogt, S. S. and Penrod, G. D.: 1983, *Astrophys. J.* **275**, 661.
Walker, G. A. H., Yang, S., and Fahlman, G. G.: 1979, *Astrophys. J.* **233**, 199.
Winget, D. E., Van Horn, H. M., and Hansen, C. J.: 1981, *Astrophys. J.* **245**, L33.
Winget, D. E., Van Horn, H. M., Tassoul, M., Hansen, C. J., Fontaine, G., and Carrol, B. M.: 1982a, *Astrophys. J.* **252**, L65.
Winget, D. E., Robinson, E. L., Nather, E., and Fontaine, G.: 1982b, *Astrophys. J.* **262**, L11.
Woodard, M. and Hudson, H.: 1983a, *Solar Phys.* **82**, 67.
Woodard, M. and Hudson, H.: 1983b, *Nature* **305**, 589.

STELLAR CHROMOSPHERES, CORONAE, AND WINDS (INVITED REVIEW)*

THEODORE SIMON

Institute for Astronomy, University of Hawaii, Honolulu, Hawaii, U.S.A.

(Received 13 June, 1985)

Abstract. Within the last few years there have been important advances in our understanding of the chromospheres, coronae, and winds of late-type stars. This progress has been the result of ultraviolet and X-ray observations obtained from space with the *International Ultraviolet Explorer* and *Einstein* satellites. In addition, ground-based astronomy has contributed much to synoptic studies of stellar activity. In this review I summarize some of the major recent developments in this field.

1. Introduction

Over the last few years, the study of the chromospheres, coronae, and winds of the late-type stars has seen many changes in our understanding of their nature and evolution. It would be very difficult to cover in this short review all of these recent, outstanding discoveries. Fortunately, there are a number of very comprehensive reviews that discuss the material that cannot be dealt with here, including Baliunas (1984), Linsky (1981), and Pallavicini (1985). For earlier results in this field, an excellent summary can be found in *Solar Phenomena in Stars and Stellar Systems* (Bonnet and Dupree, 1981).

2. The Main Sequence

Soft (0.2–4 keV) X-ray emission and far-ultraviolet (1200–3200 Å) emission lines are observed in Main-Sequence stars of spectral type F through M (e.g., Vaiana *et al.*, 1981; Ayres *et al.*, 1981b). The strength of this emission correlates with rapid rotation and youth, as is exemplified by the strong coronal X-ray fluxes of stars in the Hyades cluster (Stern *et al.*, 1981) and in the nucleus of the Ursa Major cluster (Walter *et al.*, 1984), which are up to 30 times that of the quiet Sun. There is a close resemblance between *IUE* spectra of late-type dwarfs (Figure 1) and the ultraviolet spectrum of the solar chromosphere and transition region (TR). However, in terms of surface flux, the stellar lines may be significantly brighter. Furthermore, although chromospheric emission in the Ca II H–K lines is difficult to observe in stars earlier than F5, ultraviolet chromospheric lines can be traced farther up the Main Sequence. At spectral type F2 the IUE spectrum of 78 UMa shows prominent high-temperature emission lines with surface fluxes about four times the quiet Sun's.

* Presented at the IAU Third Asian-Pacific Regional Meeting, held in Kyoto, Japan, between 30 September–6 October, 1984.

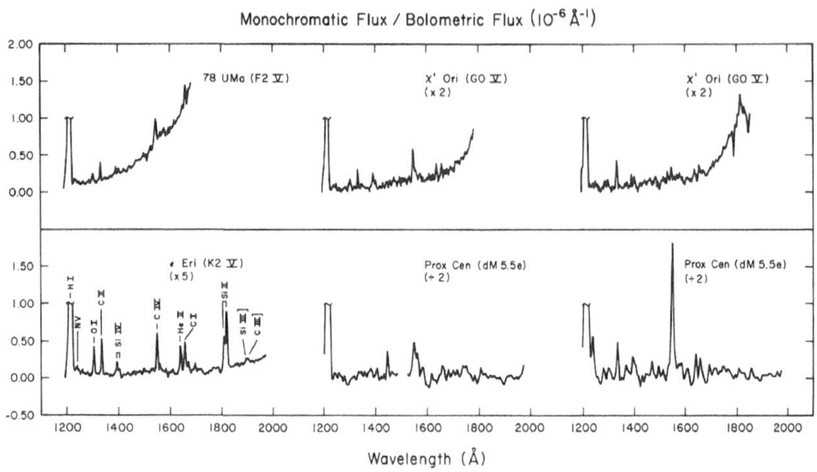

Fig. 1. Low-dispersion *IUE* spectra of Main-Sequence stars. The Lα line of H I is blended with geocoronal emission. Note the variability of the emission lines of χ^1 Ori and Proxima Centauri.

As a consequence of the large-scale organization of magnetic fields in the solar atmosphere, the ultraviolet and X-ray appearance of the Sun is very complex. Presumably the analogs of solar plages, sunspots, coronal holes, and magnetic loops are to be found on the surfaces of other stars. For many F–K stars there is already evidence from the observed modulation of calcium emission with rotation period for an uneven distribution of plages over the stellar surface (Wilson, 1978; Vaughan *et al.*, 1981). An example of rotational modulation at ultraviolet wavelengths is illustrated by the pair of spectra of the G0 V star χ^1 Orionis in Figure 1 (Boesgaard and Simon, 1984). These observations were taken one day apart, corresponding to a difference of 0.22 in rotational phase. The strength of the TR line of C IV at 1549 Å declines by a factor of 3 and the C II 1355 Å line brightens by 50%. On the following days the changes reverse. Variability on similar time scales in *IUE* spectra of other G–K stars has been reported by Hallam and Wolff (1981). Even at the limit of the photometric accuracy of *IUE*, this variability corresponds to large amounts of energy. For example, the fluctuation in power at the C IV line of χ^1 Ori in Figure 1 is 6×10^{27} erg s^{-1} and changes in the flux of the Mg II lines of χ^1 Ori, which are at the level of observational error of ~10% (Boesgaard and Simon, 1984), are equivalent in power (~3×10^{28} erg s^{-1}) with the energy release of a class 3 solar flare.

From emission-line fluxes measured in low-resolution spectra or in spectra obtained with the high-resolution echelle spectrographs of *IUE*, semi-empirical chromospheric models have been derived for some of the brightest dwarf and giant stars. These calculations have been summarized by Linsky *et al.* (1984). For the chromospherically active K2 dwarf ε Eridani (see Figure 1), Simon *et al.* (1980) derived a pressure at the top of the chromosphere of 0.5 dyn cm^{-2}, or about 3 times that of the quiet Sun. Over the 11.8 day rotation period of this star, the Ca H–K emission flux varies with a range of 10% (Vaughan *et al.*, 1981), as do the UV lines (Hallam and Wolff, 1981). Inter-

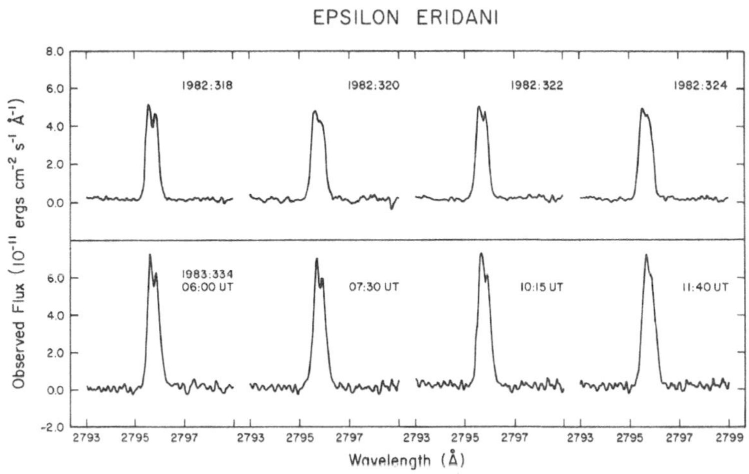

Fig. 2. Variability of Mg II k line of ε Eri (K2V). Each panel is labeled with the year and elapsed day number.

estingly, the shape of the profile of the Mg II k line (Figure 2) varies from day to day (top row) and hour to hour (bottom row), the most obvious changes occurring on the red side of the central emission peak. Although the low signal-to-noise ratio of *IUE* spectra (10–15) makes such changes suspect, similar variability is present in high signal-to-noise spectra of the Ca k line taken with Coudé Reticon of the Canada–France–Hawaii telescope on Mauna Kea. If intrinsic to ε Eri, these changes are unlikely to result from Doppler shifts of active regions at the stellar limb since the blue and red components in the central emission peak of the k line are separated by 30 km s^{-1}, while the measured stellar rotation rate is only 2.2 km s^{-1} (Gray, 1984). Similar changes (although perhaps on different time scales) in Mg II line profiles above sunspots (Lites and Skumanich, 1982) have been attributed to wave motions or downflows in the umbral chromosphere. The strength and FWHM ($\sim$0.5 Å) of the emission in the core of the stellar line is typical of solar plages (Doschek and Feldman, 1977; Kelch and Linsky, 1978), but differences in line shape within the solar network, plages, or supergranulation cells (Lemaire and Skumanich, 1973) are small enough to go undetected at the 20 km s^{-1} resolution of the echelle spectrographs of *IUE*. Therefore, while we may be assured that the surfaces of stars are as highly structured as the Sun's, we are still limited in the diagnostic techniques we can use to study the details of the structure from disk-integrated observations.

The two spectra of Proxima Cen included in Figure 1 were taken 90 min apart, the second during a large X-ray flare observed with the *Einstein* satellite (Haisch *et al.*, 1983). Coordinated observations of flare stars at X-ray, UV, optical, and radio wavelengths have been made on numerous occasions, and although there is some correlation between flare events observed at different wavelengths, the correlation is weaker than one might like. The X-ray luminosity of Proxima Cen at the time of the *IUE* observations reached a peak of 2×10^{28} erg s^{-1}, and during its 2.5 hr lifetime the flare produced 4×10^{31} erg. The energy released over the duration of the flare in the UV lines, which

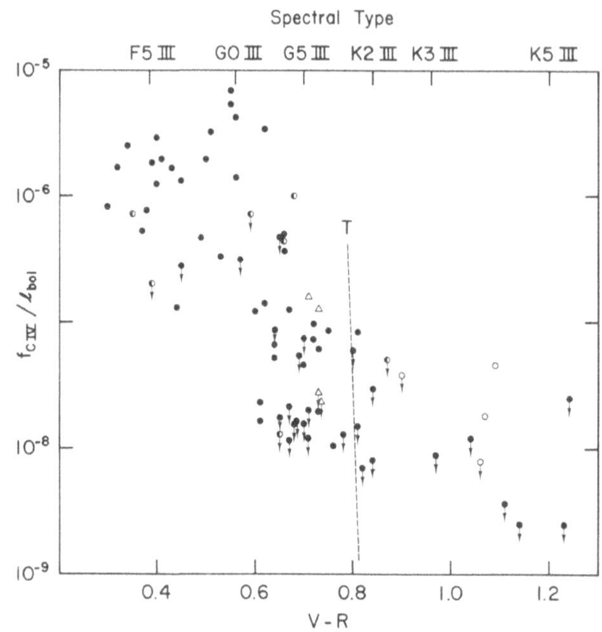

Fig. 3. Strength of the C IV 1549 Å line for late-type giants. The observed line flux is normalized by the apparent bolometric luminosity. Filled circles denote luminosity class III, triangles the Hyades giants, open circles hybrid stars, and half-filled circles luminosity class II other than hybrids. The dashed line (T) is the TR and coronal boundary line.

brightened from their preflare quiescent levels by a factor of 3, was about an order of magnitude less than in X-rays. This suggests that the flare plasma cooled by radiation rather than by conduction (Haisch *et al.*, 1983), which may be a more important mechanism for cooling in solar flares. Although the flare emission of Proxima Cen represents a large fraction of its bolometric luminosity, note that the fluctuations in C IV luminosity of χ^1 Ori are 30 times larger than the C IV luminosity of the Proxima Cen flare and are comparable with the X-ray losses of this flare.

3. Giants and Boundary Lines in the H–R Diagram

For stars several times more massive than the Sun, chromospheric activity begins after they leave the Main Sequence, enter the Hertzsprung gap, and develop hydrogen convection zones. Plotted against the $B - V$ or $V - R$ colour index, the TR emission of luminosity class III giants increases in the F stars to a peak near spectral type G0, and then declines steeply to much lower levels among the late G and K giants (Figure 3). There is simultaneously a falloff in average rotation rate with advancing spectral type (Kraft, 1966; Alschuler, 1975; Gray, 1982). Thus, if we exclude close binary stars, as stars evolve across the Hertzsprung gap, they undergo magnetic braking from coronal winds, and the effect of this spindown is to weaken dynamo action along the giant

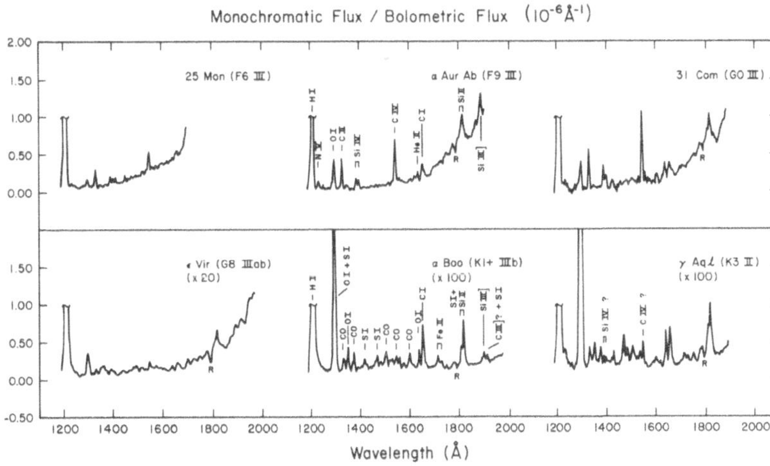

Fig. 4. Low-dispersion *IUE* spectra of giant stars.

branch. The large spread in TR emission levels among G5–K0 stars at the base of the red giant branch could be the result of an evolutionary funneling effect, which mixes first-crossing stars ascending the giant branch and He-burning post-red giants that have returned along a blue loop in their evolutionary tracks. Some first-crossing stars, e.g., the G6 III primary of Capella (Ayres and Linsky, 1980), reach this point in the H–R diagram with weak UV emission, while the second-crossing stars could have lost essentially all of their angular momentum as red giants because of mass loss in cool winds (Reimers, 1977; Stencel and Mullan, 1980). There is some evidence that rotation is not the only determinant of activity. Not only is there a spread that appears unrelated to rotation rate in UV emission levels among early F stars just entering along the blue edge of the Hertzsprung gap, but also rotation rate and TR emission are uncorrelated among the late G-early K giants (Simon, 1984). The Hyades giants, for example, have nearly identical $v \sin i$ values but differ in their UV and X-ray emission by a factor of 4 (Baliunas *et al.*, 1983).

At $V - R = 0.8$, corresponding to spectral type K1 III, there is a fundamental change in the character of *IUE* spectra of giant stars (Linsky and Haisch, 1979; Simon *et al.*, 1982). This is illustrated in Figure 4. Note that the high excitation lines of multiply-ionized metals in the spectra of yellow giants are replaced by the lines of neutral and singly-ionized atoms, molecular bands, and fluorescent lines in the spectra of red giants. Do the high-temperature lines vanish at this point because TRs (and coronae) are truly absent in the cooler stars, or do they simply weaken to the point where their emission is below the threshold for detection? If TRs and coronae disappear, to be replaced by very extended cool chromospheres, this transformation must happen over a narrow range in effective temperature because stars just to the left of the 'TR dividing line' fail to show the cool, photo-excited lines observed in stars to the immediate right. Coincident with the 'TR dividing line' there is a similar 'coronal dividing line' observed for X-ray emission (Ayres *et al.*, 1981a; Haisch and Simon, 1982), which separates the yellow

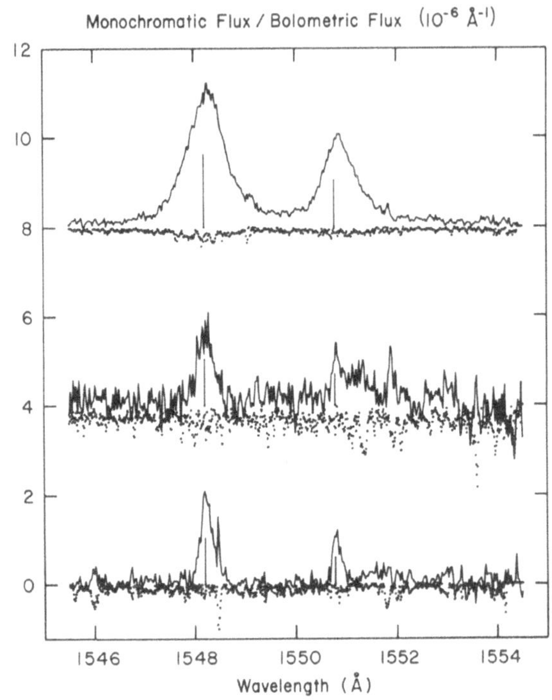

Fig. 5. High-dispersion *IUE* spectra of the C IV TR lines of α Aur (top), χ^1 Ori (middle), and ε Eri (bottom). Both the line width and redshift increase with luminosity (figure by courtesy of T. R. Ayres).

giants to the left that commonly exhibit coronal emission from the red giants to the right that show no coronal emission, in some cases to very low limits. For example, the soft X-ray flux from Arcturus is several hundred times fainter than the emission from a solar coronal hole (Ayres *et al.*, 1982). Whether these dividing lines among giant stars are real or an artefact of the sensitivity limits of *IUE* and *Einstein* remains an unsettled question at this time.

The earliest spectral type at which chromospheric and TR emission lines are detected in giant stars is F2 III (Simon, 1985). Lines of N V, Si IV, and C II are observed in low-dispersion spectra at F2 III, but C IV is difficult to see against the steep continuum at wavelengths longward of 1500 Å. At F5 III all of the usual signatures of TRs and chromospheres, including C IV, are clearly present, as in the spectrum of 25 Mon shown in Figure 4. For supergiant stars Böhm-Vitense and Dettmann (1980) suggested that the onset of chromospheric activity was bounded by the Cepheid strip. A few Cepheids have transitory emission in TR and chromospheric lines at certain phases of their light cycles (Schmidt and Parsons, 1982), but none has been detected in X-rays (Böhm-Vitense and Parsons, 1983).

High-temperature TR and chromospheric lines are most prominent in the spectra of rapidly rotating giant stars. Examples are the secondary of Capella (F9 III) and 31 Comae (G0 III). The weak emission of ε Virginis (see Figure 4) is more typical of yellow giants. A chromospheric model that gave a pressure at the top of the chromosphere of 0.017 dyn cm^{-2} was computed for β Ceti, a G9.5 III giant near the 'TR boundary' by Eriksson *et al.* (1983). This gas pressure is 10 times less than the quiet Sun's. Thus, increasing scale height, despite lower gas pressure, accounts for the activity observed in evolved stars.

The spectrum of Arcturus, also shown in Figure 4, is representative of early K giants lying to the right of the 'dividing lines'. The same cool features identified in the spectrum of Arcturus are present in the spectrum of the luminosity class II giant γ Aquilae, except that the fluorescent features of O I, S I, and CO are now stronger and there may be weak emission at the position of the Si IV lines near 1400 Å. The latter spectrum, though exposed for 8.5 hr, demonstrates why it is so difficult to determine whether TRs disappear in red giants: at low resolution, the cool emission lines formed in extended chromospheres may be easily confused with the usual indicators of TRs because of wavelength coincidences. Thus, weak emission in the γ Aql spectrum at 1549 Å may be from the Fourth Positive System of CO, which is the identification given for the same feature in the Arcturus spectrum by Ayres *et al.* (1981c), rather than from C IV. This ambiguity could be removed at high resolution, but unfortunately γ Aql is too faint to be observed with the echelle spectrographs of *IUE*.

4. Winds and Hybrid Stars

The mass loss rate in the coronal wind of a Main-Sequence star like the Sun is too low to produce observable effects in optical or *IUE* spectra. Cool, dense winds are detected in the Ca II H–K lines (Reimers, 1977; Stencel, 1978) and Mg II h–k lines (Stencel and Mullan, 1980) of red giants and supergiants, as distortions of the emission cores and also as blueshifted absorption components. The central emission features of high-luminosity G and early K stars are typically double-peaked, the peaks appearing symmetric or slightly asymmetric with the blue peak brighter than the red; at later spectral types the asymmetry reverses (red peak dominant) because of an outward acceleration of the wind in the low chromosphere. Since the onset of these winds in low-gravity stars occurs at nearly the same locus in the H–R diagram where high-temperature TR and coronal emission disappears (Linsky and Haisch, 1979; Ayres *et al.*, 1981a), it is thought that expansion of the wind may cool the outer atmospheres of red giants and supergiants. However, note in Figure 3 that the upper limits on TR emission are as low in some G giants, to the left of the dividing lines, as in K giants to the right, without any evidence for winds appearing in the Mg line profiles of the G stars. Thus, the connection between the weakness of high-temperature coronae and TRs in red giants and the presence of cool winds remains unclear.

The 'wind dividing line' for luminosity class III giants does not appear to extend to higher luminosities, since Hartmann *et al.* (1980) found evidence of cool winds in the Mg II spectra of the early G supergiants α Aqr and β Aqr. These two stars and the K4 II/III star α TrA (Hartmann *et al.*, 1981) were named 'hybrid' stars because their *IUE* spectra contain high-temperature lines such as C IV 1549 Å, low-excitation and pumped lines typical of late K stars with extended chromospheres, and also blueshifted absorption components in Mg II from a cool wind. Attributing the high-temperature lines in hybrid spectra to wind emission, Hartmann *et al.* (1981) computed wind models for the hybrids based on Hartmann and MacGregor's (1980) theory of Alfvén wave-driven winds. Their results gave wind temperatures of ~ 100000 K and expansion velocities of ~ 100 km s^{-1} within 2–3 stellar radii above the photosphere, while at greater distance expansion cooled the wind to Mg II temperatures.

A number of objections to the hybrid wind models have been raised. The outflow in the models is well underway in the low chromosphere, but the low-velocity, blue-shifted absorption component in Mg II taken as evidence of chromospheric acceleration is very likely to be interstellar rather than stellar (Drake *et al.*, 1984). The models predict that the optically thin lines of Si III] and C III] will be blueshifted by ~ 100 km s^{-1}, but the measured radial velocities are within 10 km s^{-1} of the photospheric velocity; in addition the line fluxes are 30 times less than predicted. The broad halfwidths of the high-temperature lines, which were cited by Hartmann *et al.* as evidence of a fast warm wind, are, in fact, typical of high-luminosity stars (see Figure 5) and may be caused by opacity broadening or turbulence associated with magneto-acoustic waves (Brown *et al.*, 1984). Finally, VLA observations of three luminosity class II hybrid stars proposed by Reimers (1982), namely, ι Aur, θ Her, and γ Aql, place 3 σ upper limits on the mass loss rate for ionized gas at $< 3 \times 10^{-9} M_0$ yr^{-1}, which is less than the mass loss rate required by the wind models. (The classification of these three stars as hybrids is by no means secure, because of the difficulty of identifying high-temperature lines in their spectra as explained in Section 3.)

5. Chromospheric Activity and Spindown

Rapid rotation has often been suggested as the fundamental ingredient in explaining strong chromospheric/coronal activity. Walter and Bowyer (1981) and Pallavicini *et al.* (1981) found correlations between X-ray luminosity and rotation rate for close binary stars and for single stars observed with *Einstein*. More recently, Noyes *et al.* (1984) and Simon *et al.* (1984b) showed that Main-Sequence chromospheric and TR emission-line strengths are governed by the Rossby number, which is a measure of both rotation rate and the depth of the convection zone.

It is also known from observations of nearby field stars and open clusters that chromospheric emission in the cores of the Ca H–K lines declines with age along with axial rotation rates (Wilson, 1963; Skumanich, 1972). The ultraviolet chromospheric and TR emission lines of solar-type stars observed with *IUE* decay with age (Figure 6), except that the hot TR lines decay more rapidly with age than do the cool chromospheric lines (Simon *et al.*, 1984b). This time decay of emission in Main-Sequence stars is exponential from a plateau, defined by stars younger than 1 Gyr, the *e*-folding decay time being a factor of 2 shorter for the TR lines (~ 1.4 Gyr) than for the chromospheric lines (~ 2.7 Gyr). A power law similar to Skumanich's (1972) fit to the calcium flux provides a somewhat poorer fit to the *IUE* data than does the exponential.

To extend these correlations to stars younger than ~0.1 Gyr is a challenge to the sensitivity of *IUE* because young stars are found only in distant clusters and they may be heavily reddened by circumstellar or interstellar gas and dust. For the most part, only the very luminous and, therefore, the very massive, pre-Main-Sequence T Tauri stars have been observed in the ultraviolet. *IUE* fluxes for the T Tauri-stars lie ~2 dex above an extrapolation of the Simon *et al.* (1984b) exponential decay law for chromospheric and TR lines in Main-Sequence stars (Figure 6). A power law can be fitted to the *IUE* fluxes of the T Tauris and the Main-Sequence stars, but the T Tauri fluxes spread more than 3 dex about the curve. Therefore, Simon *et al.* (1984b) argue that chromospheric activity decays by separate schemes for the T Tauri-stars and the Main-Sequence stars: in the older stars the decline of chromospheric and TR emission with age is the result of spindown caused by the braking torque of coronal winds, and in the pre-Main-Sequence stars it is related to the evolutionary retreat of the convection zone. In the first instance it is the progressive falloff in rotation rate, and in the second the diminishing size of the convection zone that accounts for the decline in dynamo strength and chromospheric/coronal heating. Stars in the cluster NGC 2264, which are intermediate in age (1–10 million years) between the T Tauri and Main-Sequence stars, have been observed with *IUE* (Simon *et al.*, 1984a), but these pre-Main-Sequence stars are very faint and so only upper limits on UV emission-line strengths were derived. As indicated by Figure 6, these limits need to be reduced by perhaps an order of magnitude in order to discriminate between alternative decay schemes.

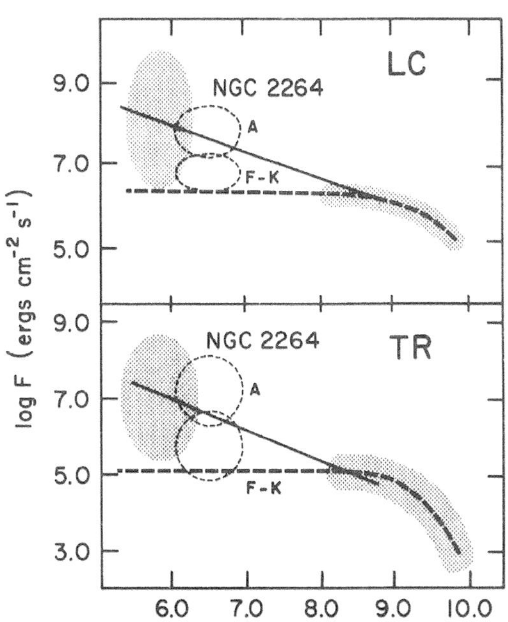

Fig. 6. Decay of chromospheric (upper panel) and transition region (lower panel) emission line surface fluxes with age. The shaded area on the left denotes the T Tauri-stars, the open circles the A and F–K pre-Main-Sequence stars in NGC 2264. The dashed line is the exponential fit to the Main-Sequence stars. The solid line is a power law fit joining the T Tauri and Main-Sequence stars.

At X-ray wavelengths there is a similar exponential trend of coronal emission with age for Main-Sequence stars older than 0.1 Gyr (Caillault and Helfand, 1985; Walter *et al.*, 1984; Stern *et al.*, 1981), but the correlation is possibly biased by the sensitivity limit of *Einstein*. T Tauri-stars are luminous X-ray sources ($\log L_x \approx 30$), in part by virtue of their large size; in terms of X-ray surface flux, their emission is comparable ($\log F_x \approx 6$) with that of the youngest Main-Sequence stars. The coronae of T Tauri-stars may be intrinsically brighter, however, if much of their X-ray emission is absorbed within their extended cool envelopes (Gahm, 1980; Walter and Kuhi, 1981).

6. The Solar-Stellar Connection

Most of the UV emission of the quiet Sun originates from the chromospheric network that outlines the supergranulation cells (Reeves, 1976). This spatial structure is the consequence of supergranulation flow patterns, which concentrate magnetic flux at the outer edge of the cells. Changes in the disk-integrated chromospheric emission of the Sun during its activity cycle are strongly correlated with the fraction of the solar surface that is covered by plages (White and Livingston, 1981; Cook *et al.*, 1980). By analogy with the Sun, plages have been assumed to be the cause of the observed rotational

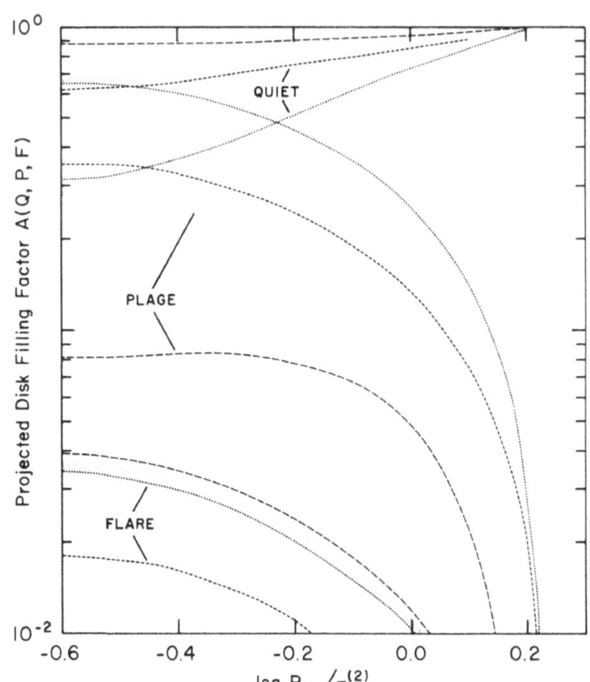

Fig. 7. Area filling factors plotted against logarithm of the Rossby number. The Rossby number is the stellar rotation period divided by the convective zone turnover time. The curves are derived from analytical expressions given by Simon *et al.* (1984b), which relate C IV flux and Mg II flux to the Rossby number. The curves are plotted for different assumed values for the emission line surface fluxes in solar plages and solar flares.

modulation of H–K emission and ultraviolet emission of stars. Generally, those stars showing the most prominent variations are much younger than the Sun and also more rapidly rotating. Since the average surface fluxes of their UV emission lines are ~25 times that of the quiet Sun or ~4 times that of a solar plage, even complete coverage of the stellar disk by solar-type plage would fall short of accounting for their bright emission. This would also leave unexplained how these stars can show rotational modulation.

Solar flares, on the other hand, have intense UV emission, and even in small subflares (Underwood *et al.*, 1978) the chromospheric and TR lines are far brighter in surface flux than the strongest stellar emission. On the reasonable assumption that the *IUE* emission of active stars may be enhanced with respect to the Sun because of a higher level of flare activity, as well as a larger area coverage of plages, Simon *et al.* (1984b) recently used a three-component model to analyze *IUE* observations of solar-type stars. Since their analysis provides a good example of the 'solar-stellar connection', it is described here in some detail. It was assumed in this model that inhomogeneities on the surface of a star can be represented as magnetically undisturbed quiet regions, plages, or flares; each component was assigned an average emission line surface flux based on solar observations of UV lines in quiet regions, plages, and flares; then the area coverage for each spatial component was derived for an individual star so as to give the best match to its chromospheric and TR line fluxes measured from *IUE* spectra.

The results of the calculations are summarized in Figures 7 and 8. Area filling factors are presented in Figure 7 as a function of the logarithm of the Rossby number. This semi-empirical parameter is the stellar rotation period divided by the convective zone turnover time (Gilman, 1980), and its value can be estimated for individual stars as described by Noyes *et al.* (1984). The curves drawn in Figure 7 were computed from the *analytical fits* given by Simon *et al.* (1984b) which relate *IUE* C IV and Mg II line strengths and Rossby number. The calculations were made for several different choices of the solar plage and flare surface fluxes. However, for further discussion let us consider only the shaded curves which denote the preferred values adopted from solar observations if we assume that stellar flares, on the average, resemble solar subflares.

For rapidly rotating stars (small Rossby number) the plage filling factor is ~30%, which is comparable with the solar plage coverage at maximum during a strong solar cycle. The flare coverage is at most 3–4%. Both of these filling factors decrease with increasing Rossby number (decreasing rotation rate), falling by an order of magnitude in slowly rotating stars like the Sun. The corresponding contribution of each spatial component of the disk-integrated flux for both Mg II and C IV is given in the following graph, Figure 8, by the solid lines and shaded areas. For comparison with these results, the observed *IUE* line fluxes for four stars ranging in age from 0.2 Gyr to 2.2 Gyr have been used in the same way to derive filling factors and the relative contributions of each surface component to the integrated flux. For a young, rapidly rotating star like π^1 UMa most of the chromospheric emission in the Mg II lines is contributed by plages; for an older star like β Com the emission comes primarily from the chromospheric network in quiet regions. Except for the very slowest rotators, more than half of the emission in TR

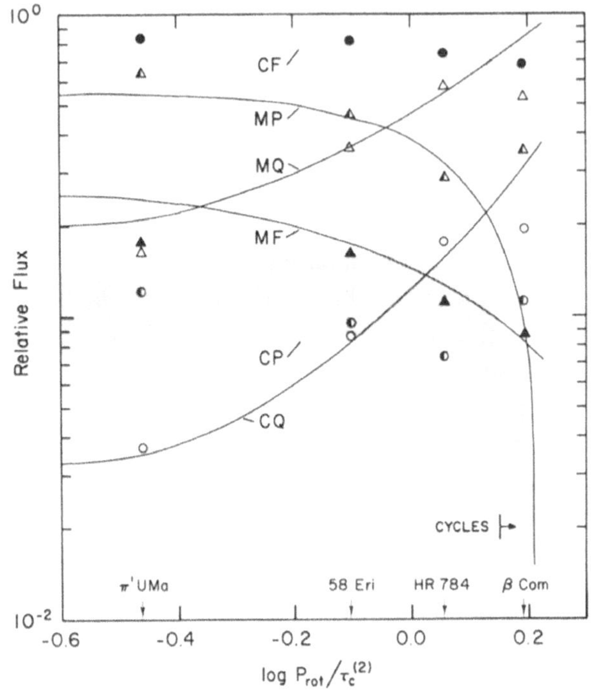

Fig. 8. The fractional contribution of each surface region to the disk-integrated flux of C IV and Mg II. C and M denote C IV and Mg II, respectively, while *F*, *P*, and *Q* denote the flare, plage, and quiet component. The results from *IUE* observations of four stars are plotted as circles and triangles for C IV and Mg II, respectively. Filled, half-filled, and open symbols indicate flare, plage, and quiet component, respectively.

lines observed with *IUE* originates in flares, which cover $<3\%$ of the surface. The more widely distributed plages contribute less than 10% to the disk-integrated light in high-temperature lines.

When compared with the plage models that have been assumed in previous discussions, this three-component model provides a very different interpretation for the bright chromospheric and TR emission of young, rapidly rotating stars. It has the advantage of reducing the plage coverage to $<30\%$, which may be more consistent with the detection of rotational modulation in both optical and UV chromospheric lines. Occasionally different modulation patterns have been observed for the TR and the chromospheric lines (Boesgaard and Simon, 1984), and this is also more consistent with expectations from the three-component analysis since the flux in these lines is shown to originate in different surface components. More importantly, for the 9 August, 1973 solar flare, which was used to model the stellar data here, Lites and Cook (1979) concluded that the excitation and ionization conditions of the flare plasma departed significantly from statistical equilibrium. Thus, attempts to model the TRs of young, active stars from ultraviolet observations may encounter the same difficulties.

References

Alschuler, W. R.: 1975, *Astrophys. J.* **195**, 649.

Ayres, T. R. and Linsky, J. L.: 1980, *Astrophys. J.* **241**, 279.

Ayres, T. R., Linsky, J. L., Vaiana, G. S., Golub, L., and Rosner, R.: 1981a, *Astrophys. J.* **250**, 293.

Ayres, T. R., Marstad, N. C., and Linsky, J. L.: 1981b, *Astrophys. J.* **247**, 545.

Ayres, T. R., Moos, H. W., and Linsky, J. L.: 1981c, *Astrophys. J.* **248**, L137.

Ayres, T. R., Simon, T., and Linsky, J. L.: 1982, *Astrophys. J.* **263**, 791.

Baliunas, S. L.: 1984, in *The Future of Ultraviolet Astronomy Based on Six Years of IUE Research*, NASA conference publication, CP-2349, p. 64.

Baliunas, S. L., Hartmann, L., and Dupree, A. K.: 1983, *Astrophys. J.* **271**, 672.

Boesgaard, A. M. and Simon, T.: 1984, *Astrophys. J.* **277**, 241.

Böhm-Vitense, E. and Dettmann, T.: 1980, *Astrophys. J.* **236**, 560.

Böhm-Vitense, E. and Parsons, S. B.: 1983, *Astrophys. J.* **266**, 171.

Bonnet, R. M. and Dupree, A. K. (eds.): 1981, *Solar Phenomena in Stars and Stellar Systems*, Proceedings of the NATO Advanced Study Institute held at Bonas, France, 25 August–5 September, 1980, D. Reidel Publ. Co., Dordrecht, Holland, p. 591.

Brown, A., Jordan, C., Stencel, R. E., Linsky, J. L., and Ayres, T. R.: 1984, *Astrophys. J.* **283**, 731.

Caillault, J.-P. and Helfand, D. J.: 1985, *Astrophys. J.* **289**, 279.

Cook, J. W., Brueckner, G. E., and Van Hoosier, M. E.: 1980, *J. Geophys. Res.* **85**, 2257.

Doschek, G. A. and Feldman, U.: 1977, *Astrophys. J.* **35**, 471.

Drake, S. A., Brown, A., and Linsky, J. L.: 1984, *Astrophys. J.* **284**, 774.

Eriksson, K., Linsky, J. L., and Simon, T.: 1983, *Astrophys. J.* **272**, 665.

Gahm, G. F.: 1980, *Astrophys J.* **242**, L163.

Gilman, P.: 1980, in D. Gray and J. Linsky (eds.), 'Stellar Turbulence', *IUA Colloq.* **52**, 19.

Gray, D. F.: 1982, *Astrophys. J.* **262**, 682.

Gray, D. F.: 1982, *Astrophys. J.* **281**, 719.

Haisch, B. M., Linsky, J. L., Bornmann, P. L., Stencel, R. E., Antiochos, S. K., Golub, L., and Vaiana, G. S.: 1983, *Astrophys. J.* **267**, 280.

Haisch, B. M. and Simon, T.: 1982, *Astrophys. J.* **263**, 252.

Hallam, K. L. and Wolff, C. L.: 1981, *Astrophys. J.* **248**, L73.

Hartmann, L. and MacGregor, K. B.: 1980, *Astrophys. J.* **242**, 260.

Hartmann, L., Dupree, A. K., and Raymond, J. C.: 1980, *Astrophys. J.* **236**, L143.

Hartmann, L., Dupree, A. K., and Raymond, J. C.: 1981, *Astrophys. J.* **246**, 193.

Kelch, W. L. and Linsky, J. L.: 1978, *Solar Phys.* **58**, 37.

Kraft, R. P.: 1966, *Astrophys. J.* **144**, 1008.

Lemaire, P. and Skumanich, A.: 1973, *Astron. Astrophys.* **22**, 61.

Linsky, J. L.: 1981, in R. Giacconi (ed.), *X-Ray Astronomy with the Einstein Satellite*, D. Reidel Publ. Co., Dordrecht, Holland, p. 19.

Linsky, J. L. and Haisch, B. M.: 1979, *Astrophys. J.* **229**, L27.

Linsky, J. L. *et al.*: 1984, in *The Future of Ultraviolet Astronomy Based on Six Years of IUE Research*, NASA conference publication, CP-2349, p. 445.

Lites, B. W. and Cook, J. W.: 1979, *Astrophys. J.* **228**, 598.

Lites, B. W. and Skumanich, A.: 1982, *Astrophys. J. Suppl.* **49**, 293.

Noyes, R. W., Hartmann, L. W., Baliunas, S. L., Duncan, D. K., and Vaughan, A. H.: 1984, *Astrophys. J.* **279**, 763.

Pallavicini, R.: 1985, in R. M. Hjellming and D. M. Gibson (eds.), *Radio Stars*, D. Reidel Publ. Co., Dordrecht, Holland, p. 197.

Pallavicini, R., Golub, L., Rosner, R., Vaiana, G. S., Ayres, T. R., and Linsky, J. L.: 1981, *Astrophys. J.* **248**, 279.

Reeves, E. M.: 1976, *Solar Phys.* **46**, 53.

Reimers, D.: 1977, *Astron. Astrophys.* **57**, 395.

Reimers, D.: 1982, *Astron. Astrophys.* **207**, 292.

Schmidt, E. G. and Parsons, S. B.: 1982, *Astrophys. J. Suppl.* **48**, 185.

Simon, T.: 1984, *Astrophys. J.* **279**, 738.

Simon, T.: 1985, in preparation.

Simon, T., Cash, W., and Snow, T. P.: 1984a, *Astrophys. J.* **293**, 542.
Simon, T., Herbig, G., and Boesgaard, A. M.: 1984b, *Astrophys. J.* **293**, 551.
Simon, T., Kelch, W. L., and Linsky, J. L.: 1980, *Astrophys. J.* **237**, 72.
Simon, T., Kelch, W. L., and Linsky, J. L.: 1980, *Astrophys. J.* **237**, 72.
Simon, T., Linsky, J. L., and Stencel, R. E.: 1982, *Astrophys. J.* **257**, 225.
Skumanich, A.: 1972, *Astrophys. J.* **171**, 565.
Stencel, R. E.: 1978, *Astrophys. J.* **223**, L37.
Stencel, R. E. and Mullan, D. J.: 1980, *Astrophys. J.* **238**, 221.
Stern, R. A., Zolcinsky, M.-C., Antiochos, S. K., and Underwood, J. H.: 1981, *Astrophys. J.* **249**, 647.
Underwood, J. H., Antiochos, S. K., Feldman, U., and Dere, K. P.: 1978, *Astrophys. J.* **224**, 1017.
Vaiana, G. S. *et al.*: 1981, *Astrophys. J.* **244**, 163.
Vaughan, A. H., Baliunas, S. L., Middelkoop, F., Hartmann, L., Mihalas, D., Noyes, R. W., and Preston, G. W.: 1981, *Astrophys. J.* **250**, 276.
Walter, F. M. and Bowyer, S.: 1981, *Astrophys. J.* **245**, 671.
Walter, F. M. and Kuhi, L. V.: 1981, *Astrophys. J.* **250**, 254.
Walter, F. M., Linsky, J. L., Simon, T., Golub, L., and Vaiana, G. S.: 1984, *Astrophys. J.* **281**, 815.
White, O. R. and Livingston, W. C.: 1981, *Astrophys. J.* **249**, 798.
Wilson, O. C.: 1963, *Astrophys. J.* **138**, 832.
Wilson, O. C.: 1978, *Astrophys. J.* **226**, 379.

NONLINEAR HYDRODYNAMICAL MODELS OF STELLAR CONVECTIVE ZONES*

WASABURO UNNO

Department of Astronomy, University of Tokyo, Bunkyo-ku, Tokyo, Japan

and

MASA-AKI KONDO

Faculty of Commerce, Sensyu University, Kawasaki-shi, Japan

(Received 13 June, 1985)

Abstract. Large-eddy simulation (LES) of turbulent convection is discussed in various versions (mixing-length theory, modal theory and spectral theory) in respect to the application to stellar convective zones. For the model construction, the non-local mixing-length formalism is suitable. However, for the determination of basic flow patterns and of mixing-length, the quasi-linear and nonlinear modal theories are useful. The eddy diffusivities are essential in these theories, and the nonlinear treatment of convection consistent with turbulent diffusivities (of effective Reynolds number of about 20 and Prandtl number of 0.4) offers a simple method of constructing stellar models without the use of the mixing-length.

1. Large-Eddy Simulation

Solar and stellar convective zones extend over many pressure scale heights and are subject to turbulence of very large-Reynolds numbers. For the construction of stellar convective zones, the large-eddy simulation (LES) is necessary to describe the turbulent state.

Large-eddy simulation has been formulated by many authors (e.g., Yoshizawa, 1982, 1983). In the astrophysical context, it offers a way to describe the eddy diffusivities and the turbulent pressure due to smaller eddies in terms of large-eddy motions. Unno *et al.* (1984) have shown that the mixing-length formalism of eddy diffusivities is recovered in the lowest order approximation. The best estimate of the eddy diffusivities may be given by the spectral theory (Nakano *et al.*, 1979).

The grain size separating the small and large eddies is a basic parameter in large-eddy simulation. If it is taken to be the limit of the smallest-eddy size, the theory will tend to the spectral theory, and if it is taken to be the limit of the largest eddy size, the theory will tend to the local mixing-length theory (Vitense, 1953). The drawback of the current spectral theory (Ledoux *et al.*, 1961; Nakano *et al.*, 1979) seems to be in the underlying assumption of homogeneity, which is not valid for the largest eddies, while the intrinsic non-local nature of convection is disregarded in the local mixing-length theory.

Xiong's (1979, 1981) non-local mixing-length theory, however, remedies the Vitense theory in that the penetrative convection at the edges of the convection zone is properly

* Paper presented at the IAU Third Asian-Pacific Regional Meeting, held in Kyoto, Japan, between 30 September–6 October, 1984.

Astrophysics and Space Science **118** (1986) 223–225.

calculated, and the turbulent pressure gives no numerical instability there. Recently, Unno *et al.* (1985) constructed a solar model in which solar granulation is predicted and found to be in good agreement with observations, and the Li–Be abundance problem is reasonably accounted for. In that study, the mixing-length was given by a semi-empirical formula implied from the study by Narasimha and Antia (1982). We believe that the local mixing-length theory should better be replaced by the non-local mixing-length theory for the construction of stellar models, since the latter theory includes the turbulent pressure, the kinetic energy transport, and other fluxes of velocity temperature correlations consistently with the concept of the large-eddy simulation.

Since the non-local mixing-length theory is not self-sustained, we must go a bit further to the largest-eddy simulation or the modal theory in which the grain size is taken fine enough to retain the largest eddies and coarse enough to average out the second (or the third) largest eddies. Van der Borght (1971) was the first to develop the modal theory of convection in deep compressible atmospheres. Some of his results are reported in this conference (Van der Borght, 1984). In the present paper, we restrict ourselves to study a Boussinesq fluid with nonlinear hydrodynamics, and then extend the interpretation of the results to find a new method for the calculation of stellar structure, which will be the subject of the following section.

2. Nonlinear LES of Turbulent Convection

Nonlinearity has already been introduced in large-eddy simulation in the form of eddy diffusivities. Therefore, the proposed linear theory in this section is actually a quasi-linear theory. We, henceforth, refer to linear modal theory, with eddy diffusivities, as the Chitre problem after the study of Antia *et al.* (1981).

Once the initial model has been calculated, eddy diffusivities can be estimated. Using these eddy diffusivities, we can attack the Chitre problem, and ask if a given mode is stable or unstable for each horizontal wavenumber. Since each horizontal wavenumber mode is representative at some depth in the convection zone (Narashima and Antia, 1982), the simplest principle to evaluate the temperature gradient at all depths is to require convective neutrality for all the horizontal wave-number modes in question. This principle, however, is too much simplified, since a homogeneity of turbulence was assumed in estimating the eddy diffusivities, and the largest eddies are directly subject to the boundary conditions.

Nonlinear Boussinesq convection (magneto-convection) has been calculated and applied to the case of fully developed turbulence, which corresponds to an effective Reynolds number of about 20, and Prandtl number of 0.4 (Rudraiah *et al.*, 1985). The results show that the Rayleigh number is adjusted by turbulence (of Reynolds number 20) to be about four times the critical Rayleigh number for stability. In other words, if the eddy viscosity and thermal conductivity were increased by a factor of 2 artificially, the Chitre problem should predict neutral stability of the model. Thus, the Chitre problem for various horizontal wavenumber modes defines the superadiabatic temperature gradient at various depths. Together with the hydrostatic condition, this neutral stability condition of the modified Chitre problem gives a simple method of constructing a stellar model. The reason that increased eddy diffusivities should be used in the Chitre problem to give the required superadiabaticity seems to be that a single largest eddy (a sine wave function in the Boussinesq fluid) should feed energy to two double subgrid eddies (sine and cosine wave functions), thereby exerting diffusivities to the largest eddy. If so, the factor 2, which increases the eddy diffusivities in the modified Chitre problem, will be valid for general cases of boundary conditions and compressibilities. Further studies are needed to check this conjecture.

References

Antia, H. M., Chitre, S. M., and Pandey, S. K.: 1981, *Solar Phys.* **70**, 67.

Ledoux, P., Schwarzschild, M., and Spiegel, E. A.: 1961, *Astrophys. J.* **133**, 184.

Nakano, T., Fukushima, T., Unno, W., and Kondo, M.-A.: 1979, *Publ. Astron. Soc. Japan* **31**, 713.

Narashima, D. and Antia, H. M.: 1982, *Astrophys. J.* **262**, 358.

Rudraiah, H., Kumudini, V., and Unno, W.: 1985, *Publ. Astron. Soc. Japan* **37**, 183 and 207.

Unno, W., Kondo, M.-A., and Chen, B.: 1984, in W. Unno and T. Asai (eds.), *Astrophysical and Geophysical Convection, Proc. Mini-Symp.*, Dept. of Astronomy, Univ. of Tokyo, p. 1.

Unno, W., Kondo, M.-A., and Xiong, D.-R.: 1985, *Publ. Astron. Soc. Japan* **37**, 235.

Van der Borght, R.: 1971, *Publ. Astron. Soc. Japan* **23**, 539.

Vitense, E.: 1953, *Z. Astrophys.* **32**, 135.

Xiong, D.-R.: 1979, *Acta Astron. Sinica* **20**, 238.

Xiong, D.-R.: 1981, *Sci. Sinica* **10**, 1406.

Yoshizawa, A.: 1982, *Phys. Fluids* **25**(9), 1532.

Yoshizawa, A.: 1983, *J. Phys. Soc. Japan* **52**, 1194.

TURBULENCE, CONVECTION, AND MIXING IN RED GIANT STARS: SOME EMPIRICAL APPROACHES BASED ON HIGH RESOLUTION SPECTROSCOPY*

TAKASHI TSUJI

Tokyo Astronomical Observatory, University of Tokyo, Mitaka, Tokyo, Japan

(Received 13 June, 1985)

Abstract. The differential velocity field in cool stars can be measured effectively on photographic plates by the use of a PDS micro-densitometer, from which it is shown that radial velocity gradients are larger for stars with larger turbulent velocities, which are determined from the high resolution echellograms. This indicates that the stellar turbulence may have something to do with the differential velocity field in stellar atmospheres. As an observational probe of the mixing processes in red giant stars, stellar abundances that are sensitive to mixing are determined on the basis of high resolution Fourier Transform spectroscopy. For example, the $^{12}C/^{13}C$ ratio has now been determined for large number of red giant stars and shows characteristic changes through the first red giant branch to the asymptotic giant branch, including both the pre-thermal pulsing and the thermal pulsing phases. This information, together with additional information on the $^{16}O/^{17}O$ ratio and on CNO abundances, provides useful constraints on the theory of mixing in red giant stars.

1. Introduction

At present, it is still difficult to have a unified picture on dynamical phenomena such as turbulence, convection, and mixing in stellar atmospheres. For improving the situation, more systematic efforts with an empirical approach should be necessaary, especially for very cool stars for which little work has yet been done. For this purpose, high resolution spectroscopy should be a promising tool and, for effectively applying such a tool to cool stars, we have been trying to prepare the physical basis for the quantitative analysis of the spectra of cool stars. This attempt shows reasonable success, as has been discussed in detail at the First Regional Meeting in Wellington (Tsuji, 1979) and we hope to show some examples of actual applications in the present paper.

2. Turbulence and Convection

We have used the notion of stellar turbulence since Struve and Elvey (1934) introduced it just fifty years ago, and we now understand that it is a very important parameter in abundance determinations, for example. However, the true nature of stellar turbulence is still not clear. For example, Nordlund (1980), from numerical simulation of the solar granular convection, has suggested that the velocity gradient of the granular velocity field

* Paper presented at the IAU Third Asian-Pacific Regional Meeting, held in Kyoto, Japan, between 30 September–6 October, 1984.

Astrophysics and Space Science **118** (1986) 227–235.

along the line-of-sight produces micro-turbulence, while spatial variation of the velocity field produces macro-turbulence. A more classical idea that the acoustic waves generated in the convective zone may be the origin of stellar turbulence has been rediscussed by Edmonds (1978). Also, Smith (1980) has suggested that non-radial oscillations may produce the effect observed as macro-turbulence.

Among these, the first possibility seems to be interesting because it showed fine success in reproducing the observed line shapes and shifts in the solar spectrum (Dravins *et al.*, 1981) and, furthermore, there is some observational evidence that a more or less similar model can be applied to red giant stars (Ridgway and Friel, 1981; Gray, 1982a). For this reason, we think it useful to examine whether any differential velocity field could be observed in cool stars other than the Sun. We have started a project to measure accurate radial velocities in cool stars by the use of the PDS micro-

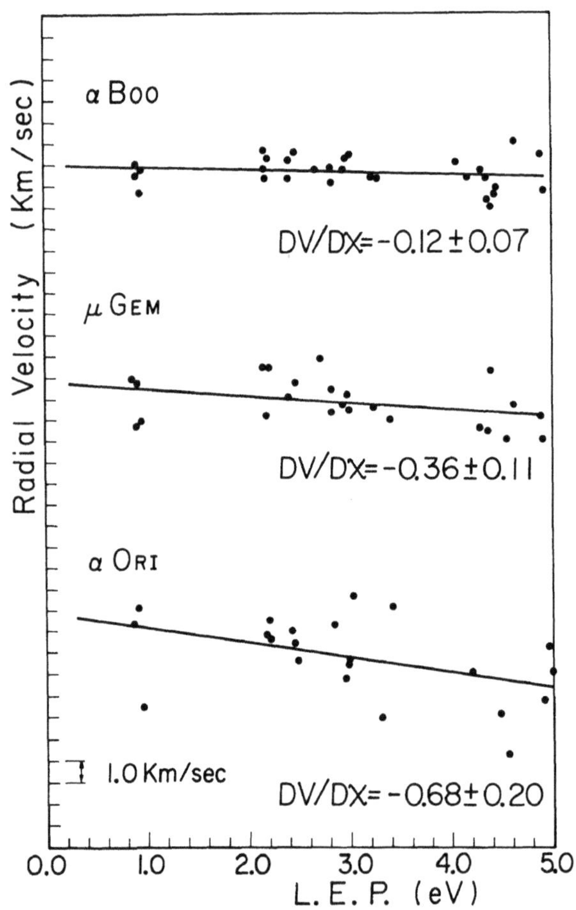

Fig. 1. Radial velocities of iron lines are plotted against lower excitation potentials (in eV) of the lines measured. Measurements are done by the PDS micro-densitometer on spectrograms of 8 Å mm^{-1} ($\lambda\lambda$ 8000–8800 Å), observed by the $F/10$ camera of the coudé spectrograph of the Okayama Astrophysical Observatory. Note that the radial velocity gradient (given in unit of km s^{-1} eV^{-1}) increases in the order of α Boo (K2IIIp), μ Gem (M3III), and α Ori (M2I$_{ab}$), as shown in the top, middle, and bottom, respectively. Note that radial velocities are measured relative to the Sun.

TABLE I

Radial velocity gradients and turbulent velocities in cool stars

Star	log g	M_{bol}	$dV/d\chi$ [a] (km s^{-1} eV^{-1})	V_{micro} (km s^{-1})
The Sun	4.4	+4.7	0.0	0.9 ± 0.1(1)
α Boo	1.5	−0.7	−0.12 ± 0.07	1.8 ± 0.2(2)
μ Gem	0.5	−3.0	−0.36 + 0.11	4.0 + 0.5
α Ori	0.0	−7.5	−0.68 ± 0.20	7.0 ± 0.5

[a] Radial velocity gradient relative to the Sun.
(1) Simmons and Blackwell (1982).
(2) Gray (1980).

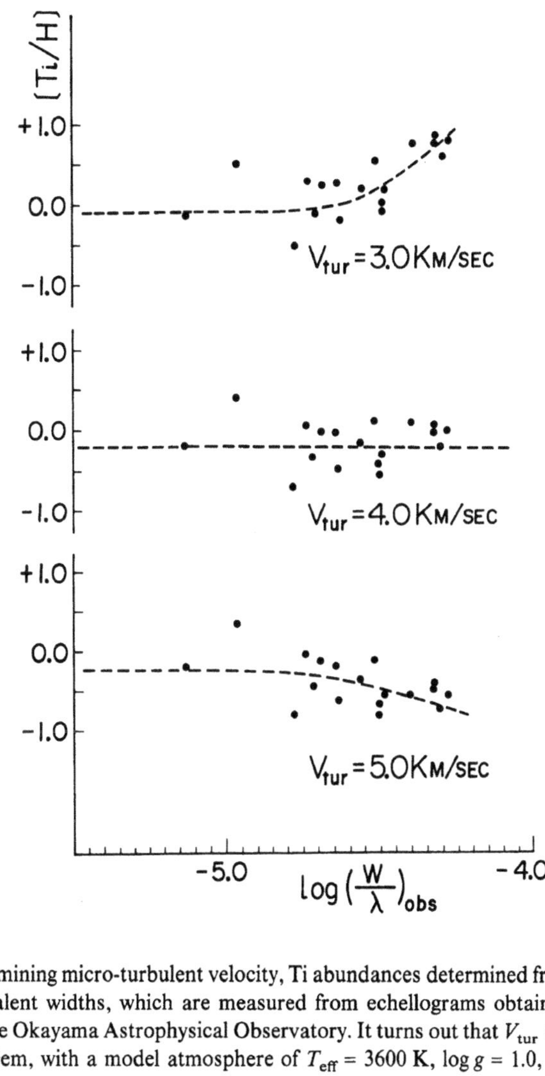

Fig. 2. For determining micro-turbulent velocity, Ti abundances determined from various lines are plotted against the equivalent widths, which are measured from echellograms obtained with the Coudé echelle spectrograph of the Okayama Astrophysical Observatory. It turns out that V_{tur} = 4 km s^{-1} gives consistent result for μ Gem, with a model atmosphere of T_{eff} = 3600 K, log g = 1.0, and V_{tur} = 3.0 km s^{-1}.

densitometer, which was introduced recently to our Observatory (Noguchi *et al.*, 1981). In fact, the measurement of radial velocities for a large number of spectral lines on the highly crowded spectra of very cool stars was first made possible by the automatic and efficient data processing of the PDS micro-densitometer. With some precautions and simple computer algorithms, we can measure radial velocities with an accuracy better than 0.3 km s^{-1} on Coudé plates of 8 Å mm^{-1} in the near infrared, for example. The accuracy of the laboratory wavelengths is a more difficult problem and, as the wavelength standard, we have used the observed solar wavelengths (Pierce and Breckinridge, 1973), which themselves show some differential velocity shifts (Dravins *et al.*, 1981).

The measured radial velocities of iron lines, relative to the Sun, are plotted against the lower excitation potentials of the measured lines in Figure 1. It is found that the radial velocity gradient of α Boo (K2III) shows little difference from that of the Sun (top off Figure 1), in agreement with the result by Dravins (1975). The more luminous and cooler giant star μ Gem (M3III) shows a differential radial velocity field somewhat larger than in the Sun (middle of Figure 1), in the sense that higher excitation lines have more negative velocities. The result for the supergiant star α Ori ($M2I_{ab}$) shows a considerable differential velocity field (bottom of Figure 1). Thus, it turns out that there are noticiable differences in the differential velocity fields among cool stars.

Recently, turbulent velocities, for a large number of G- and K-type stars, have been determined from the analysis of line profiles (e.g., Gray, 1982b). Unfortunately, to extend a similar analysis to cooler stars becomes increasingly difficult because of increasing line blending problems, and of the associated difficulty to measure accurate line profiles. This difficulty might partly be overcome by use of a very high resolution. For this purpose, we have used the Coudé echelle spectrograph, which had been installed as early as 1964 at the Okayama Astrophysical Observatory (Osawa, 1965). We are using photographic plates with this Echelle, and its advantage over modern electronic detectors is to be able to cover a wide spectral range by a single exposure (e.g., $\lambda\lambda$ 6800–8900 Å at 2 Å mm^{-1} by the use of Kodak 4N emulsion). Now, equivalent widths can be measured with reasonable accuracy, but to measure accurate line profiles is still difficult in very cool stars. Accordingly, the micro-turbulent velocity is determined so that a consistent interpretation can be given to equivalent widths of weak as well as of saturated lines. One pre-requisite in such an analysis is to assign an appropriate model atmosphere with realistic stellar parameters for such a star to be analyzed, and this problem has fully been discussed previously (Tsuji, 1979; 1981). For μ Gem, we have $T_{\mathrm{eff}} = 3600 \pm 100$ K and, with *gf*-values calibrated by the solar spectrum, the Ti abundance determined from lines of various equivalent widths gives a consistent result for $V_{\mathrm{micro}} = 4$ km s^{-1} (Figure 2). A similar analysis for α Ori ($T_{\mathrm{eff}} = 3900 \pm 150$K) gives $V_{\mathrm{micro}} = 7$ km s^{-1}.

We summarize our results in Table 1. It appears that the micro-turbulent velocity is larger for stars with larger differential radial velocity shifts. Thus, it seems that stellar turbulence may have something to do with the differential velocity field, and we hope to extend similar analysis to a larger sample of cool stars to confirm this result.

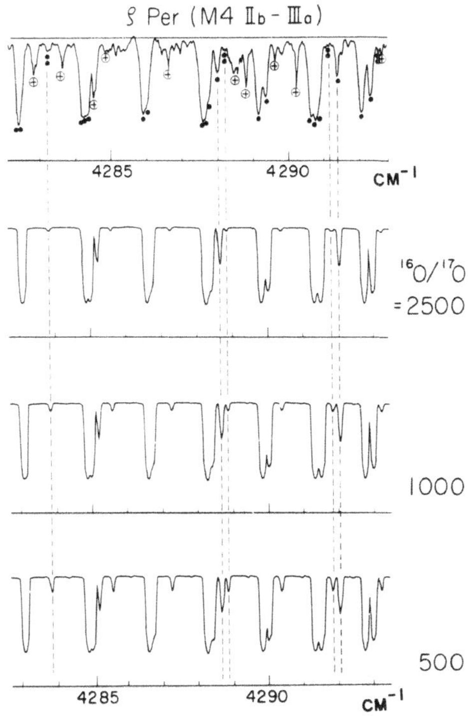

Fig. 3. A portion of the IR spectrum of ρ Per (M4III) observed by KPNO FTS is shown at the top. Dot, double dots, and plus sign indicate $^{12}C^{16}O$, $^{12}C^{17}O$, and atmospheric lines, respectively. Predicted spectra based on model atmosphere of $T_{eff} = 3500$ K, $\log g = 0.5$, and $V_{tur} = 3.0$ km s^{-1} are shown below for $^{16}O/^{17}O = 2500$, 1000, and 500. Inspection of features indicated by dashed lines suggestss that $^{16}O/^{17}O = 2500$ gives the best fit to observed spectrum.

3. Mixing

Mixing in cool stars can mostly be inferred from some abundance anomalies observed over the stellar surface, and recent studies on the CNO abundances in red giants of globular clusters revealed that mixings in these stars are far more complicated than those predicted by the standard evolutional models (e.g., Freeman and Norris, 1981). Also, recent progress in high-resolution Fourier Transform spectroscopy (e.g., Connes and Michel, 1974; Hall *et al.*, 1979) has opened a new possibility of determining C, N, O, and their isotope abundances at high precision through the analysis of molecular spectra, such as those of CO, CN, and OH, in the infrared. To realize such a possibility, the method of detailed analysis based on model atmospheres is being applied to infrared spectra of M-giant stars in the solar neighbourhood. Some results obtained so far are the followings:

(1) On the high resolution infrared spectrum of α Her (M5II, $M_{bol} \sim -5$) observed by P. Connes, we obtained: $[C/H] = -0.7 \pm 0.3$, $[N/H] = +0.8 \pm 0.4$, and $[O/H] = +0.2 \pm 0.4$ where $[M/H] = \log(M/H)_* - \log(M/H)_\odot$, but $(C+N+O)/H$ is solar to within 15%. Also, $^{12}C/^{13}C = 13 \pm 5$, $^{16}O/^{17}O = 190 \pm 80$, and $^{16}O/^{18}O$ may be not much different from the solar system value of 490 (Tsuji, 1985).

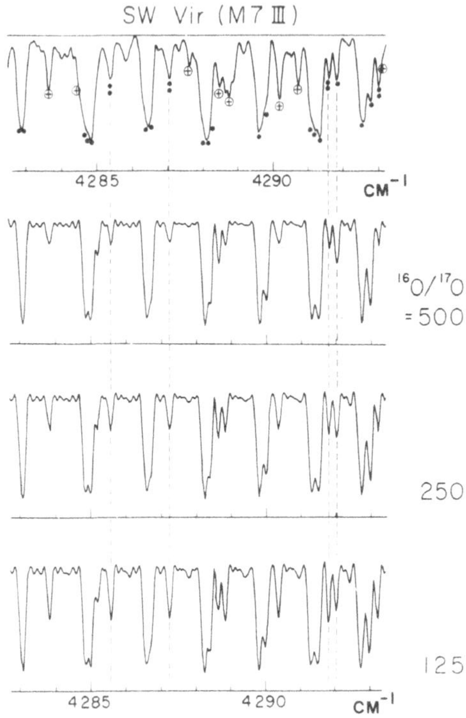

Fig. 4. A portion of the IR spectrum of SW Vir (M7III) is shown at the top. Predicted spectra based on model atmosphere of $T_{\rm eff} = 3100$ K, $\log g = 0.0$, and $V_{\rm tur} = 3.0$ km s^{-1} are shown for ^{16}O/^{17}O = 500, 250, and 125. The best fit to the observed spectrum is obtained for ^{16}O/^{17}O = 250. In comparing weak ^{12}C^{17}O lines (double dots) indicated by dashed lines with ^{12}C^{16}O lines, attention should be given to less saturated ^{12}C^{16}O line (single dot), also indicated by dashed line.

(2) From the analysis of feature due to ^{12}C^{16}O, ^{12}C^{17}O, and ^{13}C^{16}O in the infrared spectra of twelve M-giant stars (M1III–M8III, $M_{\rm bol} = -2 \sim -6$) observed by KPNO FTS, we found that ^{12}C/^{13}C ratio is below 10 in all the M-giant stars surveyed (results are shown in the middle panel of Figure 5). On the other hand, there is no evidence that the ^{16}O/^{17}O ratio is much different from the solar system value of 2700 in M-giants less luminous than $M_{\rm bol} \sim -5$; ρ Per (M4III) is shown as an example in Figure 3 and similar results are obtained for ν Vir, α Cet, β Peg, δ Vir, δ^2 Lyr, and RR UMi. Interestingly, however, ^{16}O/^{17}O ratio shows a sudden decrease to 200–1000 in M-giant stars more luminous than $M_{\rm bol} \sim -5$; for example, the ^{16}O/^{17}O ratio is estimated to be 250 for SW Vir (M7III) shown in Figure 4. Similar analyses revealed ^{16}O/^{17}O = 200, 500, 1000, and 500 for α Her, g Her, RZ Ari, and RX Boo, respectively.

These results indicate that the products of the CNO tri-cycles have been 'dredged up' in M-giant stars of high luminosity. However, the actual process of the mixing should still be identified. For this purpose, we compare our results on the carbon isotope ratio for M-giant stars with those for other red giant stars in Figure 5.

The frequency distribution of the ^{12}C/^{13}C ratio for G–K giant stars (Tomkin *et al.*, 1975, 1976; Dearborn *et al.*, 1975) is shown at the top panel of Figure 5. Generally, the rather low $^{12}C/^{13}C$ ratio in G–K giant stars, which are ascending the red giant branch for the first time, can be understood as the result of the so-called first dredge up, which mixes the products of the CN cycle at the Main-Sequence phase by extended convection at the red giant phase. However, if the initial $^{12}C/^{13}C$ ratio at the main sequence phase is as high as 90, as in the solar system, the first dredge up theory can not explain the $^{12}C/^{13}C$ ratio being below 20 in red giant stars (e.g., Dearborn *et al.*, 1976).

This difficulty of the first dredge up theory is more serious in M-giant stars, all of which show very low values of the $^{12}C/^{13}C$ ratio as noted above (middle panel of Figure 5). This fact, together with the sudden increase of ^{17}O in M-giant stars at $M_{bol} \sim -5$, suggests that the product of the new nucleo-processing at the red giant phase should be dredged up in the M-giant phase. Although a second dredge up has been predicted in high mass stars with masses larger than 4–5 $M_\odot$ (Beckers and Iben, 1979), there are some evidences that most red giant stars are low mass stars whose

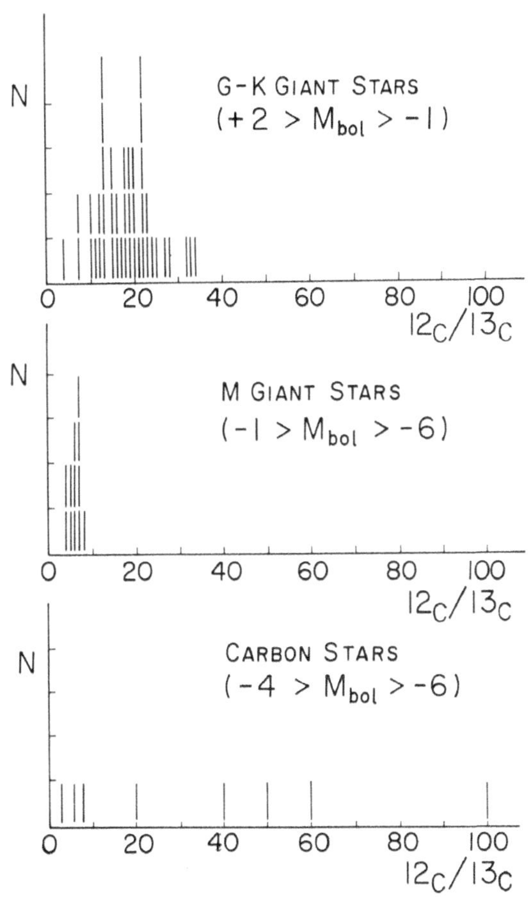

Fig. 5. The frequency distributions of $^{12}C/^{13}C$ (one bar represents one star) are shown for red giant stars in the first red giant branch (G–K giants; top panel), in the asymptotic red giant branch at pre-thermal pulsing phase (M-giants; middle panel), and in the asymptotic giant branch at thermal pulsing phase (C stars; bottom panel).

masses are near one solar mass (e.g., Scalo *et al.*, 1978). Moreover, the second dredge-up itself does not help us to understand the low $^{12}C/^{13}C$ ratio in M-giant stars. Thus, the mixing process in M-giant stars remains mostly unexplained.

Finally, the $^{12}C/^{13}C$ ratio in carbon stars shows a rather large dispersion, as was shown by Fujita and Tsuji (1977) some time ago (bottom panel of Figure 5). If carbon stars have evolved from M-giant stars with very low $^{12}C/^{13}C$ ratio (as shown in the middle panel of Figure 5) by the dredge up of the fresh ^{12}C produced in He-shell burning (e.g., Iben and Renzini, 1983), it is relatively easy to understood the rather high value of the $^{12}C/^{13}C$ ratio in some carbon stars. Thus, the so-called third dredge up theory to explain the presence of carbon stars can be consistent with the observed carbon isotope ratio. However, the presence of some carbon stars with quite low value of $^{12}C/^{13}C$ remains to be explained.

In conclusion, the evolution of isotope abundances in red giant stars from the first red giant branch (G–K giants) to the asymptotic giant branch in the pre-thermal pulsing phase (M-giants), and further to the asymptotic giant branch in the thermal pulsing phase (carbon stars) is qualitatively consistent with predictions based on the current theory of stellar evolution; but is still not fully understood on a quantitative basis.

4. Conclusions

In this paper, we showed some fragmentary information on turbulence, convection and mixing in red giant stars from observations, but we are still far from a unified understanding of these phenomena. For this purpose high resolution spectroscopy has played and will play an important role. For example, our preliminary study based on classical high resolution spectroscopy revealed rather large differential velocity fields that are correlated with micro-turbulent velocities in cool giant and supergiant stars. Furthermore, the Fourier Transform spectroscopy offers a unique opportunity to study some molecular lines that are observed only in the infrared, and we have shown that even the isotope abundance analysis alone could provide useful constraints on mixing processes in red giant stars. The information that could be recovered from high resolution infrared spectra obtained by the FTS method should still be larger if these spectra are properly analyzed, and further efforts in this direction should certainly be rewarding in promoting our understanding on physical processes and chemical evolution in red giant stars.

Acknowledgements

I am most grateful to S. T. Ridgway and K. H. Hinkle for kind advice in obtaining FTS spectra, and for invaluable discussions on these spectra. My thanks are also due to G. Ladd for efficient help in data handling and transfer. Some of the FTS spectra were observed when I was visiting astronomer at Kitt Peak National Observatory, which is operated by the Association of Universities for Research in Astronomy, Inc. under contract with the National Science Foundation.

I would like to thank T. Noguchi for helpful discussions on the PDS micro-densitometer, and the staff of the Okayama Astrophysical Observatory for helpful support in coude observations.

This work was supported in part by the Grant-in-Aid for Scientific Research No. 542002 of the Ministry of Education, Sciences and Culture.

References

Beckers, S. A. and Iben, I., Jr.: 1979, *Astrophys. J.* **232**, 831.
Connes, P. and Michel, G.: 1974, *Astrophys. J.* **190**, L29.
Dearborn, D. S. P., Lambert, D. L., and Tomkin, J.: 1975, *Astrophys. J.* **200**, 675.
Dearborn, D. S. P. Eggleton, P. P., and Schramm, D. N.: 1976, *Astrophys. J.* **203**, 455.
Dravins, D.: 1975, *Astron. Astrophys.* **36**, 143.
Dravins, D., Lindgren, L., and Nordlund, A.: 1981, *Astron. Astrophys.* **96**, 345.
Edmonds, M. G.: 1978, *Astron. Astrophys.* **64**, 103.
Freeman, K. C. and Norris, J.: 1981, *Ann. Rev. Astron. Astrophys.* **19**, 319.
Fujita, Y. and Tsuji, T.: 1977, *Publ. Astron. Soc. Japan* **29**, 711.
Gray, D. F.: 1981, *Astrophys. J.* **245**, 992.
Gray, D. F.: 1982a, *Astrophys. J.* **255**, 200.
Gray, D. F.: 1982b, *Astrophys. J.* **262**, 682.
Hall, D. N. B., Ridgway, S. T., Bell, E. A., and Yarborough, J. M.: 1979, *Proc. Soc. Photo-Opt. Instrum. Eng.* **172**, 121.
Iben, I., Jr. and Renzini, A.: 1983, *Ann. Rev. Astron. Astrophys.* **21**, 271.
Noguchi, T., Miyauchi, N., and Tsuji, T.: 1981, *Tokyo Astron. Obs. Report* **19**, 452 (in Japanese).
Nordlund, A.: 1980, in D. F. Gray and J. L. Linsky (eds.), 'Stellar Turbulence', *IAU Colloq.* **51**, 213.
Osawa, K.: 1965, *Astron. Herald* **58**, 28 (in Japanese).
Pierce, A. K. and Breckinridge, J. B.: 1973, *Kitt Peak National Observatory Contr.*, No. 559.
Ridgway, S. T. and Friel, E. D.: 1981, in C. Chiosi and R. Stalio (eds.), *Effects of Mass Loss on Stellar Evolution*, D. Reidel, Publ. Co., Dordrecht, Holland, p. 119.
Scalo, J. M., Dominy, J. F., and Pumphrey, W. A.: 1978, *Astrophys. J.* **221**, 616.
Simmons, G. J. and Blackwell, D. E.: 1982, *Astron. Astrophys.* **112**, 209.
Smith, M. A.: 1980, *Astrophys. J.* **242**, L115.
Struve, O. and Elvey, C. T.: 1934, *Astrophys. J.* **79**, 409.
Tomkin, J., Lambert, D. L., and Luck, R. E.: 1975, *Astrophys. J.* **199**, 436.
Tomkin, J., Luck, R. E., and Lambert, D. L.: 1976, *Astrophys. J.* **210**, 694.
Tsuji, T.: 1979, *New Zealand J. Sci.* **22**, 415, (Proc. First Asian-South Pacific Regional Meeting of IAU).
Tsuji, T.: 1981, *Astron. Astrophys.* **99**, 48.
Tsuji, T.: 1985, in M. Jaschek and P. C. Keenan (eds.), *Cool Stars with Excesses of Heavy Elements*, D. Reidel, Publ. Co., Dordrecht, Holland, p. 295.

HIGH SENSITIVITY SiO MASER SURVEY FOR MIRA VARIABLES*

SE-HYUNG CHO, NORIO KAIFU, NOBUHARU UKITA, MASAKI MORIMOTO, and MASAHIKO HAYASHI

Nobeyama Radio Observatory, Minamimaki, Nagano, Japan

(Received 13 June, 1985)

Abstract. Simultaneous observations of the six transitions of SiO for 106 late-type stars were made. The SiO maser emission was detected in 83 stars. The $v = 3$ maser emission was detected in eight stars, and the ^{29}SiO $v = 0$ emission in six stars. The ^{29}SiO $v = 0$ emission is stronger and narrower than that of ^{28}SiO, suggesting that the ^{29}SiO emission is masing.

1. Introduction

SiO masers are useful probes of the stellar environment because of many observable lines from various transitions. But their variation of intensity and velocity structure requires simultaneous and time-monitored observations.

In the spring of 1983, high-sensitivity and simultaneous observations of six SiO lines (^{28}SiO, $v = 0, 1, 2, 3, J = 1$–0, ^{29}SiO, $v = 0, 1, J = 1$–0) were performed with the 45-m telescope of the Nobeyama Radio Observatory. Spectra were taken with the Acousto-Optical Spectrometer, with spectral resolutions of 37 kHz (high frequency resolution) and 260 kHz (wide band resolution). Sensitivity was about 2 Jy in the case of high resolution.

TABLE I

Line	Detection	New detection
^{28}SiO ($J = 1$–0)		
$v = 0$	11	10
$v = 1$	81	24
$v = 2$	72	31
$v = 3$	12	8
^{29}SiO ($J = 1$–0)		
$v = 0$	6	5
$v = 1$	0	0

* Paper presented at the IAU Third Asian-Pacific Regional Meeting, held in Kyoto, Japan, between 30 September–6 October, 1984.

Astrophysics and Space Science **118** (1986) 237–239.

2. Observational Results

Of the 106 surveyed sources (mainly Mira variables which are bright at 4 micron), 83 sources were detected. Among them 26 were identified as new detections; 24 were M-type Mira variables, one was an S-type Mira, and one was a supergiant (i.e., W Aql, MY Cep). The results for each line are listed in Table I. The spectrum of WX Psc is presented in Figure 1 as an example of simultaneously obtained line profiles. The main results are summarized as followings.

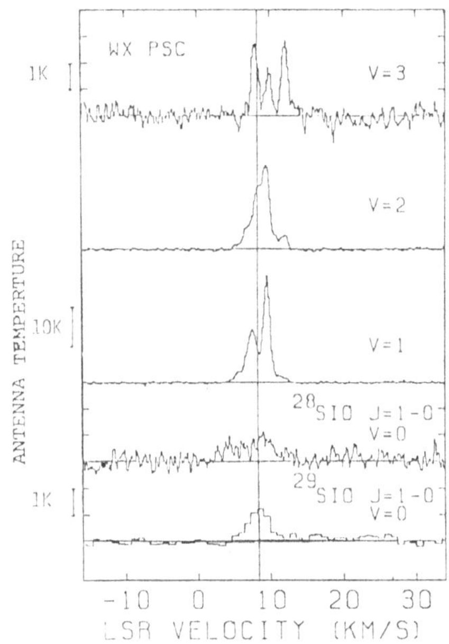

Fig. 1. Simultaneously obtained spectra of WX Psc. The vertical line indicates the stellar velocity determined by OH 1612 MHz (Silverglate *et al.*, 1979).

(1) The $v = 3$ emission; twelve stars (VY CMa, TX Cam, R Leo, W Hya, U Her, RU Her, WX Psc, Z Cas, RR Per, IRC − 20424, IRC + 10322, IRC − 30308) were detected in the third vibrational state of SiO. The latter eight stars were new detections. The optical phases of all $v = 3$ detected stars were near 0.2.

(2) Masering of the ^{29}SiO ($v = 0$, $J = 1$–0) isotope; six stars (VY CMa, VX Sgr, R Leo, V1111 Oph, WX Psc, W Hya) were detected. Figure 2 shows one of these line profiles. All of six ^{29} SiO profiles are narrower than ^{28}SiO ($v = 0, J = 1$–0) profiles, and also stronger in spite of the lesser abundance of (^{28}SiO)/(^{29}SiO) = 20 ± 5 (in VY CMa, measured by Geballe *et al.* (1979). This is some evidence of masering of ^{29}SiO.

(3) Weak masering of ^{28}SiO ($v = 0$, $J = 1$–0); eleven stars were detected. Their emissions were distinct from thermal lines of late-type stars because of the narrowness of line profiles, and the difficulty in fitting with optically thick parabolic profiles.

We also investigated the correlations between the SiO observed quantities and infrared and optical data. Meaningful results about IR pumping and $v = 3$ exitation were acquired.

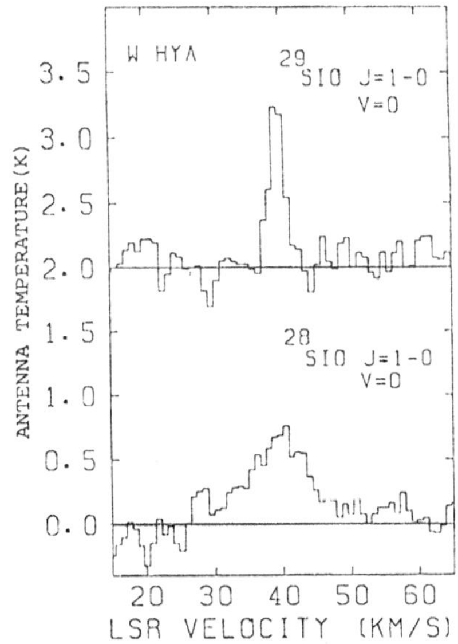

Fig. 2. Spectra of W Hya.

References

Geballe, T. R., Lacy, J. H., and Beck, S. C.: 1979, *Astrophys. J.* **230**, L47.
Silverglate, P., Zuckerman, B., Terzian, Y., Wolff, M.: 1979, *Astron. J.* **84**, 345.

CLASSICAL ALGOL SYSTEMS (INVITED REVIEW)*

EDWIN BUDDING

Carter Observatory, Wellington, New Zealand

(Received 8 March, 1985)

Abstract. Progress in the study of Algol-type binaries is looked at both in the light of more general efforts at compilation and categorization, and also some specific new and important observational results.

Results can be referred to the well known underlying model of the semi-detached configuration and Roche lobe overflow mechanism. Successes and limitations of this model will be considered.

Arising from such comparisons lines of future research are suggested. Some outstanding issues concern: (i) the evolutionary condition of the mass accreting component, (ii) instabilities and related effects in various regions of the flow, (iii) systematic outflow, (iv) the relationship of classical Algols to other evolved binaries.

1. The Classical Algol Binary Configuration

1.1. Generalities

There have been several recent reviews either devoted to, or including sections on, Algol systems, from which the author has been glad to draw ideas. These include a number of papers from the Trieste group – notably Giuricin and Mardirossian (1981), Giuricin *et al.* (1983), – Batten (1981), Olson (1983), Plavec (1982, 1983a, b, c), and more general articles such as those of Thomas (1977), Shu and Lubow (1981), Vilhu (1981), Trimble (1983), Tutukov (1982, 1983) or De Loore (1984). In that well-known compendium on close binary studies – the Proceedings of IAU Symposium No. 88 (Plavec *et al.*, 1980) – there were no fewer than 20 papers dealing specifically with classical Algol systems. In two more recent IAU Colloquia on binaries (Kopal and Rahe, 1982; or Hidayat *et al.*, 1984) Algols received special attention in a dozen or so articles. Clearly the field is alive and stimulating a good amount of astrophysical work.

Some salient facts emerge directly: classical Algols probably represent the most frequently encountered type of that informative class of variable star – the eclipsing binary system. In the catalogue of 1024 eclipsing binaries with derived parameter sets listed by Brancewicz and Dworak (1980), there were some 781 examples having 'Algol-type light curves' – a rather wider classification than that of the Algols proper, which make up somewhat more than half of this number (Budding, 1984c). Algols are distinguished by their evolved secondaries, which fill some critical surface – further expansion beyond which would entail dynamical instability of some kind. This surface has been customarily termed the Roche lobe (a term introduced into astronomical literature by Kopal in 1955); the physical configuration referred to is also associated with the description 'semi-detached' (sd) (cf. Kopal, 1955b, 1956, 1959). Algols

* Paper presented at the IAU Third Asian-Pacific Regional Meeting, held in Kyoto, Japan, between 30 September–6 October, 1984.

Astrophysics and Space Science **118** (1986) 241–255.

represent perhaps the most basic type of binary, in which the circumstance of binary nature itself is considered to have introduced a fundamental qualitative difference to the evolution of at least one of the components.

The star eclipsed at primary minimum (which, by its depth of very frequently greater than 1 mag. at visual wavelengths, is often a clue to the physical configuration) is almost invariably the more massive component, and also usually sensibly similar to a Main-Sequence star. This arrangement of the classical Algol system – the more massive, bright, earlier than mid-F type, unevolved primary accompanied by a relatively low mass, but evolved, frequently later than mid-F type secondary – at once produces the well known Algol evolution paradox, a 'prime example', according to Batten (1981), of how a superficially contradictory piece of evidence can provide a turning point in scientific thinking.

Since the time when Struve (1948) first emphasized this 'amazing discrepancy', the general view of its explanation has become sufficiently disseminated for the vocabulary of mass loss, mass transfer, Roche lobe oveflow, gaseous streams, accretion disks, conservative and non-conservative regimes to achieve widespread currency. The recognition of such a vocabulary should not, of course, imply an uncritical acceptance of the rigorous validity of any particular model, any more than a recognition of observational equipment should imply an uncritical acceptance of any particular data it may produce. This review will attempt to assess possible successes and limitations of such models in the light of newer, as well as more established, observational results.

Hall (1974, 1975b) and Hall and Neff (1979) made a number of useful critical updatings of our knowledge of a good many of the classical Algols. In particular, it does appear likely that the Algol classification is fairly stringently defined by the semi-detached qualification. The possibility of some generally similar but rather distinct subgroupings, such as the 'undersize' systems, or Algols with peculiar low mass primaries, would now appear to be very largely ruled out (Budding, 1984b).

1.2. New areas of development

In connection with subgroupings, however, Plavec (1980) drew attention to a type of semi-detached system with some peculiarities different from the great majority in the 'classical' arrangement previously described. A distinguishing feature of these systems could be seen as the presence of a 'thick disk' (see also Wilson, 1974; Wilson and Caldwell, 1978), though this in turn seems to entail a wider separation, more massive components and generally more pronounced indications of mass loss and transfer. The prototype is W Ser – hence, the designation serpentids – though perhaps the best known and most studied example is β Lyr, traditionally associated, if only from the shape of its light curve, with a rather different class of eclipsing binary system. Fresh insights into the physics of the mass loss mechanism have been made, particularly with the aid of satellite (or balloon) borne equipment, from observations of these stars. More will be said later about this.

Even with less 'exhibitionist' sources there have been interesting and important advances in observation techniques. The use of solid state spectrometric detection devices has permitted vital developments to our knowledge of basic Algol parameters (Fekel *et al.*, 1980, Popper, 1980). Time resolved vidicon spectroscopy has presented information of a type suggested by many earlier studies since the emission line discoveries of Wyse (1934) and Joy (1942), but giving much more detailed quantitative results (Kaitchuck and Honeycutt, 1982). Narrow or intermediate band photometry techniques, applied convincingly to Algol itself by Guinan *et al.* (1976), has proven very informative for the more active systems, particularly U Cep (Olson, 1976; 1982; McNamara and Feltz, 1976; Crawford, 1979). The advantages of a long-term systematic survey, by means of a reliably defined photometric system, of such stars is well shown in Olson's (1983) recent review.

U Cep also presented itself as an interesting polarimetric object in the work of Piirola (1980, 1981), whose discovered pattern of polarization variation through eclipses seems to anticipate the remarkable pointing out of the Chandrasekhar effect by Kemp *et al.* (1983) from their intensive polarization studies of β Per.

Algol itself, as the brightest example of ts class, has frequently been a target for observational innovations, noticeably, over the last two decades, in regard to an increasing range of wavelength coverage; as also with increasing resolution both spectral and spatial. Some of these developments are summarized in Table I. (Söderhjelm (1980) presented a comprehensive review of β Per.) The increasing provision of such information for various fainter members of the class can be expected.

2. The Roche Lobe Overflow Mechanism

2.1. Background

A tentative qualitative explanation for the Algol paradox in terms of what has become known as the Roche lobe overflow (RLOF) mechanism was given many years ago (Crawford, 1955; Kopal, 1955a). A considerable amount of work has since gone into the further substantiation of this hypothesis. It is possible to identify two major areas of the problem as (i) the dynamical time-scale view, or the hydrodynamic description of an overflow process, and (ii) the evolutionary time-scale view, in which the broad properties of the components are checked against the calculated evolution of models where the mass loss/transfer process, arising from the circumstance of binarity, is somehow taken into account. Paczyński (1971) proposed specific application of the term 'close' to such interacting binary systems. The evolutionary aspect will be considered more fully in the next section.

Evidence about gaseous streams, disks or rings can be traced back to the Barr effect (Barr, 1908) though the subject received a particular impetus from findings such as those of Joy (1942) and Struve (1948). Unambiguous, detailed analysis of such results is difficult, however, (cf. e.g. Batten, 1981). The discussion of Lubow and Shu (1975), apart from its clear formal treatment of the RLOF mechanism, did address such evidence and demonstrate a general consistency. Spectroscopic data were also modelled in the work of Prendergast and Taam (1974), the general agreement of whose results

TABLE I
Some advances in observational coverage of Algol

Year	Authors	Innovation	Result
1966	Chen and Reuning	IR (1.6 μ) light curve	Pronounced secondary minimum; photometric peculiarities
1966	Cristaldi *et al.*	Hα photometry	Evidence of spectal instabilities
1967	Glushneva and Esipov	IR (0.7–1.1 μ) spectroscopy	Transient emission features
1972	Hjellming *et al.*	Radio observations (2695 and 8085 MHz)	Emission with short time variations discovered
1973	Ryle and Elsmore	Radio-interferometry	Accurate location of barycentre
1974	Labeyrie *et al.*	Speckle interferometry	Resolution of Algol C.
1975	Chen and Wood	UV spectroscopy	Detailed scans of Lα at different orbital phases
1975	Eaton	UV light curves	Physical parameter values in UV
1975	Smyth *et al.*	IR (2.2 μ) light curves	Apparent discrepancies with models
1975	Longmore and Jameson	IR (2.2, 3.6, 4.8, 8.6 μ) light curves	IR excess – attributed to gas stream
1975	Bachmann and Hershey	Inclusion of accurate astrometry	More confident element specification
1975	Gibson *et al.*	Radio observations (1400, 2695, 6000, and 8085 MHz)	Large flare ($>$ 1 Jy) detected.
1976	Guinan *et al.*	Hα and Hβ index monitoring	Presence of circumstellar emission region revealed
1976	Schnopper *et al.*	X-ray observation	Emission in 2–6 keV band detected
1978	Rudy and Kemp	Polarimetry	Independent determination of orbital inclination
1978	Tomkin and Lambert	Reticon spectroscopy	Detection of Na II *D* lines of secondary
1978	Nadeau *et al.*	IR (10 μ) data	Anomalous shape of secondary minimum
1979	Bonneau	Speckle interferometry	Revised absolute elements of triple system
1979	White *et al.*	X-ray spectrometry	No eclipses. Active corona of subgiant inferred
1980	Zeilik *et al.*	IR (JHKLM) light curves	IR luminosity ratios
1979	Cugier }	UV Spectrometry	High resolution Mg II line profiles
1981	Chen *et al.* }	(Copernicus)	allow detailed modelling
1981	Kemp *et al.*	Upgraded polarimetry	Perpendicularity of AB and AC orbit planes inferred
1982	Zirin and Liggett	IR coude spectrometry	Deep absorption at λ10830 due to Algol B
1983	Kemp *et al.*	Polarimetry	Discovery of eclipse polarization effect

with the separate analytical treatment of Lubow and Shu gave force to the whole interpretation. Further similar calculations and their possible relevance to observations have been reviewed elsewhere (Budding, 1981). That article also referred to photometric data which have been interpreted along similar lines (cf. also, e.g., Hall and Walter, 1975).

2.2. Specific elements of the hypothesis

Key elements linking the proposals of a good many papers about Algols include the following:

(a) *Loser*

This is the evolved subgiant 'contact' component, a sizeable fraction of whose original mass has been lost. At first the mass loss is rapid $\dot{M} \sim M_*/(\alpha\tau_{KH})$ where $\alpha \sim 1$ on average, though perhaps $\alpha \ll 1$ during proposed episodes of 'dynamical time-scale' mass loss; later a slow phase occurs ($\alpha \gg 1$). This is the condition in which most Algols are actually observed. The existence of the slow phase is linked with the mass ratio (loser/gainer) being rather < 1. Though most losers are relatively 'cool', there appears to be a positive correlation between loser and gainer spectral types (Svechnikov, 1969; Plavec, 1982) so that a few more massive Algols are known with A type secondaries (e.g., u Her, DM Per), and there may be some still earlier (β Lyr?). Their masses, usually less than half those of the primary star, are distributed in a highly centrally condensed structure. Their envelopes develop deep convective layers below the surface during the earlier stages of mass loss, but these may later narrow down appreciably (cf., e.g., Harmanec, 1970). A CN abundance depletion in loser atmospheres was noted quite some time ago (Hall, 1967).

(b) *Gainer*

The present primary has received some fraction of the mass shed by the loser, though direct evidence about this from the gainer itself is not so obvious. Clues do exist for some gainers, however. That in the relatively active system U Cep, for example, is rotating much more rapidly than synchronously (Koch *et al.*, 1965). 'Spin-up' effects, if that is what we are observing, might be more naturally expected in the more active exchangers (Wilson and Twigg, 1980; Wilson *et al.*, 1984).

IUE data on U Cep, as also for U Sge and U CrB, reveals line profile anomalies, notably in the resonance lines of ionized carbon (Plavec, 1983a, b, c). Similar spectroscopic peculiarities have been detected in the optical and IR (Lambert, 1982). We are only at the initial stage of appreciating the full explanation of these results.

(c) *Stream*

Firmer evidence about the details of this much discussed entity seem likely to come from recent advances of observational technique (Bolton and Zubrod, 1980), perhaps especially from studying UV resonance lines (Kondo *et al.*, 1979). Polarimetry too would seem to offer a more direct indication of the variation of the stream's aspect – though there may be some complexity in the interpretation of that (see Section 4). To some extent our knowledge of the stream may also be inferred from the items it links up, i.e., the loser and accretion region (next subsection). Earlier ideas were reviewed in Kopal's (1959, 1978), Batten's (1973), or Sahade and Wood's (1978) books.

A theoretical overflow stream was formally characterized in Lubow and Shu's (1975) article. While their treatment implied a steady stream, they allowed some generalization to different astrophysical cases by having $\dot{M}$ a free parameter. Continuity through the expanding envelope will allow us to set constancy to the mass flux $4\pi r^2 s_r \rho_r$, where the density ρ_r at mean radius r attaches to a layer moving outward at a rate s_r. Such an expansion would normally be considered within the context of a stellar evolution calculation, for which the reduction to spherical geometry is usual, though towards the loser's surface, of course, the foregoing terms might be interpreted by appropriate averages.

A problem of formulation may occur here, however, in that the mass loss rate must depend on the calculated structure, which in turn depends on a specification of outer boundary conditions (OBC). In a quasi-formal way we have, in a steady state, an implicit dependance for the surface density ρ_R

$$\dot{M}\,(M, \text{Structure }(\text{OBC}(\ldots\rho_R\ldots))) \equiv 4\pi R^2 s_R \rho_R .$$

In practice, this problem was circumvented by calculating a sequence of equilibrium models, at suitably chosen time intervals Δt_i, for which (cf. Kippenhahn and Weigert, 1967)

$$R_{*i+1}\,(R_{*i}, M_{i+1}, \Delta t_i) \lesssim R_{\text{Roche } i+1}\,(R_{*i}, \Delta M_i)\,,$$

where $R_{\text{Roche } i} \simeq R_{*i}$,

$$M_{i+1} = M_i - \Delta M_i \quad \text{and} \quad \Delta M_i = 4\pi \int_{R_{*i}(M_i)}^{R_{*i+1}(M_i, \Delta t_i)} r^2 \rho_r \, \mathrm{d}r$$

so that, on average, $\Delta M_i / \Delta t_i \to \dot{M}$.

This treatment implies that slight differences in the surface conditions of the loser from those of unconfined equilibrium stars are of little net effect. (More recently, structure codes have been modified to incorporate implicitly defined OBCs (cf. Savonije, 1978)). We may note here, though, that the interface between the loser and its containing Roche lobe must be, on average, relatively dense.

Lubow and Shu (1975) demonstrate a density in the vicinity of the L_1 aperture, whose area is $\sim \varepsilon^2$ that of the loser's photosphere, of order ε^{-3} in units of $\dot{M}/\Omega\,\mathrm{d}^3$, where ε = velocity of sound/orbital velocity (Ωd). If, as we expect, $\dot{M}$ is determined largely by near-core processes (Refsdal and Weigert, 1970) and is of order $(M_*/\alpha\tau_{\mathrm{KH}})$, this density itself turns out to be, for Algol (data from Murad and Budding, 1984) $\sim 10^{-8}/\alpha$ – say, 10^{-11} g cm^{-3} in the slow phase, i.e., only about three orders of magnitude down on typical reversing layer densities, despite the fact that it is moving forward at close to the velocity of sound. This is about nine orders of magnitude faster than what could be reasonably estimated for a general radial expansion velocity of the surface (Awadalla and Budding, 1982). Hence, we may deduce that the layer of the loser at which organized bulk motion of the gas is comparable to that of the 'surface' layer for an unconfined star with a similar core structure would be at $\sim 10^3 \times$ corresponding density. The layer whose density would compare with that of a normal photosphere must already be immersed in a strong horizontal wind extending some little way beyond the critical surface itself (cf. Shu and Lubow, 1981).

The discussion so far has related to a steady stream. The possibility of instabilities in the outflow has been recognized since Morton's (1960) pioneer study. The overlow in Morton's example was stable on a dynamical time-scale – but a number of authors have pointed to the possibility of the breakdown of this (Paczyński, 1965; Bath, 1969; Webbink, 1977). The power release $\dot{E}$ associated with the loss of a fraction f of loser mass on a dynamical timescale, $\dot{E} \sim fL_* (\tau_{KH}/\tau_D)$ in the face of observational evidence on most Algols, would point against an unrestricted build-up of such an action (cf. Webbink, 1980, and subsequent *discussion* with Shu).

There are some indications of flow irregularities, however. Perhaps the best known case is that of U Cep, the consideration of whose apparent period variations led to a formulation for nonstationary effects in the mass flow (Biermann and Hall, 1973) for Algol systems, essentially similar to that of Smak (1972) for eruptive binaries. Period variation or other effects in a number of Algols have since been considered in this way (Hall and Woolley, 1973; Hall, 1975b; Hall and Stuhlinger, 1976; Hall *et al.*, 1976; Hall and Keel, 1977; Kreiner and Ziołkowski, 1978; Kreiner and Tremko, 1980; Tremko and Kreiner, 1981). However, the physical details are still not very clear. Bath (1975), did present a scheme which might give rise to quasi-periodic mass transfer episodes for stellar models whose parameters might resemble some Algol losers. Some aspects of Biermann and Hall's original picture cannot be reconciled with a closer study of the data on U Cep (Olson *et al.*, 1981), while quite different alternative pictures are also available (Matese and Whitmire, 1983).

(d) *Accretion Region*

More hydrodynamic complexity is associated with this entity, whose basic cause is attributed to retained angular momentum of the infalling material. General properties of accretion discs were reviewed by Pringle (1981). In classical Algols the intervention of the gainer's surface tends to make the accretion region a less conspicuous structure than in those situations where a deep potential well exists for its accumulation, nevertheless, interesting facts are being revealed by more incisive observational methods. The IUE data studied by Peters and Polidan (1983), for example, demonstrates high temperatures ($\sim 10^5$ K), turbulence and abundance anomalies consistent with the accretion of processed material. Basic structural changes can occur in a few orbital cycles. The ratio of primary radius to stable quasi-circular particle orbit radius (Lubow and Shu, 1975) is probably an important parameter here.

A concise review of work based on the SIT-vidicon spectrograph at Indiana and the IIDS system at Kitt Peak on the properties of accretion structures of Algol systems was given by Kaitchuck and Honeycutt (1983). Among the interesting recent discoveries has been that emission line disks definitely have an outer limiting radius at typically $1\frac{1}{2}$ times that of the primary star. The optical data also show significant structural variation over timescales comparable to an orbital period. Observed broadening effects have been associated with turbulence, though questions have been raised about the physical details (Kaitchuck and Honeycutt, 1982; Plavec and Dobiáš, 1983). For some recent theoretical appraisals see also, e.g., Fukue (1982), Cabot (1984). Electron densities in the well known 'disk' of RW Tau were put at 10^{11} cm^{-3}, and these were sensibly the same through at least one side of the structure. Velocity fields of several hundred km s^{-1} are measured, but their distribution is non-Keplerian.

Moving to the generally more massive and widely separated stars of the serpentid subgroup, it is the implied thick disk which is now thought to be responsible for many of the difficulties of interpretation. Plavec (1983b) has given good arguments, based largely on IUE data, why β Lyr should be included as an extreme example of the Algol configuration. It is the thick disk which, all along, covered over Struve's (1941) puzzling 'invisible star'. In V 356 Sgr the picture is similar, though the disk is not quite as relatively substantial (Wilson and Caldwell, 1978). If there is a continuum of types from Algol to β Lyr, as argued by Plavec (1983b), then surely ideas about disks will be better developed at the serpentid end of the range.

A special feature of the accretion region is the 'hot spot' or impact location of the stream. The importance of this can be seen in Peters and Polidan's (1983) interpretation that shocks associated with the energy release here are the primary source of heating and turbulence in the accretion region. Olson's (1980) account of what happens in this location results in a qualitatively somewhat different picture of the accretion region, at least for U Cep. He interprets an 'equatorial bulge' as a product of the stream's interaction with the gainer. Also Olson (1983) separates various elements of the mass transfer schemes, which are usually supposed linked, such as the behaviour of the loser, the mass flow, period variations, suggesting alternative possibilities from the observational evidence. Still other possibilities arise in Hall and Walter's (1974) discussion of photometry of U Cep.

(e) *Circumbinary Cloud*

Again serpentids have provided more convincing evidence about this somewhat elusive item. Batten (1973) already referred to υ Sgr, V 367 Cyg and β Lyr. Satellite, or balloon based studies such as those of Hack *et al.* (1983, 1984), Plavec (1983a, b, c), McCluskey and Kondo, 1983; Kondo *et al.* (1984) have proved very revealing.

It has long been recognized that the gainer might have difficulty in accepting at least some of the loser's output (Benson, 1970); particularly, perhaps, during the early 'rapid' stages of the transfer, which could involve a significant fraction of the total mass loss of the loser. Plavec (1983b) argues that the serpentids are nearer to this rapid phase.

The models given by Hack *et al.* (1980, 1984) for the three above mentioned stars suggest a circumbinary 'coronal' structure of radius $\sim 200\ R_\odot$ with a typical density 10^{-13} g cm^{-3} though concentrating towards the central stars and orbital plane. Matter streams out through this cloud at a rate of ~ 100 km s^{-1}, as judged by P Cyg profile comparisons; which in turn leads to a systemic mass loss rate of order $10^{-6}\ M_\odot$ yr^{-1}. This begins to be comparable to expected loser mass loss rates. Plavec (1983c) arrives at a similar conclusion from more detailed attention to β Lyr itself, judging it quite possible, in fact, that the systemic mass loss rate exceeds the accretion rate, though the accuracies of discussed numbers points up the still very preliminary nature of these findings on what are probably, in reality, complex processes (Sahade, 1980; Etzel and Meyer, 1983).

For the slower stage and generally less massive classical Algols, some indirect evidence would indicate the systemic mass loss is relatively less (Giuricin and Mardirossian, 1981; Budding, 1984a). A small proportion of matter can carry a relatively large proportion of angular momentum to great distances with an appreciable proportion of the 'cloud' ultimately coalescing towards the more condensed accretion structure closer to the gainer (cf. Lynden-Bell and Pringle, 1974). Also it should be kept in mind that known mass loss processes of stars in general of type earlier than mid B (Snow and Morton, 1976) suggest that there may be some essential difference between interaction processes in binaries containing such early type components.

3. Evolutionary Aspects

3.1. Outline

Though, of course, the long term effects in Algol systems must result from the successive developments of short term stages, the complexity implied by a completely integrated treatment has caused discussions of the evolution to take a summary approach to hydrodynamics. This concerns to what extent conservation integrals can be utlized, for example, or how to appropriately characterize, in a general way, non-conservative effects.

The pioneer investigations of the evolution of interacting binaries generally assumed conservation of mass and angular momentum of the system. Counted as successes of these studies were the predicted overluminosities of the losers, (Popper, 1973), and, to some extent, the endurance of the low mass ratio, semi-detached configuration to allow a conspicuous incidence of Algol type binaries (Giannone and Giannuzzi, 1970). A route to the appearance of low mass degenerate dwarfs in binaries could also be sought within the framework of the same theory (Refsdal and Weigert, 1971). The scale of observed period variation did not quite fit the predictions of conservative systems, (Kreiner and Ziołkowski, 1978), as neither did absolute parameter values for calculated and observed configurations (Plavec, 1973). These discrepancies might be accounted for, while not necessarily changing the essential character of the results in relation to the aforementioned strong points of the theory, by permitting some non-conservative generalization (Paczyński and Ziołkowski, 1967; Yungelson, 1972); but a clearer view of just how this works will probably develop with better observations on the hydrodynamical aspects already discussed. Some clues may be available from Algol statistics (Giuricin and Mardirossian, 1981, 1983; Budding, 1984a, Abhyankar, 1984).

A basic matter for binary evolution schemes concerns the loser's evolutionary state when RLOF first starts. The traditional terminology Case *A*, Case *B*, Case *C*, ..., is clearly exposed in Paczyński's (1971) review. Certain developments to these rather clear cut modes – e.g., Cases *AB*, *BB* – have also been discussed (Ziołkowski, 1969, 1970) (Law and Ritter, 1982). Recent studies point to Case *B* as having the most likely relevance for the great majority of low-mass Algols (cf. Giuricin and Mardirossian, 1981), though Case *A* is invoked for a few high mass and rather atypical possibles, while the relevance of Case *C* was also noted by Hack *et al.* (1980). The structure and evolution of a Case *B* loser (in its slow phase, at least) can be represented in rather a straight-forward way (Refsdal and Weigert, 1970), allowing a basic picture of the entire classical Algol process of perhaps deceptive simplicity (Awadalla and Budding, 1982). However, for doubts raised in this connection see Kopal (1984).

3.2. Specific problem areas

The question of what happens with the gainer through Algol evolution has already been mentioned. A number of investigators expect a swell-up and outflow situation to develop, particularly during the rapid stage, though this point is not entirely settled. Shu and Lubow (1981) have pointed to some different possibilities which emerge from comparable calculations on protostar inflow. This problem of just how the gainer deals with its influx connects with the perhaps surprisingly close resemblance of observed Algol primaries to normal single Main-Sequence stars (Packet, 1980).

A point of possible general significance in the context of stellar evolution theory follows from these remarks – i.e., that of ambiguity in the results of structure calculations. Specific observables permit alternative models, while similar models can lead to quite different observables. Of course, closer examination of Algol gainers reveals traces of an accretion history (cf. Section 2.2), implying more detailed observational attention as the way out of such dilemmas.

This may have already occurred with some of the previously proposed subgroupings of 'undersize' or 'R CMa' types (cf. Section 1.1). It is interesting to note, however, that some earlier models did permit an 'undersize' phase in calculated evolution tracks (Kippenhahn *et al.*, 1967; Refsdal and Weigert, 1969); while the existence of a system with such a low mass loser at relatively short period as R CMa does suggest evolutionary complications of another kind (Budding, 1984a), but here the possible role of third bodies in evolving Algol systems should be noticed (Radhakrishnan *et al.*, 1984; Van Hamme and Wilson, 1984).

There is considerable current interest in trying to link models of close binary evolution, so that the various kinds of configuration evidencing an interactive history can be brought within some general scheme. That Algols would fit somewhere near the beginning of such a scheme is reasonably expected, however the paths to subsequent stages are not necessarily well defined (Vilhu, 1981; Trimble, 1983; De Loore, 1984). Presumably, a classical Algol will move to a white dwarf/Main-Sequence star pair after the exhaustion of the loser's envelope, but more exotic circumstances can be expected when the gainer subsequently expands to become a loser – this time on to a collapsed object. Although this appears to offer a path towards a standard dwarf nova configuration, the masses of white dwarfs in such binaries support a rather different descent scenario, involving a much more massive helium burning giant and 'spiral-in'.

A common envelope state related to Algol evolution, perhaps initiated during a rapid phase, which is strongly suggested by the present parameters of at least some Algols (Budding, 1984a), brings up a possible genetic relationship which Algols might have to another kind of close binary system, viz., contact binaries. There is a certain scientific appeal about being able to link various, apparently separate, manifestations of behaviour to some uniform underlying principle, i.e., RLOF; though there are perhaps dangers in oversimplification.

There are still many avenues to follow up here. The range of model initial conditions reviewed in Thomas' (1977) paper has been, perhaps, comprehensive enough for an initial survey, at least of the conservative models, but we are still some way from a fully generalized grid against which data for the known Algols can be mapped.

4. Summary and Outlook

The theoretical explanation of the Algol paradox has by now reached the stage of being a Kuhnian paradigm, according to Batten (1981). Many groups are content to do 'normal' science within its framework. The ground seems firm, but the country is wide and there seems to be a number of little explored territories within it.

Advances in observational techniques on a variety of fronts will be indispensable to such explorations. β Per itself has been an object for most, if not all, such technical advances (Table I). Let us note here, for example, a later 'paradox' brought to light by the polarimetric studies of Kemp *et al.* (1981). The most direct interpretation of the observed polarization plane orientations is that the close (AB) and wide ($AB - C$) orbit planes are perpendicular (cf. also Harrington, 1984). This seems to be dynamically counterintuitive. What is the real meaning of this result?

Satellite-borne instruments have been profitably brought to bear on those big-time losers the serpentids. Clear results include quantitative information on carbon underabundance and nitrogen overabundance for gainer atmospheres, the presence of a hot and turbulent region surrounding the gainer, and P Cygni profiles to some high excitation resonance lines indicating systemic outflow. These results seem consistent with the basic RLOF picture, though our understanding will clearly be enhanced if we can effectively compare gainer atmosphere abundances with loser structure calculations, accretion region properties with hydrodynamic models including heating mechanisms, which can drive measurable outflows.

The steady-state model of Lubow and Shu (1975) offers a number of properties which can be compared with observational data. On the positive side, their predictions concerning likely limitations to the radial extents of disks seem to be supported, to some extent (cf. also Piotrowski and Ziołkowski, 1970). Their model of the steady-state stream was quite detailed however, and comparable quantitative observational data are still lacking. What are the influx angles, densities, ionization, and turbulence conditions of real streams? How realistic is the steady state approximation likely to be in the face of known period, photometry and spectrometry peculiarities of many Algols, especially, perhaps, U Cep?

Calculations of interactive evolution specifically to study Algols seem to have waned against the lure of more exotic objects. Indeed, until a more clear understanding of mass and angular momentum loss and transfer processes is available, parametrized approaches such as those of Paczyński and Ziołkowski (1967) may be adequate (cf. also

Popov (1970) and other non-conservative scenarios referred to in Giuricin and Mardirossian's (1981) review). Refsdal *et al.* (1974) were, however, sufficiently challenged by Popper's careful observations of AS Eri to refine their general procedures for a more intensive single study. There are probably more examples which could be studied in this way (cf. Awadalla and Budding, 1982).

A subject with a bearing on both hydrodynamic and evolutionary aspects of Algol studies was raised in Vilhu's (1981) and Tutukov's (1983) papers – namely, effects connected with the action of magnetic fields. That there are some resemblances between Algol and RS CVn subgiants has been implicit since the separating out of the latter class of binary (Hall, 1976), so that there are grounds for expecting that the well known 'activity' of RS CVn stars might have some parallel in classical Algols (Hall, 1981; Charles, 1983). A full-scale incorporation of magnetohydrodynamics into the Algol context remains to be carried out, though there are observations which would indicate a *raison d'être* for such work (cf. Walter, 1980; Söderhjelm, 1980; Gibson, 1980).

Let us recall finally our initial commendations for Algol study – they are relatively easily accessed objects with good observational coverage; they represent a first step in the range of complexity of binary evolution scenarios. Logically we should be confident of our understanding of these systems before proceeding further.

Acknowledgement

A good number of astronomers from many lands were kind enough to respond to my letter of enquiry about Algols. Unfortunately, I was not able to refer to every single item of received information and maintain a continuous, concise account. I have appreciated these responses very much, however, as indeed would continue to do so; and can only offer some apology if I have neglected to mention some potentially worthy contribution to the present review.

References

Abhyankar, K. D.: 1984, *Astrophys. Space Sci.* **99**, 355.
Awadalla, N. S. and Budding E.: 1982, in Z. Kopal and J. Rahe (eds.), *Binary and Multiple Stars as Tracers of Stellar Evolution*, D. Reidel Publ. Co., Dordrecht, Holland, p. 239.
Bachmann, P. J. and Hershey, J. L.: *Astron. J.* **80**, 836.
Barr, J. M.: 1908, *J. Roy. Astron. Soc. Canada* **2**, 70.
Bath, G. T.: 1969, *Astrophys. J.* **158**, 571.
Bath, G. T.: 1975, *Monthly Notices Roy. Astron. Soc.* **171**, 311.
Batten, A. H.: 1973, *Binary and Multiple Systems of Stars*, Pergamon Press, London.
Batten, A. H.: 1981, in T. D. Kahn (ed.), *Investigating the Universe*, D. Reidel Publ. Co., Dordrecht, Holland, p. 207.
Benson, R. S.: 1970, Ph. D. Thesis, University of California, Berkeley.
Biermann, P. and Hall, D. S.: 1973, *Astron. Astrophys.* **27**, 249.
Bolton, C. T. and Zubrod, D. J.: 1980, in M. J. Plavec, D. M. Popper, and R. K. Ulrich (eds.), 'Close Binary Stars', *IAU Symp.* **88**, 225.
Bonneau, D.: 1979, *Astron. Astrophys.* **80**, L11.
Brancewicz, H. K. and Dworak, T. Z.: 1980, *Acta Astron.* **30**, 501.
Budding, E.: 1981, in E. B. Carling and Z. Kopal (eds.), *Photometric and Spectroscopic Binary Systems*, D. Reidel Publ. Co., Dordrecht, Holland, p. 207.

Budding, E.: 1984a, *Astrophys. Space Sci.* **99**, 299.
Budding, E.: 1984b, Preprint (to be published in *Publ. Astron. Soc. Pacific*).
Budding, E.: 1984c, *Catalogue of Classical Evolved Algol-Type Binary Candidate Stars*, Bull. Inform. Centre de Données Stellaires (Strasbourg) **27**, 91.
Cabot, W.: 1984, *Astrophys. J.* **277**, 806.
Charles, P. A.: 1983, in P. B. Byrne and M. Rodono (eds.), 'Activity in Red Dwarfs', *IAU Colloq.* **71**, 415.
Chen, K. Y. and Reuning, E. G.: 1966, *Astron. J.* **71**, 283.
Chen, K. Y. and Wood, F. B.: 1975, *Astrophys. J.* **195**, L73.
Chen, K. Y., Wood, F. B., and Barker, E. S.: 1981, *Astron. J.* **86**, 258.
Crawford, J. A.: 1955, *Astrophys. J.* **121**, 71.
Crawford, R. C.: 1979, *Publ. Astron. Soc. Pacific* **91**, 111.
Cristaldi, S., Fracastoro, M. G., and Sobieski, S.: 1966, *Mem. Soc. Astron. Ital.* **37**, 347.
Cugier, H.: 1979, *Acta Astron.* **29**, 549.
De Loore, C.: 1984, *Astrophys. Space Sci.* **99**, 199.
Eaton, J. A.: 1975, *Publ. Astron. Soc. Pacific* **87**, 745.
Etzel, P. B. and Meyer, D. M.: 1983, *Publ. Astron. Soc. Pacific* **95**, 891.
Fekel, F., Lacy, C. H., and Tomkin, J.: 1980 in K. J. Plavec, D. M. Popper, and R. K. Ulrich (eds.), 'Close Binary Stars', *IAU Symp.* **88**, 53.
Fukue, J.: 1983, *Publ. Astron. Soc. Japan* **35**, 225.
Giannone, P. and Giannuzzi, M. A.: 1970, *Astron. Astrophys.* **6**, 309.
Gibson, D. M.: 1980, in M. J. Plavec, D. M. Popper, and R. K. Ulrich (eds.), 'Close Binary Stars', *IAU Symp.* **88**, 31.
Gibson, D. M., Viner, M. R., and Peterson, S. D.: 1975, *Astrophys. J.* **200**, L143.
Giuricin, G. and Mardirossian, F.: 1981, *Astrophys. J. Suppl.* **46**, 1.
Giuricin, G., Mardirossian, F., and Mezzetti, M.: 1983, *Astrophys. J. Suppl.* **52**, 35.
Glushneva, I. N. and Esipov, V. F.: 1967, *Soviet Astron.* **11**, 585.
Guinan, E., McCook, G., Bachmann, P., and Bistline, W.: 1976, *Astron. J.* **81**, 57.
Hack, M., Flora, U., and Santin, P.: 1980 in M. J. Plavec, D. M. Popper, and U. K. Ulrich (eds.), 'Close Binary Stars', *IAU Symp.* **88**, 271.
Hack, M., Sahade, J., de Jager, C., and Kondo, Y.: 1983, *Astron. Astrophys.* **126**, 115.
Hack, M., Engin, S., and Yilmaz, N.: 1984, *Astron. Astrophys.* **131**, 147.
Hall, D. S.: 1967, *Astron. J.* **72**, 301.
Hall, D. S.: 1974, *Acta Astron.* **24**, 215.
Hall, D. S.: 1975a, *Acta Astron.* **25**, 1.
Hall, D. S.: 1975b, *Acta Astron.* **25**, 95.
Hall, D. S.: 1976, in W. D. Fitch (ed.), 'Multiple Periodic Phenomena in Variable Stars', *IAU Colloq.* **29**, 287.
Hall, D. S.: 1981, in R. M. Bonnet and A. K. Dupree (eds.), *Solar Phenomena in Stars and Stellar Systems*, D. Reidel Publ. Co., Dordrecht, Holland, p. 431.
Hall, D. S. and Keel, W. C.: 1977, *Acta Astron.* **27**, 167.
Hall, D. S., Keel, W. C., and Neuhaus, G. H.: 1976, *Acta Astron.* **26**, 239.
Hall, D. S. and Neff, S. G.: 1979, *Acta Astron.* **29**, 641.
Hall, D. S. and Stuhlinger, T.: 1976, *Acta Astron.* **26**, 109.
Hall, D. S. and Walter, K.: 1974, *Astron. Astrophys.* **37**, 263.
Hall, D. S. and Walter, K.: 1975, *Astron. Astrophys.* **38**, 225.
Hall, D. S. and Woolley, K. S.: 1973, *Publ. Astron. Soc. Pacific* **85**, 507.
Harmanec, P.: 1970, *Bull. Astron. Inst. Czech.* **21**, 113.
Harrington, R. S.: 1984, *Astrophys. J.* **277**, L69.
Hidayat, B., Kopal, Z., and Rahe, J.: 1984, *Astrophys. Space Sci.* **99** (*IAU Colloq.* **80**).
Hjellming, R. M., Wade, C. M., and Webster, E.: 1972, *Bull. Amer. Astron. Soc.* **4**, 384.
Joy, A. H.: 1942, *Publ. Astron. Soc. Pacific* **54**, 35.
Kaitchuck, R. H. and Honeycutt, R. K.: 1982, *Publ. Astron Soc. Pacific* **94**, 532.
Kaitchuck, R. H. and Honeycutt, R. K.: 1983, preprint.
Kemp, J. C., Barbour, M. S., McBirney, R. E., and Rudy, R. J.: 1981, *Astrophys. J.* **243**, 557.
Kemp, J. F., Henson, G. D., Barbour, M. S., Kraus, D. J., and Collins, G. W.: 1983, *Astrophys. J.* **273**, L85.
Kippenhahn, R. and Weigert, A.: 1967, *Z. Astrophys.* **65**, 251.

Kippenhahn, R., Kohl, K., and Weigert, A.: 1967, *Z. Astrophys.* **66**, 58.
Koch, R. H., Olson, E. C., and Yoss, K. M.: 1965, *Astrophys. J.* **141**, 955.
Kondo, Y., McCluskey, G. E., and Stencel, R. E.: 1979, *Astrophys. J.* **233**, 906.
Kondo, Y., McCluskey, G. E., and Parsons, S. B.: 1984, *Astrophys. Space Sci.* **99**, 281.
Kopal, Z.: 1955a, *Mem. Roy. Soc. Liège* **15**(4), 684.
Kopal, Z.: 1955b, *Ann. Astrophys.* **18**, 379.
Kopal, Z.: 1956, *Ann. Astrophys.* **19**, 278.
Kopal, Z.: 1959, *Close Binary Systems*, Chapman and Hall, London and New York.
Kopal, Z.: 1971, *Publ. Astron. Soc. Pacific* **83**, 521.
Kopal, Z.: 1978, *Dynamics of Close Binary Systems*, D. Reidel Publ. Co., Dordrecht, Holland, Chapter VIII.
Kopal, Z.: 1984, *Astrophys. Space Sci.* **99**, 3.
Kopal, Z. and Rahe, J.: 1982, in 'Binary Stars as Tracers of Stellar Evolution', *IAU Colloq.* **69**.
Kreiner, J. M. and Ziołkowski, J.: 1978, *Acta Astron.* **28**, 497.
Kreiner, J. M. and Tremko, J.: 1980, *Bull. Astron. Inst. Czech.* **31**, 343.
Labeyrie, A., Bonneau, D., Stachnik, R. V., and Gezari, D. Y.: 1974, *Astrophys. J.* **194**, L147.
Lambert, D. L.: 1982 in Y. Kondo *et al.* (eds.), *Advances in Ultraviolet Astronomy*, NASA Conf. Publ. 2238, p. 114.
Law, W. Y. and Ritter, H.: 1982, in Z. Kopal and J. Rahe (eds.), *Binary and Multiple Stars as Tracers of Stellar Evolution*, D. Reidel Publ. Co., Dordrecht, Holland, p. 447.
Longmore, A. J. and Jameson, R. F.: 1975, *Monthly Notices Roy. Astron. Soc* **172**, 235.
Lubow, S. H. and Shu, F. H.: 1975, *Astrophys. J.* **198**, 383.
Lynden-Bell, D. and Pringle, J. E.: 1974, *Monthly Notices Roy. Astron. Soc* **168**, 603.
Matese, J. J. and Whitmire, D. P.: 1983, *Astron. Astrophys.* **117**, L7.
McCluskey, G. E. and Kondo, Y.: 1983, *Astrophys. J.* **266**, 755.
McNamara, D. H. and Feltz, K. A., Jr.: 1976, *Publ. Astron. Soc. Pacific* **88**, 688.
Morton, D. C.: 1960, *Astrophys. J.* **132**, 146.
Murad, I.M. and Budding, E.: 1984, *Astrophys. Space Sci.* **98**, 163.
Nadeau, D., Neugebauer, G., Becklin, E. E., Elias, J., Ennis, D., Matthews, K., and Sellgren, K.: 1978, *Monthly Notices Roy. Astron. Soc.* **184**, 523.
Olson, E. C.: 1976, *Astrophys. J.* **204**, 141.
Olson, E. C.: 1980, *Astrophys. J.* **241**, 257.
Olson, E. C.: 1982, *Astrophys. J.* **257**, 198.
Olson, E. C.: 1983, Preprint presented at 1983 Cambridge (UK) Workshop on Interacting Binary Stars.
Olson, E. C., Crawford, R. C., Hall, D. S., Louth, H., Markworth, N. L., and Piirola, V.: 1981, *Publ. Astron. Soc. Pacific* **93**, 464.
Packet, W.: 1980, in M. J. Plavec *et al.* (eds.), 'Close Binary Stars', *IAU Symp.* **88**, 211.
Paczyński, B.: 1965, *Acta Astron.* **15**, 89.
Paczyński, B.: 1971, *Ann. Rev. Astron. Astrophys.* **9**, 183.
Paczyński, B. and Ziołkowski, J.: 1967, *Acta Astron.* **17**, 7.
Peters, G. J. and Polidan, R. S.: 1984, *Astrophys. J.* **283**, 745.
Piirola, V.: 1980, *Astron. Astrophys.* **90**, 48.
Piirola, V.: 1981, *Astron. Astrophys. Suppl.* **44**, 461.
Piotrowski, S. L. and Ziołkowski, J.: 1970, *Astrophys. Space Sci.* **8**, 66.
Plavec, M. J.: 1973, in A. H. Batten (ed.), 'Extended Atmospheres and Circumstellar Matter', *IAU Symp.* **51**, 216.
Plavec, M. J.: 1980, in M. J. Plavec *et al.* (ed.), 'Close Binary Stars', *IAU Symp.* **88**, 251.
Plavec, M. J.: 1982, in Z. Kopal and J. Rahe (eds.), *Binary and Multiple Stars as Tracers of Stellar Evolution*, D. Reidel Publ. Co., Dordrecht, Holland, p. 159.
Plavec, M. J.: 1983a, UCLA preprint No. 157 (to be published in *Astrophys. J.*).
Plavec, M. J.: 1983b, UCLA preprint No. 159 (to be published in *J. Roy. Astron. Soc. Canada*).
Plavec, M. J.: 1983c, UCLA preprint No. 160.
Plavec, M. J. and Dobiáš, J. J.: 1983, *Astrophys. J.* **272**, 206.
Popov, M. V.: 1970, *Perem. Zvezdy* **17**, 412.
Popper, D. M.: 1973, *Astrophys. J.* **185**, 265.
Popper, D. M.: 1980, *Ann. Rev. Astron. Astrophys.* **189**, 125.
Prendergast, K. H. and Taam, R. E.: 1974, *Astrophys. J.* **189**, 125.

Pringle, J. E.: 1981, *Ann. Rev. Astron. Astrophys.* **19**, 137.
Radhakrishnan, K. R., Sarma, M. B. K., and Abhyankar, K. D.: 1983, *Astrophys. Space Sci.* **99**, 229.
Refsdal, S. and Weigert, A.: 1969, *Astron. Astrophys.* **1**, 167.
Refsdal, S. and Weigert, A.: 1970, *Astron. Astrophys.* **6**, 426.
Refsdal, S. and Weigert, A.: 1971, *Astron. Astrophys.* **13**, 367.
Refsdal, S., Roth, M. L., and Weigert, A.: 1974, *Astron. Astrophys.* **36**, 113.
Rudy, R. U. and Kemp, J. C.: 1978, *Astrophys. J.* **221**, 200.
Ryle, M. and Elsmore, B.: 1973, *Monthly Notices Roy. Astron. Soc.* **164**, 223.
Sahade, J.: 1980, *Space Sci. Rev.* **26**, 349.
Sahade, J. and Wood, F. B.: 1978, *Interacting Binary Stars*, Pergamon Press, London.
Savonije, G. J.: 1978, *Astron. Astrophys.* **62**, 317.
Schnopper, H. W., Delvaille, J. P., Epstein, A., Helmken, H., Murray, S. S., Clark, G., Jernigan, G., and Doxsey, R.: 1976, *Astrophys. J.* **210**, L75.
Shu, F. H. and Lubow, S. H.: 1981, *Ann. Rev. Astron Astrophys.* **19**, 277.
Smak, J.: 1972, *Acta Astron.* **22**, 1.
Smyth, M. J., Dow, M. J., and Napier, W.D., *Monthly Notices Roy. Astron. Soc.* **172**, 235.
Snow, T. P., Jr. and Morton, D. C.: 1976, *Astrophys. J. Suppl.* **32**, 429.
Söderhjelm, S.: 1980, *Astron. Astrophys.* **89**, 100.
Struve, O.: 1941, *Astrophys. J.* **93**, 104.
Struve, O.: 1948, *Ann. Astrophys.* **11**, 117.
Svechnikov, M. A.: 1969, *Katolog Orbitalnyich Elementov, Mass i Svetimostei Tesnyich Dvoinyich Zvezd*, Gorky Univ., Sverdlovsk.
Thomas, H. C.: 1977, *Ann. Rev. Astron. Astrophys.* **15**, 127.
Tomkin, J. and Lambert, D. L.: 1978, *Astrophys. J.* **222**, L119.
Trimble, V.: 1983, *Nature* **303**, 137.
Tremko, J. and Kreiner, J. M.: 1981, *Bull. Astron. Inst. Czech.* **32**, 242.
Tutukov A. V.: 1982, in D. Sugimoto *et al.* (ed.), 'Fundamental Problems in the Theory of Stellar Evolution', *IAU Symp.* **93**, 137.
Tutukov, A. V.: 1983, Preprint Univ. Illinois, IAP 83–25.
Van Hamme, E. and Wilson, R. E.: 1984, Preprint (submitted to *Astron. Astrophys.*).
Vilhu, O.: 1981, *Astrophys. Space Sci.* **78**, 401.
Walter, K.: 1980, in M. J. Plavec *et al.* (ed.), 'Close Binary Stars', *IAU Symp.* **88**, 305.
Webbink, R. F.: 1977, *Astrophys. J.* **211**, 486.
Webbink, R. F.: 1980, M. J. Plavec *et al.* (ed.), 'Close Binary Stars', *IAU Symp.* **88**, 127.
White, N. E., Becker, R. H., Holt, S. S., Muschlotzky, R. F., Boldt, E. A., and Serlemitsos, P. J.: 1979, *Bull. Am. Astron. Soc.* **11**, 782.
Wilson, R. E.: 1974, *Astrophys. J.* **189**, 319.
Wilson, R. E. and Caldwell, C. N.: 1978, *Astrophys. J.* **221**, 917.
Wilson, R. E. and Twigg, L. W.: 1980, in M. J. Plavec *et al.* (ed.), 'Close Binary Stars', *IAU Symp.* **88**, 263.
Wilson, R. E., Van Hamme, W., and Pettera, L.: 1984, Preprint (Univ. Florida).
Wyse, A. B.: 1934, *Lick Obsd. Bull.* **17**, 37.
Yungelson, L. R.: 1972, *Astron. Zh.* **49**, 1059.
Zeilik, M. Bayliss, L., and Heckert, P.: 1980, *Inf. Bull. Var. Stars*, No. 1790.
Ziołkowski, J.: 1969, *Astrophys. Space Sci.* **3**, 14.
Ziołkowski, J.: 1970, *Acta Astron.* **20**, 213.
Zirin, H. and Liggett, M. A.: 1982, *Astrophys. J.* **259**, 719.

MASSIVE X-RAY BINARIES: THEIR PHYSICS AND EVOLUTION*

W. SUTANTYO

Bosscha Observatory, Department of Astronomy, Institute of Technology of Bandung, Indonesia

(Received 8 May, 1985)

Abstract. We review various aspects of the evolutionary history of massive X-ray binaries. It is expected that moderately massive close binaries evolve to Be X-ray binaries, while very massive systems evolve to standard X-ray binaries.

The compact objects are formed through supernova explosions. The fairly low galactic latitudes of those systems indicate that the explosion should, in general, not have accelerated the system to a velocity larger than ~ 50 km s^{-1}. This implies that the mass of the exploding stars is in general less than ~ 5 to $6\,M_\odot$.

After the explosion, tidal forces will circularize the orbit of short period systems. Even if the tidal evolution has been completed, the expansion of the optical star during the course of its evolution will continuously disturb the stability of the orbit. Short period systems with large mass ratio may eventually become tidally unstable. Cen X-3 may be an example of such a system. The predicted rate of the orbital period decrease of Cen X-3 is in agreement with the observed rate.

A way to represent the rotational and magnetic evolution of neutron stars in close binary systems is presented. The observed distribution of the pulsation periods of X-ray pulsars with Be companions is consistent with initial magnetic fields of 10^{12}–10^{13} G of the neutron stars. We suggest that the fast X-ray pulsars 4U 0115 + 63 and A 0538 − 66 are young neutron stars, while Cen X-3 and SMC X-1 are 'recycled' pulsars.

The evolutionary relationship between massive X-ray binaries, binary pulsars, and millisecond pulsars is also discussed.

1. Introduction

Massive X-ray binaries form a group of distinct, well-studied systems. Some 25 such systems have been identified (Bradt and McClintock, 1983). The optical radiation from the systems is dominated by the intrinsic light from the massive and bright optical components. The effects of the X-ray sources on the optical radiation are small. X-ray eclipses are observed in some systems. These X-ray sources are characterized by hard X-ray spectra ($kT \sim 15$ keV; Jones, 1977). The X-ray sources often exhibit periodic X-ray pulsations (X-ray pulsars). Combination of X-ray and optical studies has provided very reliable data on the masses and the orbital parameters (Rappaport and Joss, 1983; Van Paradijs, 1983). In this paper, we review some aspects of their evolutionary history.

2. Nature of the Optical Components

Comparison of the position of the optical stars in the HR diagram with the theoretical HR diagram indicates that the stars could still be in the stage of core hydrogen burning,

* Invited paper presented at the IAU Third Asian-Pacific Regional Meeting, held in Kyoto, Japan, between 30 September–6 October, 1984.

Astrophysics and Space Science **118** (1986) 257–270.

with the possible exception of HD 77581 (Conti, 1978; Rappaport and Joss, 1983).

Massive X-ray binaries can be divided into two groups, the standard and the Be X-ray binaries (Rappaport and Van den Heuvel, 1982). In standard X-ray binaries the optical component is an early-type giant or supergiant which is nearly filling its Roche lobe. Except for 2U 1223 – 62 (Wra 977; of which $P_{orb} = 41$ d); the orbital periods are < 10 d (cf. Table 8.1 of Van den Heuvel, 1983).

As one can judge from their location in the HR diagram (Figure 1) the Be components are in general less massive and less evolved than the standard optical components in massive X-ray binaries (Rappaport and Van den Heuvel, 1982; De Loore *et al.*, 1982). The orbital periods are fairly large ($P_{orb} \gtrsim 15$ d) and the Be components are deep inside their Roche lobe.

A cut-off at $\sim 10\, M_\odot$ in the mass distribution of the optical components separates this group from the low mass X-ray binaries (of which $M_{opt} < 2\, M_\odot$).

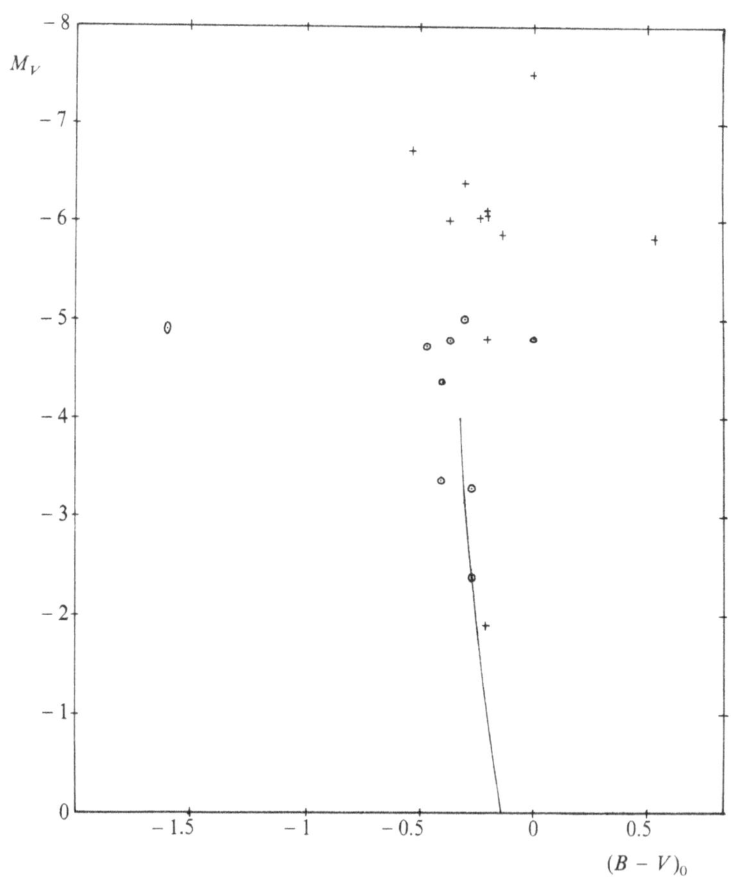

Fig. 1. Colour magnitude diagram of the optical components of standard (+) and Be (⊙) massive X-ray binaries based on the data compiled by Bradt and McClintock (1983).

3. Nature of the X-Ray Sources

Most X-ray sources in the massive X-ray binaries are X-ray pulsars (cf. Bradt and McClintock, 1983; Rappaport and Joss, 1983). The pulsation periods are in the range of 0.7 to 835 s. It is believed that these X-ray pulsars are rotating, magnetized neutron stars. The existence of periodic pulsations eliminate black holes as candidates. White dwarfs are also eliminated as some of the pulsars are rotating faster than the break up speed of white dwarfs, and the luminosity is higher than expected from accretion onto a white dwarf (Kylafis and Lamb, 1979). Furthermore, Rappaport and Joss (1977, 1983) indicate that a plot of $\dot{P}/P$ vs $PL_x^{6/7}$ (where L_x is the X-ray luminosity) follows the relation expected if the X-ray pulsars are neutron stars.

In standard massive X-ray binaries, the X-ray sources are more or less permanent. Two types of Be X-ray sources are identified, i.e., permanent and transient sources. The X-ray luminosities of the permanent sources are low, i.e., 10^{33}–10^{34} erg s^{-1} (e.g., X Per; γ Cas). On the other hand, the transient sources are among the brightest X-ray sources during their active phases. In particular, the transient Be X-ray binary A 0538 – 66 in the LMC reached $L_x \sim 10^{39}$ erg s^{-1} during outbursts, which puts it as the brightest known stellar X-ray source (White and Carpenter, 1978; Skinner *et al.*, 1980). Sudden variations in the rate of mass ejection from Be stars in the presence of compact companions has been suggested as the cause of the transient X-ray emission (Maraschi *et al.*, 1976).

4. Evolutionary History: The Basic Idea

Beginning from the work of Van den Heuvel and Heise (1972), efforts have been made to understand the origin and the evolution of massive X-ray binaries in terms of close binary evolution. The basic idea prevailing in all such hypotheses is that the compact star is formed through a supernova explosion, which happens after an extensive mass transfer. Table I summarizes the evolution of a massive close binary. (We will always call the originally more massive star as the primary, and the companion as the

TABLE I

Scheme of the evolution of a massive close binary

Physical processes	Primary		Secondary		Orbit	
Mass transfer + mass loss						
beginning	evolved	M_{10}	Main-Seq.	M_{20}	P_{0i}	0
end	WR star	M_1	Main-Seq.	M_2	P_0	0
Supernova	compact	M_{1f}	Main-Seq.	M_2	P_f	e
Tidal evolution	compact	M_{1f}	Main-Seq.	M_2	P_n	0
X-ray stage	compact	M_{1f}	Evolved	M_2	P_n	0

secondary, regardless of their masses at each stage.) We discuss in the following sections several important stages in the evolution of massive close binary systems.

5. Stellar Wind Mass Loss and First Stage of Mass Transfer

Massive stars are known to emit strong stellar winds, with a rate from 10^{-8} to $10^{-5}\ M_{\odot}\ \mathrm{yr}^{-1}$ (cf. Conti and Garmany, 1980; Lamers, 1981). If the wind is strong it is quite possible that the mass losing star loses most of its hydrogen-rich envelope before it has a chance to reach its Roche lobe. Rappaport and Van den Heuvel (1982) suggest that this wind dominated evolution will form short period WR binaries, as stellar wind mass loss is not expected to increase the orbital period drastically.

Mass transfer will happen when the primary fills up its Roche lobe and starts to spill matter to its companion through the inner Lagrangian point. In the early studies it was generally assumed that such a mass transfer happens conservatively, i.e., the total mass and orbital angular momentum are conserved (Plavec, 1968; Paczyński, 1971). However, there is evidence indicating that in some systems mass transfer should not be conservative (Kopal, 1971). Meyer-Hofmeister (1974) presented an argument that the orbital period of Cen X-3 is too short to be accounted for by conservative evolution.

Van den Heuvel (1976, 1981, 1983) suggests that some systems may undergo a so-called quasi-conservative evolution. Although some mass and angular momentum may be lost, the system behaves like a conservative system. This kind of evolution is characterized by a fast decrease of the orbital period at first followed by a slower expansion of the orbit as the mass ratio is reversed. The important consequence of a quasi-conservative evolution is that the final period is usually large (> 10 d; De Grève and Vanbeveren, 1980; Van den Heuvel, 1983). The subsequent supernova explosion will increase the orbital period even more.

The mass transfer is terminated when core helium burning commences and the primary becomes detached from its Roche lobe. Since the star loses most of its hydrogen rich envelope it becomes a helium star (originally the helium core of the primary) which can be identified as a Wolf–Rayet star (Paczyński, 1967; Van den Heuvel, 1973). During the Wolf–Rayet phase some additional mass loss may occur as Wolf–Rayet stars are known to emit strong stellar winds (Barlow *et al.*, 1981).

A massive close binary may evolve quasi-conservatively if the system can avoid contact during the mass transfer, i.e., the secondary is always detached from its Roche lobe during the course of mass transfer. This can be accomplished (Van den Heuvel, 1981) if (i) the mass ratio $q = M_2/M_1$ is not too low ($q > 3$); (ii) the orbital separation is not too short; (iii) the primary is in radiative equilibrium.

6. Common Envelope Evolution

If the system evolves to a contact configuration, both stars are embedded in a common envelope. The available angular momentum is not sufficient to co-rotate the common envelope with the binary motion. The system will then consist of two dense objects rapidly revolving inside a slow rotating envelope. The large friction generated by the relative motion will transport angular momentum from the orbital motion to the envelope. Large amounts of mass and angular momentum are expected to flow out through the outer Lagrangian points (or the equivalent points, as corotation is not likely to be maintained out to such a large distance). Common envelope evolution of close binary systems has been discussed by Paczyński (1976); Taam *et al.* (1978); and Meyer and Meyer-Hofmeister (1979). It is expected that the outcome of such evolution will be a binary system consisting of a helium star and a normal star with very short orbital period. If the mass of the helium star is $\lesssim 3\ M_{\odot}$, it will collapse into a white dwarf. Such an evolution has been suggested to explain the origin of very short binary systems consisting of a degenerate and a normal star, such as cataclysmic variables (Paczyński, 1976). On the other hand, if the mass of the helium star is large ($M \gtrsim 3\ M_{\odot}$; Van den Hcuvcl, 1981, 1983) a supernova explosion is expected to happen.

7. Supernova Explosion

We consider the case that the helium star explodes as a supernova and leaves behind a compact remnant. Since the exploding star is less massive than its companion, the explosion will in general not disrupt the system. At the end of the explosion the system becomes an eccentric binary consisting of a compact and a normal star.

Since a supernova explosion might give an additional space velocity to the system due to matter ejection, the system may 'run away' from the place where it originated, i.e.,

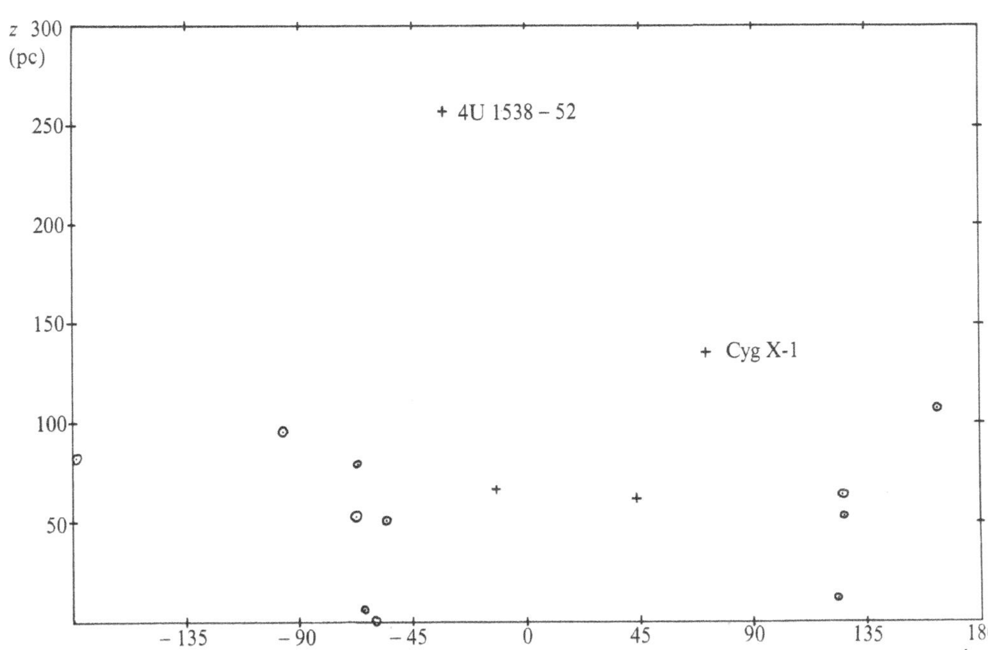

Fig. 2. Galactic z-distribution of massive X-ray binaries based on the data compiled by Bradt and McClintock (1983). (+ standard; ⊙ Be)

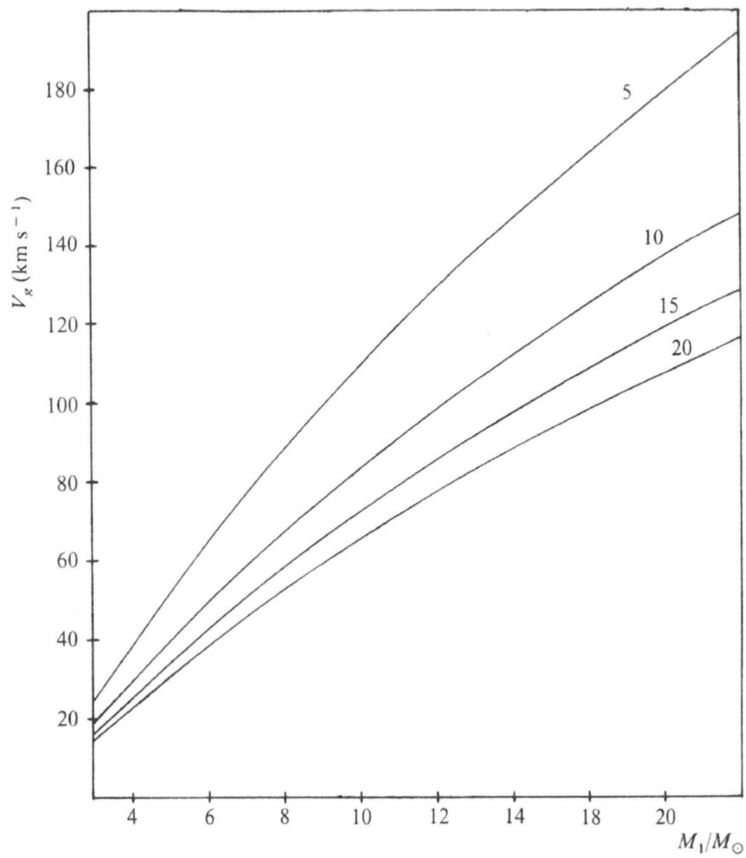

Fig. 3. Runaway velocity as a function of the mass of the exploding star. The mass of the unexploding star is assumed to be 20 $M_\odot$ and the final mass of the exploding star is 1.4 $M_\odot$. The label on each curve indicates the pre-explosion orbital period in days.

the galactic plane. We call this extra velocity the runaway velocity. Sutantyo (1975, 1981) suggested that the galactic z-distribution of massive X-ray binaries may give a constraint on the mass of the exploding star. Figure 2 shows the galactic z-distribution of massive X-ray binaries. Notice that except for 3U 1538 – 522 (z = 257 pc) and Cyg X-1 (z = 135 pc), all other massive X-ray binaries are located close to the galactic plane ($z < 100$ pc). The mean z of the known massive X-ray binaries is 70 pc. If the lifetime of the system is 5×10^6 yr, the runaway velocity should, in general, not be larger than 50 km s^{-1}.

Figure 3 shows the runaway velocity V_g as a function of the mass of the exploding star (M_1) for various initial (pre-explosion) orbital periods (P_0) in the case of a symmetric supernova explosion. We assume that after the explosion the star becomes a neutron star with mass (M_1^f) of 1.4 $M_\odot$. The mass of the unexploding star (M_2) is assumed to be 20 $M_\odot$. Notice that for $P_0 < 10$ d a runaway velocity $\lesssim 50$ km s^{-1} can be obtained if $M_1 \lesssim 6\, M_\odot$. If we assume $M_2 = 10\, M_\odot$ the upper limit of M_1 is reduced to 5 $M_\odot$. We conclude, therefore, that the galactic z-distribution of massive X-ray binaries is consistent with the mass of the exploding star being $\lesssim 6\ M_{\odot}$.

Adopting the statistical distribution of mass ratios and orbital periods of unevolved massive binaries, as obtained by Garmany *et al.* (1980; cf. Van den Heuvel, 1983), we can derive the statistical distribution of V_g. Figure 4 depicts the results for conservative mass transfer. Note that in most cases the runaway velocity is < 50 km s^{-1} as expected. An examination of non-conservative mass transfer in the way described by Vanbeveren *et al.* (1979), does not alter the above results significantly.

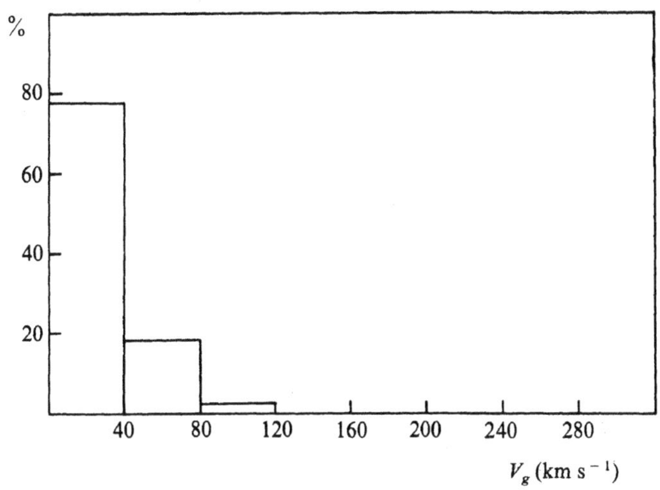

Fig. 4. Statistical distribution of runaway velocities derived for conservative evolution based on the distribution of mass ratios and orbital periods of unevolved binaries as obtained by Garmany *et al.* (1980).

8. X-Ray Stages

X-ray radiation can be produced by the accretion of matter from binary stellar companions onto compact components. Two modes of mass transfer have been suggested, i.e. stellar wind accretion and Roche lobe overflow (Davidson and Ostriker, 1973; Savonije, 1983). Stellar wind accretion is much too low to power the intrinsically bright massive X-ray binaries such as Cen X-3 and SMC X-1 (Lamers *et al.*, 1976; Bonnet-Bidaud and Van der Klis, 1979, 1981). On the other hand fully developed Roche-lobe overflow in massive binaries would certainly extinguish the X-ray source, as the accretion would soon exceed the critical rate set by the Eddington critical luminosity. In this case the accumulating matter expelled by radiation pressure in the neighborhood of the compact star will degrade the X-rays into low energy photons. Savonije (1979, 1983) solves these conflicting arguments by indicating that a sufficient, but not too much, rate of mass transfer could produce X-rays from such systems on time-scale 10^4 to 10^5 yr if the stellar companion is undergoing atmospheric Roche-lobe overflow, i.e., the star is filling its Roche lobe during the core hydrogen burning. Only massive systems with short orbital periods (<4–5 d) can produce X-rays powered by atmospheric Roche-lobe overflow. The X-ray lifetime drops very sharply for longer orbital periods, as Roche-lobe overflow happens at the post Main-Sequences stages when the star expands its outer envelope rapidly. Wide systems like Vela X-1 and 4U 1223 − 62 are more likely powered by stellar wind accretion.

9. What Systems Become X-Ray Binaries

The high X-ray luminosities of transient Be X-ray binaries are very likely caused by rotationally-induced mass ejection from the equatorial regions (Maraschi *et al.*, 1976; Rappaport and Van den Heuvel, 1982). Two conditions are, therefore, required for the formation of Be X-ray binaries, (1) the optical star should be rapidly rotating, (2) the post mass transfer, and post supernova, orbital period should be sufficiently large such that tidal braking is not operative within the lifetime of the system.

Van den Heuvel (1983) suggests that quasi-conservative evolution of moderately massive B-type close binaries ($10 < M_{10} < 20\ M_{\odot}$) seems to fulfil those requirements for two reasons. First, mass transfer may induce rapid rotation of the accreting star as the infalling matter has large specific angular momentum. Second, mass transfer and a SN explosion in moderately massive close binaries will, in most cases, end up with long orbital period systems.

In very massive close binaries mass transfer and a supernova explosion will, in general, not result in very wide systems. Such systems are expected to evolve to standard massive X-ray binaries.

In the following we summarize the behaviour of various type of binary systems consisting of a normal and a compact star.

(a) *Very massive binaries* ($M_2 > 20\ M_{\odot}$) *with short orbital periods* (<5 d). Such systems can be observed as permanent standard massive X-ray binaries. The X-ray is powered by atmospheric Roche-lobe overflow. Those systems are very likely formed through common envelope evolution. Examples: Cen X-3, SMC X-1.

(b) *Very massive binaries with long orbital periods* ($P > 5$ d). Those systems may belong to standard massive X-ray binaries in which the X-ray source is powered by stellar wind accretion. Examples: Vela X-1, 4U 1223 − 62.

(c) *Moderately massive binaries* ($10 < M_2 < 20\ M_{\odot}$). As mass transfer and the supernova explosion in such systems will, in general, increase the orbital period drastically, only systems with long orbital periods ($P > 10$ d) are expected to exist. Such systems may show up as transient Be X-ray binaries. Examples: A 0538 − 66, 4U 0115 + 63.

(d) *Intermediate mass systems* ($2 < M_2 < 10\ M_{\odot}$). The orbital periods of such systems are expected to be large (> 15 d). Rotationally induced mass ejection is probably not operative in these systems. Stellar winds are too weak while Roche-lobe overflow is too large to produce X-rays (Van den Heuvel, 1975). We do not expect X-rays, therefore, from such systems. This can explain the observed gap between ~ 2 and $10\ M_{\odot}$ in the mass distribution of the optical components of X-ray binaries.

(e) *Low mass systems* ($M_2 < 2\ M_{\odot}$). Among these systems belong the low mass

X-ray binaries in which the X-rays are produced by fully developed Roche-lobe overflow. Examples: Her X-1, Sco X-1.

10. Tidal Evolution

The orbital eccentricities of short period ($P < 5$ d) massive X-ray binaries are almost exactly zero. This is contrary to the expectation that the orbit should be eccentric as a consequence of the supernova explosion which has formed the compact stars. The most likely mechanism to have circularized the orbit is tidal interaction between the two components. At the end of the tidal evolution the system will reach the most stable configuration, i.e., the secondary rotates synchronously with the orbital motion and the system has a circular orbit. Whether this state can be attained within the lifetime of the system depends on the ratio R/a (the ratio of the secondary radius to the orbital semi-major axis) and on the rate of the energy dissipation process. Zahn (1977) indicates that the dissipative process in stars with a convective envelope is mainly due to the turbulent friction in the convective region of the star, while in stars

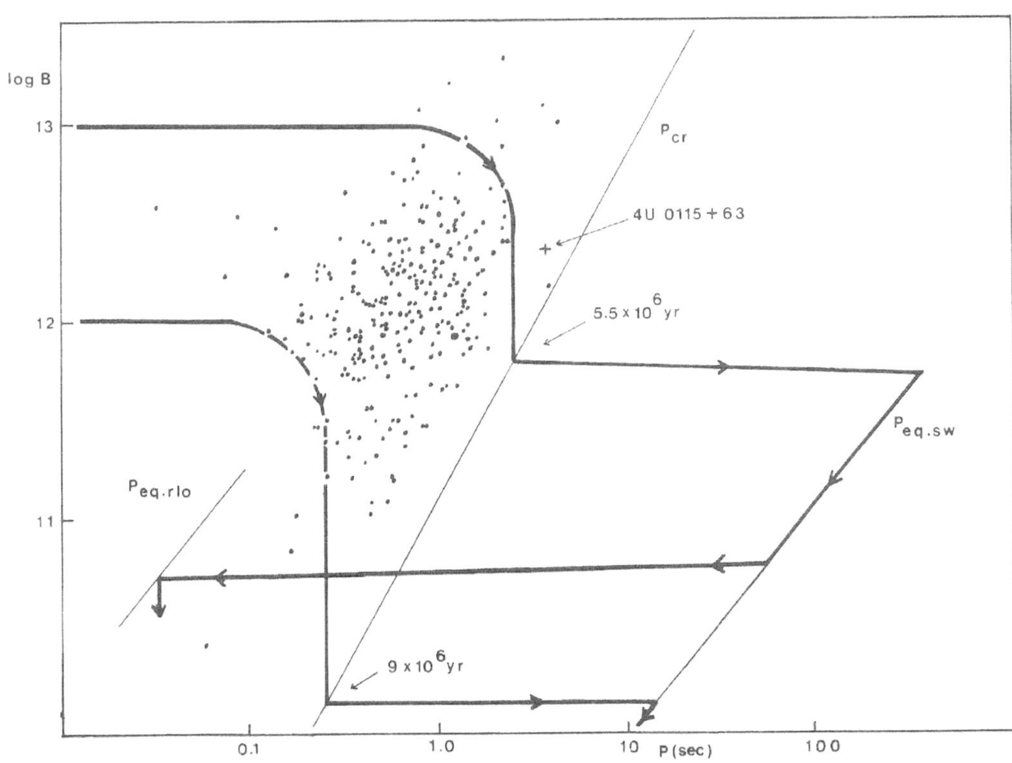

Fig. 5. Evolutionary tracks of neutron stars in close binaries in the *BP* diagram. The dots represent radio pulsars. After a pulsar crosses P_{cr} a strong braking mechanism, due to the interaction between the neutron star and stellar wind from the companion, will slow down the pulsar to its equilibrium period ($P_{eq,\,sw}$). When the companion is filling its Roche lobe the pulsar will be recycled to its new equilibrium period ($P_{eq,\,rlo}$) by mass transfer.

possessing radiative envelopes and convective cores the dissipative process is mainly caused by the radiative damping. The role of the energy dissipation process on the orbital evolution can conveniently be expressed in terms of a mean dynamic viscosity, without referring to the real (known) physical processes (Kopal, 1968; Alexander, 1973). From the fact that some massive X-ray binaries have an almost perfectly circular orbit, and that the lifetime of such systems (after the supernova explosion) should be about 5×10^6 yr, and that Vela X-1 has not yet reached a circular orbit, Sutantyo (1974) derived that the mean dynamic viscosity of the systems should be of the order of $10^{12} - 10^{13}$ g cm^{-1} s^{-1}. This is the same conclusion as obtained by Alexander (1973) for the AG Per system.

Although a system may have completed its tidal evolution, a static equilibrium cannot be maintained. The expansion of the secondary during and after the Main-Sequence stages will continuously disturb the stability of the orbit. Due to the increase of its moment of inertia, the rotation of the secondary is slowed down. On the other hand, the tidal forces try to maintain the synchronism. Therefore, there will be transfer of angular momentum from the orbital motion to the rotation. As a result the orbit will decay. A system may become tidally unstable if the ratio of the orbital to rotational angular momentum becomes smaller than 3 (Counselman, 1973; Hut, 1981; Shibazaki *et al.*, 1983). De Grève *et al.* (1975) considered the tidal evolution of Cen X-3 and concluded that the system may at present be tidally unstable. They obtained that, assuming the same viscosity as would be required for the circularization of the orbit, the orbit of Cen X-3 is now decaying on time-scale of 10^5–10^6 yr. Investigations by Kelley *et al.* (1983) and Nagase *et al.* (1983) do indicate a decay of the orbit of Cen X-3 on time-scale 5×10^5 yr.

11. X-Ray Pulsars

Neutron stars can be observed either as radio pulsars or X-ray pulsars. One of the basic difference between those two kinds of pulsars is that most radio pulsars are single, while most (if not all) X-ray pulsars are members of binary systems.

Radhakrishnan and Srinivasan (1981) have suggested that a *BP* diagram (B = the surface magnetic field and P = the pulsation period) may serve as a suitable representation for studying the evolution of pulsars, in a similar way to the use of the *HR* diagram in stellar evolution. They show that the best fit to the distribution of radio pulsars in the *BP* diagram can be obtained if the magnetic field of pulsars decays on a timescale of $\sim 2 \times 10^6$ yr. In this respect single pulsars would follow evolutionary tracks parallel to B = constant when they are young and after a few million years go asymptotically to P = constant as shown in Figure 5. In this figure the dots represent radio pulsars.

We can apply the same diagram to study the evolution of neutron stars in close binary systems. X-ray pulsars in Be X-ray binaries may represent the early evolution of such neutron stars, as the Be components are relatively unevolved. A neutron star in such a system will at first the same evolution as a single neutron star. The neutron star will not, however, show up as a radio pulsar, as f–f transitions in the stellar wind from its companion will obscure any radio pulse (Illarionov and Sunyaev, 1975; Maraschi and Treves, 1979). When the neutron star becomes less energetic, stellar wind matter from the nondegenerate star may enter the magnetosphere of the neutron star. This is expected to happen when the magnetospheric radius becomes equal or smaller than the light cylinder radius (Van den Heuvel, 1977) – i.e., $P > P_{cr}$ where,

$$P_{cr} = 0.0503(B/10^{12}\ \text{G})^{4/7}\,(R/10\ \text{km})^{12/7}\,(M_a/10^{17}\ \text{g s}^{-1})^{-2/7} \times \\ \times (M_x/M_\odot)^{-1/7}\ \text{s}\,,$$

where B, R, M_x are, respectively, the magnetic field, the radius, and the mass of the neutron star, while M_a is the accretion rate on to the neutron star. In the *BP* diagram this will happen when the neutron star crosses the straight line labeled by P_{cr} (in drawing this line we have adopted $M_a \sim 6 \times 10^{10}\ \text{g s}^{-1}$ as a typical accretion rate in Be X-ray binaries; cf. Lamers *et al.*, 1975). At this stage a strong braking mechanism will rapidly spin down the neutron star to an equilibrium period (Davidson and Ostriker, 1973; Van den Heuvel, 1977; Henrichs, 1983)

$$P_{eq} = 1.6(B/10^{12}\ \text{G})^{6/7}\,(R/10\ \text{km})^{18/7}\,(M_x/M_\odot)^{-5/7} \times \\ \times (M_a/10^{17}\ \text{g s}^{-1})^{-3/7}\ \text{s}\,.$$

The equilibrium period for stellar wind accretion in a typical Be X-ray binary is indicated by the straight line labeled by $P_{eq,\,sw}$.

After crossing P_{cr}, a neutron star will move very rapidly to the right and then settle down on the equilibrium period defined by the accretion rate at this stage. The neutron star will stay on this equilibrium period for about the Main-Sequence lifetime of the Be companion ($\sim 10^7$ yr). Note if the initial magnetic field of a neutron star is $\sim 10^{12}$–10^{13} G (the typical values for radio pulsars) the equilibrium period would be a few ten to a few hundred seconds, just in the range of the observed values (Henrichs, 1983; see Kelley *et al.*, 1983, for additional data).

After some 10^7 yr Roche-lobe overflow will occur. The accretion rate on to the neutron star may approach the critical rate set by the Eddington critical luminosity ($\sim 10^{-8}\ M_\odot\ \text{yr}^{-1}$). The neutron star will be spun up to the new equilibrium period (indicated by a straight line labeled by $P_{eq,\,rlo}$ in Figure 5). Those 'recycled' pulsars are spinning with periods of a few tenths to a few seconds (Radhakrishnan and Srinivasan, 1981; Srinivasan and Van den Heuvel, 1982; Van den Heuvel, 1984).

Only one direct estimate of the magnetic field of an X-ray pulsar with a massive companion is available. Wheaton *et al.* (1979) observed an absorption feature in the hard X-ray spectrum of 4U 0115 + 63 at 20.1 keV. If the feature arises by cyclotron resonance absorption near the magnetic poles of the neutron star, it implies a magnetic field of ~ 1.8–2.5×10^6 G. The X-ray pulsar has a pulsation period of 3.61 s. Its location in the *BP* diagram is shown in Figure 5. The companion of the X-ray pulsar is a Be star. We conclude that the X-ray pulsar is a young neutron star which has not been spun down to its equilibrium period. We suggest that the fastest X-ray pulsar A 0538 – 66 ($P = 0.069$ s), which also has a Be companion, is also at an early stage of its evolution. Other fast X-ray pulsars such as Cen X-1 ($P = 4.84$ s) and SMC X-1 ($P = 0.71$ s), are 'recycled' pulsars as evidenced by their more evolved companions and the presence of an accretion disk. Those 'recycled' X-ray pulsars are characterized by intrinsically bright and permanent X-ray emission.

12. Binary and Millisecond Radio Pulsars

Four binary radio pulsars are known. Except for PSR 0820 + 02 those pulsars have peculiar combinations of magnetic field and pulsation period, i.e., low B and short P (one of them, PSR 1953 + 29, is a millisecond pulsar). This fact indicates that they are 'recycled' pulsars (Smarr and Blandford, 1976; Damashek *et al.*, 1982; Srinivasan and Van den Heuvel, 1982). It is believed that the short orbital period binary radio pulsars (PSR 1913 + 16 and PSR 0655 + 64) are descendants of massive X-ray binaries (Flannery and Van den Heuvel, 1975). Following Van den Heuvel and Taam (1984) we will discuss the origin of those systems. (The long orbital period binary radio pulsars PSR 0820 + 02 and PSR 1953 + 29 have evolved from low mass systems (see Van den Heuvel and Taam, 1984), we will not discuss those systems.)

The orbital periods of PSR 1913 + 16 and PSR 0655 + 64 are, respectively, 0.32 and 1.03 d. In PSR 1913 + 16 the companion of the pulsar is most likely a neutron star, while in PSR 0655 + 64 it is a white dwarf (Blandford and De Campli, 1981; Van den Heuvel, 1981).

It is believed that those systems have originated from long period binaries ($P_{orb} >$ a few weeks) consisting of a neutron star and a more massive normal companion. In such systems mass transfer occurs when the companion is climbing the asymptotic giant branch. At this stage the star possesses a deep convective envelope. Runaway mass transfer will happen and the system will undergo a common envelope evolution.

Suppose the original mass of the mass losing star is > 8–$10\ M_{\odot}$. As the orbit is wide the star has a chance to develop a massive CO core before filling its Roche lobe. The star will eventually explode as a supernova leaving behind a neutron star. In case the system is not disrupted by the explosion, the result will be an eccentric short period binary consisting of two neutron stars. PSR 1913 + 16 is probably formed in this way. Note that in PSR 1913 + 16 the observed pulsar is the old 'recycled' pulsar (Smarr and Blandford, 1976). We do not see the young pulsar probably due to the beaming effect. Good candidates for this type of evolution are Be X-ray binaries.

Due to the emission of gravitational radiation the two neutron stars in PSR 1913 + 16 are approaching each other. A coalescence of the system will happen in $\sim 3 \times 10^8$ yr. The orbital angular momentum will be transformed into rotational angular momentum of the merging pulsars, leading to the formation of a single millisecond pulsar like PSR 1937 + 214 (Henrichs and Van den Heuvel, 1983).

If at the end of the common envelope phase the mass of the core is $< 3\ M_{\odot}$, the star will collapse into a CO white dwarf. A short period system consisting of a white dwarf and a neutron star is formed. This may explain the formation of the binary pulsar PSR 0655 + 64.

Acknowledgements

The author would like to thank the Japan Society for the Promotion of Science for the support which has enabled them to attend this meeting.

References

Alexander, M. E.: 1973, *Astrophys. Space Sci.* **23**, 459.
Barlow, M. J., Smith, L. J., and Willis, A. J.: 1981, *Monthly Notices Roy. Astron. Soc.* **196**, 101.
Blandford, R. D. and De Campli, W. M.: 1981, in W. Sieber and R. Wielebinski (eds.), 'Pulsars', *IAU Symp.* **95**, 371.
Bonnet-Bidaud, J. M. and Van der Klis, M.: 1979, *Astron. Astrophys.* **73**, 90.
Bonnet-Bidaud, J. M. and Van der Klis, M.: 1981, *Astron. Astrophys.* **97**, 134.
Bradt, H. V. and McClintock, J. E.: 1983, *Ann. Rev. Astron. Astrophys.* **21**.
Conti, P. S.: 1978, *Astron. Astrophys.* **63**, 225.
Conti, P. S. and Garmany, C. D.: 1980, *Astrophys. J.* **238**, 190.
Counselman, C. C.: 1973, *Astrophys. J.* **180**, 307.
Damashek, M., Backus, P. R., Taylor, J. H., and Burkhardt, R.: 1982, *Astrophys. J.* **253**, L57.
Davidson, K. and Ostriker, J. P.: 1973, *Astrophys. J.* **179**, 583.
De Grève, J. P. and Vanbeveren, D.: 1980, *Astrophys. Space Sci.* **68**, 433.
De Grève, J. P., De Loore, C., and Sutantyo, W.: 1975, *Astrophys. Space Sci.* **38**, 301.
De Loore, C., Burger, M., van Dessel, E. L., and Mouchet, M.: 1982, in M. Jaschek and H. G. Groth (eds.), 'Be Stars', *IAU Symp.* **98**, 347.
Flannery, B. P. and Van den Heuvel, E. P. J.: 1975, *Astron. Astrophys.* **39**, 61.
Garmany, C. D., Conti, P. S., and Massey, P. S.: 1980, *Astrophys. J.* **242**, 1063.
Henrichs, H. F.: 1983, in W. H. G. Lewin and E. P. J. Van den Heuvel (eds.), *Accretion Driven Stellar X-Ray Sources*, Cambridge University Press, Cambridge, p. 393.
Henrichs, H. F. and Van den Heuvel, E. P. J.: 1983, *Nature* **303**, 213.
Hut, P.: 1981, *Astron. Astrophys.* **99**, 126.
Illarionov, A. F. and Sunyaev, R. A.: 1975, *Astron. Astrophys.* **39**, 185.
Jones, C.: 1977, *Astrophys. J.* **214**, 856.
Kelley, R. L., Rappaport, S., Clark, G. W., and Petro, L. D.: 1983, *Astrophys. J.* **268**, 790.
Kelley, R. L., Rappaport, S., and Ayasli, S.: 1983, *Astrophys. J.* **274**, 765.
Kopal, Z.: 1968, *Astrophys. Space Sci.* **1**, 411.
Kopal, Z.: 1971, *Publ. Astron. Soc. Pacific* **83**, 521.
Kylafis, N. D. and Lamb, D. Q.: 1979, *Astrophys. J.* **228**, L105.
Lamers, H. J.: 1981, *Astrophys. J.* **245**, 593.
Lamers, H., Van den Heuvel, E. P. J., and Petterson, J.: 1975, *Astron. Astrophys.* **49**, 327.
Maraschi, L. and Treves, A.: 1979, *Nature* **279**, 401.
Maraschi, L., Treves, A., and Van den Heuvel, E. P. J.: 1976, *Nature* **259**, 292.
Meyer-Hofmeister, E.: 1974, *Astron. Astrophys.* **36**, 361.
Meyer, F. and Meyer-Hofmeister, E.: 1979, *Astron. Astrophys.* **78**, 167.
Nagase, F., Sato, N., Makishima, K., Kawai, N., and Mitani, K.: 1983, *ISAS Research Note* **234**, 40.
Paczyński, B.: 1967, *Acta Astron.* **17**, 355.
Paczyński, B.: 1971, *Ann. Rev. Astron. Astrophys.* **9**, 183.
Paczyński, B.: 1976, in P. Eggleton, S. Mitton, and J. Whelan (eds.), 'Structure and Evolution of Close Binary Systems', *IAU Symp.* **73**, 75.
Plavec, M.: 1968, *Adv. Astron. Astrophys.* **6**, 201.
Radhakrisnan, V. and Srinivasan, G.: 1981, in B. Hidayat and M. W. Feast (eds.), *Proc. of the Second Asian-Pacific Regional Meeting on Astronomy*, Tira Pustaka, Jakarta, p. 423.
Rappaport, S. A. and Joss, P. C.: 1977, *Nature* **266**, 683.
Rappaport, S. A. and Joss, P. C.: 1983, in W. H. G. Lewin and E. P. J. Van den Heuvel (eds.), *Accretion Driven Stellar X-Ray Sources*, Cambridge University Press, Cambridge, p. 1.
Rappaport, S. A. and Van den Heuvel, E. P. J.: 1982, in M. Jaschek and H. G. Groth (eds.), 'Be Stars', *IAU Symp.* **98**, 327.

Savonije, G. J.: 1978, *Astron. Astrophys.* **62**, 317.
Savonije, G. J.: 1983, in W. H. G. Lewin and E. P. J. Van den Heuvel (eds.), *Accretion Driven Stellar X-Ray Sources*, Cambridge University Press, Cambridge, p. 343.
Shibazaki, N., Makishima, K., Inoue, H., Murakami, T.: 1983, *ISAS Research Note* **234**, 55.
Skinner, G. K., Shulman, S., Share, G., Evans, W. D., McNutt, D., Meekins, J., Smathers, H., Wood, K., Yentis, D., Byram, E. T., Chubb, T. A., and Friedman, H.: 1980, *Astrophys. J.* **240**, 619.
Smarr, L. L. and Blandford, R. D.: 1976, *Astrophys. J.* **207**, 574.
Srinivasan, G. and Van den Heuvel, E. P. J.: 1982, *Astron. Astrophys.* **108**, 143.
Sutantyo, W.: 1974, *Astron. Astrophys.* **35**, 251.
Sutantyo, W.: 1975, *Astron. Astrophys.* **41**, 47.
Sutantyo, W.: 1981, in P. W. Sanford, P. Laskarides, and J. Salton (eds.), *Galactic X-Ray Sources*, John Wiley, Chichester, p. 27.
Taam, R. E., Bodenheimer, P., and Ostriker, J. P.: 1978, *Astrophys. J.* **222**, 269.
Vanbeveren, D., De Grève, J. P., van Dessel, E. L., and de Loore, C.: 1979, *Astron. Astrophys.* **73**, 19.
Van den Heuvel, E. P. J.: 1973, *Nature Phys. Sci.* **242**, 71.
Van den Heuvel, E. P. J.: 1975, *Astrophys. J.* **198**, L109.
Van den Heuvel, E. P. J.: 1976, in P. Eggleton, S. Mitton, and J. Whelan (eds.), 'Structure and Evolution of Close Binary Systems', *IAU Symp.* **73**, 35.
Van den Heuvel, E. P. J.: 1977, *Ann. N.Y. Acad. Sci.* **302**, 14.
Van den Heuvel, E. P. J.: 1981, in W. Sieber and R. Wielebinski (eds.), 'Pulsars', *IAU Symp.* **95**, 379.
Van den Heuvel, E. P. J.: 1983, in W. H. G. Lewin and E. P. J. Van den Heuvel (eds.), *Accretion Driven Stellar X-Ray Sources*, Cambridge University Press, Cambridge, p. 303.
Van den Heuvel, E. P. J.: 1984, *J. Astrophys. Astron.* (in press).
Van den Heuvel, E. P. J. and Heise, J.: 1972, *Nature Phys. Sci.* **239**, 67.
Van den Heuvel, E. P. J. and Taam, R. E.: 1984, *Nature* **309**, 235.
Van Paradijs, J.: 1983, in W. H. G. Lewin and E. P. J. Van den Heuvel (eds.), *Accretion Driven Stellar X-Ray Sources*, Cambridge University Press, Cambridge, p. 189.
Wheaton, Wm. A., Doty, J. P., Primini, F. A., Cooke, B. A., Dobson, C. A., Goldman, A., Hecht, M., Hoffman, J. A., Howe, S. K., Scheepmaker, A., Tsiang, E. Y., Lewin, W. H. G., Matteson, J. L., Gruler, D. E., Baity, W. A., Rotschild, R., Knight, F. K., Nolan, P., and Peterson, L. E.: 1979, *Nature* **282**, 240.
White, N. E. and Carpenter, G. F.: 1978, *Monthly Notices Roy. Astron. Soc.* **183**, 11 p.
Zahn, J. P.: 1977, *Astron. Astrophys.* **57**, 383.

RAPID OSCILLATIONS IN CATACLYSMIC VARIABLES*

BRIAN WARNER

Department of Astronomy, University of Cape Town, Rondebosch, South Africa

(Received 8 May, 1985)

Abstract. Rapid coherent and quasi-periodic oscillations in the optical and X-ray fluxes of cataclysmic variable stars are described.

1. Introduction

It is thirty years since the discovery of coherent 71 s periodic variations in the brightness of the nova remnant DQ Her (Walker, 1954) and a dozen years since the discovery that such a phenomenon is at least intermittently present in other classes of cataclysmic variables (Warner and Robinson, 1972). During the past decade the observational record of rapid oscillations in cataclysmic variables (CVs) has grown steadily richer, but the nature of the underlying mechanisms still defies certain understanding. We review here some of the major observational conclusions to be kept under consideration while wrestling with the problems of interpretation.

2. The Data Base

We first look at the range of objects that show rapid optical oscillations (Table I). Previous tabulations have been given by Warner (1979), Patterson (1981), and Cordova and Mason (1982, 1983), in which most references to work prior to 1982 may be found; here we update these with additional references. Rapid oscillations in polars have been excluded from consideration.

Two types of rapid oscillation are generally recognised: those in which a high degree of coherence is maintained for at least hundreds of cycles and 'quasi-periodic' oscillations in which the coherence length may be as short as a few cycles (Robinson and Nather, 1979). Because of their stability, the coherent oscillations can be detected to amplitudes as low as 0.0002 mag. (e.g., SS Cyg: Patterson *et al.*, 1978a). Such low signals, in the presence of flickering noise from the star and atmospheric scintillation, require Fourier transform techniques for their detection. Occasionally the coherent oscillations are of sufficient amplitude (> 0.01 mag.) to be visible directly in the light

* Invited paper presented at the IAU Third Asian-Pacific Regional Meeting, held in Kyoto, Japan, between 30 September–6 October, 1984.

Astrophysics and Space Science **118** (1986) 271–277.

curve (e.g., VW Hyi: Warner and Harwood, 1973). The quasi-periodic oscillations, on the other hand, generally have their power spread over such a wide range of frequencies that they are not detectable in power spectra unless readily visible in the light curve. Low amplitude quasi-periodic oscillations may, therefore, be a common but undetectable phenomenon.

A few CVs show rapid periodic modulation of their X-ray flux (Table II). The amplitude of modulation is generally large (20–100%) and both coherent and quasi-periodic oscillations are known, although the coherence is often less than that of the corresponding optical oscillations (Cordova and Mason, 1982, 1983).

TABLE I

Optical oscillations in cataclysmic variables

Object	Type	Orbital period (hr)	Coherent periods (s)	Quasi periods (s)	Additional references
SS Cyg	DN	6.60	7.5–9.7	32–36	
RTU Peg	DN	8.99	11.6–11.8	~51	
TT Ari	NL	3.2		~12, ~32, ~40, 50–1100	Jensen *et al.* (1983) Mardirossian *et al.* (1980) Sztanjno (1979)
EM Cyg	DN	6.98	14.6–21.2		Steining *et al.* (1982)
Z Cam	DN (Z)	6.96	16.0–18.8		
V436 Cen	DN (SU)	1.50	19.5–20.1		
VW Hyi	DN (SU)	1.78	20–32	23, 88, 253, 413	Robinson and Warner (1984) Warner (unpublished)
HT Cas	DN	1.77	20.2–20.4	~100	
RR Pic	N	3.48		20–40	Schoembs and Stolz (1981)
KT Per	DN	–	22.0–29.2	82–147	
SY Cnc	DN	–	23.3–33.0		
AH Her	DN	5.93	24.0–38.8	~100	
CN Ori	DN	3.91	24.3–25.0		Schoembs (1982)
CDP −48°1577	NL	4.5:	24.6–29.1		Warner *et al.* (1984)
PS74	DN (SU)	2.0:	27–29	248	Warner (unpublished)
Z Cha	DN (SU)	1.79	24.8, 27.7		Warner (unpublished)
WZ Sge	DN	1.36	27.87, 28.97		
UX UMa	NL	4.72	28.5–30.0		
V3885 Sgr	NL	4.94	29–32		Warner (unpublished)
AE Aqr	NL	9.88	33.08	~36	
RX And	DN	5.08		36	
V2051 Oph	DN (SU)	1.50	40		O'Donoghue and Warner (unpublished)
V533 Her	N	–	63.63		Robinson and Nather (1979)
DQ Her	N	4.65	71.07		
U Gem	DN	4.25		73–146	
YZ Cnc	DN (SU)	2.21		75–95	
X Leo	DN	–		~160	
GK Per	N, DN	1.99 d		360–390	Watson *et al.* (1984)
RW Sex	NL	5.93		620, 1280	
V442 Cen	DN	11:		925	Marino and Walker (1984)

TABLE II

X-ray oscillations in cataclysmic variables

Object	Type	Orbital period (hrs)	Coherent period (s)	Quasi periods (s)	References
SS Cyg	DN	6.60		9–12	Cordova *et al.* (1984)
TT Ari	NL	3.2		9, 12, 32	Jensen *et al.* (1983)
VW Hyi	DN (SU)	1.78	14.06		Heise *et al.* (1984)
U Gem	DN	4.25		20–30	Cordova *et al.* (1984)
AE Aqr	NL	9.88	33		Patterson (1980)
YZ Cnc	DN (SU)	2.21		227	Cordova and Mason (1984)
GK Per	N, DN	1.99 d	351		Watson *et al.* (1984)

3. Temporal Behaviour

There is a wide range of temporal behaviour seen in the oscillating CVs. The prototype, DQ Her, has 71 s oscillations always visible (though of differing amplitude around the orbit and from year to year) and of high stability (Balachandran *et al.*, 1983). V533 Her seemed to be in the same class until its oscillations disappeared (Robinson and Nather, 1983). WZ Sge has one, sometimes two, oscillations always present (Robinson *et al.*, 1978) and AE Aqr usually shows a 33 s oscillation and its first harmonic (Patterson, 1979). In the other systems the oscillations are in various ways transient.

The nova-like objects V3885 Sgr (CD $-42°$14462), CDP $-48°$1577 and UX UMa all behave similarly in possessing slowly varying coherent periods near 29 s that are occasionally visible. In V3885 Sgr the amplitude of oscillation can reach as large as 0.33% but is less than 0.01% when apparently absent (Warner, 1973; and unpublished). There is no correlation of the presence or absence of the oscillations with the (small) variations in mean brightness of the system. In CPD $-48°$1577 the oscillations grow and decline in amplitude on timescales of hours (Warner *et al.*, 1984). The cause of these variations in visibility is not known but it does suggest that the coherent oscillations are *easily* obscured and therefore their absence in some similar systems (e.g. RW Tri, RW Sex) is not evidence for absence of the underlying phenomenon in these stars. Similarly, the cessation of observable oscillations in the nova remnant V533 Her could arise from the same cause and should not therefore be used as a compelling argument to differentiate between V533 Her and DQ Her (Robinson and Nather, 1983).

The dwarf novae (with the exception of the atypical WZ Sge) only show oscillations during outburst. In a given dwarf nova the behaviour of the oscillations is generally similar from outburst to outburst, covering a range of periods characteristic of the individual and appearing and disappearing during the outburst at approximately the same stages, even in ordinary maxima and supermaxima (Warner, 1976). There are exceptions, however: for example, an eruption of KT Per was extensively observed by Nevo and Sadeh (1978) and by Robinson and Nather (1978) without detecting the oscillations seen in other comparable outbursts of that star. In several well-observed dwarf novae (e.g., YZ Cnc: Patterson, 1981) coherent oscillations have never been detected.

There are some large differences in behaviour from star to star. In most dwarf novae the oscillations first appear just before or at maximum and are present for a few days (e.g., Patterson, 1981). In AH Her, however, the oscillations disappear for two days at maximum (Hildebrand *et al.*, 1980) and it may be significant that the same type of behaviour appears in the hard X-ray emission of SS Cyg (Ricketts *et al.*, 1979). In VW Hyi rapid coherent optical oscillations are seen only towards the end of outburst (Warner, 1976).

4. The Periods

The shortness of most of the coherent periods implies an origin close to the white dwarf component of the CV systems. In DQ Her the long-term stability of its period has been used to argue in favour of a rotational origin (e.g., Bath *et al.*, 1974) but the discovery of even greater stability in the ZZ Cet pulsating white dwarfs (e.g., Robinson and Kepler, 1980) eliminates this as a convincing argument. The rate of change of period, $dP/dt = -8.1 \times 10^{-13}$ (Balachandran *et al.*, 1983) is, however, readily understood as a result of acquisition of angular momentum by a rotating, accreting white dwarf (Lamb and Patterson, 1983). In addition, the phase variations of the 71 s oscillations, seen around the orbit and through eclipse (Patterson *et al.*, 1978b), are at least moderately well modelled by the effects of a beam of radiation emerging from a rotating white dwarf (Chester, 1979; Petterson, 1980).

In the nova-like and dwarf novae the situation is less clear – the large variations of period appear to preclude purely rotation of the white dwarf as the principal clock controlling the oscillations. However, the phase shifts during eclipse in UX UMa (Nather and Robinson, 1974; Petterson, 1980) and HT Cas (Patterson, 1981), although opposite in sign from those observed in DQ Her, show that a beam mechanism is probably required. The phase shifts in eclipses of Z Cha (Warner and Brickhill, 1978; Warner, unpublished) show quite different behaviour, which apparently depends on the orbital phase at which superhump maxima occur; analysis of these is currently underway and should aid in elucidating both the nature of the coherent oscillations and the structural changes that cause the superhumps (Warner, 1984).

The period variations that occur during outbursts of dwarf novae are of great significance (Warner, 1976; Patterson, 1981). On the rising branch of the outburst $dP/dt < 0$, with minimum period occurring about one day after visual maximum brightness (Patterson, 1981). On the falling branch of the outburst $dP/dt > 0$ but the same period is not observed at identical visual luminosities on the two branches. However, as shown by Bailey (1980) and Hassall *et al.* (1983), the ultraviolet flux (especially the far ultraviolet) lags behind the visible flux on the rising branch, reaching maximum about one day after visual maximum. It is probable, therefore, that there is a monotonic period-ultraviolet luminosity relationship for the coherent oscillations. This and the X-ray oscillations show that the coherent oscillations originate in the inner regions of the accretion disc – white dwarf structure, where the high-energy flux is generated.

The nova-like variables probably show a similar period-luminosity relationship (Warner *et al.*, 1984) and it is therefore possible that the oscillation periods could provide a monitor of the variations in far ultraviolet flux of the type seen in UX UMa (Holm *et al.*, 1982).

From Tables 1 and 2 it can be seen that, with the exception of AE Aqr (in which coherent, identically phased oscillations appear in both optical and X-ray regions: Patterson, 1980), there is only a loose correlation between the optical and X-ray periods. There is a striking difference in the coherence of the SS Cyg optical oscillations (Patterson *et al.*, 1978a) and its X-ray oscillations (Cordova *et al.*, 1984) but simultaneous observations have yet to be made. The apparent coherence of the 351 s X-ray oscillations seen during the dwarf nova-like outburst of GK Per (Watson *et al.*, 1984) contrast strangely with the variable quasi-periodic optical oscillations in that star. The 14.06 s X-ray oscillation in VW Hyi (Heise *et al.*, 1984), which appeared coherent for 10 days during the outburst, is of shorter period than anything yet detected in the optical region but occurred earlier in the outburst, when VW Hyi was brighter (and therefore should generate shorter period oscillations) than the phase at which coherent optical oscillations are normally present. If the X-ray and optical oscillations in VW Hyi originate in the same mechanism, then the period-luminosity relationship discussed above would require that the high-energy flux remain remarkably constant despite the general decline in visible flux.

Table I shows that the quasi-periodic oscillations generally have periods 3–5 times larger than the coherent oscillations in the same stars. On this basis, the quasi-periodic variations evidently originate further out in the accretion disc than the coherent ones. That they are not associated with the shock region (bright spot), where the inter-star stream collides with the accretion disc, is demonstrated by the fact that quasi-periodic oscillations in U Gem are seen throughout eclipse of the bright spot (Robinson and Nather, 1979).

There is as yet no detailed study of any correlation between the periods of the coherent and the quasi-periodic oscillations as they vary during a dwarf nova outburst.

5. Models

There are many qualitative models proposed for the region of high energy flux which is responsible for the beam that illuminates the accretion disc and causes the optical oscillations. Even though the X-ray modulations provide a more direct view of the physical processes there is still no identified mechanism. The various models include inhomogeneities in the inner accretion disc, non-radial pulsations of the accretion disc or the surface layers of the white dwarf, instabilities of mass transfer from the inner edge of the disc via magnetic field lines to the surface of the white dwarf. These are reviewed by Cordova and Mason (1982) who conclude that the observations are insufficient to distinguish between the models.

Mathematical models of the oscillations, aimed at quantifying the varying amounts of coherence, have been carried out (Robinson and Nather, 1979; Cordova and Mason, 1982; Cordova *et al.*, 1984; Robinson and Warner, 1984). These include descriptions in terms of amplitude, frequency or phase modulation and the superposition of many oscillations of similar periods. To this could be added the possibility of chaos in the form of Kolmogorov Instability (Perdang and Blacher, 1982); but it is far from clear that the results of such mathematical modelling will provide any physical insight into the underlying mechanisms.

References

Bailey, J.: 1980, *Monthly Notices Roy. Astron. Soc.* **120**, 119.
Balachandran, S., Robinson, E. L., and Kepler, S. O.: 1983, *Publ. Astron. Soc. Pacific* **95**, 653.
Bath, G. T., Evans, W. D., and Pringle, J. E.: 1974, *Monthly Notices Roy. Astron. Soc.* **166**, 113.
Chester, T. J.: 1979, *Astrophys. J.* **230**, 167.
Cordova, F. A. and Mason, K. O.: 1982, in J. P. Cox and C. J. Hansen (eds.), *Pulsations in Classical and Cataclysmic Variable Stars*, JILA, Boulder, p. 23.
Cordova, F. A. and Mason, K. O.: 1983, in W. H. G. Lewin and E. P. J. Van den Heuvel (eds.), *Accretion-Driven Stellar X-Ray Sources*, Cambridge, p. 147.
Cordova, F. A. and Mason, K. O.: 1984, *Monthly Notices Roy. Astron. Soc.* **206**, 879.
Cordova, F. A., Chester, T. J., Mason, K. O., Kahn, S. M., and Garmire, G. P.: 1984, *Astrophys. J.* **278**, 729.
Hassall, B. J. M., Pringle, J. E., Schwarzenberg-Czerny, A., Wade, R. A., Whelan, J. A. J., and Hill, P. W.: 1983, *Monthly Notices Roy. Astron. Soc.* **203**, 865.
Heise, J., Paerels, F., and van der Woerd, H.: 1984, *IAU Circ.*, No. 3939.
Hildebrand, R. H., Spillar, E. J., Middleditch, J., Patterson, J., and Stiening, R. F.: 1980, *Astrophys. J.* **238**, L145.
Holm, A. V., Panek, R. J., and Schiffer, F. H.: 1982, *Astrophys. J.* **25**, L35.
Jensen, K. A., Cordova, F. A., Middleditch, J., Mason, K. O., Grauer, A. D., Horne, K., and Gomes, R.: 1983, *Astrophys. J.* **270**, 211.
Lamb, D. Q. and Patterson, J.: 1983, *IAU Colloq.* **72**, 229.
Mardirossian, F., Mezzeti, M., Pucillo, M., Santin, P., Sedmak, P., Sedmak, G., and Giuricin, G.: 1980, *Astron. Astrophys.* **85**, 29.
Marino, B. F. and Walker, W. S. G.: 1984, *Southern Stars* **30**, 389.
Nather, R. E. and Robinson, E. L.: 1974, *Astrophys. J.* **190**, 637.
Nevo, I. and Sadeh, D.: 1978, *Monthly Notices Roy. Astron. Soc.* **182**, 595.
Patterson, J.: 1979, *Astrophys. J.* **234**, 978.
Patterson, J.: 1980, *Astrophys. J.* **140**, L133.
Patterson, J.: 1981, *Astrophys. J. Suppl.* **45**, 517.
Patterson, J. *et al.*: 1978a, *Astrophys. J.* **226**, L137.
Patterson, J., Robinson, E. L., and Nather, R. E.: 1978b, *Astrophys. J.* **224**, 570.
Perdang, J. and Blacher, S.: 1982, *Astron. Astrophys.* **112**, 35.
Petterson, J. A.: 1980, *Astrophys. J.* **241**, 247.
Ricketts, M. J., King, A. R., and Raine, D. J.: 1979, *Monthly Notices Roy. Astron. Soc.* **186**, 233.
Robinson, E. L. and Kepler, S. O.: 1980, *IAU Colloq.* **58**, 613.
Robinson, E. L. and Nather, R. E.: 1979, *Astrophys. J. Suppl.* **39**, 461.
Robinson, E. L. and Nather, R. E.: 1983, *Astrophys. J.* **273**, 255.
Robinson, E. L. and Warner, B.: 1984, *Astrophys. J.* **277**, 250.
Robinson, E. L., Nather, R. E., and Patterson, J.: 1978, *Astrophys. J.* **219**, 168.
Schoembs, R.: 1982, *Astron. Astrophys.* **115**, 190.
Schoembs, R. and Stolz, B.: 1981, *Inf. Bull. Var. Stars*, No. 1986.
Steining, R. F., Dragovan, M., and Hildebrand, R. H.: 1982, *Publ. Astron. Soc. Pacific* **94**, 672.
Sztanjno, M.: 1978, *Inf. Bull. Var. Stars*, No. 1710.

Walker, M. F.: 1954, *Publ. Astron. Soc. Pacific* **66**, 230.
Warner, B.: 1973, *Monthly Notices Roy. Astron. Soc.* **163**, 25P.
Warner, B.: 1976, *IAU Symp.* **73**, 85.
Warner, B.: 1979, in H. van Horn and V. Weidemann (eds.), *White Dwarfs and Variable Degenerate Stars*, Rochester, p. 417.
Warner, B.: 1984, in P. P. Eggleton and J. E. Pringle (eds.), *Interacting Binaries*, Cambridge, p. 367.
Warner, B. and Brickhill, A. J.: 1978, *Monthly Notices Roy. Astron. Soc.* **182**, 777.
Warner, B. and Harwood, J. M.: 1973, *Inf. Bull. Var. Stars*, No. 756.
Warner, B. and Robinson, E. L.: 1972, *Nature Phys. Sci.* **239**, 2.
Warner, B., O'Donoghue, D. and Allen, S.: 1985, *Monthly Notices Roy. Astron. Soc.* **212**, p. 9.
Watson, M. G., King, A. R., and Osborne, J.: 1985, *Monthly Notices Roy. Astron. Soc.* **212**, 917.

SHORT-PERIOD NONCONTACT BINARIES*

ATSUMA YAMASAKI

Department of Earth Science and Astronomy, University of Tokyo, Japan

AKIRA OKAZAKI

Tsuda College, Kodaira, Tokyo, Japan

and

MASATOSHI KITAMURA

Tokyo Astronomical Observatory, Mitaka, Tokyo, Japan

(Received 15 May, 1985)

Abstract. Five short-period noncontact binaries (GR Tau, UU Lyn, BV Eri, RU Eri, and BW Eri) are studied. It is found from photometric and spectroscopic observations that (1) both components are noncontact to each other but are very close to the respective Roche lobes; (2) period change does not appear or it is very small; and (3) component stars of UU Lyn and RU Eri are normal main sequence stars while those of GR Tau and BV Eri are smaller and less massive components indicating in an advanced evolutionary state having experienced mass reversal.

We have made optical observations of some close binaries (short-period noncontact binaries) in which (1) the orbital period is less than ~ one day, (2) both components are noncontact, but (3) neither of the components is a degenerate star. The aim of the present study is to obtain accurate physical quantities on the components and to clarify the evolutionary state of these binaries.

So far, we have studied five binaries: GR Tau, UU Lyn, BV Eri, RU Eri, and BW Eri. Photometric and spectroscopic observations were made for GR Tau, UU Lyn, and RU Eri by us at Okayama and at Dodaira, and for BV Eri and BW Eri by H. W. Duerbeck (University Bonn) and D. Baade (ESO) at ESO (La Silla) and at CTIO. (For RU Eri, light curves of Sarma and Sanwal (1981) were used in the photometric analysis.)

Spectral types were determined from spectroscopic and photometric (colour) observations. They are A and F as expected from their orbital periods. The spectral types are considered to be those of the primary components since most of the light comes from the primary components.

No emission lines were detected in any phase in these binaries. Also, no lines of the secondary components were detected because of their faintness. Thus, from the radial velocity measurements, only the mass-function values were obtained. It was found that UU Lyn and RU Eri have reasonable values of mass functions while GR Tau and BV Eri have very small mass functions.

* Paper presented at the IAU Third Asian-Pacific Regional Meeting, held in Kyoto, Japan, between 30 September–6 October, 1984.

Astrophysics and Space Science **118** (1986) 279–282.

TABLE I

Short-period noncontact binaries studied

System (period)	Spectral type	Mass functions ($M_\odot$)	Light curve	Mass ratio	Mass ($M_\odot$)		Radius ($R_\odot$)		Reference
					Primary	Secondary	Primary	Secondary	
GR Tau ($0^d.430$)	A5 V	0.0101	Asymmetry	(a) 0.32	0.54	0.17	1.03	0.49	Yamasaki *et al.* (1984)
				(b) 0.20	2.0 (assumed)	0.39	1.69	0.74	
UU Lyn ($0^d.468$)	F3 V	0.0496	Symmetry	(a) 0.41	1.40	0.57	1.40	0.90	Yamasaki *et al.* (1983)
BV Eri ($0^d.508$)	F2–3	0.0079	Asymmetry	(a) 0.30	0.50	0.15	1.10	0.63	Baade *et al.* (1983)
				(b) 0.20	1.5 (assumed)	0.30	1.55	0.88	
RU Eri ($0^d.632$)	F3–4 IV–V	0.0853	Almost symmetry	(a) 0.55	1.38	0.76	1.60	1.20	Nakamura *et al.* (1984)
BW Eri ($0^d.638$)	A8 V	–	Symmetry	(a) 0.36	–	–	–	–	Baade *et al.* (1982)

Three systems, UU Lyn, RU Eri, and BW Eri, have symmetrical light curves while two systems, GR Tau and BV Eri, show marked asymmetry. For example, GR Tau is faint at phase around 0.65 by 0.06 mag. (V) and 0.08 mag. (B). If the photometric disturbance is distributed symmetrically, the disturbance must be seen from $0^p.4$ to $0^p.9$ (Yamasaki *et al.*, 1984). In BV Eri, the photometric disturbance is seen between $0^p.7$ and $0^p.9$ (Baade *et al.*, 1983). Thus in both binaries, disturbances were seen mainly in the last half phase.

From times of minima observations, the periods of GR Tau and RU Eri were found to be stable for the last 50 yr and 70 yr, respectively. We may say that a period change does not appear, or it is very small.

Light curves of the five systems were analyzed by the synthesis method developed by one of us (AY). Mass ratio (photometric), inclination, fractional radii, and luminosities were determined. Derived photometric mass ratios are given in rows (a) in Table I. For GR Tau and BV Eri, unperturbed parts of the light curves were analyzed.

It was found that either component or both are very close to their respective Roche lobes, but no component is in contact with it. No contact solutions were possible.

Combining the results of spectroscopic and photometric analyses, we derived masses and radii as given in rows (a) in Table I. For UU Lyn and RU Eri, the derived quantities are consistent with the spectral type, while for GR Tau and BV Eri these values are not consistent with Main-Sequence components of the same spectral type.

On the other hand, if we assume that the masses of the primary components of GR Tau and BV Eri can be given by those of the Main Sequence of the same spectral types, then larger masses and radii of the components are obtained (shown in rows (b) in Table I). However, theoretical light curves computed with these elements no longer satisfy the observations.

If the masses and radii are taken at face value as given in rows (a) in Table I, the positions of the components of these short-period noncontact binaries in the mass-radius diagram indicate that:

(1) Component stars of UU Lyn and RU Eri, in which no light curve disturbances were observed, are Main Sequence, or near Main-Sequence stars; while (2) component stars of GR Tau and BV Eri, in which light curves show asymmetry and radial velocity curves give very small amplitude, have low mass, small size, and low luminosity, and they cannot be Main-Sequence stars.

Therefore, it can be said that short-period noncontact binaries may evolve like

[UU Lyn / RU Eri]	→	Roche overflow	→	Mass transfer, mass loss	→	[GR Tau / BV Eri]
Main-Sequence components				(mass reversal)		Low mass, low luminosity components.

References

Baade, D., Duerbeck, H. W., Karimie, M. T., and Yamasaki, A.: 1982, *Astrophys. Space Sci.* **88**, 197.
Baade, D., Duerbeck, H. W., Karimie, M. T., and Yamasaki, A.: 1983, *Astrophys. Space Sci.* **93**, 69.
Nakamura, Y., Yamasaki, A. and Kitamura, M.: 1984, *Publ. Astron. Soc. Japan* **36**, 277.
Sarma, M. B. K., and Sanwal, N. B.: 1981, *Astrophys. Space Sci.* **74**, 41.
Yamasaki, A., Okazaki, A., and Kitamura, M.: 1983, *Publ. Astron. Soc. Japan* **35**, 131.
Yamasaki, A., Okazaki, A., and Kitamura, M.: 1984, *Publ. Astron. Soc. Japan* **36**, 175.

THE EFFECT OF BACK PRESSURE ON THE CONTACT EVOLUTION OF A CLOSE BINARY SYSTEM*

MASAOMI NAKAMURA
Astronomical Institute, Tohoku University, Sendai, Japan

and

YASUHISA NAKAMURA
Komaba High School, Ohashi, Tokyo, Japan

(Received 8 May, 1985)

Abstract. The contact condition which includes the effect of pressure in common envelope is derived and employed in the evolutionary calculation of a close binary system in contact phase. The result indicates that the effect is negligibly small and the equality of surface potentials of both components is a sufficiently good approximation for the contact condition.

1. Introduction

In evolutionary calculations of close binary systems in the contact phase, the equality of surface potentials of both components is usually assumed as the contact condition. However, there are some investigators who question the results of evolutionary calculations which use such a contact condition. They insist on the following evolutionary scenario. A primary fills its critical Roche lobe and begins to transfer its mass to the secondary. The secondary drastically increases in its radius and fills its Roche lobe. At the onset of contact the back pressure of the secondary might prevent further mass transfer, and the evolution after that might be quite different from what usual evolutionary calculations indicate. We think that this is not a correct evolutionary picture. We consider this problem as follows. As pointed out by Neo *et al.* (1977), although the radius of the accreting star increases as the matter falls onto the surface and 'piles up', each layer of accumulated matter is contracting and sinking when it is seen in the Lagrangian mass coordinate. Then if the transfer of mass is prevented, the radius of the secondary begins to decrease and the mass is then transferred from primary to secondary in the thermal time-scale of the interior region.

In any case, since the effect of pressure is neglected in the usual contact condition, then we would now generalize the contact condition so as to incorporate this effect and examine how large it is.

* Paper presented at the IAU Third Asian-Pacific Regional Meeting, held in Kyoto, Japan, between 30 September–6 October, 1984.

Astrophysics and Space Science **118** (1986) 283–286.

2. Procedures

We assume that in the contact phase the transfer flow is described by Bernoulli's law. Also, assuming a polytropic relation $P = K\rho^\gamma$ for the gas, we obtain the relation

$$\psi + \frac{\gamma}{\gamma - 1}\frac{P}{\rho} + \frac{1}{2}v^2 = \text{constant}, \tag{1}$$

along a stream line. Here, ψ, P, ρ, v, and γ are the potential, pressure, density, velocity, and ratio of specific heats, respectively. The flow velocity may be considered to be small in all its regions except the vicinity of the L_1 point. Therefore, relation (1) may be approximated, almost everywhere in the common envelope, by

$$\psi_1 + \frac{\gamma}{\gamma - 1}\left(\frac{P}{\rho}\right)_1 = \psi_2 + \frac{\gamma}{\gamma - 1}\left(\frac{P}{\rho}\right)_2, \tag{2}$$

where the subscript 1 refers to the primary and 2 to the secondary. In order to evaluate the total effect of pressure we take an average of Equation (2) over the common envelope and obtain the relation

$$\frac{1}{M_1^{\text{ce}}}\int_{R_1^{\text{cr}}}^{R_1}\left(\psi_1 + \frac{\gamma}{\gamma - 1}\frac{P_1}{\rho_1}\right)4\pi r^2\rho_1\,\mathrm{d}r = \frac{1}{M_2^{\text{ce}}}\int_{R_2^{\text{cr}}}^{R_2}\left(\psi_2 + \frac{\gamma}{\gamma - 1}\frac{P_2}{\rho_2}\right)4\pi r^2\rho_2\,\mathrm{d}r, \tag{3}$$

where M_i^{ce} is the mass of the common envelope which belongs to each component, i.e.,

$$M_i^{\text{ce}} = \int_{R_i^{\text{cr}}}^{R_i} 4\pi r^2\rho_i\,\mathrm{d}r; \quad i = 1, 2 \tag{4}$$

where R_i and R_i^{cr} are the radii of stellar surface and of inner critical surface, respectively.

We adopt Equation (3), with Equation (4), as the new contact condition. Then we follow the evolution of a close binary system, whose component stars' initial masses are $M_1^0 = 1.50\,M_\odot$ and $M_2^0 = 0.75\,M_\odot$ and initial orbital separation is $A^0 = 4.50\,R_\odot$.

3. Results

Figure 1 shows the evolution of both components in the mass-radius diagram. Both components are on the Zero-Age Main-Sequence at point 1. The primary fills its critical Roche lobe at point 2 and mass transfer begins. The secondary grows rapidly in its radius and fills its critical Roche lobe at point 3. At the onset of contact the mass-transfer rate begins to decrease, but the direction of the flow is still from the primary to the secondary. Passing through the mass reversal, the secondary detaches from its critical surface at point 4. At point 5 both components reach a thermal equilibrium state and the system is semidetached.

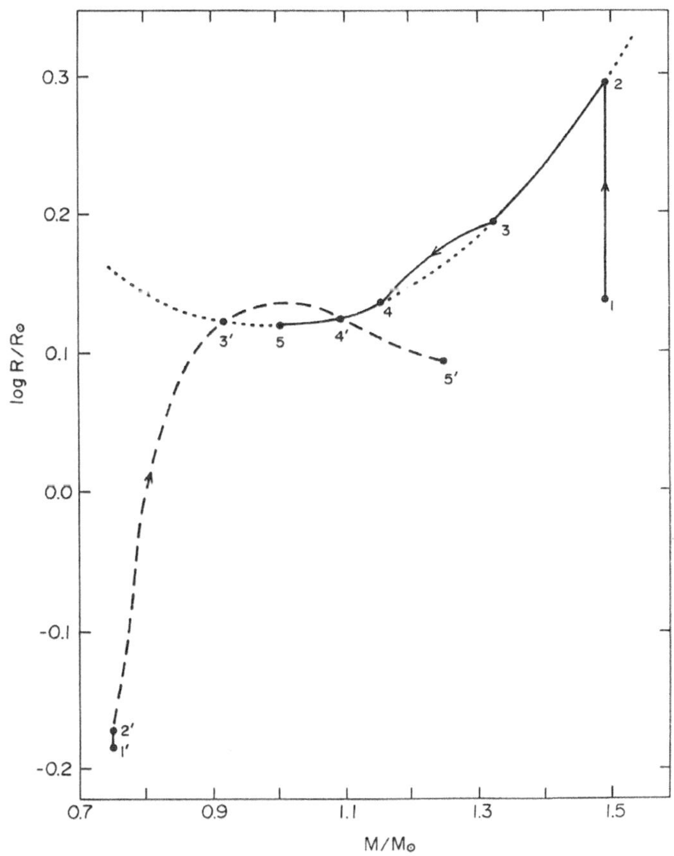

Fig. 1. Evolutionary paths of the components in mass-radius diagram. The solid and dashed curves refer to the primary and the secondary components, respectively. The numerals are the stage numbers described in the text. The dotted curve shows the radius of the inner critical Roche lobe.

For the sake of comparison with this result, we also calculated the evolution of the same system but by using the usual contact condition, which assumes the equality of surface potentials of both components. The result is shown in Figure 2. As can be seen, there arises no appreciable difference.

4. Conclusions

We have calculated the evolution of a close binary system in the contact phase by taking account of the effect of pressure in the common envelope. As can be seen in Figure 2, the primary component is slightly undersized by the introduction of the new contact condition. This describes the effect of pressure which is very small.

We can conclude that the effect of pressure on the evolution of the system is negligibly small and that the equality of the surface potentials of both components is a sufficiently good approximation for the contact condition.

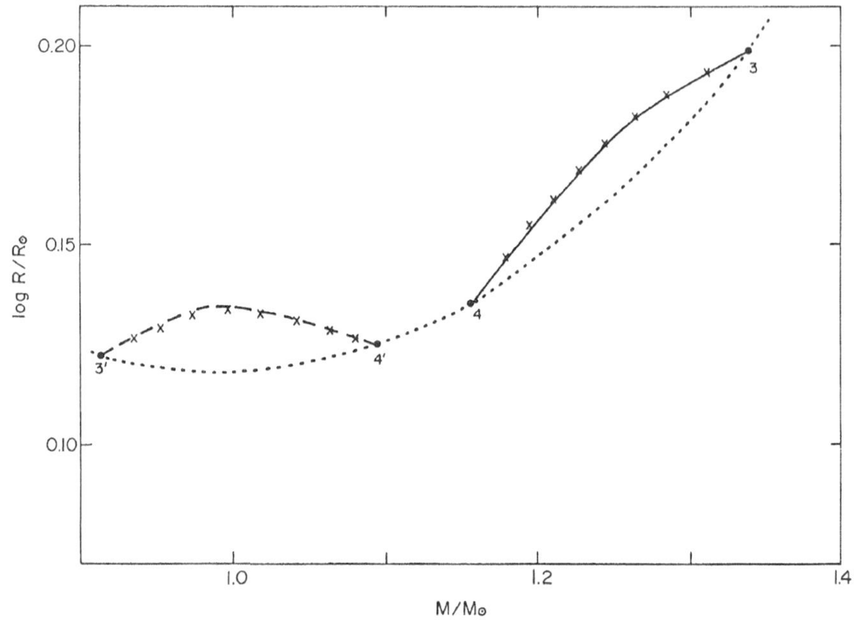

Fig. 2. Evolutionary changes in radii of the primary component (solid curve) and of the secondary component (dashed curve) for the case computed with new contact condition (Equation (3) in the text). The symbols × indicate the radii of the evolutionary models with the conventional contact condition.

Acknowledgements

We wish to express our thanks to Professor K. Suda for encouragement and discussions. We also thanks to Professor D. Sugimoto for useful comments. One of us (Y.N.) is grateful to Professor M. Kitamura for encouragement. Numerical computations were performed on the ACOS 1000 computer at the Computer Center of Tohoku University.

Reference

Neo, S., Nomoto, K., Miyaji, S., and Sugimoto, D.: 1977, *Publ. Astron. Soc. Japan* **29**, 249.

THE ELECTRON TEMPERATURE OF AM HERCULIS TYPE SYSTEMS*

D. T. WICKRAMASINGHE and S. MEGGITT

Department of Mathematics, Australian National University, Canberra A.C.T., Australia

(Received 8 May, 1985)

Abstract. We argue that the polarised optical emission from AM Herculis-type systems arises from a compact high temperature ($T \gtrsim 10$ keV) region closely associated with the accretion shock.

It is well established that the optical radiation emitted by AM Herculis type systems has a cyclotron component that is strongly polarised and originates from the vicinity of the magnetic poles of the accreting white dwarf. The polar field strength has been estimated directly from the cyclotron spectrum in VV Puppis (Wickramasinghe and Visvanathan, 1980) and estimated from the Zeeman spectrum in several other systems (Latham *et al.*, 1984). The estimates range of polar field strengths $B_p \sim 1\text{–}3 \times 10^7$ G corresponds to cyclotron harmonic numbers $\omega/\omega_c \sim 5\text{–}15$ in the optical band.

Given the magnetic field strength and the wavelength dependence of intensity and polarisation, it should in principle be possible to estimate the electron temperature T_e and the size parameter Λ of the cyclotron emission region. The presently available analyses have yielded widely different results for T_e and Λ. For instance, Wickramasinghe and Meggitt (1982) have estimated $T_e = 10$ keV, $\Lambda \simeq 10^5$ from wavelength dependent optical observations of VV Puppis. On the other hand, Barrett and Chanmugam (1984) deduced $T_e \sim 0.1\text{–}2$ keV, $\Lambda \sim 10^7\text{–}10^8$ from broad band (V) polarisation observations of AM Herculis. Thus it is at present unclear if the polarised optical radiation arises from a compact hot region closely associated with the shock or an extended cool preshock region. A resolution of this discrepancy is important for theories of the structure of the accretion column.

We believe that it is impossible to determine unambiguously the electron temperature from observations in a single broad band. However, a distinction between the high- and low-temperature models can be made by the use of multiwavelength (preferably spectropolarimetric) observations. The characteristics of low and high temperature models which enable this distinction to be made are summarised in Figure 1. The calculations have been carried out using the theory and notation outlined by Meggitt and Wickramasinghe (1982). We note that:

(i) Low-temperature models show a strong modulation of intensity and polarisation (circular and linear) with frequency while the high-temperature models do not show such structure. With the exception of VV Puppis, which sometimes shows weak cyclotron

* Paper presented at the IAU Third Asian-Pacific Regional Meeting, held in Kyoto, Japan, between 30 September–6 October, 1984.

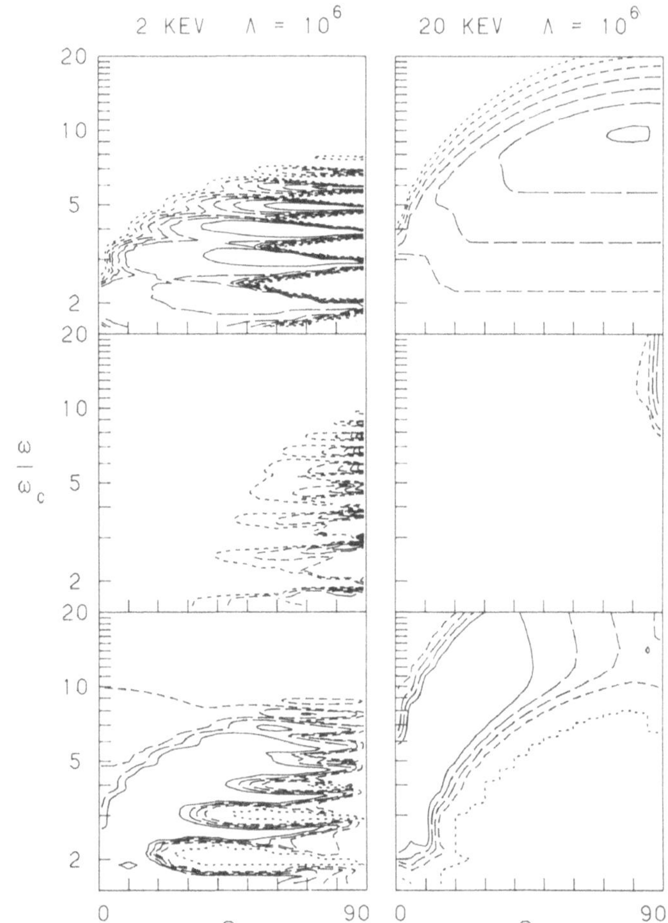

Fig. 1. Contour plots of intensity (top panel), linear polarisation (centre panel), and circular polarisation (bottom panel) for uniform temperature models with T_e = 2 and 20 keV, $\Lambda = 10^6$, and electron number density $N_e = 10^{16}$ cm^{-3}. ω/ω_c is the cyclotron harmonic number and θ is the angle between the line of sight and the direction of the magnetic field. The intensity I_ω is represented by 1 mag. contours, the solid curve corresponding to the brightest intensity. The linear polarisation contours correspond to 80 (solid) 60, 40, 20, and 10%. The circular polarisation contours correspond to 80 (solid), 60, 40, 20, and 0 (dotted) %.

harmonic features corresponding to $T_e \approx 10$ keV, harmonic features are not seen in any other AM Herculis type system.

(ii) The linear polarisation pulse is sharper, more pronounced, and extends to higher harmonics at high temperatures. The available observations indicate sharp pulses in the frequency range $\omega/\omega_c \sim 5$–15 again supporting the high-temperature models.

We conclude that the optical emission region in AM Herculis type systems is compact and hot ($T \gtrsim 10$ keV) probably associated with the shock front.

References

Barrett, P. E. and Chanmugam, G.: 1984, *Astrophys. J.* **278**, 298.
Latham, D. W., Liebert, J. and Steiner, J. E.: 1981, *Astrophys. J.* **246**, 919.
Meggitt, S. M. A. and Wickramasinghe, D. T., 1982, *Monthly Notices Roy. Astron. Soc.* **198**, 71.
Stockman, H. S., Foltz, C. B., Schmidt, G. D., and Tapia, S.: 1983, *Astrophys. J.* **271**, 725.
Wickramasinghe, D. T. and Meggitt, S. M. A.: 1982, *Monthly Notices Roy. Astron. Soc.* **198**, 975.
Wickramasinghe, D. T. and Visvanathan, N. V.: 1980, *Monthly Notices Roy. Astron. Soc.* **191**, 589.

PHOTOMETRY, POLARIMETRY, AND SPECTROSCOPY OF AM HERCULIS VARIABLES*

I. R. TUOHY, N. VISVANATHAN, and D. T. WICKRAMASINGHE

Australian National University, Canberra, Australia

(Received 8 May, 1985)

Abstract. We report AAT observations of two southern AM Herculis variables, H0139–68 and E1405–451. H0139–68 was found to be in an unusually faint state ($V \gtrsim 17$), and this has permitted the first measurement of the magnetic field of the white dwarf, $B_p \sim 3 \times 10^7$ G. Linear polarimetry of E1405–451 has revealed a substantial position angle variation with phase, corresponding to an orbital inclination of $\sim 60°$ and a colatitude of $\sim 20°$ for the magnetic pole of the white dwarf.

1. H0139–68

H0139–68 was observed in white light on 5 June and 7 September, 1983 using the Single Channel Automatic Polarimeter (SCAP; Visvanathan and Tuohy, 1983) at the Cassegrain Focus of the Anglo-Australian Telescope (AAT). Figures 1 and 2 show the sky-subtracted light curves, plotted against extrapolated binary phase. The deep eclipsing behaviour observed previously during the bright state of H0139–68 (Pickles and Visvanathan, 1983) was absent on each occasion; instead, the source was found to be in a faint ($V \sim 17$ mag.) and relatively inactive state. The only pronounced variability that is evident is a series of narrow minima in the 7 September data. These have a depth of up to 1 mag. and all occur near $\phi = 0$. However, none of the observed minima are separated by *exactly* the binary period of 113.6 min suggesting, therefore, that they are aperiodic in nature, rather than being true eclipse phenomena. Nevertheless, their occurrence near the phase at which the shock front is expected to be viewed near 90° to the direction of the magnetic field suggests that they are the result of optical depth variations within the shock front.

Polarimetric data acquired simultaneously with the photometry of Figures 1 and 2 show that the strong ($\sim 20\%$) linear polarization pulse observed previously near $\phi = 0$ (Visvanathan and Tuohy, 1983) was absent on both occasions. Vector addition of the data yields an upper limit to the polarization of 2% (note that our preliminary result of $P \sim 6\%$ for 5 June given by Visvanathan *et al.* (1983) is incorrect). The absence of significant linear polarization from H0139–68 is not surprising in view of the weakness of the cyclotron component implied by the faintness of the system.

Our detection of the first faint state of H0139–68 (IAU Circular, No. 3833), subsequently noted by other observers (IAU Circulars 3846 and 3895), provided an important opportunity to conduct a spectroscopic study of the underlying white dwarf

* Paper presented at the IAU Third Asian-Pacific Regional Meeting, held in Kyoto, Japan, between 30 September–6 October, 1984.

Astrophysics and Space Science **118** (1986) 291–294.

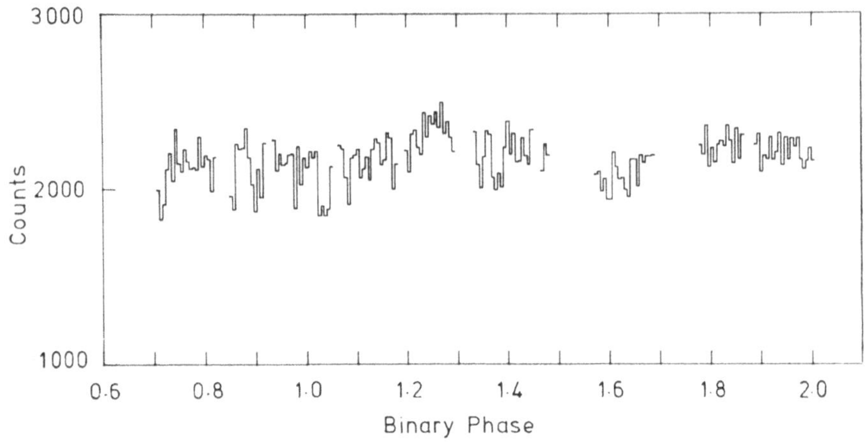

Fig. 1. The light curve obtained for H0139–68 on 5 June, 1983, 16 : 58 to 19 : 26 UT. Binary phase has been calculated using the ephemeris of Pickles and Visvanathan (1983), and is accurate to $\Delta\phi \sim 0.1$. Note the absence of a deep eclipse preceding $\phi = 1.0$.

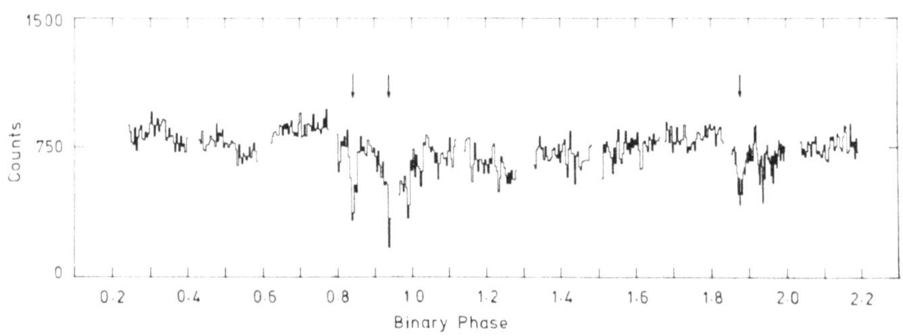

Fig. 2. The light curve obtained for H0139–68 on 7 September, 1983, 13 : 54 to 17 : 38 UT. Arrows indicate prominent narrow minima preceding $\phi = 1.0$ and $\phi = 2.0$.

(which is usually invisible when the accretion column dominates the luminosity). With this in mind, we obtained a phase-averaged spectrum of H0139–68 on 29 September, 1983 using the AAT. The spectrum is shown in Figure 3, and clearly reveals the presence of several broad features attributable to Zeeman absorption in a hydrogen atmosphere. Modelling of these features in terms of a centered-dipole field distribution indicates a polar field strength of $B_p = 3 \pm 0.3 \times 10^7$ G (Figure 3; see also Wickramasinghe *et al.*, 1984). There are now four measurements of field intensity for AM Herculis systems, and, interestingly, all lie in the narrow range $2\text{–}3 \times 10^7$ G.

2. E1405–451

We observed E1405–451 with the AAT and SCAP on 5 June, 1983 and 28 March, 1984 for the specific purpose of measuring the behaviour of the linear polarization position

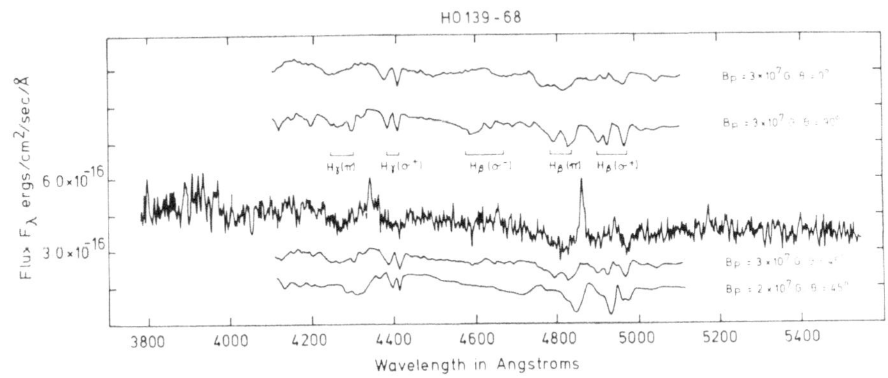

Fig. 3. A phase-averaged spectrum of H0139−68 obtained on 29 September, 1983. Zeeman hydrogen absorption components are indicated, together with theoretical spectra for different polar magnetic field strengths (B_p) and viewing angles (θ). Closest agreement with the data is for $B_p \sim 3 \times 10^7$ G and $\theta \sim 45°$.

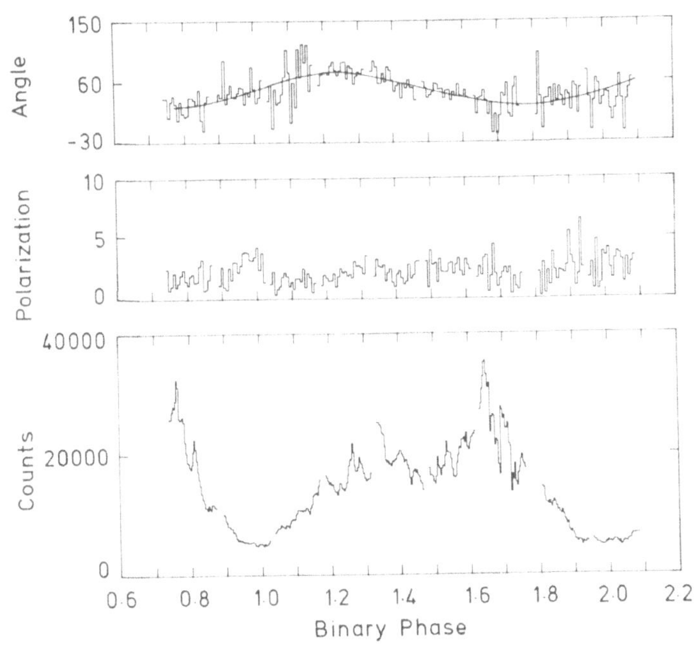

Fig. 4. Light curve, linear polarization, and position angle for E1405−451 on 5 June, 1983, 13 : 58 to 16 : 14 UT. Binary phase has been computed using the ephemeris of Tuohy *et al.* (1985). Note the significant residual linear polarization and associated position angle variation, which is fitted by the model discussed in the text.

angle as a function of binary phase. The 5 June data were taken in white light for maximum statistics, and the sky-subtracted results for one of the observed binary cycles are shown in Figure 4. Significant linear polarization ($P \sim 2\%$) is evident throughout the light curve, and exhibits a systematic variation in position angle ($\Delta\theta \sim 70°$) with phase. A similar variation is present in earlier cycles on 5 June, and in *V*-band data acquired on 28 March, 1984. The 28 March data also show the presence, for the first time, of a pronounced linear polarization pulse ($P \sim 8\%$) at exactly $\phi = 0.5$ (Tuohy *et al.*, 1985). This latter result supports a geometry for E1405–451 in which the light-emitting magnetic pole is closest to the limb of the white dwarf at $\phi = 0.5$.

On the basis of this geometry, we have modelled the observed phase variation of the polarization position angle using the theory outlined by Meggitt and Wickramasinghe (1982). Our best-fit model, plotted in Figure 4, corresponds to an orbital inclination of $60 \pm 5°$, and a colatitude for the dominant magnetic pole of $20 \pm 5°$. These parameters imply that the emitting pole is not eclipsed as the white dwarf rotates, in agreement with the circular polarization data of Tapia (1982) and Bailey *et al.* (1983), which show no change of sign with phase.

3References

Bailey, J., Axon, D. J., Hough, J. H., Watts, D. J., Giles, A. B., and Greenhill, J. G.: 1983, *Monthly Notices Roy. Astron. Soc.* **205**, 1P.
Meggitt, S. M. A. and Wickramasinghe, D. T.: 1982, *Monthly Notices Roy. Astron. Soc.* **198**, 71.
Pickles, A. and Visvanathan, N.: 1983, *Monthly Notices Roy. Astron. Soc.* **204**, 463.
Tapia, S.: 1982, *IAU Circ.*, No. 3685.
Tuohy, I. R., Visvanathan, N., and Wickramasinghe, D. T.: 1985, *Astrophys. J.* **289**, 721.
Visvanathan, N. V. and Tuohy, I.: 1983, *Astrophys. J.* **275**, 709.
Visvanathan, N. V., Tuohy, I., and Wickramasinghe, D. T.: 1983, *IAU Circ.*, No. 3833.
Wickramasinghe, D. T., Visvanathan, N., and Tuochy, I. R.: 1984, *Astrophys. J.* **286**, 328.

LINE PROFILES IN SYMBIOTIC STARS*

B. LOUISE WEBSTER, L. H. TAAFFE

School of Physics, The University of New South Wales, Kensington, Australia

and

A. J. KALNAJS

Mount Stromlo and Siding Spring Observatory, Australian National University, Canberra, Australia

(Received 8 May, 1985)

Abstract. High resolution profiles of permitted and forbidden lines in symbiotic stars are described, with particular reference to the broad wings. An interpretation of the data in terms of an optically thin accretion disc around a hot subdwarf is presented.

Symbiotic star line profiles in the optical region are characterized by broad wings on the hydrogen lines, and usually by a relation between ionization potential and width in lines from other ions. This paper summarizes the results of high dispersion spectroscopy on hydrogen, [O III] and other lines in a few symbiotic stars, and on Hα in a larger number, with particular reference to the broad wings. Line profiles predicted for an optically-thin accretion disc are then compared with the observations.

The line profiles were observed with the coudé echelle spectrograph and 2DPCA on the Mount Stromlo 1.9 m telescope and with the intermediate dispersion spectrograph and IPCS on the 3.9 m Anglo-Australian Telescope. The resolutions at Hα were 5 and 27 km s^{-1}, respectively. Reductions were performed with the computing facilities at the AAO. Some results on individual objects of particular interest follow.

RX Puppis: In its low excitation state in 1973, the Hβ line had particularly strong wide wings (Webster and Allen, 1975) and P Cygni profiles (Swings and Klutz, 1976). The wings had weakened by 1983 when the ionization level had increased. In the latter stage the line width was positively correlated with the ionization potential.

He 2–177: (Hen 1242, PK 326 − 10° 1). The profiles are illustrated in Figure 1 and the development of the wings between 1965 and 1972 and the line widths have been described by Webster (1973). No wings comparable to those on the Balmer lines are present in [O III]. The wings are stronger on Hα than on Hβ.

RR Telescopii: The line widths in 1983 were relatively constant at about 60 km s^{-1}, irrespective of ionization stage, and the hydrogen lines were weak. Friedjung (1966) found that between 1952 and 1965 the widths of all lines were decreasing with time,

* Paper presented at the IAU Third Asian-Pacific Regional Meeting, held in Kyoto, Japan, between 30 September–6 October, 1984.

Astrophysics and Space Science **118** (1986) 295–298.

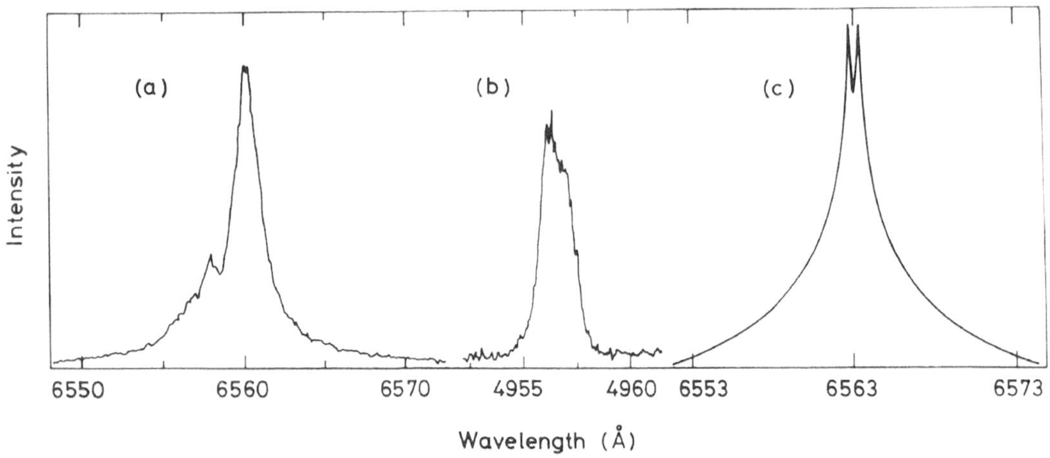

Fig. 1. A comparison between observed line profiles and one calculated for an optically thin accretion disc. (a) He 2–177, observed Hα profile, (b) He 2–177, observed [O III] 4959 profile, (c) calculated Hα profile.

although lines from high ionization states were the widest, and the trend has apparently continued. Several lines ([O III], Hα, Fe II) have a satellite line about 100 km s^{-1} blueward, which may be a displaced component since Thackeray (1977) lists no other lines at these positions.

He 2–390: (PK 5 – 5°2). This is one of the rare symbiotics with [N II] lines and these have similar profiles, with wings, to the Hα profiles.

The following general statements about the broad wings to the line profiles can be made, based on assorted observations of more than fifteen stars:

(1) Generally in the optical region very broad wings are found only on the hydrogen Balmer lines. Similar profiles are seen in some early type binaries and other emission-line stars.

(2) The Balmer decrement for the wings is steeper than that for the central component. That is, the wings are most prominent on Hα.

(3) No obvious relation could be discovered between the shape of the Hα line (e.g., the presence of an absorption dip, or the proportion of wing to central component) and any other features of the spectrum. In particular, there is no clear distinction between the profiles from D and S class symbiotics (Webster and Allen, 1975) even though the ions represented in the nebular spectrum are quite different.

(4) The Full Width at Zero Intensity (FWZI) of the Hα wings, though hard to define in noisy spectra, varies only from 20 to 40 Å. (Estimates from the profiles published by Oliversen and Anderson (1982) for ten other symbiotics vary from 17 to 30 Å.)

Possible origins of the wings are (a) electron scattering, (b) stellar wind/s, and (c) an accretion disc. Although both (a) and (b) are quite plausible, the ubiquitous nature of the wings and lack of correlation with the forbidden line spectrum suggest a connection with a dense region, such as the accretion disc which would be expected to exist if the system is an interacting binary. The line profiles emerging from an optically thin disc have been computed by Unewisse *et al.* (1984), and the agreement in shape between the calculated profiles and the observed outer wings is excellent. The centre of the line is expected to be affected by self-absorption, distortion of the orbits, the nebular contribution and other effects. The steep Balmer decrement is consistent with collisional excitation at electron temperatures below 16000 K, as predicted by the models. Unewisse *et al.* (1985) find a high density in the disc which conforms with the general lack of wings in the forbidden lines. The [N II] profiles in He 2–390 are a mystery, whatever the origin of the wings, and a mechanism for keeping the density of an accretion disc low may be required. Two puzzles concern the changes in CI Cygni through eclipse (Oliverson *et al.*, 1982) and the behaviour and width of He II. It may be that the eclipse is of a He II hotspot, as is found for cataclysmic variables.

The measurements of the FWZI point to a mass to radius ratio of about 10 $M_{\odot}/R_{\odot}$ for the hot central object in the majority of symbiotics. This value is similar to that of a planetary nebula nucleus or hot subdwarf. The source of ionization is unlikely to be the transition zone between the disc and the star, as the temperature here depends on the mass accretion rate (Pringle, 1977) and there is no apparent relation between the accretion rate deduced from the wings and the ionizing temperature. The unseen companion has been variously proposed in the literature as being a main-sequence star, a white dwarf or a subdwarf (see Plavec, 1982). A subdwarf has been suggested from luminosity or evolutionary considerations. Our accretion disc interpretation of the wings supports this type of ionizing object for a different reason – the determination of mass to radius ratio. The most plausible model would thus appear to be one of the type proposed by Tutukov and Yungel'son (1976) or Paczyński and Rudak (1980) in which mass is accreted onto a hot subdwarf. The luminosity depends primarily on the core mass, while the temperature depends on the mass in a hydrogen-rich surface layer which, in turn, varies with the accretion rate and with luminosity. By analogy with planetary nebulae, the subdwarf mass needs to be above about 1 $M_{\odot}$ to reach the high temperatures observed, and this is not inconsistent with the luminosities and galactic distribution of symbiotic stars.

Acknowledgements

We are grateful to Mount Stromlo and Siding Spring Observatory for observing time on the 1.9 m telescope and to the Anglo-Australian Observatory for the use of the VAX computing facilities.

References

Friedjung, M.: 1966, *Monthly Notices Roy. Astron Soc.* **133**, 401.

Oliversen, N. A. and Anderson, C. M.: 1982, in M. Friedjung and R. Viotti (eds.), *The Nature of Symbiotic Stars*, D. Reidel Publ. Co., Dordrecht, Holland, p. 71.

Oliversen, N. A., Anderson, C. M., and Nordsieck, K. H.: 1982, in M. Friedjung and R. Viotti (eds.), *The Nature of Symbiotic Stars*, D. Reidel Publ. Co., Dordrecht, Holland, p. 153.

Paczyński, B. and Rudak, B.: 1980, *Astron. Astrophys.* **82**, 349.
Plavec, M. J.: 1982, in M. Friedjung and R. Viotti (eds.), *The Nature of Symbiotic Stars*, D. Reidel Publ. Co., Dordrecht, Holland, p. 231.
Pringle, J. E.: 1977, *Monthly Notices Roy. Astron Soc.* **178**, 195.
Swings, J. P. and Klutz, M.: 1976, *Astron. Astrophys.* **46**, 303.
Thackeray, A. D.: 1977, *Mem. Roy. Astron. Soc.* **83**, 1.
Tutukov, A. V. and Yungel'son, L. R.: 1976, *Astrofizika* **12**, 521.
Webster, B. L.: 1973, *Monthly Notices Roy. Astron. Soc.* **164**, 381.
Webster, B. L. and Allen, D. A.: 1975, *Monthly Notices Roy. Astron Soc.* **171**, 171.
Unewisse, M. H., Neilson, D., Taaffe, L. H., and Webster, B. L.: 1985, in preparation.

LIGHT CURVES OF V711 TAURI (HR 1099)*

IL-SEONG NHA and JOON-YOUNG OH

Yonsei University Observatory, Seoul, Korea

(Received 8 May, 1985)

Abstract. *V* and *B* observations of V711 Tauri (HR 1099) for the year 1982–1983 are presented. The light curves made with the data of this season are compared with those of the previous observations of the 1981–1982 season. The single-peaked maximum light had shifted slightly and the peak amplitude had decreased to 0.04 mag. in *V*. No clear sign of colour, $B - V$, variation in comparison with the low temperature dark-spot is detected.

V711 Tau (HR 1099) is one of the most observed RS CVn stars, and that over a wide range of wavelengths, from X-rays to radio. Nevertheless, the peculiar nature of V711 Tau is far from being comprehensively understood, given the present knowledge of stellar evolution. Among the many peculiarities that V711 Tau exhibits are light variations which are not associated with the classical binary interactions, such as 'reflection' and 'ellipticity' effects. The most powerful hypothesis for the explanation of these light variations is a starspot, covering mostly one hemisphere of the more active component. This has been postulated by many investigators. The two-spot model developed by Dorren *et al.* (1981) and extended later by Dorren and Guinan (1982) is the latest and perhaps the most promising working hypothesis to describe the light variations of the *V*-light curves of other observers, and their own λ6585 light curves. They assumed the two spots to be circular and the same size, and cooler than the surrounding photosphere by about 1800 K.

Soon after this two-spot model was anounced, this hypothesis was questioned by Mekkaden *et al.* (1982) who had made a *UBV* photometric study of V711 Tau over two years, the 1979–1980 and 1980–1981 seasons. With their empirical relation for the wave amplitude vs the light maximum, shown in their Figure 3, Mekkaden *et al.* (1982) found that a decrease in the light amplitude is followed by a decrease in the brightness of the star, a result that is in disagreement with the prediction of Dorren *et al.* (1981) that the brightness of a spot-free component should correspond to the maximum observed brightness when the wave amplitude tends to zero.

All the light curves of V711 Tau known before the 1980–1981 season, since the discovery of the light variation in 1963 (Cousins, 1963), may be characterized as having a single peak followed by a curved minimum whose total shape resembles that of the δ Cep stars. But a dramatic change happened in the V711 Tau system in the course of its long-term light curve evolution. This change was reported by Blanco *et al.* (1981) who

* Paper presented at the IAU Third Asian-Pacific Regional Meeting, held in Kyoto, Japan, between 30 September–6 October, 1984.

Astrophysics and Space Science **118** (1986) 299–303.

first noticed it in their light curve made in 1980–1981 in which the single-peaked curve suddenly changed to a double-peaked curve, which appeared again in the light curve of Mekkaden *et al.* (1982).

In order to check V711 Tau for its light variations after this unexpected change, *B* and *V* observations were made during two observation seasons, 1981–1982 and 1982–1983, with the 40 cm reflector at the Yonsei University Observatory. Instrumentation and observational techniques adopted for these observations are presented elsewhere (Nha and Kang, 1982). The standardized *B* and *V* data made in the first season, 1981–1982, were reported by Sarma *et al.* (1984) and those of the latter season, 1982–1983, are given in Table I.

TABLE I

Normal points in standard *V* and $\Delta(B-V)$ made at the Yonsei University Observatory in 1982–1983

J.D. Hel. 2445000 +	ΔV	$\Delta(B-V)$	n	J.D. Hel. 2445000 +	ΔV	$\Delta(B-V)$	n
253.1275	1.500	+0.370	3	344.0602	1.497	+0.352	3
.1563	1.480	0.326	2	.1982	1.499	0.340	3
258.1099	1.461	0.380	3	.1322	1.514	0.333	3
.1678	1.483	0.374	3	.1617	1.499	0.325	3
.2168	1.484	0.339	3	.1865	1.503	0.326	2
.2682	1.491	0.372	3	364.9886	1.460	0.355	3
267.0489	1.510	0.346	3	365.0235	1.459	0.359	3
.0865	1.522	0.345	3	.0542	1.452	0.373	3
.1350	1.520	0.368	4	.0837	1.453	0.334	3
.1752	1.510	0.366	3	371.9386	1.478	0.333	3
308.9695	1.486	0.369	2	.9570	1.497	0.343	3
309.0390	1.501	0.368	3	.9820	1.503	0.323	3
.1069	1.509	0.351	3	.9940	1.484	0.343	3
.1601	1.507	0.348	3	372.0400	1.488	0.368	4
315.9112	1.448	0.371	3	.0590	1.482	0.353	3
.9389	1.467	0.355	3	.0710	1.469	0.363	4
.9695	1.454	0.353	3	.0930	1.456	0.357	4
316.0011	1.468	0.369	3	378.9296	1.446	0.343	2
.0466	1.456	0.363	3	.9500	1.432	0.358	3
.0972	1.466	0.348	3	.9763	1.452	0.352	3
.1336	1.457	0.347	3	379.0166	1.436	0.320	3
.1732	1.462	0.346	3	.0515	1.446	0.334	3
.2141	1.461	0.339	3	.0813	1.434	0.334	3
.2543	1.428	0.363	3	385.9321	1.484	0.350	3
336.9327	1.453	0.347	3	.9607	1.499	0.346	3
.9828	1.466	0.335	3	.9893	1.504	0.339	3
337.0300	1.467	0.344	3	386.0156	1.496	0.351	3
.0818	1.482	0.374	3	392.9420	1.468	0.337	3
.1237	1.492	0.340	3	.9693	1.439	0.362	3
.1554	1.493	0.321	3	.9956	1.450	0.346	3
343.9377	1.494	0.351	3	393.0209	1.452	0.337	3
.9684	1.501	0.349	3	.0425	1.431	+0.352	3
344.0131	1.478	+0.396	3				

Since this dramatic light change in 1980–1981, V711 Tau has exhibited puzzling light variations. In the first year, 1981–1982, as reported by Nha and Oh (1983), a single-peak of amplitude $0^m.09(V)$ appeared at a phase of about $0^p.75$ computed from the ephemeris of Bopp and Fekel (1976); but the smaller peak of Blanco *et al.* (1981) at a phase of about $0^p.3$ was replaced by constant light over a wide phase range $0^p.95$–$0^p.55$. In our second year, 1982–1983, the single-peaked maximum light had shifted slightly to about $0^p.70$, but the peak amplitude had decreased drastically to $0^m.04(V)$. The constant light is still present in the light curve made in this second year but the phase interval is somewhat decreased, $0^p.0$–$0^p.5$.

In dealing with a scenario for the light variations in V711 Tau it is convenient to designate the 1980–1981 season as an epoch separating the light curves of V711 Tau made before and after. These variations are best seen in Figure 1 in which four representative light curves (one for pre-, one during-, and two post-change) are plotted. The upper two light curves in Figure 1 are made with the data of Mekkaden *et al.* (1982) and the lower two from the present work. The solid curves are free-hand curves which represent the trend of the observations in each season. The 1979–1980 curve shows the typical shape for V711 Tau during pre-change time. The light curve during the change (1980–1981) is definitely different in shape with a significant secondary peak, but this

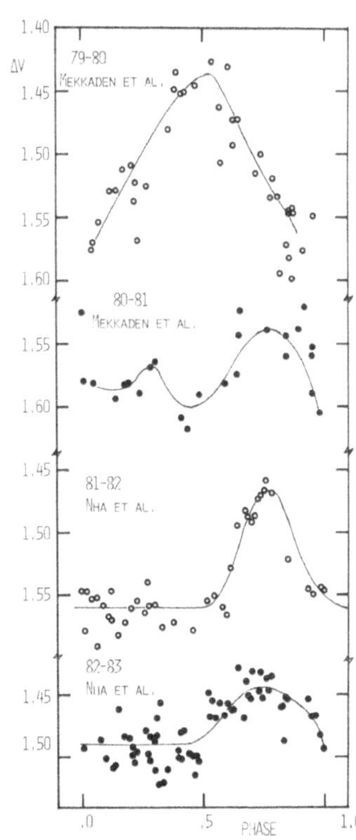

Fig. 1. V-light curves of V711 Tau. From top to bottom, one pre-change (open circles), one during-change (close circles), and two post-change (open and closed circles).

secondary peak is totally absent in the light curves made post-change, and is replaced by the flat minimum.

It is our purpose here to test with our photometric data, using also limb darkening values and $B - V$ variations, the consistency of the spot model which has been such a powerfull working theory for the light variations of the pre-change light curves.

We present our two year observations in Figure 2 for comparison. In this figure the V light curve of the check star, 12 Tau, is shown at the top, and the V and $B - V$ light curves of V711 Tau are, respectively, at center and bottom. Symbols used in the figure are open circles for 1981–1982 and closed circles for 1982–1983 observations for both stars. The ΔV (12 Tau–10 Tau) data demonstrate that the light from both check and comparison stars did not change in the two years, showing a straight line distribution with a mean light of about $1\overset{m}{.}28$ and an observational error deduced from the check star observations of $\pm 0\overset{m}{.}005$.

The V-light curves of V711 Tau in Figure 2 are, on the other hand, significantly dissimilar between the two years. V711 Tau increased in brightness in the 1982–1983 season, at the flat minimum by $0\overset{m}{.}07$ and at the peak by $0\overset{m}{.}02$ over the previous year, but the peak V amplitude has changes from $0\overset{m}{.}09$ in the first year only 0.05 in the second year, nearly one-half of the previous year's. A significant feature of these two light curves is that both have long flat minima lasting for about one-half of the orbital period. If one assumes limb darkening to some degree on one hemisphere of the more active star, then this spotted surface would be seen with unequal brightness as the star rotates. As a consequence of this, the light minima should not be uniformly flat. These flat and relatively long minima observed in two seasons by the present investigators

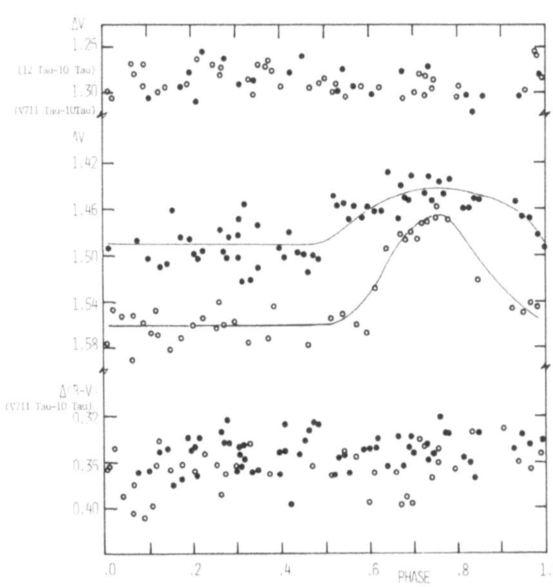

Fig. 2. Light curves of 12 Tau and V711 Tau and the colour curves of V711 Tau made in 1981–1982 (open circles) and in 1982–1983 (closed circles).

cannot be described by the two-spot model as a star with limb-darkening coefficients of 0.5–0.7, as given by Dorren *et al.* (1981).

The supporting evidence, which is contrary to the dark-spot hypothesis for the light variations of the V711 Tau system, is the $B - V$ variation shown at the bottom of Figure 2. The average $B - V$ values are $0^{m}.375$ and $0^{m}.345$ for the 1981–1982 and 1982–1983 seasons, respectively. The phase of maximum brightness, if this is assumed to be attributable to the unspotted photosphere, should show corresponding $B - V$ values bluer than those of the minimum phases, but in each season in Figure 2 no clear sign of color variation is detected.

There is, therefore, no realistic explanation as yet for the peculiar light behavior of the photometrically complicating V711 Tau system. Although the spot model has been known as a powerful working hypothesis to most investigators for V711 Tau and for RS CVn stars in general, we need to search for other mechanisms as well, and meantime these stars should be systematically monitored in anticipation of new clues.

Acknowledgement

This work was supported by grants from the Ministry of Education, Republic of Korea.

References

Blanco, C., Catalano, S., Marilli, E., Rodono, M., and Scaltriti, F.: 1981, *Inf. Bull. Var. Stars*, No. 2000.
Bopp, B. W. and Fekel, F.: 1976, *Astron. J.* **81**, 771.
Cousins, A. W. J.: 1963, *Monthly Notices Astron. Soc. S. Afr.* **22**, 58.
Dorren, J. D. and Guinan, E. F.: 1982, *Astrophys. J.* **252**, 296.
Dorren, J. D., Siah, M. J., Guinan, E. F., and McCook, G. P.: 1981, *Astron. J.* **86**, 572.
Mekkaden, M. V., Raveendran, A. V., and Mohin, S.: 1982, *J. Astrophys. Astron.* **3**, 27.
Nha, I.-S. and Kang: 1982, *Publ. Astron. Soc. Pacific* **94**, 496.
Nha, I.-S. and Oh, J.-Y.: 1983, in 'Activity in Red-Dwarf Stars', *IAU Colloq.* **71**, 409.
Sarma, M. B. K., Ausekar, B. D., Prakasa Rao, B. V. N. S., Oh, J.-Y., and Nha, I.-S.: 1985, *Publ. Astron. Soc. Japan* **37**, 107.

CARBON DEFLAGRATION MODELS FOR TYPE-I SUPERNOVAE AND THEORETICAL OPTICAL SPECTRA*

KEN'ICHI NOMOTO

Dept of Earth Science and Astronomy, University of Tokyo, Japan

FRIEDRICH-K. THIELEMANN

Max-Planck-Institut für Physik und Astrophysik, Institut für Astrophysik, Garching, F.R.G.

KOICHI YOKOI

Institut für Kernphysik III, Karlsruhe, F.R.G.

and

DAVID BRANCH

Dept of Physics and Astronomy, University of Oklahoma, Norman, OK, U.S.A.

(Received 14 June, 1985)

Abstract. A carbon deflagration model in the accreting C + O white dwarf is presented as a plausible model for type-I supernovae (SN I). The evolution of the white dwarf is calculated from the beginning of accretion through complete disruption. The propagation of a carbon deflagration wave from center to the surface and associated nucleosynthesis are calculated. About 0.6 $M_{\odot}$ ^{56}Ni and substantial amounts of Ca, Ar, S, Si, Mg, and O are synthesized.

The synthetic spectrum at maximum light for this carbon deflagration model is calculated and compared to the observed spectrum of SN 1981b in NGC 4536 at maximum light. If the radial stratification of composition in the outer layers of the model is removed by mixing, the theoretical spectrum is in good agreement with the observed spectrum.

1. Introduction

Type-I supernovae (SN I) are quite distinct from type-II supernovae (SN II) because of hydrogen-deficiency in their spectra near maximum light. SN I are observed in all types of galaxies including elliptical galaxies and not concentrated in spiral arms (see, e.g., Trimble, 1982, for a review). These facts can be naturally accounted for by the white dwarf model of SN I. Although this model is currently most popular (e.g., Nomoto, 1984), whether it is consistent with other observational constraints needs more investigations.

In the present study, we calculate the evolution of a C + O white dwarf from the beginning of accretion through explosion. In particular, nucleosynthesis associated with the carbon deflagration is investigated in detail. We also calculate a theoretical optical spectrum for this model and compare it with observations.

* Paper presented at the IAU Third Asian-Pacific Regional Meeting, held in Kyoto, Japan, between 30 September–6 October, 1984.

Astrophysics and Space Science **118** (1986) 305–309.

2. Propagation of the Carbon Deflagration Wave

The initial model is a white dwarf of $M = 1\,M_\odot$ that has been cooled down for 5.8×10^8 yr. The accretion rate was assumed to be $\dot{M} = 4 \times 10^{-8}\,M_\odot\,\mathrm{yr}^{-1}$. The accretion of hydrogen-rich matter at a rate of $\dot{M} \gtrsim 4 \times 10^{-8}\,M_\odot\,\mathrm{yr}^{-1}$ results in a steady or weakly unstable hydrogen shell burning. Accretion of helium with $\dot{M} \gtrsim 4 \times 10^{-8}\,M_\odot\,\mathrm{yr}^{-1}$ also leads to a weak flash (e.g., Nomoto, 1982a, b). Therefore, the accretion increases the mass of the C + O white dwarf through recurrence of the shell flashes.

We approximated such an increase in M by assuming the steady hydrogen-helium shell burning, where the accreted material is processed into C + O at the same rate as the accretion. As matter is accreted, both ρ_c and T_c increase by compression and heat conduction from the outer hot layers. When ρ_c reaches $2.6 \times 10^9\,\mathrm{g\,cm}^{-3}$ ($M = 1.378\,M_\odot$), carbon is ignited at the center. The central carbon flash grows into thermonuclear runaway because of strong electron degeneracy, and the material is incinerated into nuclear statistical equilibrium (NSE) composition.

Afterwards, the explosive carbon burning front propagates outward on the time-scale for convective energy transport across the front (Nomoto *et al.*, 1976; Muller and Arnett, 1982; Woosley *et al.*, 1984). A burning front that propagates at a subsonic velocity with respect to the unburnt material is called a *deflagration wave* in contrast to a detonation wave that propagates at a supersonic velocity.

To simulate the propagation of a convective carbon deflagration wave with a 1D hydrodynamic code, the time-dependent mixing-length theory of convection by Unno (1967) was employed. The mixing-length was taken to be $l = 0.7\,H_p$ where H_p is a pressure scale-height.

In the present model, it takes 1.2 s for the deflagration wave to reach the shell at $M_r = 1.3\,M_\odot$ which is six times longer than the propagation of a detonation wave (Arnett, 1969). During such a slow propagation of the deflagration wave, the white dwarf gradually expands to decrease the density and temperature. Such an expansion weakens the explosive nuclear burning at the deflagration wave and eventually quenches the carbon burning when the deflagration wave reaches $M_r = 1.3\,M_\odot$ where $\rho \sim 10^7\,\mathrm{g\,cm}^{-3}$.

The total nuclear energy release exceeds the initial binding energy of the white dwarf so that the star is disrupted completely leaving no compact star behind. The explosion energy is 1.3×10^{51} erg. The expansion velocities at $M_r = 0.9$–$1.2\,M_\odot$ are $v_{\mathrm{exp}} = 10\,000$–$14\,000\,\mathrm{km\,s}^{-1}$ and in good agreement with the observed velocities of SN I deduced from the spectra near maximum light.

3. Explosive Nucleosynthesis in the Carbon Deflagration Wave

The white-dwarf material undergoes explosive burning of carbon, neon, oxygen, and silicon at the passage of the deflagration wave in a generalized sense according to final burning products. The nuclear products of such explosive burning depends mainly on the peak temperature T_p at the deflagration wave. The final composition structure based on the network is shown in Figure 1.

For the inner layer at $M_r < 0.7\,M$, $T_{p,9} \gtrsim 6$ so that the nuclear reactions are rapid enough to incinerate the material into almost NSE composition ($T_{p,9} \equiv T_p/10^9$ K). As seen in Figure 1, the central layers are composed of neutron-rich iron-peak elements (mostly ^{56}Ni).

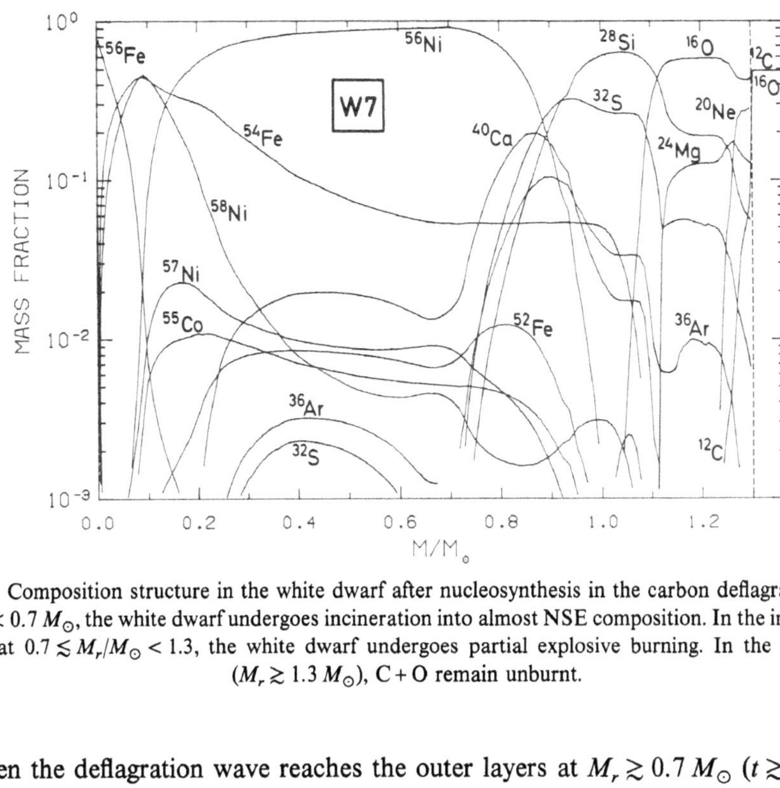

Fig. 1. Composition structure in the white dwarf after nucleosynthesis in the carbon deflagration wave. At $M_r < 0.7\,M_\odot$, the white dwarf undergoes incineration into almost NSE composition. In the intermediate region at $0.7 \lesssim M_r/M_\odot < 1.3$, the white dwarf undergoes partial explosive burning. In the outer layer ($M_r \gtrsim 1.3\,M_\odot$), C + O remain unburnt.

When the deflagration wave reaches the outer layers at $M_r \gtrsim 0.7\,M_\odot$ ($t \gtrsim 1$ s), the white dwarf has already expanded and ρ_c has decreased by a factor of more than 10. Accordingly the peak temperature attained during explosive burning T_p is lower because of larger heat capacity. Then the nuclear reactions are slowed down so that the material undergoes explosive burning but is not incinerated into NSE composition.

For the layers at $0.7 \lesssim M_r/M_\odot < 0.9$, T_p is as low as $6 > T_{p,9} \gtrsim 5$. The reactions are not rapid enough to process the material into ^{56}Ni; in other words, this layer undergoes partial Si-burning whose products are Si, S, Ar, Ca, Fe, Ni, etc. For $0.9 \lesssim M_r/M_\odot < 1.1$, $5 > T_{p,9} \gtrsim 4$. Explosive carbon and oxygen burning produces Si, S, and Ar. For $1.1 \lesssim M_r/M_\odot < 1.25$, $4 > T_{p,9} \gtrsim 3$ so that this layer undergoes explosive carbon and neon burning which produces O, Mg, Si, etc., but oxygen burning is too slow to proceed appreciably. Finally, in the quenching phase of the deflagration wave, only the carbon burning products appear at $1.25 \lesssim M_r/M_\odot < 1.3$. In the outermost layer at $M_r \gtrsim 1.3\,M_\odot$, the original C + O remains unburnt.

4. Light Curves and Hubble Constant

The carbon deflagration supernova models presented here ejected 0.58 $M_\odot$ ^{56}Ni. The approximate light curves powered by the radio-active decays of ^{56}Ni and ^{56}Co (e.g., Axelrod, 1980) were calculated by employing Chevalier's (1981) approximation. Qualitative characteristics of the SN I light curves are well reproduced. The peak bolometric luminosities are 1.3×10^{43} erg s^{-1}. Since the observed value for typical SN I is 2×10^{43} erg s^{-1} $(H_0/50$ km s^{-1} Mpc$^{-1})^{-2}$ (Wheeler *et al.*, 1980), the present model is consistent with observations of SN I if $H_0 \simeq 60$. For $H_0 > 70$, it is difficult to get a carbon deflagration model that is consistent with SN I observations (Arnett *et al.*, 1985).

5. Theoretical Optical Spectra

The early-time spectra of SN I show the composition structure in the outer layer which is expanding at $\sim 10^4$ km s^{-1} (Branch, 1984). We calculate the synthetic spectrum for the present model by taking the time from explosion to maximum light to be 15 days (Figure 2, lower). Spectral lines are assumed to be formed by resonant scattering, and are treated in the escape-probability approximation. A fit to the observed spectra of SN I 1981b at maximum light (Figure 2, upper) is fairly good provided that the outer layers at $M_r \gtrsim 0.7\, M_\odot$ are mixed. On the other hand, the spectrum for the model with complete composition stratification does not show a good fit.

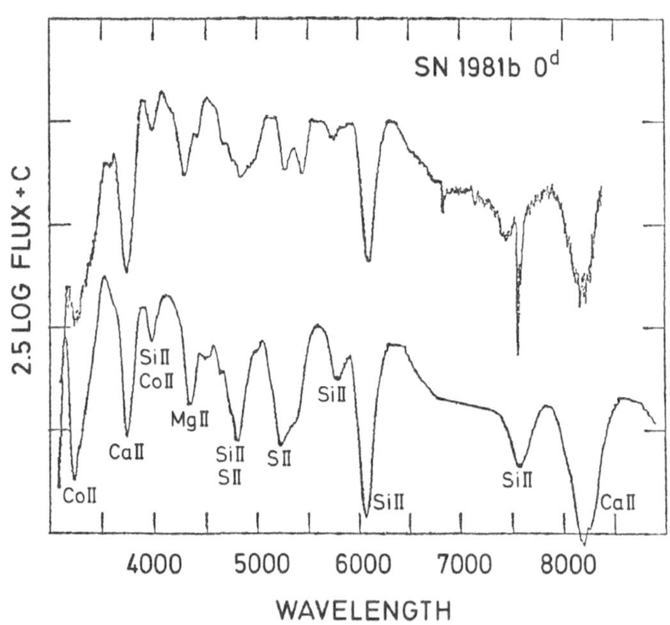

Fig. 2. The maximum-light spectrum of SN 1981b (upper) is compared to a synthetic spectrum based on the carbon deflagration model (lower).

Although a mixing process after passage of the deflagration wave was not included in the present calculation, convective mixing in the outer layers is expected to occur because of the decaying nature of the deflagration wave. This point needs more study.

In conclusion, many of the observed features of SN I, in particular, near maximum light, can be accounted for by the carbon deflagration model in accreting white dwarfs.

Acknowledgement

This work is supported in part by the Japanese Ministry of Education, Science, and Culture through the Research Grant Nos. 58340032, 5938001, and 60540152.

References

Arnett, W. D.: 1969, *Astrophys. Space Sci.* **5**, 180.
Arnett, W. D., Branch, D., and Wheeler, J. C.: 1985, *Nature* **314**, 337.
Axelrod, T. S.: 1980, in J. C. Wheeler (ed.), *Type I Supernovae*, University of Texas, Austin, p. 80.
Branch, D.: 1984, in W. D. Arnett (ed.), *Nucleosynthesis: Challenges and New Developments*, University of Chicago Press, Chicago.
Chevalier, R. A.: 1981, *Astrophys. J.* **246**, 267.
Muller, E. and Arnett, W. D.: 1982, *Astrophys. J.* **261**, L107.
Nomoto, K.: 1982a, *Astrophys. J.* **253**, 798.
Nomoto, K.: 1982b, *Astrophys. J.* **257**, 780.
Nomoto, K.: 1984, in C. Chiosi and A. Renzini (eds.), *Stellar Nucleosynthesis*, D. Reidel Publ. Co., Dordrecht, Holland, p. 205.
Nomoto, K., Sugimoto, D., and Neo, S.: 1976, *Astrophys. Space Sci.* **39**, L37.
Trimble, V.: 1982, *Rev. Mod. Phys.* **54**, 1183.
Unno, W.: 1967, *Publ. Astron. Soc. Japan* **19**, 140.
Wheeler, J. C., Branch, D., and Falk, S. W., Jr.: 1980, in J. C. Wheeler (ed.), *Type-I Supernovae*, Univ. of Texas, Austin, p. 199.
Woosley, S. E., Axelrod, T. S., and Weaver, T. A.: 1984, in C. Chiosi and A. Renzini (eds.), *Stellar Nucleosynthesis*, D. Reidel Publ. Co., Dordrecht, Holland, p. 263.

A SURVEY OF EMISSION-LINE STARS IN THE CMa STAR FORMATION REGION*

SUHARDJA D. WIRAMIHARDJA**, TOMOKAZU KOGURE, MAKOTO NAKANO, SHIGEOMI YOSHIDA, and KEN'ICHI TATEMATSU

Department of Astronomy, Faculty of Science, Kyoto University, Kyoto, Japan

(Received 13 June, 1985)

Abstract. A survey of emission-line stars was carried out in the CMa star formation region, based on spectral observations with the Kiso Schmidt telescope. In total, 128 emission-line stars were detected, ranging from $V = 6$ to 15, and the majority are found with V between 11 and 14. A comparison with the existing catalogues shows that some emission-line stars are common with ours, while some exhibit no detectable emission line on our plates, suggesting the variability of emission-line strength. The two-colour ($U - B$, $B - V$) diagram is also shown for the detected stars.

1. Introduction

The CMa star formation region is one of the well-known sites of star formation, which was first discovered by Ambartsumian (1949) as an O association. Since then this region has been the subject of many investigations.

Kopylov (1958) studied the distribution of several aggregates of early-type stars, and Ruprecht (1966) defined its boundary to be from 222° to 226° in galactic longitude, and from $-3\overset{\circ}{.}4$ to $+0\overset{\circ}{.}7$ in latitude. He described it as CMa OB1. Claria (1974a) made photoelectric measurements in the *UBV* system for 247 stars, most of which are of early types. Based on the photometric and spectroscopic data of these OB stars, Claria (1974b) confirmed the existence of the stellar association at a mean distance of 1150 parsec from the Sun, and with the mean visual absorption of 0.81 mag. In addition, van den Bergh (1966) noted that within the limit of CMa OB1 there is a conglomerate of reflection nebulae and called this group CMaR1. It is claimed by Claria that this group is physically associated with the CMa OB1.

We now report a survey for emission-line stars in this CMa star formation region undertaken by us with the Kiso Schmidt telescope. Results are compared with the previous surveys of Wackerling (1970), Sanduleak and Stephenson (1973), and Stephenson and Sanduleak (1977a, hereafter referred to as SS).

2. Observations

For this survey the observations were made using the 105/150/330 cm Schmidt telescope of the Kiso Observatory, University of Tokyo. The Schmidt plate covers a sky area of $5\overset{\circ}{.}2 \times 5\overset{\circ}{.}2$. The plate-filter combination of IIIaF (hypersensitized by baking in forming

* Paper presented at the IAU Third-Asian-Pacific Regional Meeting, held in Kyoto, Japan, between 30 September–6 October, 1984.

** On leave from the Bosscha Observatory, Institute of Technology Bandung, Indonesia.

gas) and Schott RG 615 is mainly used to detect emission-line stars. The 4°-objective prism attached to the telescope yields a dispersion of 700 Å mm^{-1} at Hα.

Besides the program region, we also surveyed a reference region adjacent to it in the S–E direction. The photographed region, therefore, consists of three overlapping areas, which have the following centers:

Area	R.A. (1950)	Dec. (1950)
A-1052	07^h00^m	− 10°00′
A-7052	07 08	− 12 00
A-1125	07 20	− 15 00

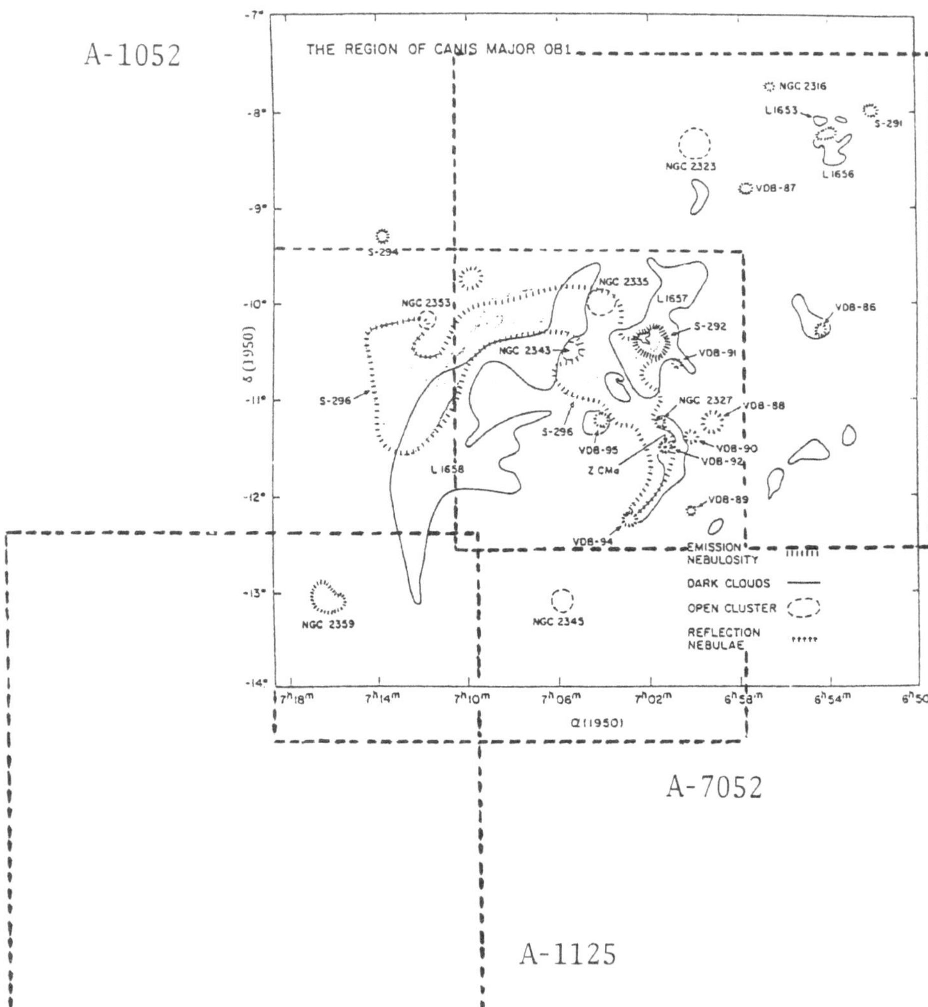

Fig. 1. The photographed regions (dashed line), superimposed on the schematic map of CMa OB1/R1 reproduced from Blitz (1978). A-1052 and A-7052 are the programme regions and A-1125 is the reference region.

The whole area covered by the present survey is about fifty square degrees. Ten spectral plates, with exposure times ranging from 3 to 90 min, were used for this survey. The programme and reference regions are shown in Figure 1, superimposed on the schematic map of CMa OB1/R1 reproduced from Blitz (1978).

Photometric plates were collected with the 40/70/120 cm Ouda Schmidt telescope of Kyoto University. The CMa region was direct-photographed in *U*, *B*, and *V* with the plate-filter combination as follows: Kodak 103aO + U-360 for *U*, Kodak 103aO + L-39 for *B*, and Kodak 103aD + L-50 for *V*.

3. Results and Discussion

The discoveries of Hα-emission-line stars were made visually using magnifiers. On our plates an Hα-emission line of medium strength can usually be easily detected without much doubt. But this is not the case when the intensity of the emission line is weak with respect to the continuum. Only when the emission line appears on two or more different plates, do we define that emission to be real.

The *U*, *B*, *V* magnitudes of Hα-emission-line stars were derived using the standard stars of Hoag *et al.* (1950) for NGC 2323 and NGC 2353, and of Kazanasmas *et al.* (1982) for NGC 2345 and Haffner 8. The accuracy is estimated to be ± 0.15 (p.e.). The limiting magnitude of the present survey is somewhat brighter than $V = 15$, although a few emission-line stars with magnitude fainter than this value are still visible.

The distribution of the stars with respect to the *V* magnitude shows that the number of stars increases as the magnitude goes fainter, and the majority of the emission-line stars are found with *V* between 11 and 14. A sudden decrease occurs at the beginning of $V = 14$, which might be caused by incompleteness of the survey for fainter magnitudes.

From the characteristics of the surveyed region, it was expected to find T Tauri stars associated with dark clouds. However, no strong grouping of faint Hα-emission-line stars was found. Instead, a total of 128 Hα-emission-line stars were detected somewhat scattered over the region.

After examining existing lists of Hα-emission objects, covering the CMa region, it was found that in our surveyed region about 51 Hα-emission-line stars were previously known (Wackerling, 1970). Among them 17 are common with ours, while the rest exhibit no detectable Hα-emission on our plates. A comparison with Stephenson and Sanduleak (1977a, b) was also made. Out of 34 emission-line stars contained in their catalogues 27 could be detected and 7 showed no trace of emission lines. The undetectability of previously known emission-line stars can probably be considered as the result of variation in emission-line strength.

The possibility of detecting T Tauri stars in this region can be described as follows. If we adopt the distance and absorption for the CMa association from Claria (1974b), i.e., 1150 parsecs and 0.81 mag., respectively, and the magnitude of typical T Tauri stars to be $M_v = 5$ (Allen, 1973); it is expected that T Tauri stars associated with this association would be found beginning at $V = 16$. Since this is just fainter than our limiting magnitude, it cannot be expected that there will be many T Tauri stars among the Hα-emission-line stars found in the present survey. It is, therefore, believed that the majority of these Hα-emission-line stars belong to early types. The two-colour diagram of these stars in Figure 2, including the OB stars of Claria, shows that most of the emission-line stars are reddened B stars.

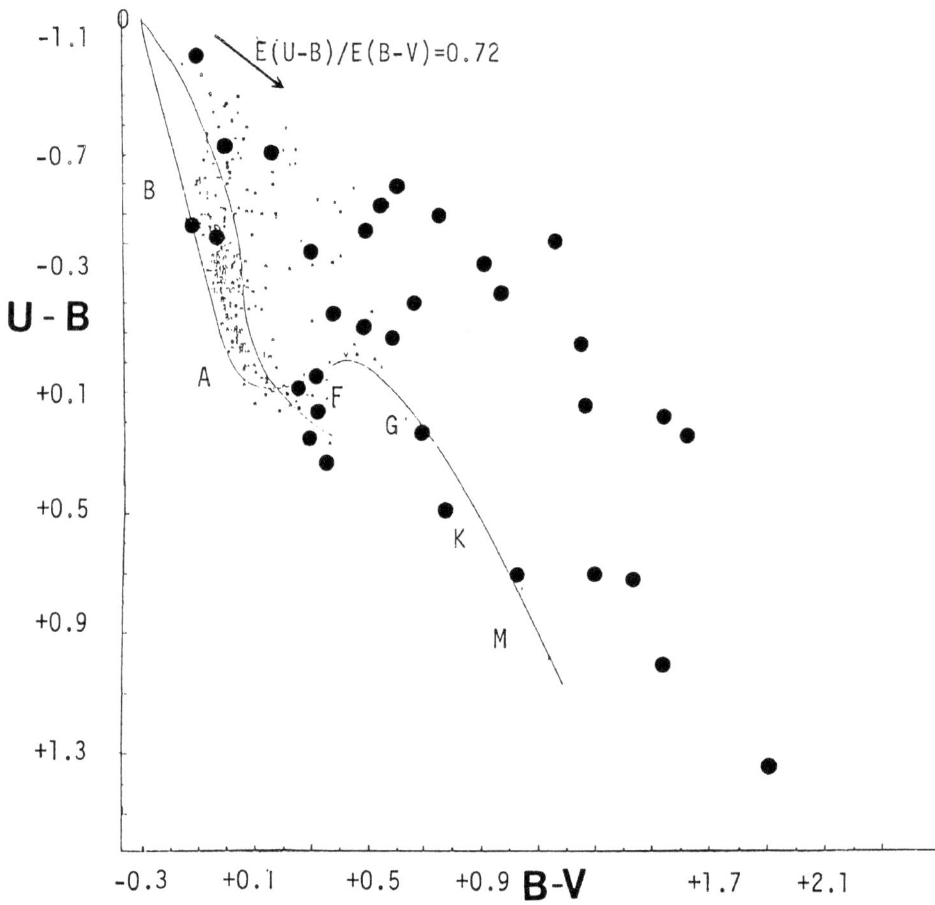

Fig. 2. The two-colour diagram for the emission-line stars in the CMa region, including the OB stars of Claria (1974b). Small dots are OB stars and large dots are emission-line stars.

Acknowledgements

The authors wish to express their gratitude to the staff members of Kiso Observatory and Mr T. Tsujimura of the Ouda Station, Kyoto University.

References

Allen, C. W.: 1973, *Astrophysical Quantities*, 3rd ed., The Athlone Press, London.
Ambartsumian, V. A.: 1949, *Dokl. Akad. Nauk USSR* **68**, 22.

Blitz, L.: 1978, Doctor's Thesis, NASA Technical Memorandum 70708.
Claria, J. J.: 1974a, *Astron. J.* **79**, 1022.
Claria, J. J.: 1974b, *Astron. Astrophys.* **37**, 229.
Hoag, A. A., Johnson, L. L., Iriarte, B., Mitchell, R. I., Hallam, K. L., and Sharpless, S.: 1950, *Publ. of US Naval Obs.* **XVII**, 334.
Kazanasmas, M. S., Zavershneva, L. A., and Tomak, L. F.: 1982, *Atlas of Photometric Charts of Stellar Fields*, Academy of Science of the Ukrainian SSR.
Kopylov, I. M.: 1958, *Izv. Astrophys. Obs. Crimean* **20**, 156.
Ruprecht, J.: 1966, *Trans. IAU* **XII**, 350.
Sanduleak, N. and Stephenson, C. B.: 1973, *Astrophys. J.* **185**, 899.
Sharpless, S.: 1959, *Astrophys. J. Suppl.* **4**, 257.
Stephenson, C. B. and Sanduleak, N.: 1977a, *Astrophys. J. Suppl.* **33**, 459.
Stephenson, C. B. and Sanduleak, N.: 1977b, *Publ. Warner Swasey Obs.* **2**, 71.
Van den Bergh, S.: 1966, *Astron. J.* **71**, 990.
Wackerling, L.: 1970, *Mem. Roy. Astron. Soc.* **73**, 153.

X-RAY SPECTRA OF VARIOUS BURSTS FROM THE RAPID BURSTER*

YUZURU TAWARA, SATIO HAYAKAWA, HIDEYO KUNIEDA, and WANG DE-YU

Department of Physics, Nagoya University, Nagoya, Japan

(Received 13 June, 1985)

Abstract. About 20 type-I bursts and 3000 type-II bursts were observed with Tenma and Hakucho in 1983. Spectral evolutions were analyzed for both types of bursts and results are presented.

1. Introduction

The 'rapid burster' (MXB 1730–335) is a peculiar X-ray burst source, which lies in the globular cluster Liller I near the galactic center (Lewin and Joss, 1981, and references therein). This source shows a recurrent activity of the burst and persistent X-ray emission with a period from half a year to a few years (Grindlay and Gursky, 1977; or Kunieda *et al.*, 1984). This burster emits two kinds of bursts (Hoffman *et al.*, 1978). One is the type-I burst, considered to be a similar phenomenon to those observed in many other X-ray burst sources and be caused by thermonuclear flashes on the surface of the accreting neutron star. Another is the type-II burst considered to be different from the type-I burst because of the peculiarities of its burst activity (rapid recurrence etc.), and of no clear spectral evolution compared with that of the type-I burst. Although no clear evidence has been discovered, the type-II bursts are considered to be caused by some instability of mass accretion on to a neutron star.

In order to investigate the mechanism of the type-II burst, the energy spectrum of a type-II burst should give important information. The occurrence of a type-I burst indicates that the rapid burster should have some similar situation to other bursters, so we can investigate it by comparing the properties with other bursters. In this paper, we present some properties of the energy spectra discovered by Tenma and Hakucho observations in August 1983 (Kunieda *et al.*, 1984).

* Paper presented at the IAU Third Asian-Pacific Regional Meeting, held in Kyoto, Japan, between 30 September–6 October, 1984.
** On leave from Purple Mountain Observatory, Academia Sinica, Nanjing, China.

Astrophysics and Space Science **118** (1986) 317–320.

2. Observations and Results

Tenma and Hakucho observed about 20 type-I bursts and about 3000 type-II bursts, from 5 August to 31 August in 1983, with gas scintillation proportional counters (total effective area of 640 cm^2) on board Tenma and with proportional counters (effective area of 80 cm^2) on board Hakucho.

For 18 type-I bursts, time resolved energy spectra were analyzed. They can be well fitted by black-body spectra with $N_H = (4 \pm 2)10^{22}$ H cm^{-2}. The derived properties are summarized in Table I, being roughly consistent with those of bursts from other bursters.

TABLE I

Burst properties of type-I bursts from the rapid burster

Rise time (s)	5
Total energy (ergs)	$(2-3) \times 10^{39} (D/10 \text{ kpc})^2$
Burst time interval (min)	85
Peak luminosity (ergs s^{-1})	$(0.7-1.0) \times 10^{38} (D/10 \text{ kpc})^2$
Decay time (s)	37
Maximum colour temperature (keV)	1.9–2.0
Colour radius (km)	$7 \times (D/10 \text{ kpc})$
α-value	30
γ-value	0.15

Some interesting properties different from other bursters are (i) small fluctuation of the burst peak fluxes and maximum colour temperatures, (ii) a relatively large γ-value, that is, the ratio of the persistent flux to the burst peak flux, as compared with those of other bursters. Such properties are similar to those of bursts from MXB 1636–536 observed by EXOSAT in June 1983 (Turner and Breedon, 1984).

We observed about 2000 type-II bursts with Hakucho and about 800 with Tenma. Here, we classify the observed type-II bursts into three subgroups, morphologically, as follows (a) short bursts with a duration from less than 1 to 50 s, and showing a clear pattern of flux variation associated with damped oscillations, (b) long, flat-top bursts with a duration of > 50 s, and showing flux saturation, (c) exotic bursts, which show a very slow rise and slow decay. These have been newly discovered in the present observations.

We analyzed the spectral evolution of each event for all classes of type-II bursts by fitting the observed data to a black-body spectrum. For all classes of bursts ((a), (b), (c)), the colour temperatures averaged over corresponding durations are nearly the same; and also less spectral evolution is found for each event in contrast to that of type-I bursts, which show the clear spectral softening in the luminosity decay.

However, we can find weak a correlation between luminosities and colour temperatures for short bursts of class (a) as follows; (i) colour temperature decreases at the first dip of the flux (ii) the averaged colour temperature of the first half of each event is lower than that of the second half of the event.

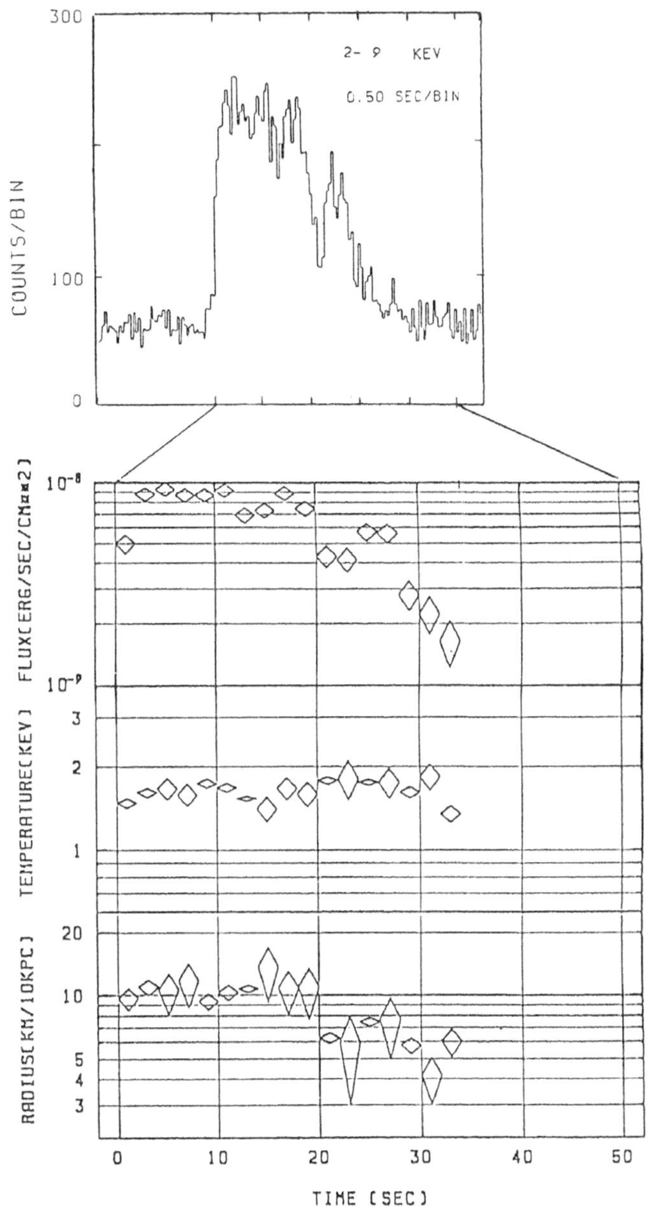

Fig. 1. Results of spectral analysis for short type-II (class (a) – see text) bursts, best-fit energy fluxes, colour temperatures and apparent radii, assuming a spherical blackbody emission and a source distance of 10 kpc.

3. Discussion

For type-I bursts from the rapid burster, a relatively low colour temperature of about 2.0 keV at the burst peak is consistent with the absence of double peak structure in colour temperature, as observed from MXB 1636–536 (Inoue *et al.*, 1984). Since the event with double peak structure shows the highest colour temperature of about 3.0 keV at its maximum, while the highest luminosity may be considered as the Eddington limit, the observed maximum luminosity of the type-I burst from the rapid burster is lower than this Eddington luminosity by a factor of 3–4, if we use the relation between maximum colour temperature and maximum luminosity from the case of MXB 1636–536.

For the type-II bursts, the fact that nearly the same averaged colour temperatures is observed for all classes of type-II bursts indicates that they originate by the same mechanism.

References

Grindlay, J. E. and Gursky, H.: 1977, *Astrophys. J.* **218**, L117.

Hoffman, J. A., Marshall, H. L., and Lewin, W. H. G.: 1978, *Nature* **271**, 630.

Inoue, H., Waki, I., Koyama, K., Matsuoka, M., Ohashi, T., Tanaka, Y., Miyamoto, S., and Tsunemi, H.: 1984, *Publ. Astron. Soc. Japan* **36**, 831.

Kunieda, H., Tawara, Y., Hayakawa, S., Masai, K., Nagase, F., Inoue, H., Koyama, K., Makino, F., Makishima, K., Matsuoka, M., Murakami, T., Oda, M., Ogawara, Y., Ohashi, T., Shibazaki, N., Tanaka, Y., Kondo, I., Miyamoto, S., Tsunemi, H., and Yamashita, K.: 1984, *Publ. Astron. Soc. Japan* **36**, 807.

Lewin, W. H. G. and Joss, P. C.: 1981, *Space Sci. Rev.* **28**, 3.

Turner, M. J. L. and Breedon, L. M.: 1984, *Monthly Notices Roy. Astron. Soc.* **208**, 29.

TRAPEZIUM-TYPE MULTIPLE SYSTEMS*

G. N. SALUKVADZE

Abastumani Astrophysical Observatory, Byurakan, U.S.S.R.

(Received 13 June, 1985)

Abstract. Trapezium type multiple systems are one of the youngest systems in the galaxy. Ambartsumian (1954) regards many of the Trapezium-type multiple systems as having positive total energy.

The examination of Table I, compiled from the *Abastumani Catalogue of Trapezia* (Salukvadze, 1978; Poveda *et al.*, 1977) and the *Index Catalogue of Visual Binary Stars*, shows that real trapezia are met among multiple stars with their primaries of O and B spectral types.

TABLE I

Spectral class of the primary	Total number of multiple stars	Number of trapezia observed	Calculated number of pseudo trapezia	Probable number of real trapezia	Relative number of trapezia (%)
O–B2	59	39	5	34	58
B3–B5–B	72	23	6	17	24
B8–B9	118	25	11	14	12
A	394	60	35	25	6
F	309	41	28	13	4
G	224	33	20	13	6
K	153	37	14	23	15
M	11	8	1	7	64
Unknown spectrum	526	146	47	99	19
In all	1867	412	167	245	

Table II lists the results of a special investigation of the problem of relation of Trapezium-type systems with associations and galactic clusters (Salukvadze, 1979).

TABLE II

Spectral classes	Number of real trapezia	Number of trapezia belonging to		Percent of trapezia belonging to	
		Associations	Clusters	Associations	Clusters
O–B2	33	27	21	82	64
B3–B5 + 5	16	12	6	89	39
B8–B9	13	8	2	61	15
Unknown spectrum	96	32	10	34	13

* Paper presented at the IAU Third Asian-Pacific Regional Meeting, held in Kyoto, Japan, between 30 September–6 October, 1984.

Astrophysics and Space Science **118** (1986) 321–323.

Trapezium-type multiple systems were also searched for in T-associations (Salukvadze, 1980a, b, 1981a). In 13 of them 120 trapezia were revealed. Out of these 85% are real physical systems. The majority of trapezia are members of nebulae, and they are concentrated mainly in groups.

A great deal of papers, from various investigators with quite different methods, have been devoted to the problem of trapezia stability. We shall not stop to consider them, but give only the results of our own investigations concerning this problem (Salukvadze, 1981b, 1982, 1985).

Kinematics of 15 Trapezium-type multiple systems from the *Abastumani Catalogue* with the primaries of O–B2 spectral classes have been specially studied. They are: ADS 719, 2783, 2843, 3709, 4241, 4728, 5322, 5977, 13374, 13626, 14526, 14831, 15184, 16095, 16381. It is found that out of these 15 studied Trapezia no less than 14 show expansion.

As an example Figure 1 shows expansion of the Trapezium-type multiple system ADS 2843.

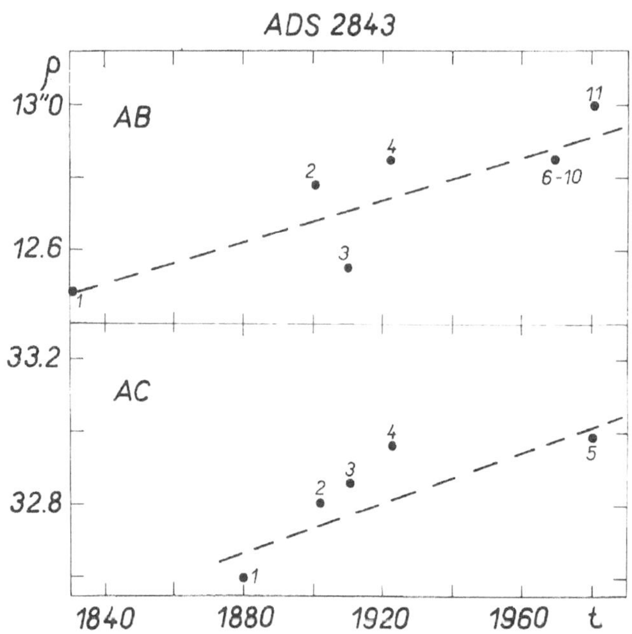

Finally it can be concluded that at present the statistical properties of trapezia are quite reliably determined. Consequently, it is of great importance to investigate individual trapezia. From this standpoint first of all it is necessary to estimate the photometric and spectral properties of stars making trapezia; and to determine the relative positions and radial velocities by which the problem of number of trapezia with positive total energy could be finally established.

References

Ambartsumian, V. A.: 1954, *Bull. Bjurakan Obs.* **15**, 3.
Poveda, A., Allen, C., and Warman, L.: 1977, *Rev. Mex. Astron. Astrophys.* **3**, 127.
Salukvadze, G. N.: 1978, *Bull. Abastumani Astrophys. Obs.* **49**, 39.
Salukvadze, G. N.: 1979, *Astrophysics* **15**, 311.
Salukvadze, G. N.: 1980a, *Astrophysics* **16**, 505.
Salukvadze, G. N.: 1980b, *Astrophysics* **17**, 689.
Salukvadze, G. N.: 1981a, *Bull. Georgian SSR Acad. Sci.* **101**, 573.
Salukvadze, G. N.: 1981b, *Bull. Georgian SSR Acad. Sci.* **104**,1, 41.
Salukvadze, G. N.: 1982, in Z. Kopal and J. Rahe (eds.), 'Binary and Multiple Stars as Tracers of Stellar Evolution', *IAU Colloq.* **69**, 109.
Salukvadze, G. N.: 1985, *Astrophysics* **22**, 97.

AGE-METALLICITY RELATION FOR F-STARS*

HONG BAE ANN

Department of Earth Sciences, Pusan National University, Korea

and

YONG HEE KANG

Department of Earth Sciences, Kyungpook National University, Korea

(Received 2 July, 1985)

Abstract. We derived the age-metallicity relation for the solar neighbourhood from the *uvby*, Hβ photometry of F-stars using the evolutionary models of Hejlesen (1980). It is shown that disk formation is divided into two phases according to the fast and slow collapse which accompany the initial rapid enrichment and slow increase of the metallicity of the galactic disk, respectively.

The age-metallicity relation (AMR) of stars in the galactic disk (Twarog, 1980b) and the metallicity distribution of G-dwarfs (i.e., the G-dwarf problem) (Schmidt, 1963) are the most important observational constraints on the models of chemical evolution of the solar neighbourhood. Most of the theoretical models have been developed to interpret the latter constraint, and some of these have been successful without introducing *ad hoc* assumptions (Lee and Ann, 1981; Tinsley and Larson, 1978). Recent publication of *uvby*, Hβ photometry of a large number of intermediate Population II stars (Olsen, 1983) made it possible to investigate the AMR of the solar neighbourhood in more detail, although there is no suitable calibration of the photometric indices in terms of the physical parameters for Population II stars.

The observational data for our derivation of the AMR for the solar neighbourhood is provided by the catalogues of Olsen (1983) and Hauck and Mermilliod (1980). For direct comparison with the AMR of Twarog (1980b), we carried out the same procedures with the data of Twarog (1980a). The definition of F-stars is the same as that of Crawford (1975) and Twarog (1980a, b), and the calibration of the photometric indices in terms of the physical parameters by Crawford (1975) was used, except for the δm1 calibration. Kang and Ann's (1984) δm1 calibration, in terms of [Fe/H], was used for the metallicity of the stars; however, Kang and Ann's calibration is not much different from that of Crawford.

For the determination of the age of a star, we used the isochrones of Hejlesen (1980), in combination with the observational physical parameters of a star: [Fe/H], T_{eff}, and M_v. This method of age determination is the same as that of Twarog (1980b). In Figure 1 we plotted our derived AMRs, as well as that of Twarog (1980b). The theoretical

* Paper presented at the IAU Third Asian-Pacific Regional Meeting, held in Kyoto, Japan, between 30 September–6 October, 1984.

Astrophysics and Space Science **118** (1986) 325–327.

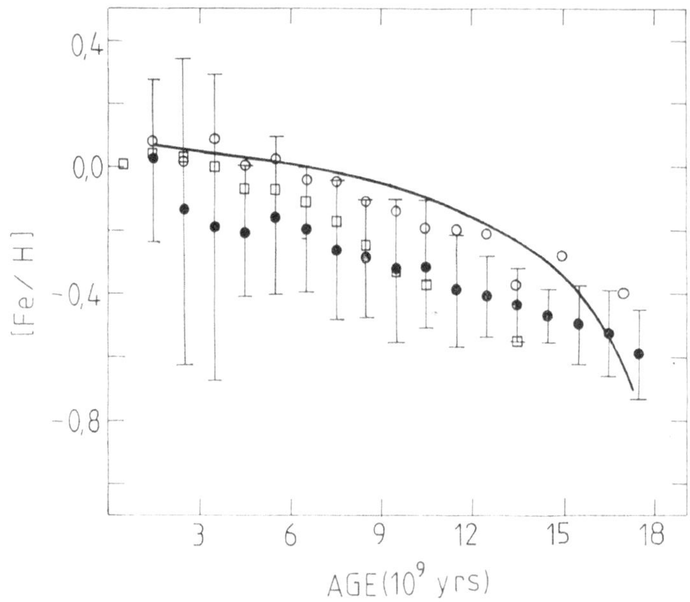

Fig. 1. The age-metallicity relation (AMR) of F-stars in the solar neighbourhood. The open square represents the AMR of Twarog (1980b) and the open and closed circle indicate the AMR from the data of Twarog (1980a) and from the data of Olsen (1983), respectively. The vertical bars on the closed circles are dispersions of [Fe/H] in each age bin. The theoretical prediction of Lee and Ann (1981) is shown by heavy line.

predictions of Lee and Ann's (1981) two-zone disk-halo model is also given for comparison. Considering the fact that Twarog took the age of our Galaxy as 15×10^9 yr, while we have taken it as 18×10^9, the AMR of Twarog and that of the present study, using the Twarog's (1980a) data, fairly well coincide with each other. The AMR from Olsen's data shows systematically small metallicities compared with that from Twarog's data. This difference may be caused by some intrinsic difference in chemical properties of the two groups of stars, i.e., intermediate Population II and Population I stars. The AMR from Olsen's data strongly suggests that in the vicinity of the Sun there are few or no extremely metal deficient stars, and at the very early stages of galactic evolution there should have been some events that produced an initial enrichment of the metallicity of the galactic disk. These are consistent with the model predictions of Lee and Ann (1981).

The basic property of Lee and Ann's model is an unclosed two-zone disk-halo feature; and the model assumes the time dependent yield P_z and fraction of remnant gas α. The star formation rate (SFR) is assumed to be a power law, and the inflow rate and the removal rate are assumed to be proportional to the SFR. The good agreement of Lee and Ann's model and the derived AMRs suggests that the predictions of this model on the history of the chemical evolution of the solar neighbourhood is not incorrect. This implies that disk formation was divided into two phases of fast and slow collapse, which accompanied an initial rapid enrichment and subsequent slow increase of metallicity of the galactic disk, respectively.

References

Crawford, D. L.: 1975, *Astron. J.* **80**, 955.
Hauck, B. and Mermilliod, M.: 1980, *Astron. Astrophys. Suppl.* **40**, 1.
Hejlesen, P. M.: 1980, *Astron. Astrophys. Suppl.* **39**, 347.
Kang, Y. H. and Ann, H. B.: 1986, in preparation.
Lee, S.-W. and Ann, H. B.: 1981, *J. Korean Astron. Soc.* **14**, 55.
Olsen, E. H.: 1983, *Astron. Astrophys. Suppl.* **54**, 55.
Schmidt, M.: 1963, *Astrophys. J.* **137**, 758.
Tinsley, B. M. and Larson, R. B.: 1978, *Astrophys. J.* **221**, 554.
Twarog, B. A.: 1980a, *Astrophys. J. Suppl.* **44**, 1.
Twarog, B. A.: 1980b, *Astrophys. J.* **242**, 242.

THE PULSAR VELOCITIES AND THEIR BINARY ORIGINS (INVITED REVIEW)*

V. RADHAKRISHNAN and C. S. SHUKRE
Raman Research Institute, Bangalore, India

(Received 2 July, 1985)

Abstract. The observed proper-motions are shown to indicate that most pulsars come from binary systems. Due to a remarkable relationship between the speed of a pulsar and that of its progenitor in orbit, it is concluded that these binaries went through a common-envelope phase. This provides independent support to similar conclusions drawn from stellar evolution considerations.

For several years after pulsars were discovered, very little attention was paid to possible binary origins for them; there was enough excitement trying to understand their observed properties to worry about such questions. But because the majority of stars are in double or multiple systems, the possibility of finding pulsars in binary systems did get discussed a little, even before the discovery of the first binary pulsar PSR 1913 + 16 (Hulse and Taylor, 1975). The way things work, discoveries are rarely made in the right order so that our understanding increases step by step with each of them. In the case of PSR 1913 + 16, we got not one neutron star in a binary, but two. If one was going to jump to general conclusions from this one example, it seemed to suggest that a binary system could survive not only one but two explosions without disruption. Today, we have four radio pulsars in binary systems and a large number of binary X-ray sources each of which is believed to contain a neutron star. What we have learnt in the last decade or so seems to indicate that a double neutron star binary system is a rather rare thing, and this is what was meant by discoveries being in the wrong order.

The question we wish to address today is not how many observable radio pulsars there are in binary systems, but whether all of the 300 or so single radio pulsars we see originated in binary systems. The reason for even suspecting that this may be true is twofold. As mentioned earlier, most stars seen in an evolutionary stage today seem to be in binary or multiple systems (Abt and Levy, 1976, 1978). The difficulty of getting rid of the angular momentum in gas clouds which are collapsing to form stars appears to strongly favour the formation of binary or larger systems. Many O and B stars which could be expected to later explode as supernovae and leave behind a pulsar are in 'runaway' binaries which acquired their velocities on the explosion of their companions (Stone, 1979). Indeed, according to Stone, there is little or no evidence that single stars of this category exist.

* Paper presented at the IAU Third Asian-Pacific Regional Meeting, held in Kyoto, Japan, between 30 September–6 October, 1984.

Astrophysics and Space Science **118** (1986) 329–336.

On the basis of what has been said above, one could take the stand that, as far as we know, every single pulsar that we see could have had an origin in a binary system. But it would be much more satisfactory to find properties or characteristics of the now single pulsars which point to such a past. We shall show the importance in this respect of the proper motion or space velocity on which considerable information has been obtained in recent years.

Even in the early stage of pulsar research it was suggested by Gunn and Ostriker (1970) that the Z-distribution of pulsars indicated that they were born in the galactic plane in binary systems and acquired high velocities at the time of their birth. They should, therefore, have a scale height commensurate with their velocities, and we should also tend to see older pulsars at greater distances from the plane. The considerable amount of data collected over the years have confirmed fairly well this picture originally advanced with very meagre data.

Direct measurement of the proper motion in recent years, using interferometric techniques, have yielded the best data we have today on the velocities of pulsars. Lyne *et al.* (1982) have obtained accurate velocities for 26 pulsars ranging from approximately 10 to 450 km s^{-1}. These authors have noted that the distribution of velocities does not seem to be Maxwellian, there being excesses of both high- and low-velocity objects. This distribution of velocities is interestingly different from the corresponding distribution for early-type stars in general, whose velocities are much lower, and is clearly trying to tell us something. An important clue to understanding this distribution comes from the remarkable correlation observed by Anderson and Lyne (1983) between the velocities of the 26 pulsars, and their magnetic dipole moments derived from P and $\dot{P}$. These authors came to no firm conclusions on the origin of the pulsar velocities; however, they suggested that the velocities are probably due to asymmetric explosions, and that the asymmetry could be influenced by the magnetic dipole moment.

A recent paper (Radhakrishnan, 1984) has advanced arguments to show that the above interesting correlation is not causal, but accidental, and that pulsar velocities are most probably governed by their binary histories and not in any way by their magnetic fields. As an appreciation of this reasoning is essential for the development of te present theme, it is briefly recounted below:

(1) The energy stored in the dipole fields is many orders of magnitude lower than the kinetic energy of the pulsar.

(2) No pure dipole field, or indeed any single multipole component, can cause such an asymmetry, but only a combination of two or more multipole field components.

(3) Even if such multipole components exist, and with the enormous strength required, the correlation of the velocities should be with them and not the dipole components derived from the slow-down of the pulsars.

(4) The correlation worsens after correction for magnetic field decay instead of improving, as should be the case for velocities imparted at birth. Nevertheless, the correlation is there and a satisfactory explanation is required which will also account for why it is better for 'observed' fields rather than 'initial' fields.

At the last regional meeting held in Bandung three years ago, it was proposed

(Radhakrishnan and Srinivasan, 1981) that there must exist a class of recycled pulsars which were spun-up in binary systems through accretion, and later released as single pulsars on the disruption of the binary by the second explosion. Such a pulsar would have a low field, due to its decay during the evolution time of the companion star, but the neutron star should still be observable as a radio pulsar due to its anomalously high spin (for the field). In the three years since then a very large number of papers by numerous authors have appeared on the same theme stimulated by, among other things, the discovery of millisecond pulsars.

A necessary but not sufficient condition, which is an excellent diagnostic for such recycled pulsars, is that they must all have characteristic ages ($P/2\dot{P}$) of $\geq 10^7$ yr. The reason for this can be seen from Figure 1, which is a $\log B - \log P$ plot showing the field

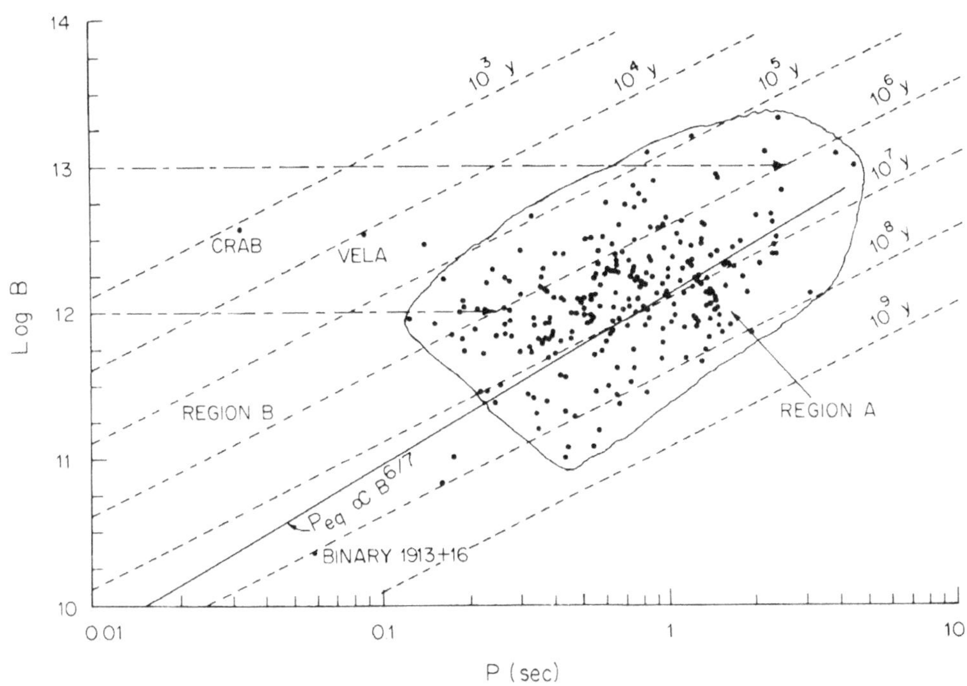

Fig. 1. Plot of derived magnetic field vs period for pulsars showing characteristic ages and the field-period equilibrium condition for spin-up (solid line) from Srinivasan and Van den Heuvel (1982). Figure taken from V. Radhakrishnan: *Contemp. Phys.* **23**, 207.

period equilibrium condition or 'spin-up' line (Srinivasan and van den Heuvel, 1982) for accretion in binaries. This line has almost the same slope as the characteristic age lines in this plot, and its position practically coincides with the age line for 10^7 yr. Therefore, this provides us with the diagnostic referred to above. It is important to note that this condition is *independent* of the decay time-scale for the magnetic fields. Figure 2

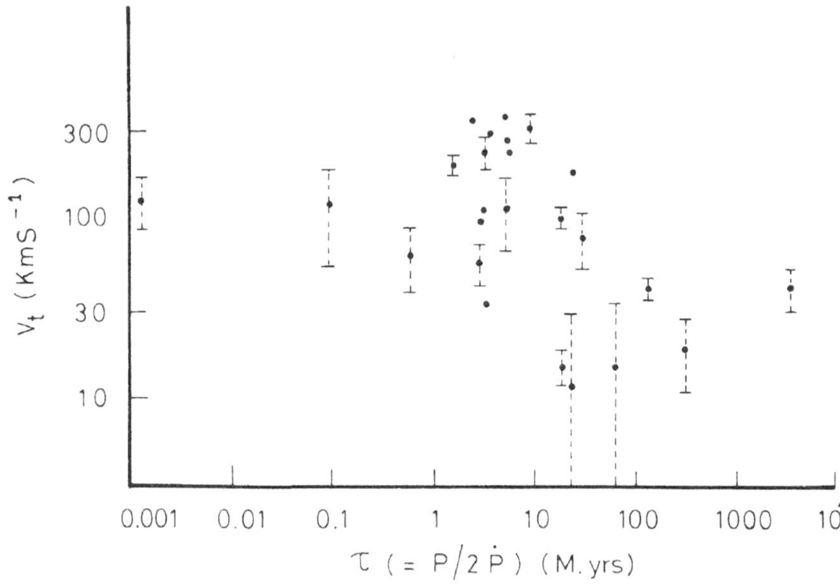

Fig. 2. Transverse velocities from Lyne *et al.* (1982) plotted against the characteristic age $\tau(\frac{1}{2}P/\dot{P})$. Small error bars of some of the pulsars have been omitted from the figure for clarity.

taken from Radhakrishnan (1984) shows a plot of the velocities against characteristic ages and suggests strongly a division of the pulsars into two groups. The young high-field ones have high velocities, and the smaller group – all with characteristic ages $> 10^7$ yr – have low velocities. It was on the basis of this correlation with characteristic age that the apparent correlation with magnetic field found by Anderson and Lyne (1983) was explained. Further, the fact that the decrease in correlation with 'initial fields' was only partial, and not total, is well explained as due to incorrect extrapolation backwards for recycled pulsars. The spin-up moves the pulsars out of their original evolutionary tracks to anomalous positions; extrapolation backwards will then be along different tracks leading to apparent lower initial fields (Radhakrishnan, 1984).

While this exercise provided persuasive evidence for the existence of a minority population of recycled pulsars with binary histories, it made the understanding of their velocities even more difficult. These velocities as seen earlier were low (approximately 30 km s^{-1}), and it is easy to show that after the second supernova explosion in a massive close binary system the velocity of the first-born neutron star is expected to be greater than 100 km s^{-1}. One plausible way out of this difficulty (Radhakrishnan and van den Heuvel, 1985) is that the binary system involves a companion whose core has not fully evolved before a common envelope phase sets in. The companion gets disrupted into a disk and is eventually dispersed leaving behind a spun-up old neutron star. The velocity of this neutron star will be simply the 'runaway' velocity of the whole binary system after the first explosion, which for such systems is estimated to be only 20 to 30 km s^{-1} (van den Heuvel and Bonsema, 1984).

The 26 pulsars chosen for the proper motion measurements by Lyne *et al.* (1982), were selected on the basis of positional nearness on the sky to extragalactic point sources, and not on the basis of any pulsar characteristics. Therefore, the 26 pulsars observed must constitute a representative sample. It would appear, therefore, that of all single pulsars observed, if we leave out recycled ones which form a minority, the rest have high velocities ($\sim 250\ \mathrm{km\ s^{-1}}$), and high fields. The possibility that the velocities could result from an asymmetric explosion governed by the magnetic field has already been dealt with. Another good reason for doubting asymmetric explosions in general is the fine-tuning required. The amount of energy in a supernova explosion is many orders of magnitude higher than the kinetic energy of the pulsar. Thus, while it is true that a minute asymmetry could provide the energy, it would be extraordinary if the spread in asymmetry were so small as to produce the relatively narrow distribution of velocities of the young high-field pulsars.

A binary origin for the velocities has been proposed by many authors. Van den Heuvel concluded at the International Astronomical Union Symposium on Pulsars that the majority of them were likely to come from binaries (Van den Heuvel, 1981). As there is no other satisfactory explanation to date for the high velocities observed, we shall proceed on the assumption that all pulsars had an origin in binary systems, and that the explosions which released them were symmetric.

We shall trace backwards to derive constraints on binary systems that produced the pulsar with the velocity as observed today. This is similar to the attempts, mentioned earlier, to constrain the evolution time in the binaries, based on the present magnetic fields and periods of pulsars for which a binary history was responsible (Radhakrishnan and Srinivasan, 1981). Thus, if the pulsar velocities observed today are a consequence of their having been in binary systems, what can we conclude about the binary systems themselves?

Towards this aim, we must digress to mention a remarkable property of binaries which are disrupted by symmetric explosions. To an observer situated in the same frame of reference as the centre of mass of a binary system with circular orbits, a sudden expulsion of mass from one of the stars in a symmetrical way leaves the speed of the remnant of this star *unchanged* for all time. Remarkably, this is true irrespective of the amount of mass lost, and whether as a consequence, the system remains bound or is disrupted. The reason for this surprising property, can be understood in terms of a hodograph, which is a representation of the motion of a particle in velocity space just as the orbit is the representation of its motion in coordinate space. It is well known that this path in velocity space is a circle in the Keplerian case, even though both the speed and direction are changing continuously as in an elliptic orbit (Goldstein, 1980). Except for circular orbits, the centre of the hodograph is not at the origin. For a parabolic orbit, the hodograph is still a full circle but it passes through the origin. For a hyperbolic orbit it shrinks to an arc of a circle whose center is displaced from the origin by more than its radius. If the observer now puts himself in a frame which is at the centre of the hodograph, clearly the speed is always the same as seen by him. The remarkable thing is that when one of the stars in a binary system suddenly loses a certain amount of mass, the centre of mass of the remaining two-body system (whether bound or not) moves in such a way with respect to the original centre of mass, that the observer continues to remain at the centre of *this* star's hodograph whose radius also remains unchanged.

If we return to our main objective, the implication of this extraordinary property is that the progenitor of a pulsar had precisely the same speed in orbit (assumed to be circular) just before the explosion, as the pulsar has today far away in space and time from this event. If we ignore for the moment the possibility of orbits being elliptical, we can conclude that the distribution of speeds observed for the young high-field pulsars was exactly the distribution of their orbital speeds prior to the explosions which liberated the pulsars (note that this is not true for the other object in the binary system whose mass did not undergo any sudden change). Here we are neglecting the motion of the whole binary system before the explosion, which is a relatively minor perturbation.

We can now use this to learn something about the binary systems from which the pulsars originated. The explosion which releases a pulsar as a single object can either be the first, or the second explosion in the binary system. We have already seen that there are many systems containing a neutron star which must have clearly survived the first explosion, and at least one case which survived two (we are ignoring, possible neutron star formation by accretion on to white dwarfs although the amount of energy released has to be about the same). Still, let us consider release of the pulsar in the first explosion.

For Keplerian motion we can write

$$V = \frac{438 M_2}{\sqrt{(M_1 + M_2)a}} \text{ km s}^{-1}, \tag{1}$$

where M_1 is the mass of the pre-supernova and M_2 that of the companion, in $M_\odot$. The separation between M_1 and M_2 in solar radii is a, and V is the velocity of M just before the explosion which, due to the above considerations, would be the observed pulsar velocity. Since more than one half of the total mass has to be lost in the explosion to disrupt the binary, and assuming a mass for the neutron star of 1.4 $M_\odot$, we have

$$M_1 \gtrsim 2.8 + M_2 . \tag{2}$$

leading to

$$a \lesssim \left(\frac{438}{V\,(\text{km s}^{-1})}\right)^2 \frac{M_2^2}{2.8 + 2M_2} R_\odot . \tag{3}$$

It is clear that the separation required to give the observed velocities is very small (e.g., consider a typical $M_2 = 5$; then $a \lesssim 2\,R_\odot$ for $V = 450$ km s^{-1}). We also know that it is only in the case of very wide binaries that the individual stars can evolve just as if they were single. In such wide binaries, the velocities would be clearly too low. On the other hand, if the stars were as close as required by the high velocities, then mass transfer is inevitable, and the primary will become the less massive of the two before it reaches the supernova stage (Paczyński, 1971; Van den Heuvel, 1976). The ensuing explosion cannot unbind the system and we will have a binary system containing one neutron star which can then become an X-ray source and/or a recycled pulsar.

If we then conclude that all young high-field high-velocity pulsars come from the second explosions in binaries, we have the additional constraint that the mass of the other star is now $\sim 1.4\,M_\odot$. We can now go to Figure 3 which depicts, for $M_2 = 1.4$, the dependence of the separation of the stars on the total mass as given by Equation (1). Considering that speeds as high as 400 km s^{-1} have been found, we see that the

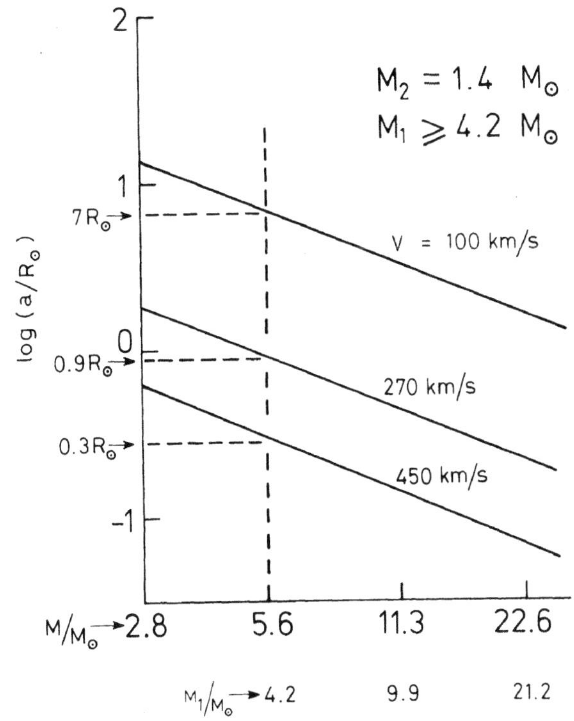

Fig. 3. Plot of the separation a, between the stars vs the total mass M of the binary, as derived from Equation (1) assuming the mass of the neutron star companion $M_2 = 1.4\,M_\odot$. Numbers with arrows along the Y-axis indicate maximum separations appropriate for the observed range of velocities, and correspond to the minimum value of M (the vertical broken line) required to disrupt the binary.

separations have to be very small, and the collapsing stars or at least their dense central regions have to be very compact indeed. This implies that the evolution of the binary system after the first SN must go through a common envelope phase. If the orbit just prior to the explosion were elliptical, and not circular as we had assumed, our conclusions would not hold exactly. However, the observed high velocities would still require extremely small separations as before. The consequent tidal interactions would inescapably lead to circularisation of the orbits, thus fully justifying our assumption.

We seem to be led to the conclusion that in all cases where a young pulsar is produced – if it is produced in binary systems – a common envelope phase is inevitable. It is interesting that this conlusion drawn here purely on the basis of pulsar proper motions is consistent with the conclusions of numerous studies undertaken from the point of view of stellar evolution in binaries.

Acknowledgement

We thank Rajaram Nityananda for interesting discussions on hodography.

References

Abt, H. A. and Levy, S. G.: 1976, *Astrophys. J. Suppl.* **30**, 277.
Abt, H. A. and Levy, S. G.: 1978, *Astrophys. J. Suppl.* **36**, 241.
Anderson, B. and Lyne, A. G.: 1983, *Nature* **303**, 597.
Goldstein, H.: 1980, *Classical Mechanics*, 2nd ed., Addison-Wesley, New York.
Gunn, J. E. and Ostriker, J. P.: 1970, *Astrophys. J.* **160**, 979.
Hulse, R. A. and Taylor, J. H.: 1975, *Astrophys. J.* **195**, L51.
Lyne, A. G., Anderson, B., and Salter, M. J.: 1982, *Monthly Notices Roy. Astron. Soc.* **201**, 503.
Paczyński, B.: 1971, *Ann. Rev. Astron. Astrophys.* **9**, 183.
Radhakrishnan, V.: 1984, Proceedings of the Workshop on Millisecond Pulsars, Green Bank, 6–8 June, 1984.
Radhakrishnan, V. and Srinivasan, G.: 1981, Proceedings of the Second Asian-Pacific Regional Meeting of the IAU, Bandung, Indonesia, 24–29 August, 1981.
Radhakrishnan, V. and Van den Heuvel, E. P. J.: 1985, in preparation.
Srinivasan, G. and Van den Heuvel, E. P. J.: 1982, *Astron. Astrophys.* **108**, 143.
Stone, R. C.: 1979, *Astrophys. J.* **232**, 520.
Van den Heuvel, E. P. J.: 1976, in P. Eggleton, S. Mitton, and J. Whelan (eds.), 'Structure and Evolution of Close Binary Systems', *IAU Symp.* **73**, 35.
Van den Heuvel, E. P. J.: 1981, in W. Sieber and R. Wielebinski (eds.), 'Pulsars', *IAU Symp.* **95**, 379.
Van den Heuvel, E. P. J. and Bonsema: 1984, *Astron. Astrophys.* **139**, L16.

DYNAMICS OF DISK GALAXIES (INVITED REVIEW)*

K. C. FREEMAN

Mount Stromlo and Siding Spring Observatories, Research School of Physical Sciences, Australian National University, Canberra, Australia

(Received 2 July, 1985)

Abstract. In order to detect the gravitational effect of the dark corona component of disk galaxies, it is necessary to have surface photometry and rotation data that extend well beyond three disk-scale lengths. Such data exists now for several pure disk galaxies. There is very strong evidence for dark coronas in these galaxies. It has been possible to estimate the scale length and density for their dark components, which can be well represented as isothermal spheres. The pure disk galaxies that have been studied so far have absolute magnitudes in the range $-17 > M_B > -19.5$. It would be very interesting to know whether the fainter dwarf irregular galaxies (which are also disk systems) have dark coronas: are these coronas necessary for disk galaxies to form?

1. Introduction

Disk galaxies have two major visible components, the disk and the bulge. There is strong evidence now for a third dark component, which is dynamically very important. This dark component is the subject of this review.

We should like to know the characteristic parameters of the dark components of disk galaxies, such as the scale lengths and densities, because these are important for understanding the dynamics of the galaxies and the nature of the dark matter itself. The most direct way to estimate these parameters is from the galactic rotation curves, so I will discuss the problems and recent advances in using the rotation curves to measure the parameters of galactic dark coronas. I will not discuss here the problem of dark matter in clusters of galaxies, nor the other less direct evidence for the existence of dark matter in individual galaxies.

The disks of disk galaxies are supported in the radial direction mainly by their rotation. The equilibrium of disk matter is then given approximately by

$$GM(r)/r^2 = V^2(r)/r\,,$$

where $M(r)$ is the mass within radius r and $V(r)$ is the rotational velocity at radius r. At large r, the integrated luminosity $L(r)$ approaches the total luminosity L_T. If $M(r)$ similarly tends to a limit, then, at large r, we would expect $V(r) \propto r^{-1/2}$. This is not seen. For most disk galaxies, the rotational velocity is approximately constant in the outer parts (see for example Rubin *et al.*, 1982, and references therein). Constant rotational

* Paper presented at the IAU Third Asian-Pacific Regional Meeting, held in Kyoto, Japan, between 30 September–6 October, 1984.

Astrophysics and Space Science **118** (1986) 337–342.

velocity suggests that $M(r) \propto r$, like an isothermal sphere, and this is the most direct evidence for the presence of the massive dark corona component.

However, there has been some argument about whether a flat rotation curve (i.e., V = const.) necessarily implies the presence of a dark corona. We will see that the answer depends on how far out in radius the rotation measurements extend. Some recent work by Kalnajs (1983) illustrates this very clearly.

2. Flat Rotation Curves

Kalnajs considered disk galaxies with measured rotation curves $V(r)$ and surface brightness distribution $I(r)$. He assumed that the surface density distribution was proportional to the surface brightness distribution (i.e., constant M/L ratio). He then devised a neat algorithm to calculate the expected rotation curve from the observed surface brightness distribution, assuming, as above, that the disk is in centrifugal equilibrium. The question was: using the observed $I(r)$ distribution, can one reproduce the observed (flat) rotation curve without requiring a massive dark corona component? In other words, can the gravitational field of the luminous components (bulge and disk) alone produce a flat rotational curve?

For several examples shown by Kalnajs, the rotation curve predicted from the surface brightness profile was indeed in excellent agreement with the observed rotation curve. However, as pointed out by Haud in the discussion of Kalnajs's paper, the rotation data extends only to about three disk scale lengths (for disks, the surface brightness profile is typically exponential: $I(r) \propto \exp(-r/h)$, where h is the disk scale length). Kalnajs's result warns us that rotation data out to only three scale lengths is not sufficient to show unambiguously the gravitational effects of the dark corona. Out to a radius of about three scale lengths, the disk and the bulge together can produce a fairly flat rotation curve. A simple model shows why this is so.

We represent a disk galaxy as a bulge + disk. The bulge has the typical $r^{1/4}$ surface brightness law

$$\log I = a - br^{1/4};$$

and the disk has the exponential surface brightness law

$$I = I_0 \exp(-r/h).$$

We then calculate the rotation curve for the resulting gravitational potential, assuming a constant mass-to-light ratio for each component. For illustration, this model is applied to M31, which has a typical flat rotation curve. The scale lengths for the disk and the bulge are taken from de Vaucouleurs's (1958) surface photometry. We assume that the bulge and the disk have intrinsic axial ratios of 0.6 and 0.1, respectively; these are typical values. The M/L ratios for the two components are not known a priori, so we adjust them to make the resulting rotation curve as flat as possible, at the observed value. Figure 1 shows the outcome: for the bulge M/L = 11 and the disk M/L = 8, the rotation curve from these two luminous components alone is approximately flat, at the observed value, from near the galactic center out to a radius of about three scale lengths. These M/L values are typical of those for old stellar populations. However, Figure 1 also shows that the rotation curve calculated for the two luminous components falls beyond three disk scale lengths, so we cannot explain rotation curves that are flat beyond about three scale lengths in this way.

Another interesting fact emerges from Figure 1. As shown by Rubin *et al.* (1982), rotation curves for Sb galaxies (these have significant bulge components) are usually flat between a few kpc from the centre and the outer parts. From Figure 1, we see that the flat rotation curve in the inner parts of the model comes about because the peak rotational velocities for the disk and the bulge potentials separately are fairly similar. It is not obvious why this should be so. For example, consider again the simple model for M31. Take the bulge and disk as above, with the same M/L ratios but with the bulge's scale length reduced by a factor 2.5. The bulge's mass is the same as before, so the peak rotational velocity for the bulge potential is significantly increased. The total rotation curve then has the form shown in Figure 2. This kind of rotation curve, rising towards the galactic center, is rarely observed.

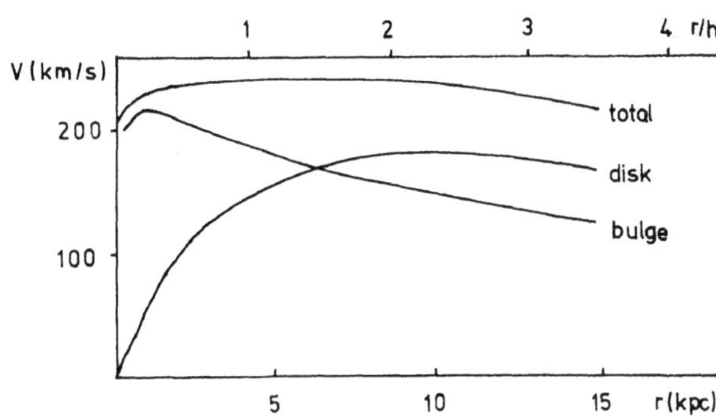

Fig. 1. The rotation curve for the simple (bulge plus disk) model described in the text. The curves labelled 'disk' and 'bulge' show the rotation curves for the gravitational fields of these components separately. The rotation curve for the bulge plus disk is labelled 'total'. The parameters of the model have been scaled to represent M31. This shows how the bulge and disk together can produce a flat rotation curve, from near the galactic center out to about three disk scale lengths.

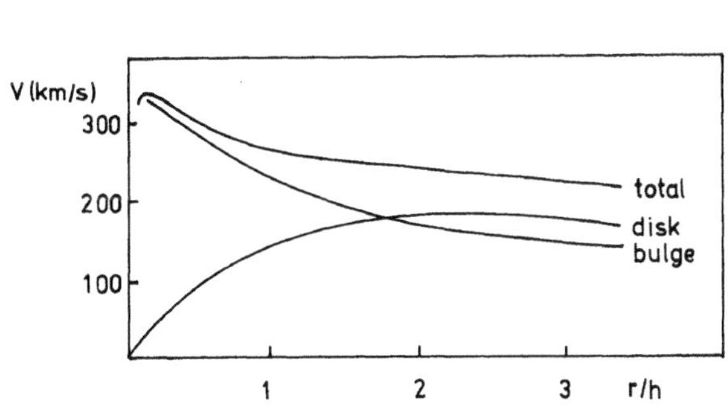

Fig. 2. The rotation curve for another bulge plus disk model. The disk is the same as in Figure 1. The scale length of the bulge is 2.5 times smaller than in Figure 1; the bulge mass is the same. See the text for explanation.

What is the implication? The formation of the bulge component is probably dissipative: while the bulge matter is mainly in the form of gas, it contracts and dissipates energy until it reaches a state when large-scale star formation occurs. At this stage, the dissipation ceases, so the total energy of the bulge is fixed and, therefore, the characteristic density and length-scale are fixed. These parameters determine the peak rotational velocity for the bulge potential. Then the fact that most rotation curves are flat means that, for some reason, the dissipation continues until the peak rotational velocity for the bulge has risen to be about equal to that for the disk alone. Then the bulge undergoes large-scale star formation. It would be interesting to know why star formation in the bulge occurs precisely at this point. Maybe, as several authors have suggested, the formation history of both the bulge and the disk are linked to the dynamical properties of the dark corona.

We have seen how the bulge and the disk together can produce flat rotation curves out to about three disk scale lengths, and that rotation data out to only three disk scale lengths tells us little about the presence of a dark corona component. Now I will turn to recent work by Carignan (see Carignan and Freeman, 1984) and van Albada *et al.* (1984) on pure disk galaxies (systems with negligible bulges), which provides more solid evidence.

3. Pure Disk Galaxies

Pure disk galaxies are ideal systems in which to study the properties of the dark coronas. Firstly, they are structurally simple: there are only the two components (the disk and the dark corona), so it is easier to identify the contribution of each component to the rotation curve $V(r)$. Secondly, these systems are of late Hubble types, so $V(r)$ can be measured by optical or H I techniques out to many disk scale lengths: such $V(r)$ data is essential if we are to estimate the parameters for the dark coronas. Thirdly, there is indirect evidence (e.g., Tinsley, 1981) from population models that later type galaxies contain more dark matter.

Carignan studied the nearby late-type galaxies NGC 247, NGC 300, and NGC 3109. For these systems, he obtained surface brightness distributions $I(r)$ and rotation curves $V(r)$ out to several disk scale lengths. He then used Kalnajs's algorithm to derive the rotation curve that would be produced by the luminous matter alone, assuming a constant M/L ratio. In the central regions of each galaxy, he assumed that the disk dominates the gravitational field, so he fitted the calculated rotation curve to the observed $V(r)$. This gives the value of the M/L ratio for the disk.

In each case, the rotation curve calculated from the surface photometry fits the data well in the inner parts of the system. Also, in two out of three cases, the estimated M/L ratio for the disk agrees well with values expected from stellar population models. Together, these facts support the assumption that the disk dominates the gravitational field in the inner parts. However, in the outer parts the observed $V(r)$ curve rises well above the calculated curve (see Carignan and Freeman, 1984, for details). This mismatch in the outer parts is presumably due to the additional gravitational field of the dark corona.

The detailed structure of the dark corona is not known, so it was necessary to adopt a simple model. Carignan represented the dark corona as an isothermal sphere. The scale parameters for an isothermal sphere are its core radius r_c, the central density ρ_0, and the velocity dispersion σ. These parameters are related by

$$4\pi G\rho_0 r_c^2 = 9\sigma^2 ;$$

so that only two of them are independent. The circular velocity in an isothermal sphere tends to the asymptotic value $\sqrt{2}\,\sigma$ at large radii. To represent the observed rotation curves, Carignan added the gravitational field of an isothermal sphere to the field of the disk. The parameters of the isothermal sphere (r_c and σ) were adjusted to give the best fit to the observed $V(r)$ curve. In each case, the rotation curve calculated for the photometric disk plus the isothermal corona was an excellent fit to the observed $V(r)$ curve.

Although the rotation data extended to several disk-scale lengths for these three galaxies, this is not far enough to define clearly the 'turnover' radius for the contribution that the isothermal sphere makes to the rotation curve. In the region where there is data, this contribution is not much different from solid body. Therefore the values of r_c and σ derived from the fits (and given in Table I below) are just lower limits. However, their ratio σ/r_c is well determined; this ratio is a measure of the central density ρ_0, which is, therefore, also well determined. Table I also gives the ratio of the dark mass to the luminous mass within the Holmberg radius.

TABLE I

Parameters for dark coronas

	r_c (kpc)	σ (km s^{-1})	ρ_0 ($M_\odot$ pc^{-3})	M_D/M_L
NGC 247	22.5	100	0.0033	0.85
NGC 300	12.0	60	0.0042	0.70
NGC 3109	10.5	40	0.0024	1.52

Recently, van Albada *et al.* (1984) have studied the disk galaxy NGC 3198, for which the rotation data extends out to 11 disk scale lengths. The results are generally similar to those above. The inner parts of the rotation curve are well fit by the calculated rotation curve, with a disk M/L value of 3.6, which is consistent with population models. Again, the observed $V(r)$ curve is much higher than the calculated curve in the outer parts of the galaxy. Although their analysis is somewhat different from Carignan's, the parameters for the dark corona are clearly comparable with those given in the table above. Within a radius of 30 kpc, the ratio of dark mass to luminous mass is about 4.

References

Carignan, C. and Freeman, K. C.: 1984, preprint.
de Vaucouleurs, G.: 1958, *Astrophys. J.* **128**, 465.

Kalnajs, A.: 1983, in E. Athanassoula (ed.), *Internal Kinematics and Dynamics of Galaxies*, D. Reidel Publ. Co., Dordrecht, Holland, p. 87.
Rubin, V. C., Ford, W. K., Thonnard, N., and Burstein, D.: 1982, *Astrophys. J.* **261**, 439.
Tinsley, B.: 1981, *Monthly Notices Roy. Astron. Soc.* **194**, 63.
van Albada, T. S., Bahcall, J. N., Begeman, K., and Sancisi, R.: 1984, preprint.

THE GALACTIC CENTRE (INVITED REVIEW)*

A. R. HYLAND

Mount Stromlo and Siding Spring Observatories, Research School of Physical Sciences, Australian National University, Canberra, Australia

(Received 2 July, 1985)

Abstract. Recent observations of the galactic centre at radio and infrared wavelengths are reviewed. Particular emphasis is placed on the new large scale radio structures – the galactic centre lobe, and the filamentary ribbon structures in the radio arc – which have just been reported. New observations of the two parsec ring of molecular hydrogen, and the relationship of the ring to the interior distribution of ionised gas are discussed. The kinematics and distribution of neutral material, shocked molecular hydrogen and plasma in the inner few parsecs of the nucleus are considered in terms of a central luminous mass-losing object, together with rotation and infall of the neutral material.

1. Introduction

Studies of galactic nuclei hold the key to the understanding of many important and energetic astrophysical phenomena. These include high velocity gas jets, nuclear activity, BL Lac objects, quasars, and the very nature of the sources which power these phenomena. All external galaxies are at such distances that small scale details cannot be studied with present technology. However, despite its obscuration by ~ 30 mag. of visual absorption, our own galactic centre is close enough for radio and infrared techniques to provide an extremely detailed view of physical processes there.

Although clearly not in the same class as energetic active nuclei, our galactic nucleus contains many indications of past energetic events, which may have relevance to nuclear activity in general. In this rapidly developing subject, there is now such a wealth of data, that one cannot cover all aspects in a review of this length. The galactic centre evokes such interest that the subject has been reviewed many times in the last few years (Oort, 1977; Wollmann, 1978; Lacy, 1980; Gatley and Becklin, 1981; Townes *et al.*, 1983; Gatley, 1984; Brown and Liszt, 1984).

Thus, while giving an overview of some commonly held beliefs of the galactic centre region, emphasis in this paper will be placed on those new results which impact significantly on ideas relating to large scale ($\gg 10$ pc) and small scale (< 10 pc) structures within the region.

2. Large Scale Structure of the Galactic Centre Region

2.1. Radio line observations

Much of our present knowledge of the large scale structure in the vicinity of the galactic centre comes from the interpretation of radio line data, in particular, the emission and

* Paper presented at the IAU Third Asian-Pacific Regional Meeting, held in Kyoto, Japan, between 30 September–6 October, 1984.

absorption line observations of H I, and the many molecules (OH, CO, H_2O, HCO^+, CN, and CS) abundant in the direction of the galactic centre.

All early work on these subjects has been reviewed in detail in the treatise of Oort (1977). Since that time much new molecular data has been accumulated (Fukui *et al.*, 1977; Liszt and Burton, 1978, 1980; Whiteoak and Gardner, 1979; Bania, 1980; Güsten and Downes, 1980; Cohen and Few, 1976; Linke *et al.*, 1981; Güsten, 1982; Güsten and Henkel, 1983) and new high quality H I data has been obtained covering the region within $\pm 1\overset{\circ}{.}5$ of the galactic nucleus (Rohlfs and Braunsfurth, 1982). These last authors are among the latest to attempt a detailed interpretation of the line data.

The interpretation is generally based on longitude-velocity plots of the line data, which reveal a number of important features in the region (see the schematic representation given in Figure 1). These may be interpreted in terms of dynamical phenomena, which commonly show the presence of non-circular motions relative to the nucleus.

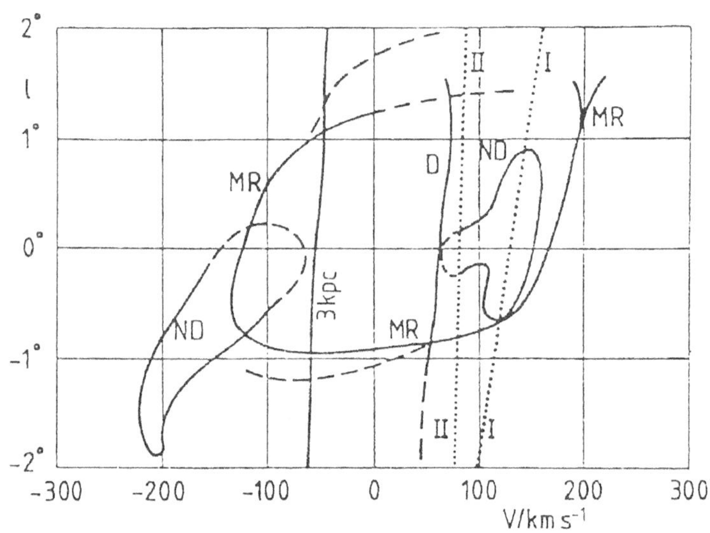

Fig. 1. A schematic representation of the $T(l, V)$ diagram for the galactic centre, showing the major features: ND, nuclear disk; MR, molecular ring; 3 kpc, 3 kpc arm; I and II possible spiral features (from Rohlfs and Braunsfurth, 1982).

Among the important features are the following:

(1) The 3 kpc arm, an intense feature at ~ -53 km s^{-1} first described by Rougoor and Oort (1960) and Rougoor (1964).

(2) The $+65$ km s^{-1} feature, which Sinha (1979) suggests may be two spiral structures.

(3) The nuclear disk, prominent in both H I and molecular lines, which exhibits line widths $\gamma \sim 25$ km s^{-1}, and which appears warped in its outer regions (Burton and Liszt, 1978).

(4) The $+135$ km s^{-1} feature, which is distinctly noticeable in the $T(l, V)$ diagrams.

It has been considered to be an expanding arm (Cohen, 1978), possibly superposed on the nuclear disk (Cohen and Davies, 1979).

(5) The molecular ring (Kaifu *et al.*, 1972; Scoville, 1972), which was first seen in OH and H_2CO, but is probably best seen in the CO observations of Liszt and Burton (1978). It is also present in neutral hydrogen, but only weakly. It is asymmetric in both radius and velocity, and is rotating at an angular velocity much slower than that of galactic rotation.

From the available data in H I and molecular lines, Rohlfs and Braunsfurth (1982) derive a model of the physical locations of the features within the central few hundred parsecs, as shown in Figure 2. This interpretation has many aspects in common with earlier similar analyses (Oort, 1977; Güsten and Downes, 1980) but differs in certain aspects such as the detailed positions of individual molecular clouds, and must be regarded as just one point of view. Many aspects of such analyses are still controversial (see, e.g., Brown and Liszt, 1984).

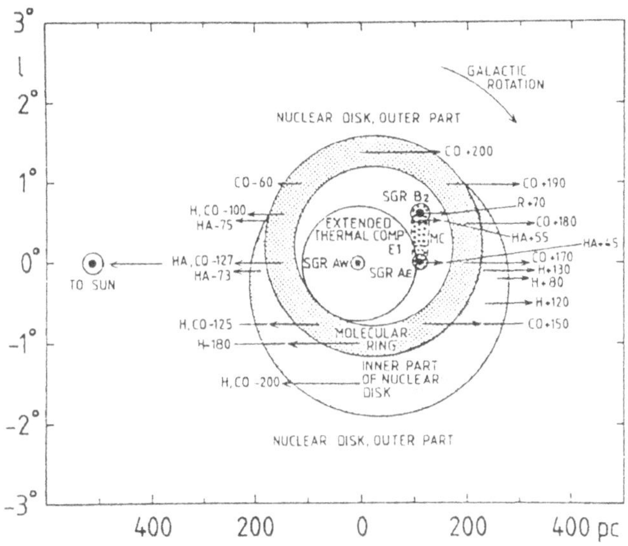

Fig. 2. Model for the galactic centre region (from Rohlfs and Braunsfurth, 1982). The arrows indicate the radial velocities for the marked positions while the letters denote the source of the line radiation (H, 21 cm emission; HA, 21 cm absorption; CO, CO emission; R, H recombination line emission).

The important question naturally relates to the mechanism by which these large scale features have been produced. While there is no established consensus, Oort (1977) discussed at some length two competing hypotheses advanced to explain the large scale dynamics and distribution of matter near the galactic centre, viz., the expulsion hypothesis, and the field hypothesis. Although the effects of a possible bar like structure on the kinematics of material near the galactic centre are still a possible explanation, and need further investigation (Brown and Liszt, 1984), the expulsion model is currently favoured as the explanation for the non-circular velocities and expansion velocities of many species near the galactic centre.

In this context, the large mass of ejected material (requiring a constant mass loss of $1\ M_{\odot}\ \mathrm{yr}^{-1}$), the velocities and the derived timescales suggest that approximately 10^7 yr ago the galactic nucleus was the centre of some extremely energetic ejection process, and that it is now merely in a comparatively quiescent phase.

2.2. Radio continuum observations

The large-scale distribution of 3.75 cm continuum radio emission is shown in Figure 3 (Downes and Maxwell, 1966). In large part this emission may be associated with typical

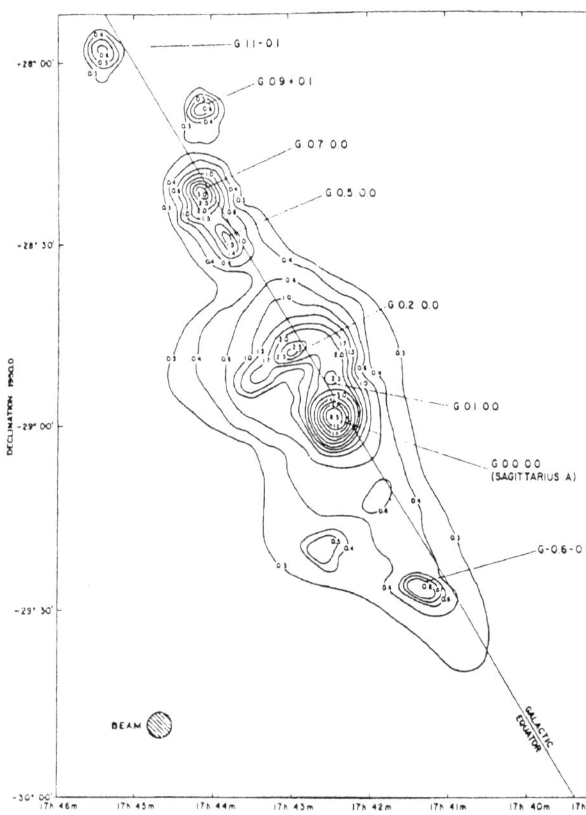

Fig. 3. The distribution of thermal radio emission around the galactic centre, taken from the 3.75 cm survey of Downes and Maxwell (1966). The beam was 4.2 arc min, and the contours represent antenna temperature.

H II regions along the galactic plane, themselves associated with molecular clouds and regions of star formation. Details of the relationship between the neutral atomic and molecular species and the ionized gas are, however, highly complex and difficult to interpret; an excellent overview has been given by Brown and Liszt (1984).

Of particular interest for many years has been the prominent arc (spur) of continuum emission perpendicular to the galactic plane and crossing it at $l = 0°.18$ (see, e.g., Pauls *et al.*, 1976; Mills and Drinkwater, 1984) as shown in Figure 4. There is no clear association of the arc with far infrared emission, but it does contain several compact radio sources (Downes *et al.*, 1978) and shows a remarkable alignment with the edge of the dense 50 km s^{-1} molecular CO cloud (Liszt and Burton, 1985), suggesting that a causal relationship exists.

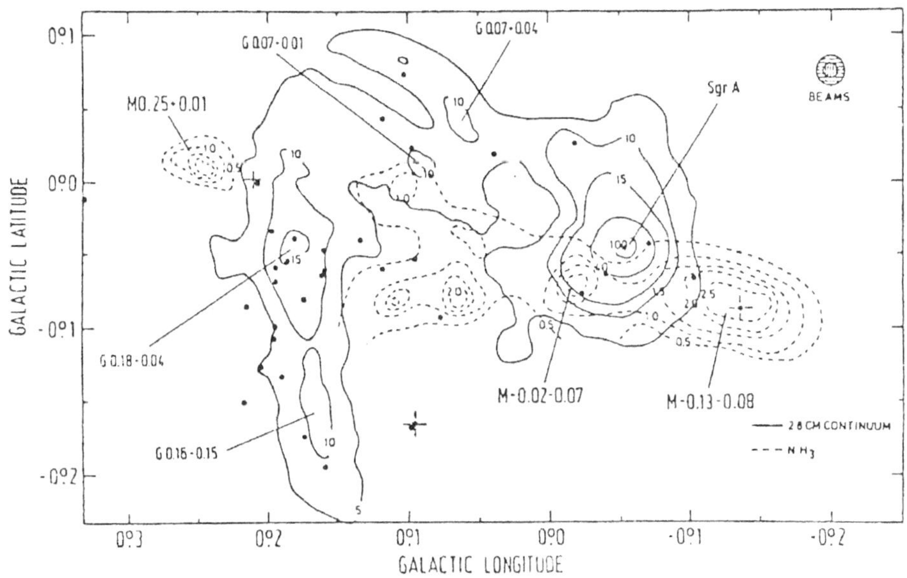

Fig. 4. The distribution of ammonia intensity in the (1, 1) transition compared with that of the 2.8 cm continuum. Dots represent compact radio sources (Downes *et al.*, 1978) and crosses are H_2O sources. From Güsten (1982).

Crucial new observations with the Very Large Aperture (VLA) reveal remarkably ordered structures within the arc, as discussed below, and provide the rationale for completely new interpretations of the origin of the arc.

2.2.1. *Filamentary Structure of the Radio Arc*

Some of the most exciting new observations of the galactic centre region are those of Yusef-Zadeh *et al.* (1984a, b) made with the VLA CD array, giving a 25 arc min field of view at 20 cm, with a half power beam width of 4×9 arc sec. The observations show that the radio arc consists of a system of narrow filamentary structures, with a typical width of ~ 1 pc and length of ~ 30 pc (Figure 5). These highly ordered large-scale filaments are clearly related to some galactic scale phenomenon, rather than being associated with local areas of star formation, and the new data brings into question any ideas that the galactic centre region is currently one of active star formation.

The authors identify two classes of filaments, (i) the vertical filaments, located at

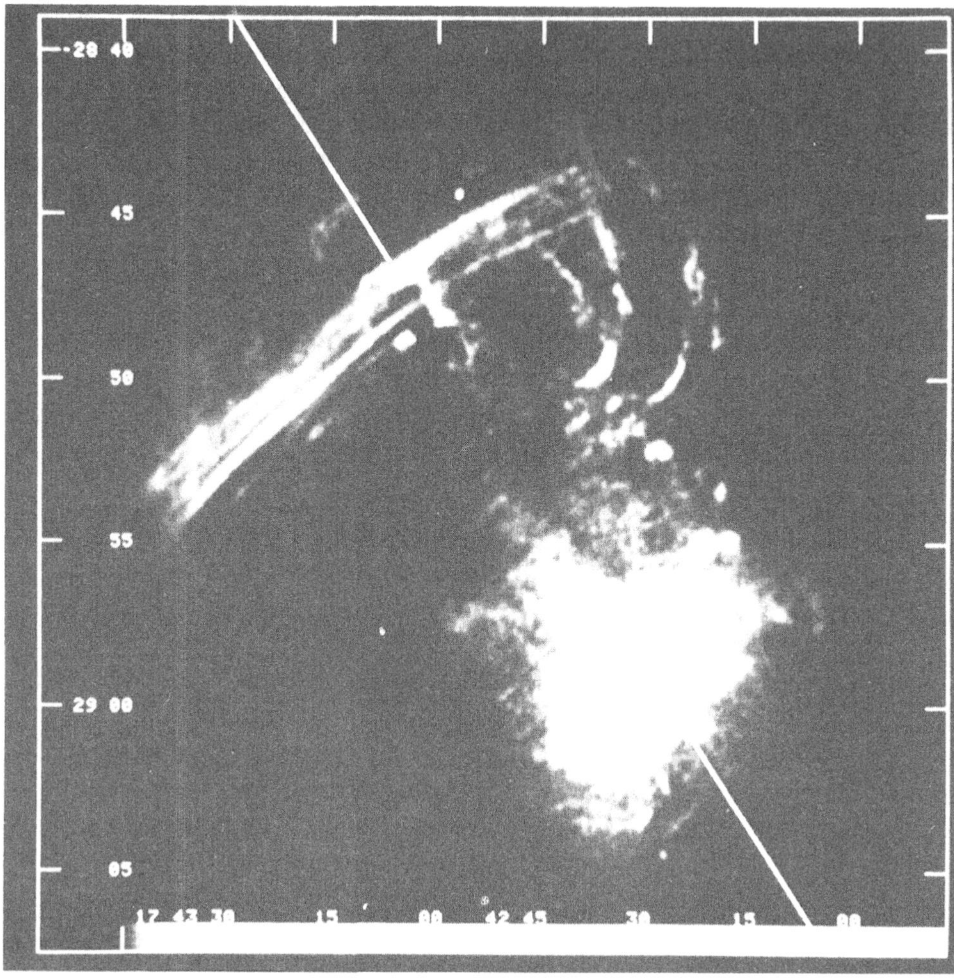

Fig. 5. A radiograph of the large scale 20 cm radio emission features within 60 pc of the galactic nucleus. The line shown is parallel to the galactic plane at $b = -0^\circ.05$. The half power beam width is 4×9 arc sec, and the coordinates are 1950 (from Yusef-Zadeh *et al.*, 1984).

$l = 0^\circ.16$–$0^\circ.18$, perpendicular to the galactic plane, parallel to each other, and extremely regular and unbroken over their lengths, and (ii) the arched filaments emanating in a tight curve from the northern region of the low surface brightness halo of Sgr A, and eventually joining the vertical filaments. In addition, there appear to be some large scale helical structures encircling the vertical filaments.

The vertical and arched features differ in several characteristics. The latter have a counterpart in the far infrared and are probably thermal, while the vertical filaments have no such counterpart, have uniform brightness temperature and significant polarization (15–25% at 6 cm) at several positions along their length, notably where the helical structures cross the filaments. These may be non-thermal, and the presence of synchrotron radiation and its associated relativistic particles is indicated. Unfortunately, the spectral index information from the 6 and 20 cm observations, while being consistent with such a hypothesis, is not sufficiently discriminating to prove the case.

New VLA recombination line (H110α) observations by Morris *et al.* (1984) are also consistent with the view that the arched filaments are thermal, and the vertical filaments non-thermal. The authors argue that such unique ordered phenomena must be occurring within the unique environment of the galactic centre, and suggest that magnetic phenomena related to a significant galactic poloidal field must be a dominant contributor to the structure. They also note that the structure is in some ways analogous to that of solar prominences, which is interesting in the light of the galactic centre lobe observations discussed below.

2.2.2. *The Galactic Centre Lobe*

A second large scale radio structure, possibly related to the 'arc', has also recently been identified. Sofue and Handa (1984) and Sofue *et al.* (1984) working with the Nobeyama Radio telescope at 10.5 GHz have demonstrated the existence of an extensive thermal radio lobe structure (called the galactic centre lobe, GCL), perpendicular to the galactic plane, stradding the galactic centre and extending to $b > +1.^{\circ}2$. While evidence for such a structure may be found in earlier low frequency surveys (Altenhoff *et al.*, 1970, 1978; Haynes *et al.*, 1978) its importance to the physics of the galactic nucleus has not previously been considered. Elsewhere in this conference Sofue *et al.* (1984) consider three possible mechanisms for the production of the GCL. Of particular interest is the spatial (and possibly causal) relationship of the eastern edge of the GCL near $b = 0°$ with the filamentary structure of the radio arc described above, and the fact that, in common with proposed explanations of the arc structure, the presence of a magnetic field of 20 μG, perpendicular to the galactic disk, is required by the non-explosive mechanisms proposed for the GCL. These include (i) a magnetically produced channelled exhaust, and (ii) the inflation of a giant loop of magnetic tube filled with ionised gas, similar to a solar loop prominence.

2.3. Far infrared emission

Observations of far infrared emission (40–350 μm) along the galactic plane in the vicinity of the galactic nucleus have provided important data complementary to the radio observations (see Alvarez *et al.*, 1974). The radiation is thermal emission from dust heated by starlight, and may arise from fairly low density regions, because, as long as $A_v \gtrsim 1$, virtually all the power will be thermalised into the far infrared.

More recently, Odenwald and Fazio (1984) have undertaken an extensive study in the 40–250 μm band with a 1 arc min beam (see Figure 6). This study, covering $\pm 1.^{\circ}5$ from the galactic centre, warrants attention as an investigation of the giant H II regions (and associated star formation regions) in the inner region of the Galaxy not penetrated by spiral density waves. While they find a large number of discrete far infrared sources coincident with thermal radio sources, and that these are remarkably similar to H II regions in spiral arms in terms of their far IR luminosities, Lyman continuum fluxes, dust properties and low dust to gas ratios, there are important differences. Many of the

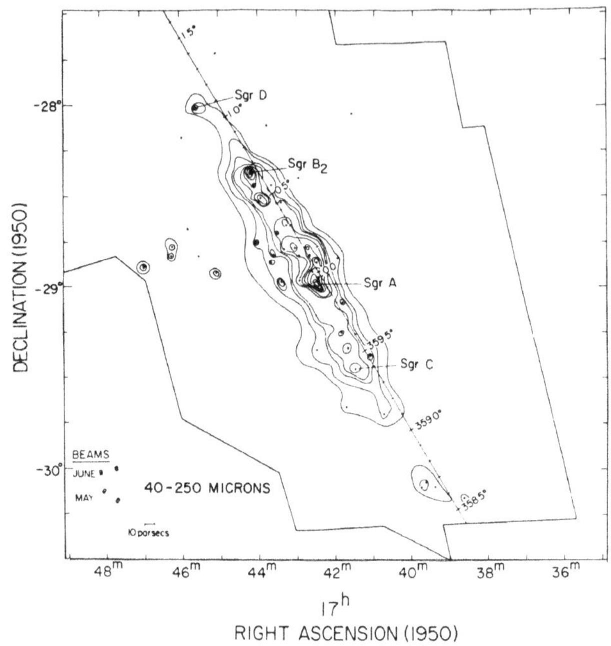

Fig. 6. The large scale 40–250 μm map from Odenwald and Fazio (1984). The lowest contour is 400 Jy/beam, and successive contours are at multiples of the lowest contour, at intervals of 2, 3, 4, 6, 8, 10, 12, 14, 16, 20, and 40. The boundaries of the survey, and the orientation of the chopped beams are shown.

discrete sources do not appear to be embedded within the dense H_2CO, HCN, and NH_3 molecular clouds found within the region. The molecular clouds appear to be cold ($T \sim 25$ K), and are certainly not currently regions of active star formation. This conclusion adds to the growing body of evidence that significant star formation activity does not appear to be occurring within 300 pc of the galactic centre.

An extended diffuse component of far infrared emission dominates the total luminosity of the far IR ($7 \times 10^8 L_\odot$) over a 6°.5 × 2°.0 area (Hoffmann and Frederick, 1969). The discrete sources account for only ~15% of the total luminosity. The extended component is best seen in the large beam surveys of Nishimura *et al.* (1980) and Hoffmann *et al.* (1971). The large infrared excess (> 50) derived when this luminosity is combined with the Lyman continuum flux of the extended radio component (at least 3.6×10^{51} s^{-1}, Mezger *et al.*, 1974), requires that the stellar population responsible for the ionisation be much later than usually found in H II regions. A viable suggestion is that there is an extremely large population of early B stars (e.g., 30 000 B0 stars, Odenwald and Fazio, 1984) producing the ionisation and luminosity. The consequences of the existence of such large numbers of B stars are that either there is a very large number of evolved red supergiants in the region, or the initial luminosity function in no way resembles the Salpeter function, such that the formation of stars with masses greater than $20\,M_\odot$ is inhibited. Since the available data on the stellar population within the central degree of the Galaxy (Becklin *et al.*, 1974) does not appear to show the required number of red supergiants, the latter suggestion is presently to be favoured. It is of interest to compare this suggestion with the processes of star formation occurring within the 30 Doradus complex in the LMC. There, star formation appears to be taking place in a relatively low density environment (not in molecular clouds), and the luminosities of the protostars indicate that the brightest are less than 20 $M_\odot$ (Hyland *et al.*, 1985).

2.4. Near infrared emission and the stellar population

The stellar population near the galactic centre is dominated by late-type giant stars, which emit a major fraction of their luminosity in the near infrared. Furthermore, since the interstellar extinction at 2 μm is approximately $\frac{1}{15}$ of that in the visual, radiation in the 2 μm window provides the best means of studying the stellar distribution, and the disk and bulge contributions in the galactic centre region.

For the central degree, the 1.2 arc min observations of Becklin *et al.* (1974) remains the best available information, showing as it does the definite alignment of the stellar distribution along the galactic plane, peaking at the position of Sgr A. This is presumably the dynamical centre of the Galaxy.

Large scale observations covering areas up to 10° × 10° centred on the galactic nucleus are due in large part to the valuable contribution of Japanese groups who have undertaken balloon-borne observations at 2.4 μm (a minimum in the airglow). There have been a large number of similar observations (see Matsumoto *et al.*, 1982; Hiromoto *et al.*, 1984 and references therein), which have been analysed in terms of axisymmetric models of stellar and dust components of the disk and bulge. Despite being fairly small at the wavelength used, modelling of the interstellar extinction has been the biggest problem. Hiromoto *et al.* (1984) have undertaken the most recent observations and analysis covering an area of 10° × 10°. Using a new model for the local extinction (Ichikawa *et al.*, 1982), they have derived a distribution of the 2.4 μm emission which basically reveals the contributions from stars and dust in the vicinity of the galactic centre. They find three regions of significantly decreased 2.4 μm emission. One region corresponds with a known prominent CO complex (Inatani, 1982), but the others do not yet have known molecular counterparts. It will be of considerable interest to know whether these too may be explained as due to extinction within molecular clouds, or are real regions of low star density near the nucleus (see Figure 7).

3. Components of the Central 10 Parsecs

While the large-scale phenomena described above provide evidence for extensive non-circular mass motions, and lead to an expulsive hypothesis for physical phenomena near the galactic centre, it is clear that the origin of these effects lies in the very core region of the nucleus. In this section some recent observations which appear crucial to our understanding of the central source will be discussed. These all pertain to the inner region, within a few parsecs of the nucleus.

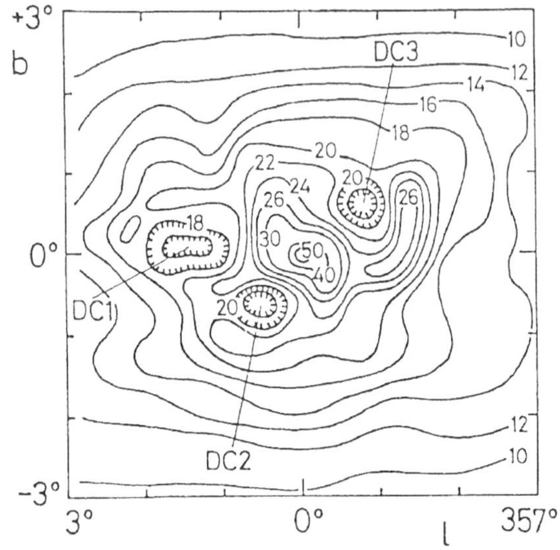

Fig. 7. The 2.4 μm contour map of the central bulge of the Galaxy derived by Hiromoto *et al.* (1984) after correction for local extinction. The contour units are 10^{-10} W cm^{-2} m^{-1} sr^{-1}. Three conspicuous dips in the 2.4 μm emission are denoted by DC1, DC2, and DC3.

3.1. The Distribution of the Dust and Neutral Gas

Among the observations which have had a critical impact on our understanding of the physical conditions within a few parsecs of the galactic centre, none surpass the far infrared observations at 30, 50, and 100 μm with 30 arc sec resolution (Becklin *et al.*, 1982) obtained using the Kuiper Airborne Observatory. Power absorbed by dust grains near the galactic centre is reradiated as optically thin radiation at these wavelengths. The observations show a gradual change in the morphology of the emission with wavelength, from a centrally peaked source at 30 μm to a double lobed source at 100 μm, with the galactic centre lying between the lobes as shown in Figure 8. Both the luminosity and temperature peak at the position of the galactic nucleus, and decrease smoothly in all directions. The explanation for these observations follows from the fact that the volume emissivity of the dust is proportional to both the heating flux and the number density of grains. Thus, a maximum in surface brightness may correspond either to a peak in the dust temperature, or to a peak in the dust density. Becklin *et al.* (1982), therefore, attribute the observations to the presence of a central luminous object, surrounded by a cavity of low density material, and encircled by a thick ring structure of neutral material (corresponding to a dust density peak), at about 2 pc radius from the centre, and aligned along the galactic plane. Within this ring $A_v \geq 1$, the mean gas density is $\sim 10^{3.5}$ cm^{-3}, and its total mass is $\sim 10^3\ M_\odot$. These observations also lead to the best present estimate of the luminosity of the central source as 1–3 × $10^7\ L_\odot$.

The above picture is supported by observations of the 63 μm fine structure line of neutral oxygen (Lester *et al.*, 1981; Genzel *et al.*, 1984) which is spatially extended and well correlated with the far infrared continuum. Velocity measurements of the oxygen line also show that the proposed ring structure is participating in galactic rotation.

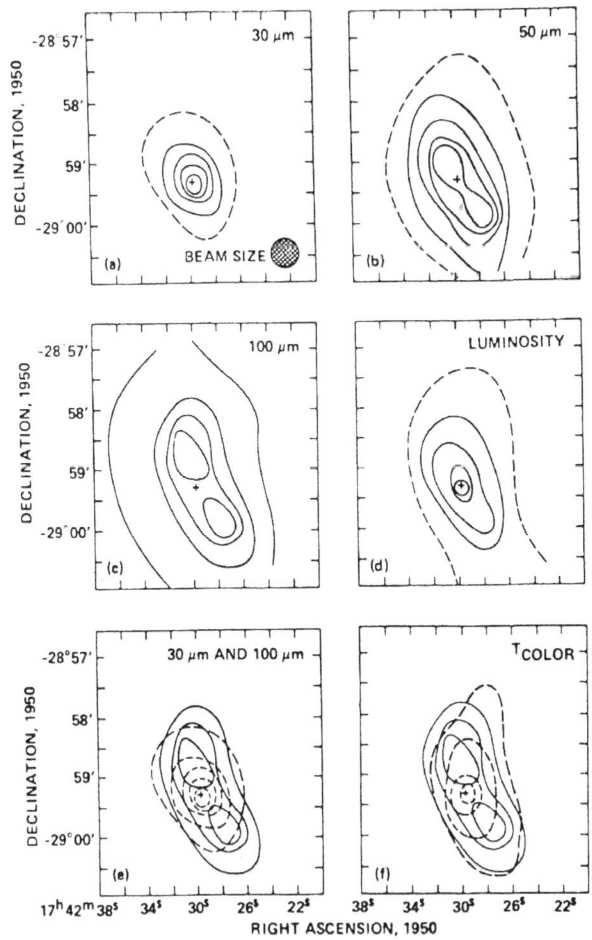

Fig. 8. The far infrared surface brightness at 30 μm (a), 50 μm (b), and 100 μm (c) of the galactic centre region measured with 30 arc sec (1.3 pc) resolution (from Becklin *et al.*, 1982). The position of the galactic centre is shown by a cross (+). The distributions of luminosity (d) and colour temperature (f) are deduced from the data in a, b, and c.

Further ground based observations from Mauna Kea, at 20 and 34 μm with 10 arc sec resolution (Gatley *et al.*, 1984), establish the near infrared source IRS 16 as the location of the highest temperature grains, and hence as the major source of luminosity (Gatley, 1984).

3.2. The Distribution of Shocked Molecular Hydrogen

Another set of observations which have provided critical insight into the nature of the galactic centre, is that of shocked molecular hydrogen (Gatley *et al.*, 1984) located at the inner edge of the neutral ring. From limited spectra of the 2.122 μm $v = 1 \rightarrow 0$ $S(1)$ and 2.248 μm $v = 2 \rightarrow 1$ $S(1)$ lines of H_2 they infer that the galactic centre is encircled by a thin ring of shocked gas. The ring has a radius of ~ 2 pc, a thickness of $\sim 10^{13}$ cm, and is symmetric about the notional centre of the Galaxy, IRS 16. The observations are also consistent with the ring rotating in the sense of galactic rotation.

The momentum needed to shock the gas at the required rate ($\sim 10^{-2}\,M_\odot\,\mathrm{yr}^{-1}$) cannot be provided radiatively by the available central luminosity ($\sim 3 \times 10^7\,L_\odot$). Gatley *et al.* (1984) suggest that mass loss from a central object is a natural explanation for both the central 'cavity' and the shocked gas, requiring an isotropic mass loss rate of $3 \times 10^{-3}\,M_\odot\,\mathrm{yr}^{-1}$. Such a rate suggests that a large fraction of the luminosity within the central parsec should be concentrated in a single luminous object, on the basis of mass loss/luminosity laws derived for young hot stars.

Time-scale arguments suggest that observable molecular hydrogen is a long-lived phenomenon, and lead to interesting speculations regarding the history of the galactic centre. An attractive picture is that there was an energetic nuclear outburst some 10^7 yr ago, which cleared the central volume of neutral gas, and was responsible for the large-scale dynamical features described earlier. The present $10^3\,M_\odot$ of neutral material

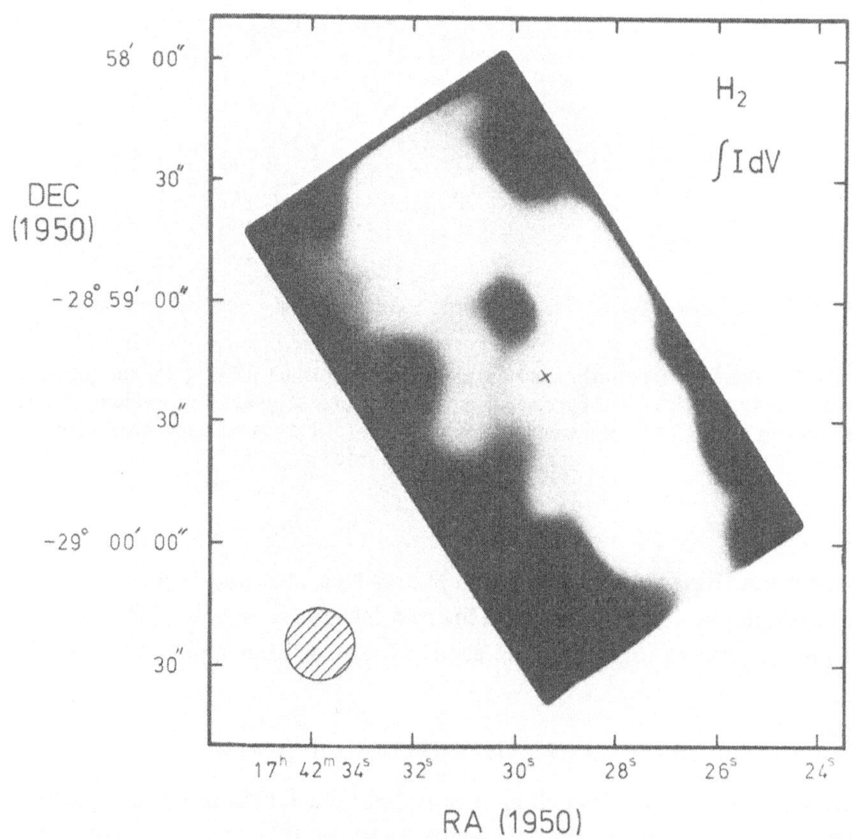

Fig. 9. A photographic representation of the integrated emission in the $v = 1 \rightarrow 0\ S(1)$ line of molecular hydrogen in the galactic centre region. This is from a fully sampled map of the region made at UKIRT with an 18 arc sec beam. The position of IRS 16 is shown by a cross (+). The morphology resembles that of a tilted disk (Gatley *et al.*, 1985).

within the central few parsecs must then be accumulated mass loss from giant stars in the intervening period, the observed molecules have formed, in situ, and the plasma in Sgr A is infalling.

The mechanism for possible periodic outbursts is suggested as accretion onto a central massive object (possibly a black hole), which is at present in a low level of activity.

These ideas were based on extremely limited data and have, therefore, to be treated cautiously. However, comprehensive new data acquired earlier this year on UKIRT (Gatley *et al.*, 1985) vindicates the previous speculations, and clarifies the morphology of the H_2 distribution substantially. This data was obtained using an 18 arc sec beam, and with 90 km s^{-1} resolution. It covers an area of $\sim 3 \times 1.5$ arc min, and was fully sampled every 9 arc sec. A photographic representation of the observed integrated intensity of the $v = 1 \rightarrow 0$ $S(1)$ line is given in Figure 9. The new velocity resolved data also provides conclusive evidence that the ring is rotating in the sense of galactic rotation. This is clearly seen in the contour maps shown in Figures 10 and 11, and the (l, V) diagram (Figure 12).

The new data do, however, suggest a number of changes to the initial picture: (i) the ring is not axisymmetric about IRS 16, as the SW lobe is slightly closer. It is possible

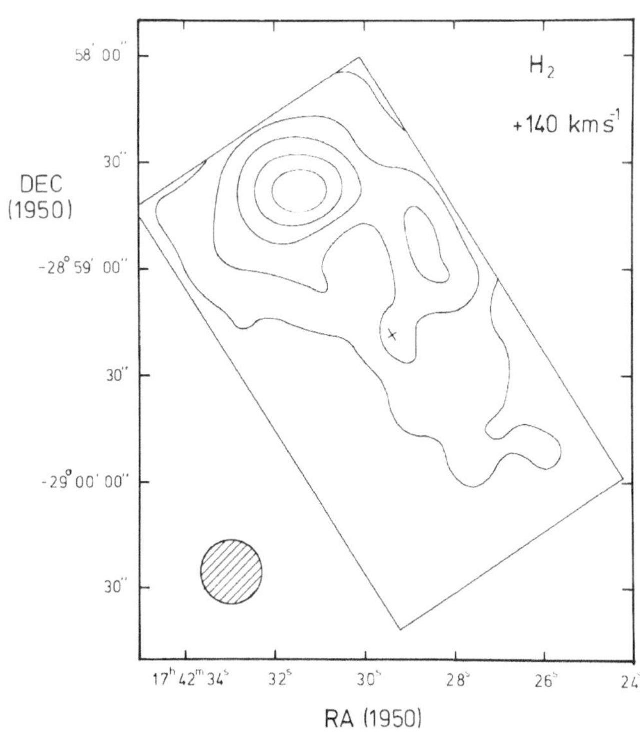

Fig. 10. Contour map of the intensity of the $v = 1 \rightarrow 0$ $S(1)$ line of molecular hydrogen in the $+140$ km s^{-1} channel. Comparison with the contour map in the next figure shows the velocity difference between the two lobes.

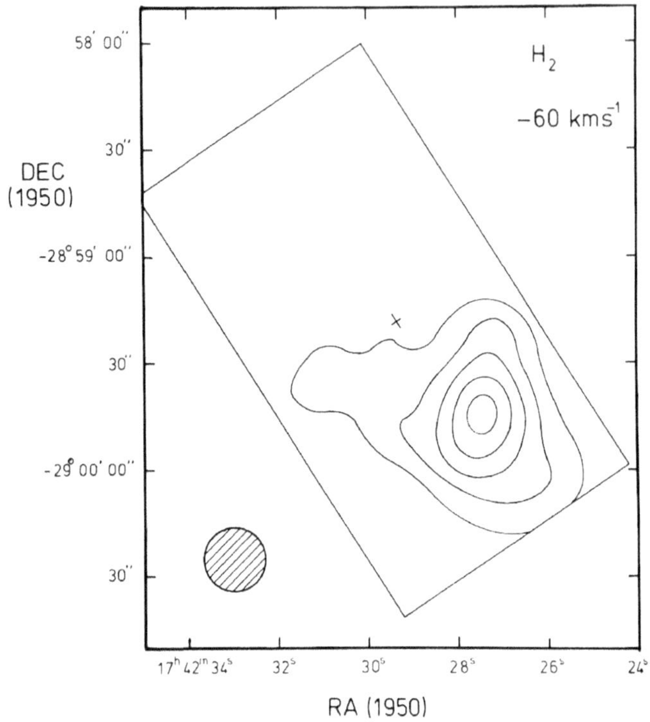

Fig. 11. As for Figure 10 but in the -60 km s^{-1} channel.

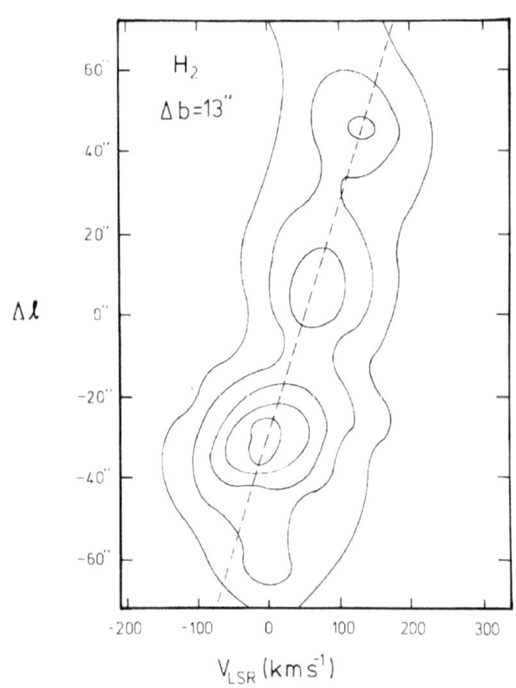

Fig. 12. A velocity-longitude diagram of the intensity of the $v = 1 \rightarrow 0$ $S(1)$ line of molecular hydrogen, taken at a latitude offset of 13 arc sec from IRS 16.

that this morphology is related to a spiralling infall of neutral material towards the centre. Certainly gas just interior to the ring of shocked H_2 is ionised, and apparently also has similar rotation characteristics (see below). (ii) The ring is slightly offset (~ 13 arc sec) in latitude from IRS 16, and gives the appearance of being a tilted ring. There is considerable emission between the peaks of the lobes at mid velocity (~ +40 km s^{-1}). The interpretation of this emission and the effects of extinction from intervening molecular clouds on the appearance of the ring need further investigation. Also, the apparent velocity offset of the midpoint of the ring (+40 km s^{-1}) remains a puzzle.

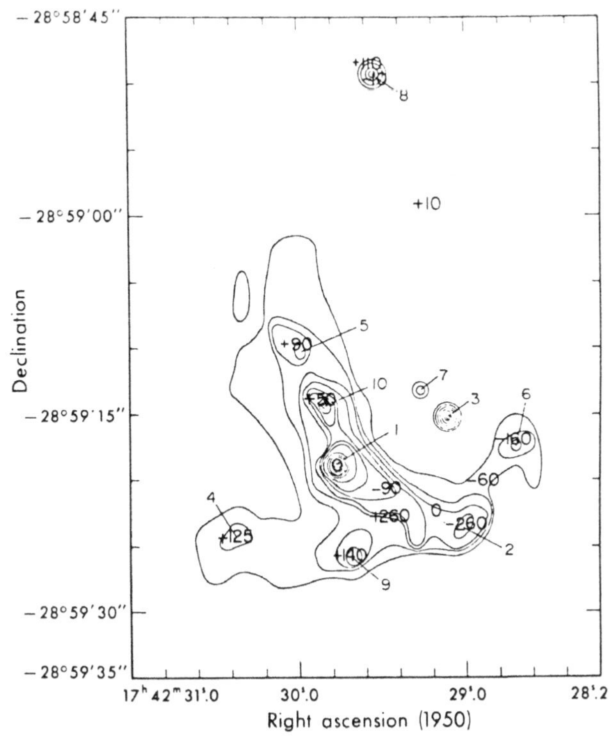

Fig. 13. Velocities of the ionised gas clumps of Ne II (bold face numbers) superimposed on the 10 μm continuum map of Becklin *et al.* (1978) (from Lacy *et al.*, 1980).

3.3. The Distribution of Ionised Gas

Inside the ring of shocked molecular hydrogen the gas is ionised. The total mass of the ionised gas is small (~ 10 $M_{\odot}$), but its distribution and kinematics provide important constraints on galactic centre models.

The distribution of ionised material is revealed by three independent sets of observations:

(a) 10 μm continuum radiation (Becklin *et al.*, 1978) traces the distribution of hot dust, presumably heated by trapped Lyman radiation.

(b) 12.8 μm observations of Ne II emission (Lacy *et al.*, 1979, 1980) correlate well with the 10 μm continuum, and also provide important information on the kinematics of the ionised gas. The gas appears to be concentrated in transient high density clumps (or possibly streamers), with $M \sim 1\ M_\odot$ and lifetimes $\sim 10^4$ yr. If the velocity dispersion of the clumps is attributed to the gravitational field of the stellar density peak at the galactic centre, Lacy *et al.* (1980) infer the presence of $\sim 10^7\ M_\odot$ within the central parsec.

(c) High resolution radio continuum observations such as those of Ekers *et al.* (1983), Lo and Claussen (1983) and Brown and Johnston (1983) define the morphology of the gas extremely well. This is dominated by three bright thermal arms as is shown in Figures 14 and 15, together with the non-thermal compact source Sgr A*. Ekers *et al.* (1983) initially proposed that the arms might be part of a spiral pattern, while Brown (1982) proposed precessing twin-beam ejection.

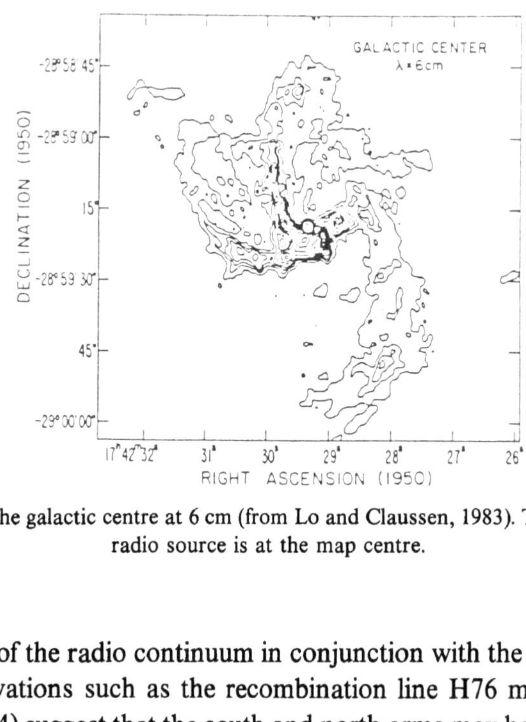

Fig. 14. VLA map of the galactic centre at 6 cm (from Lo and Claussen, 1983). The compact nonthermal radio source is at the map centre.

Careful analysis of the radio continuum in conjunction with the Ne II velocities, and more recent observations such as the recombination line H76 measurements of van Gorkom *et al.* (1984) suggest that the south and north arms may be part of an elliptical, rotating, infalling stream of plasma just inside the ring of shocked H_2 (see Lo and Claussen, 1983; or Brown and Liszt, 1984). It is interesting to compare the morphology of the plasma with that of the H_2. There are certain similarities, including the apparent tilt to the structure.

Any discussion of the ionised gas is confronted by several observations which indicate that the plasma clumps are not typical H II regions (see Gatley and Becklin, 1981). The most widely discussed of these is the low ionisation state of the clumps, indicative of a low temperature of the ionising radiation $T \sim 35\,000$ K. Furthermore, in addition to

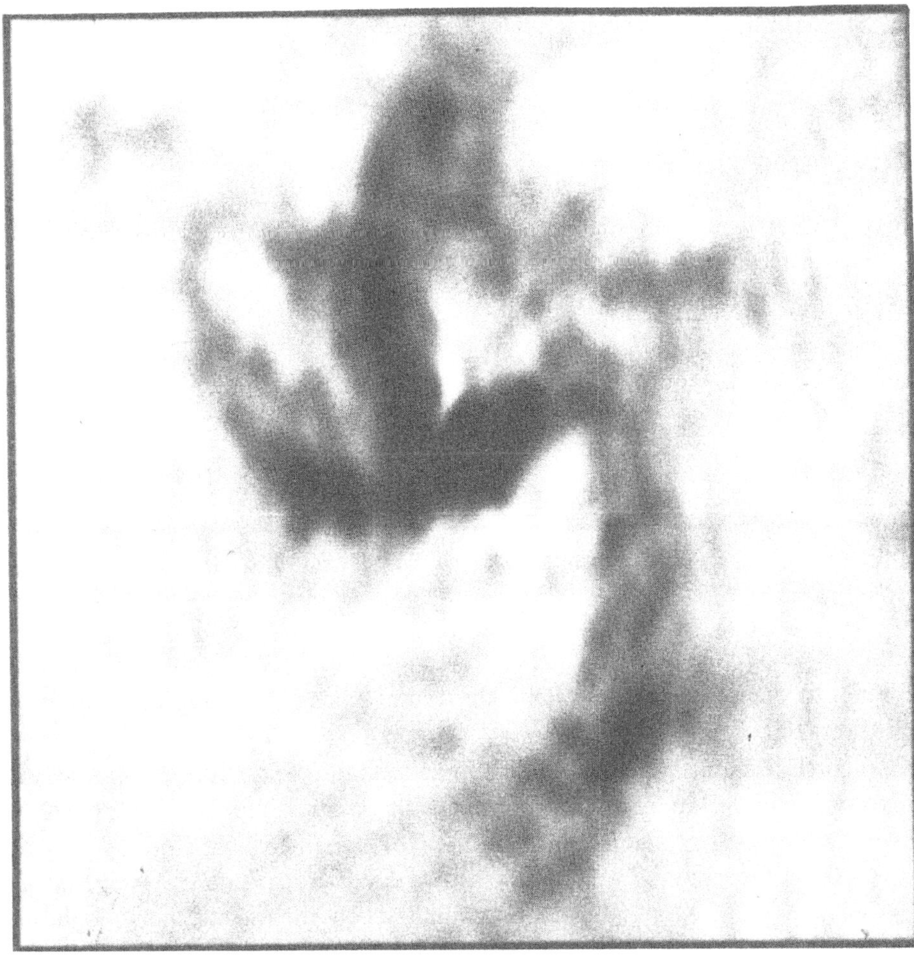

Fig. 15. Radiograph of the galactic centre at 2 cm. North is up, East to the left. The total north–south extent is 80 arc sec (from Ekers *et al.*, 1983).

the ordered rotation exhibited by the plasma ring, there are peculiar velocities in a bar-like structure extending from SE–NW through the centre of the radio maps. The gas in this region shows large velocity widths and high temperature gas compared with the ring (Brown and Liszt, 1984). No model at present adequately accounts for all the observations.

3.4. The Stellar Distribution and IRS 16

The dynamical centre of the Galaxy is defined by the stellar mass distribution, and is expected to peak at IRS 16. High resolution images at 2.2 μm of the central 45 arc sec have been made by Allen *et al.* (1983) to determine the radial density distribution of the inner core. They find that the distribution follows a $\rho \propto r^{-1.68}$ law and is a smooth continuation of the larger scale distribution. They were unable to find any flattening of

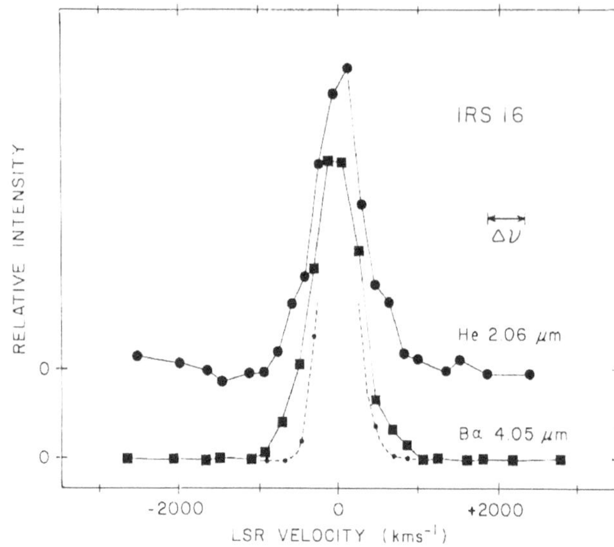

Fig. 16. Spectra of Brα at 4.05 μm and the 2.058 μm line of He I in IRS 16. The extensive wings on both lines in IRS 16 compared with the unresolved Bα line in NGC 6572 are evident (from Geballe *et al.*, 1984).

the curve to their observational limit, thus emphasising the point-like nature of the central source.

The source **IRS** 16 is elongated with multiple component structure, having some three components within 3 arc sec. It has unusually blue colours, and Hall *et al.* (1982) have shown that it possesses and extremely broad He I line, inferring the presence of a high mass loss rate. Geballe *et al.* (1984) have shown that Brα also has large wings extending to ± 700 km s^{-1}. They also have limited data which shows that the strength of the 2.06 μm line of He decreases more rapidly than those of Ne II or H as one moves away from IRS 16. This is consistent with the idea that IRS 16 is the dominant source of ionisation of the plasma clumps, and adds weight to proposals invoking a central engine within the galactic centre.

A crucial question in the discussion of IRS 16 is that of its relationship to the compact radio source Sgr A*. The best available positions of the two sources are not identical. The best position for the near infrared source comes from the 0.98 μm CCD picture of Henry *et al.* (1984), which overcomes the usual astrometric problems of finding positions from single detector measurements in the infrared. As seen in Figure 17, the positions are separated by less than 1 arc sec. This demonstrates the pressing need for the development of 2-D detectors further into the infrared. As an example of what we may expect within the next few years, Figure 18 presents an image obtained with a 3.5 μm In-Si array with a Si CCD, developed by Rockwell/U of Hawaii (Capps *et al.*, 1984). This image covers the innermost 18 × 18 arc sec of the galactic nucleus, with a resolution of 1 arc sec. All the well-known sources can be identified, and their positions can be accurately measured.

Certainly, IRS 16 and Sgr A* are so close that a causal relationship most probably

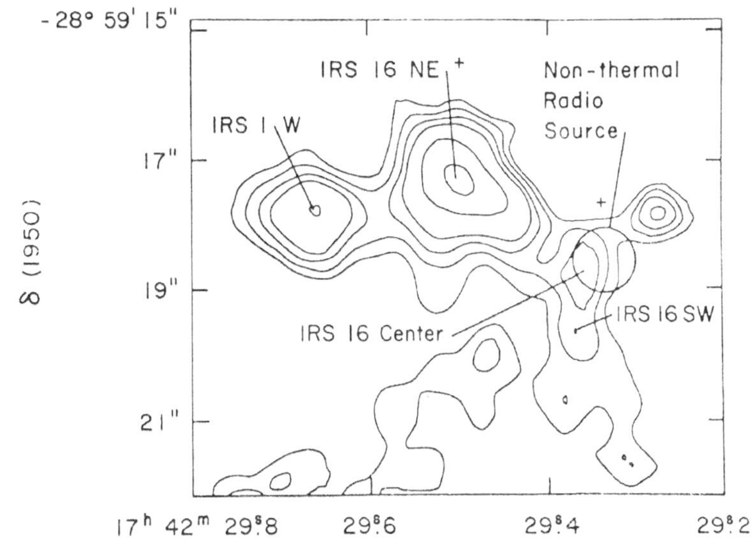

Fig. 17. Contour diagram of the 0.98 μm image around IRS 16 smoothed to a resolution of 1.36 arc sec (FWHM). The two crosses mark the locations of the near IR sources IRR1 and IRR2, which have been subtracted off. The error circle for the position of the nonthermal radio source corresponds to the combined uncertainties in the radio and 0.98 μm maps (from Henry *et al.*, 1984).

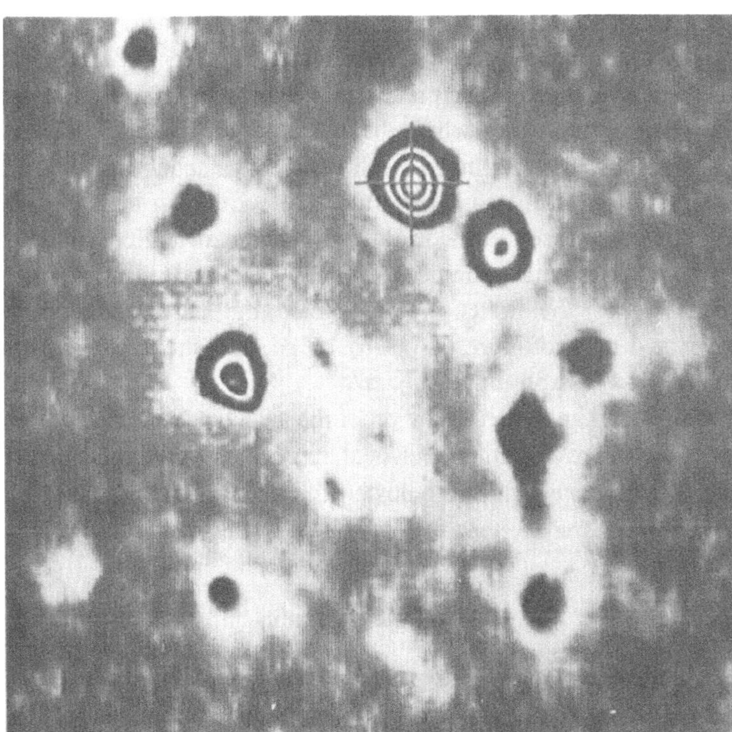

Fig. 18. Photographic representation of an 18 × 18 arc sec image of the galactic centre, acquired using the IRTF with a 3.5 μm Rockwell/UH In-Si plus Si CCD array. The cross (+) marks the position of the bright infrared source IRS 7. The resolution is 1 arc sec. Astrometry from new 2-D detectors such as this promise great advances in the positioning of infrared sources, and the consequent understanding of the phenomena involved (from Capps *et al.*, 1984).

exists. They lie at the dynamical centre of the Galaxy, are clearly related to mass loss and other more complicated dynamical phenomena. The compact radio source is variable (Brown and Lo, 1982), and is probably related to the positron-annihilation gamma ray source identified with the galactic centre (see, e.g., MacCallum and Leventhal, 1983). We thus expect the exact source of galactic centre activity, past and present, to be at least causally related to IRS 16 and/or Sgr A*.

4. Summary

The task of revealing and understanding the nature of the galactic centre remains challenging and rewarding. Any future models will clearly be influenced by the variety of intriguing new results described above, a few of which are summarised here.

(i) The new large scale galactic centre lobe and filamentary arc radio structures provide the first clear indications for the presence of a significant poloidal galactic magnetic field near the galactic centre.

(ii) Radio and far infrared observations show that the dense molecular clouds within 300 pc of the nucleus are very cool, and not active in the formation of massive stars. The conditions of the extended plasma and thermal radiation in the far IR can be interpreted in terms of a peculiar Initial Mass Function, such that very few stars more massive than $\sim 20\, M_\odot$ have been formed within the past 10^7 yr.

(iii) The distribution of dust, neutral atoms and shocked molecules within a few parsecs of the centre all indicate the presence of a massive ($10^3\, M_\odot$) rotating toroidal cloud. The inner edge of this ring, where the molecules are shocked (arguably by mass loss from a central object), is rotating in the direction of galactic rotation at 100 km s^{-1}. Inside this ring is a cavity of low density material containing ionised gas clumps or streamers. Careful comparison of the radio continuum distribution with the molecular ring suggests that a clear connection between the plasma and the ring exists, providing the best evidence yet for the existence of significant material spiralling in towards the nucleus. This would be photo-ionised out to 1.7 pc, and shock excited beyond, by a single luminous, mass loss object.

While demonstrating the clearest view yet of the galactic centre, the present observations remain unable to provide an unequivocal description of the central energy source. However, as new techniques and more powerful telescopes are brought to bear on the problem, the day when the mystery will be solved comes ever closer.

Acknowledgements

I would like to thank Drs Gatley, Jones, Capps, and Becklin, and Dr Ron Ekers and his colleagues, for providing me with much data (some as yet unpublished), for inclusion in this review.

References

Allen, D. A., Hyland, A. R., and Jones, T. J.: 1983, *Monthly Notices Roy. Astron. Soc.* **204**, 1145.
Altenhoff, W. J., Downes, D., Goad, L., Maxwell, A., and Rinehart, R.: 1970, *Astron. Astrophys. Suppl.* **1**, 319.
Altenhoff, W. J., Downes, D., Pauls, T. A., and Schraml, J.: 1978, *Astron. Astrophys. Suppl.* **35**, 23.
Alvarez, J. A., Furniss, I., Jennings, R. E., King, K. J., and Moorwood, A. F. M.: 1974, in A. F. M. Moorwood (ed.), 'H II Regions and the Galactic Centre', *Proc. 8th ESLAB Symposium*, ESTEC, Noordwijk, The Netherlands, p. 69.
Bania, T. M.: 1980, *Astrophys. J.* **242**, 95.
Becklin, E. E., Gatley, I., and Werner, M. W.: 1982, *Astrophys. J.* **258**, 134.
Becklin, E. E., Matthews, K., Neugebauer, G., and Willner, S. P.: 1978, *Astrophys. J.* **219**, 121.
Becklin, E. E., Neugebauer, G., and Early, D.: 1974, in A. F. M. Moorwood (ed.), 'H II Regions and the Galactic Centre', *Proc. 8th ESLAB Symposium*, ESTEC, Noordwijk, The Netherlands, p. 227.
Brown, R. L.: 1982, *Astrophys. J.* **262**, 110.
Brown, R. L. and Johnston, K. J.: 1983, *Astrophys. J.* **268**, L85.
Brown, R. L. and Liszt, H. S.: 1984, *Ann. Rev. Astron. Astrophys.* **22**, 223.
Brown, R. L. and Lo, K. Y.: 1982, *Astrophys. J.* **253**, 108.
Burton, W. B. and Liszt, H. S.: 1978, *Astrophys. J.* **225**, 815.
Capps, R., Tollestrup, E., and Becklin, E. E.: 1986, in press.
Cohen, R. S.: 1977, *Monthly Notices Roy. Astron. Soc.* **178**, 547.
Cohen, R. S. and Davies, R. D.: 1979, *Monthly Notices Roy. Astron. Soc.* **186**, 453.
Cohen, R. S. and Few, R. W.: 1976, *Monthly Notices Roy. Astron. Soc.* **176**, 495.
Downes, D. and Maxwell, A.: 1966, *Astrophys. J.* **146**, 653.
Downes, D., Goss, W. M., Schwarz, U. J., and Wouterloot, J. G. A.: 1978, *Astron. Astrophys. Suppl.* **35**, 1.
Ekers, R. D., van Gorkom, J. H., Schwarz, U. J., and Goss, W. M.: 1983, *Astron. Astrophys.* **122**, 143.
Fukui, Y., Iguchi, T., Kaifu, N., Chikada, Y., Morimoto, M., Nagane, K., Miyazawa, K., and Miyaji, T.: 1977, *Publ. Astron. Soc. Japan* **29**, 643.
Gatley, I.: 1984, in M. F. Kessler and J. P. Phillips (eds.), *Galactic and Extragalactic Infrared Spectroscopy*, ESA, Paris, p. 347.
Gatley, I. and Becklin, E. E.: 1981, in C. G. Wynn-Williams and D. P. Cruikshank (eds.), *Infrared Astronomy*, D. Reidel Publ. Co., Dordrecht, Holland, p. 281.
Gatley, I., Becklin, E. E., and Telesco, C. M.: 1986, (in press).
Gatley, I., Jones, T. J., Hyland, A. R., Beattie, D. H., and Lee, T. J.: 1984, *Monthly Notices Roy. Astron. Soc.* **210**, 565.
Gatley, I., Jones, T. J., Hyland, A. R., Wade, R., Krisciunas, K., and Geballe, T. R.: 1985, in preparation.
Geballe, T. R., Krisciunas, K., Lee, T. J., Gatley, I., Wade, R., Duncan, W. D., Garden, R., and Becklin, E. E.: 1984, *Astrophys. J.* **284**, 118.
Genzel, R., Watson, D. M., Townes, C. H., Dinerstein, H. L., Hollebbach, D., Lester, D. F., Werner, M. W., and Storey, J. W. V.: 1984, *Astrophys. J.* **276**, 551.
Güsten, R.: 1982, in G. R. Riegler and R. D. Blandford (eds.), *The Galactic Center*, AIP, New York, p. 9.
Güsten, R. and Downes, D.: 1980, *Astron. Astrophys.* **87**, 6.
Güsten, R. and Henkel, C.: 1983, *Astron. Astrophys.* **126**, 136.
Hall, D. N. G., Kleinmann, S. G., and Scoville, N. Z.: 1982, *Astrophys. J.* **260**, L53.
Haynes, R. F., Caswell, J. L., and Simons, L. W. J.: 1978, *Australian J. Phys. Suppl.* **45**, 1.
Henry, J. P., DePoy, D. L., and Becklin, E. E.: 1984, *Astrophys. J.* **285**, L27.
Hiromoto, N., Maihara, T., Mizutani, K., Takami, H., Shibai, H., and Okuda, H.: 1984, *Publ. Astron. Soc. Japan* (in press).
Hoffmann, W. F. and Frederick, C. L.: 1969, *Astrophys. J.* **155**, L9.
Hoffmann, W. F., Frederick, C. L., and Emery, R. J.: 1971, *Astrophys. J.* **164**, L23.
Hyland, A. R., Jones, T. J., Gatley, I., and Straw, S.: 1985, in preparation.
Ichikawa, T., Hamajima, K., Ishida, K., Hidayat, B., and Roharto, M.: 1982, *Publ. Astron. Soc. Japan* **34**, 231.
Inatani, J.: 1982, *Publ. Astron. Soc. Japan* **34**, 515.
Kaifu, N., Kato, T., and Iguchi, T.: 1972, *Nature* **238**, 105.
Lacy, J. H.: 1980, in P. A. Wayman (ed.), *Highlights of Astronomy* **5**, D. Reidel Publ. Co., Dordrecht, Holland, p. 163.

Lacy, J. H., Baas, F., Townes, C. H., and Geballe, T. R.: 1979, *Astrophys. J.* **227**, L17.
Lacy, J. H., Townes, C. H., Geballe, T. R., and Hollenbach, D. J.: 1980, *Astrophys. J.* **241**, 132.
Lester, D. F., Werner, M. W., Storey, J. W. V., Watson, D. M., and Townes, C. H.: 1981, *Astrophys. J.* **248**, L109.
Linke, R. A., Stock, A. A., and Frerking, M. A.: 1981, *Astrophys. J.* **243**, 147.
Liszt, H. S. and Burton, W. B.: 1978, *Astrophys. J.* **226**, 790.
Liszt, H. S. and Burton, W. B.: 1980, *Astrophys. J.* **236**, 779.
Liszt, H. S. and Burton, W. B.: 1986, *Astrophys. J.* (in press).
Lo, K. Y. and Claussen, M. J.: 1983, *Nature* **306**, 647.
MacCallum, C. J. and Leventhal, M.: 1983, in M. L. Burns, A. K. Harding, and R. Ramett (eds.), *Positron-Electron Pairs in Astrophysics*, AIP, New York, p. 211.
Matsumoto, T., Hayakawa, S., Koizumi, H., Murakami, H., Uyama, K., Yamagami, T., and Thomas, J. A.: 1982, in G. R. Riegler and R. D. Blandford (eds.), *The Galactic Center*, AIP, New York, p. 48.
Mezger, P. G., Churchwell, E. B., and Pauls, T. A.: 1974, in L. N. Mauridis (eds.), 'Stars and the Milky Way System', *Proc. EAS Meeting*, Springer-Verlag, Berlin, p. 140.
Mills, B. Y. and Drinkwater, M. J.: 1984, *J. Astrophys. Astron.* **5**, 43.
Morris, M., Yusef-Zadeh, F., and van Gorkom, J. H.: 1986, (in press).
Nishimura, T., Low, F. J., and Kurtz, R. F.: 1980, *Astrophys. J.* **239**, 101.
Odenwald, S. F. and Fazio, G. G.: 1984, *Astrophys. J.* **283**, 601.
Oort, J. H.: 1977, *Ann. Rev. Astron. Astrophys.* **15**, 295.
Pauls, T., Downes, D., and Mezger, P. G.: 1976, *Astron. Astrophys.* **46**, 407.
Rohlfs, K. and Braunsfurth, E.: 1982, *Astron. Astrophys.* **113**, 237.
Rougoor, G. W.: 1964, *Bull. Astron. Inst. Neth.* **17**, 381.
Rougoor, G. W., and Oort, J. H.: 1960, *Proc. Acad. Natl. Sci.* **46**, 111.
Scoville, N. Z.: 1972, *Astrophys. J.* **175**, L127.
Sinha, R. P.: 1979, *Astron. Astrophys. Suppl.* **37**, 403.
Sofue, Y. and Handa, T.: 1984, *Nature* **310**, 568.
Sofue, Y., Handa, T., Nakai, N., Hirabayashi, H., Innoue, M., and Akabane, K.: 1986, *Publ. Astron. Soc. Japan* (in press).
Townes, C. H., Lacy, J. H., Geballe, T. R., and Hollenbach, D. J.: 1983, *Nature* **301**, 661.
van Gorkom, J. H., Shwarz, U. J., and Bregman, J. D.: 1986, (in press).
Whiteoak, J. B. and Gardner, F. F.: 1979, *Monthly Notices Roy. Astron. Soc.* **188**, 445.
Wollmann, E. R.: 1978, in W. B. Burton (ed.), 'The Large-Scale Characteristics of the Galaxy', *IAU Symp.* **84**, 367.
Yusef-Zadeh, F., Morris, M., and Chance, D.: 1984, *Nature* **310**, 557.
Yusef-Zadeh, F., Morris, M., and Chance, D.: 1985, (in press).

EXPECTED AND OBSERVED LATE GIANT STAR COUNTS IN THE MILKY WAY BULGE*

V. M. BLANCO and B. M. BLANCO

National Optical Astronomy Observatories, Cerro Tololo Inter-American Observatory[†], *La Serena, Chile*

(Received 2 July, 1985)

Abstract. In this paper, late M giants star counts in the Milky Way central bulge are compared with those expected from the Hubble and de Vaucouleurs surface brightness distributions.

The integrated light of central galaxy spheroids shows late giants to be a common component (Morgan and Mayall, 1957; Morgan and Osterbrock, 1969). Among others, Whitford (1978) and de Vaucouleurs and Pence (1978), have shown that the Milky Way central bulge is similar to that of other galaxies. For the Milky Way the relevant late giants have been identified by Nassau and Blanco (1958), and Blanco *et al.* (1984) have shown that these stars can be identified in the clearer areas of the central bulge to distances well beyond that the galactic center. Star count estimates for such area can be made, as shown by Bahcall and Soneira (1980), if one assumes within the galactic bulge stellar space densities based on Hubble's (1930) or de Vaucouleurs' (1953) laws for the distribution of surface brightness in galaxy spheroids. It is, therefore, interesting to compare such estimates with actually observed star counts.

Here, giants of type M5 or later were identified and counted in thirteen relatively clear areas near $l = 0°$ and at $b = -2°$ to $-12.6°$. The CTIO 4 m telescope was used for the observations, and the observations were as described by Blanco *et al.* (1984). In Figure 1 are plotted (solid dots with circles) the observed numbers of late giants after correction for expected foreground objects estimated from galactic disc models by Mikami and Ishida (1981) and Hidayat and Blanco (1968).

Theoretical star count estimates must use effective radii and space density distributions corresponding to the Hubble and de Vaucouleurs laws. For distributions with spherical symmetry and constant mass-luminosity ratios, space densities have been published by Hubble (1930) and Young (1976). Here, two values of the effective radius, $a = 0.1$ and 1.0 kpc, were used with Hubble's distribution and two values of $\alpha(= R_0/r_e) = 3$ and 10, r_e being the effective radius, with the de Vaucouleurs' law. These values are close to those found by de Vaucouleurs and Pence (1978). For R_0 we used 8 kpc. This value, and also those of a and α, within reasonable limits, did not alter appreciably the estimated star counts. These are presented by smooth curves in Figure 1

* Paper presented at the IAU Third Asian-Pacific Regional Meeting, held in Kyoto, Japan, between 30 September–6 October, 1984.

[†] Operated by the Association of Universities for Research in Astronomy, Inc., under contract with the U.S. National Science Foundation.

Astrophysics and Space Science **118** (1986) 365–366.

with the a and α values for Hubble's (H) and de Vaucouleurs' (dV) laws. All our integrations of space densities were up to 15 kpc.

Clearly the calculated star counts fail to reproduce even approximately the observed ones. This suggests, for the late Milky Way bulge giants, a space density distribution which is very different (possibly highly flattened) from the symmetrical one assumed in the models.

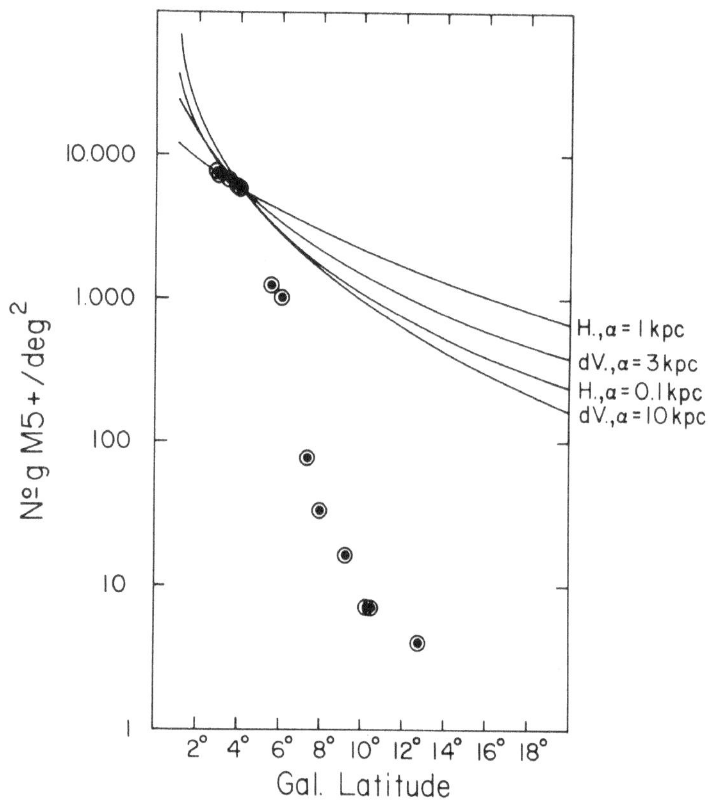

Fig. 1. Observed (solid dots with circles) and estimate (smooth curves) total star counts for Milky Way bulge late M giants. See text for other symbols.

References

Bahcall, J. and Soneira, R.: 1980, *Astrophys. J. Suppl.* **44**, 73.
Blanco, V. M., McCarthy, M. F., and Blanco, B. M.: 1984, *Astron. J.* **89**, 636.
de Vaucouleurs, G.: 1953, *Monthly Notices Roy. Astron. Soc.* **113**, 134.
de Vaucouleurs, G. and Pence, W. D.: 1978, *Astron. J.* **83**, 1163.
Hidayat, B. and Blanco, V. M.: 1968, *Astron. J.* **73**, 712.
Hubble, E.: 1930, *Astrophys. J.* **71**, 231.
Mikami, T. and Ishida, K.: 1981, *Publ. Astron. Soc. Japan* **33**, 135.
Morgan, W. W. and Mayall, N. U.: 1957, *Publ. Astron. Soc. Pacific* **69**, 291.
Morgan, W. W. and Osterbrock, D. E.: 1969, *Astron. J.* **74**, 15.
Nassau, J. J. and Blanco, V. M.: 1958, *Astrophys. J.* **128**, 46.
Whitford, A. E.: 1978, *Astrophys. J.* **226**, 777.
Young, S.: 1976, *Astron. J.* **81**, 807.

POST-COLLAPSE EVOLUTION OF SMALL-N CLUSTERS*

SHOGO INAGAKI

Department of Astronomy, University of Kyoto, Japan

(Received 2 July, 1985)

Abstract. N-body simulations were made to study post-collapse evolution of small-N clusters. The qualitative behaviour of the dynamical evolution is the same as in the Fokker–Planck approximation. The flatter is the mass spectrum, the more violent is the activity of binaries and so are the expansion and contraction of the cluster core.

Theories based on the Fokker–Planck equation predict that the central densities of globular clusters become infinite in finite time. According to the prediction, the central density becomes infinite at about 330 times the initial relaxation time (Cohn, 1980). Several clusters have central relaxation times of less than ten million years. This means that those clusters will collapse completely in about one thousand million years. Since one thousand million years is much shorter than the age of the Universe, some clusters may have collapsed completely. This fact has stimulated the researches investigating the evolution of globular clusters after complete core collapse.

In real globular clusters, the Fokker–Planck approximation will break down before the central density of the cluster becomes infinite. This is because the number of stars contained in the cluster core becomes too small. In such a situation, hard binaries are formed by three-body encounters and 'heat' the core (Heggie, 1975). Due to this energy input from the binaries, the core turns into expansion. The expansion phase is called the post-collapse phase and was recently reviewed by Heggie (1985) in detail.

In a realistic cluster, the number of stars contained in the core becomes less than one hundred when the central density becomes very high. This number is too small to apply the Fokker–Planck approximation. So N-body techniques must be applied to investigate the behaviour of the core.

In this paper direct N-body integrations are adopted, using as large a value of N as possible. The numerical code used here is NBODY5 developed by Aarseth (1984). This code can regularize violent three-body encounters, as well as close two-body encounters. Thus the code is ideal to treat the post-collapse evolution. The purpose of this paper is to examine the effects of a spectrum of masses. The first problem is to see whether equipartition can be achieved or not. The second problem is to see how the evolution is accelerated. In N-body calculations, detailed information about binaries is also available.

Figure 1 shows the change of the radius containing a fixed fraction of the total mass for a single component cluster. The initial condition is the Plummer model. The core

* Paper presented at the IAU Third Asian-Pacific Regional Meeting, held in Kyoto, Japan, between 30 September–6 October, 1984.

Astrophysics and Space Science **118** (1986) 367–370.

turns into expansion at about 21 t_{rh}. Unfortunately, a numerical instability occured at the end of the simulation and I could not follow the evolution in the post-collapse phase for a long time. The density profile at the most collapsed phase is proportional to $r^{-2.2}$.

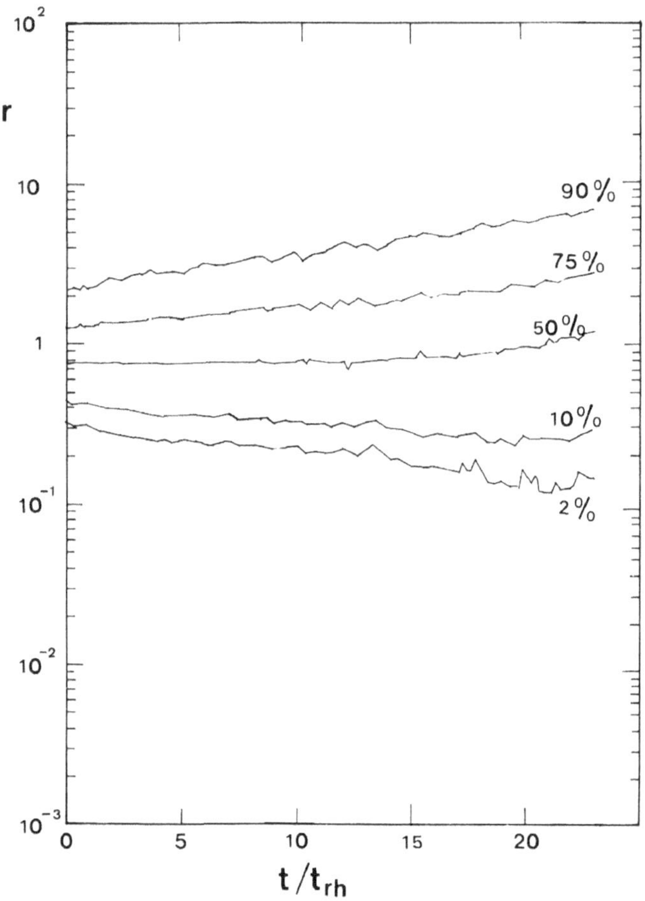

Fig. 1. The time evolution of radii containing 2, 10, 50, 75, 90% of the total mass of the cluster. All stars have the same mass. The initial state is Plummer's model with an isotropic velocity distribution.

We see that the outer region is expanded if we compare it with the initial model. At the end of the simulation the central density has slightly decreased.

Figure 2 shows the change of the radius containing a fixed fraction of the total mass for the cluster with the parameter α set to be 2.5. Here the mass spectrum is taken to be of the form, $dM \propto m^{-\alpha}\, dm$, where dM is the mass of stars in the mass range $(m, m + dm)$. The radius containing the central 2% of the mass fluctuates rather conspicuously. So the radius containing 10% of the mass will be a better measure to see the behaviour of the core. It is seen that the radius containing 10% of the mass performs contractions and expansions several times. This change of the radius is more marked than in the single-component cluster. We also see that the epoch to turn to

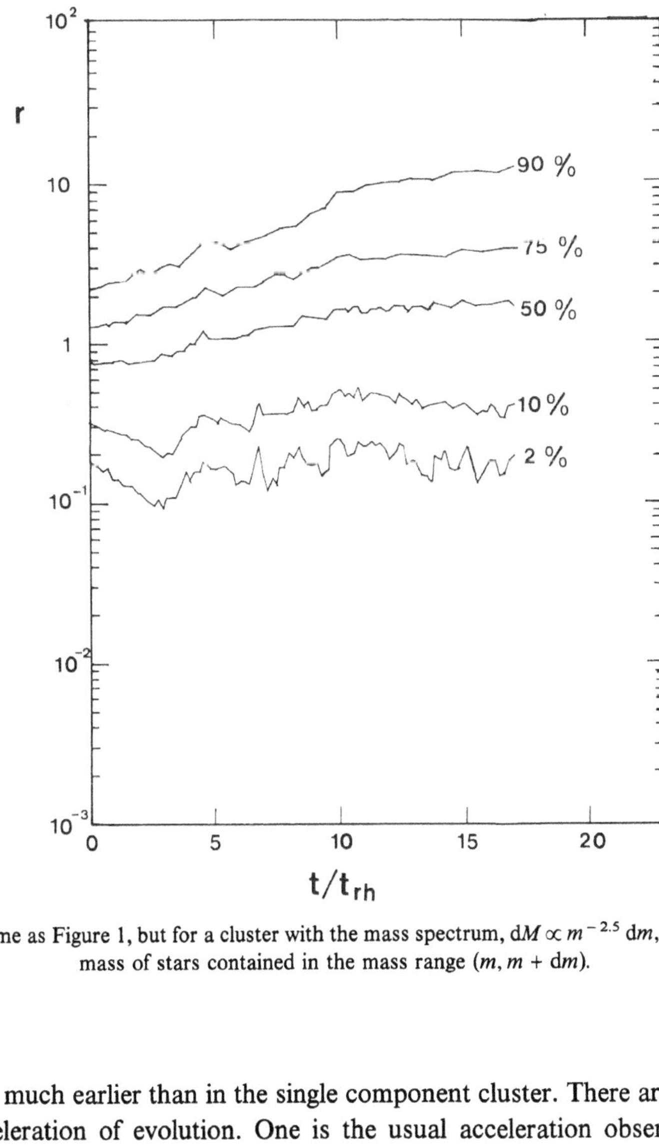

Fig. 2. The same as Figure 1, but for a cluster with the mass spectrum, $dM \propto m^{-2.5}\,dm$, where dM is the mass of stars contained in the mass range $(m, m + dm)$.

expansion is much earlier than in the single component cluster. There are two reasons for this acceleration of evolution. One is the usual acceleration observed in multi-component clusters within the Fokker–Planck approximation (Inagaki and Sastaw, 1985). The other is a more efficient binary formation mechanism which can operate in multi-mass clusters (Heggie, 1975). If we compare similar figures with different α, we see that the oscillation of the radius is more violent as α becomes smaller.

All the expansions are related with hardening of the binaries. Therefore, we can conclude that the expansions are caused by an energy input from binaries. Unfortunately, the fluctuations are too large to see whether equipartition is achieved.

Acknowledgements

I am grateful to Dr S. J. Aarseth for providing his N-body programme and for helpful advice. Numerical computations were carried out on FACOM M382 at the Data Processing Center of Kyoto University.

References

Aarseth, S. J.: 1984, in J. U. Brackbill and B. I. Cohen (eds.), *Multiple Time-Scales*, Academic Press, New York.
Cohn, H.: 1980, *Astrophys. J.* **242**, 765.
Heggie, D. C.: 1975, *Monthly Notices Roy. Astron. Soc.* **173**, 729.
Heggie, D. C.: 1985, in P. Hut and J. Goodman (eds.), 'Dynamics of Star Clusters', *IAU Symp.* **113**.
Inagaki, S. and Saslaw, W. C.: 1985, *Astrophys. J.* **292**, 339.

THE GALACTIC CENTER RADIO LOBE – A COSMIC JET IN OUR GALAXY?*

Y. SOFUE, T. HANDA

Nobeyama Radio Observatory, Minamimaki, Nagano, Japan

and

I. SUWA and Y. FUKUI

Dept. Physics, Nagoya University, Nagoya, Japan

(Received 2 July, 1985)

Abstract. The galactic center radio lobe found in the radio continuum is interpreted as due to a channeled exhaust of material perpendicular to the galactic plane driven by an activity in the nuclear region. A common characteristic of the lobe to extragalactic more energetic jets is pointed out.

A radio continuum survey of the galaxy center region of $4° \times 4°$ at 10 GHz has revealed an off-plane, Ω-shaped lobe of diameter 200 pc (Sofue and Handa, 1984). The lobe emerges from the nuclear disk perpendicularly (Figure 1). A cross section of the lobe indicates that it is a result of an explosive or energetic outflow phenomenon (Figure 2). As seen from Figure 2 the radio spectrum of the lobe component is flat, which suggests a thermal gas origin with a total mass of ionized hydrogen of $4 \times 10^5\, M_\odot$. A 2.6 mm CO line survey has shown that high-velocity CO gas with $V_{\rm LSR} = -150$ to 150 km s^{-1} is associated with the lobe ridges and the total mass of molecular hydrogen is about several times $10^5\, M_\odot$ (Figure 3).

We propose the following possible formation mechanisms:

(a) A shock wave produced by an explosion at the nucleus of the Galaxy: a model computation shows an explosion of energy 10^{54} ergs some 10^6 yr ago produces an Ω-shaped shock front with an expansion velocity of about 100 km s^{-1}, at present, which resembles the observed lobe structure (Sofue, 1984).

(b) A channelled exhaust of gas produced by a twisted magnetic field due to an accretion of gas in the nuclear disk (Uchida and Shibata, 1984): the accreting gas twists the field lines and is accelerated along a cylinder perpendicular to the disk. A model calculation by Uchida and Shibata (1984, in this issue) reproduces well the observed feature of the radio lobe.

(c) An inflation of a giant loop of magnetic tube filled with ionized gas, like as a solar prominence: the inflation may be triggered by a Parker-type instability in a magnetized nuclear disk with cosmic rays.

* Paper presented at the IAU Third Asian-Pacific Regional Meeting, held in Kyoto, Japan, between 30 September–6 October, 1984.

Astrophysics and Space Science **118** (1986) 371–374.

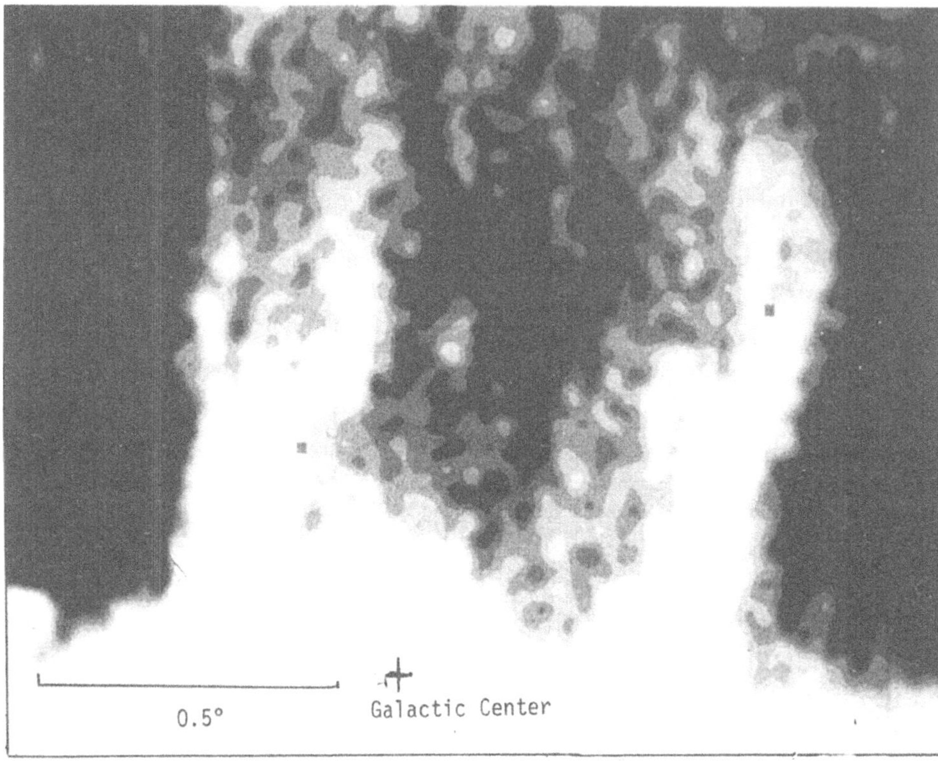

Fig. 1. The galactic center radio lobe at 10 GHz as reproduced from Sofue and Handa (1984).

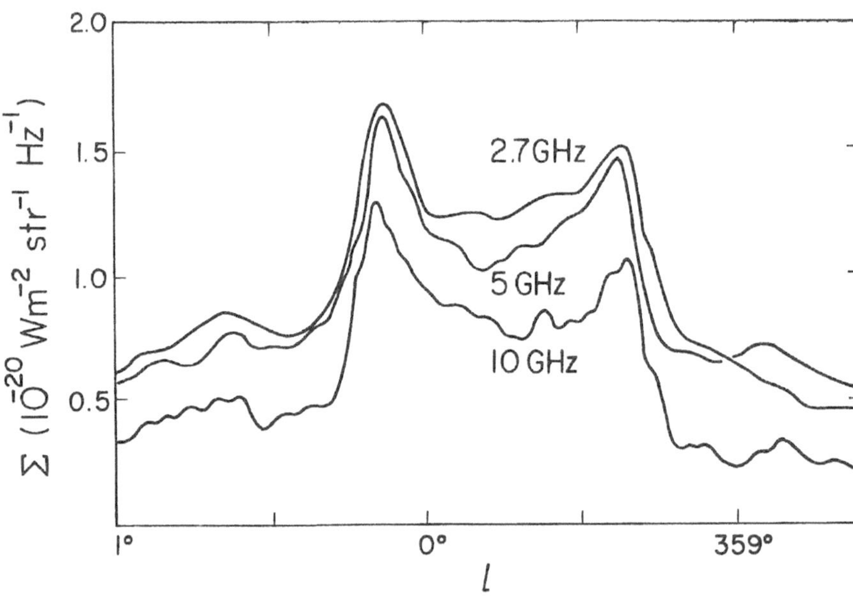

Fig. 2. Cross-sections of the radio lobe at 2.7, 5, and 10 GHz. The absolute zero levels are not correctly calibrated, due to a large-scale galactic background radiation. The data are from the Bonn 2.7 and 5 GHz, and Nobeyama 10 GHz surveys.

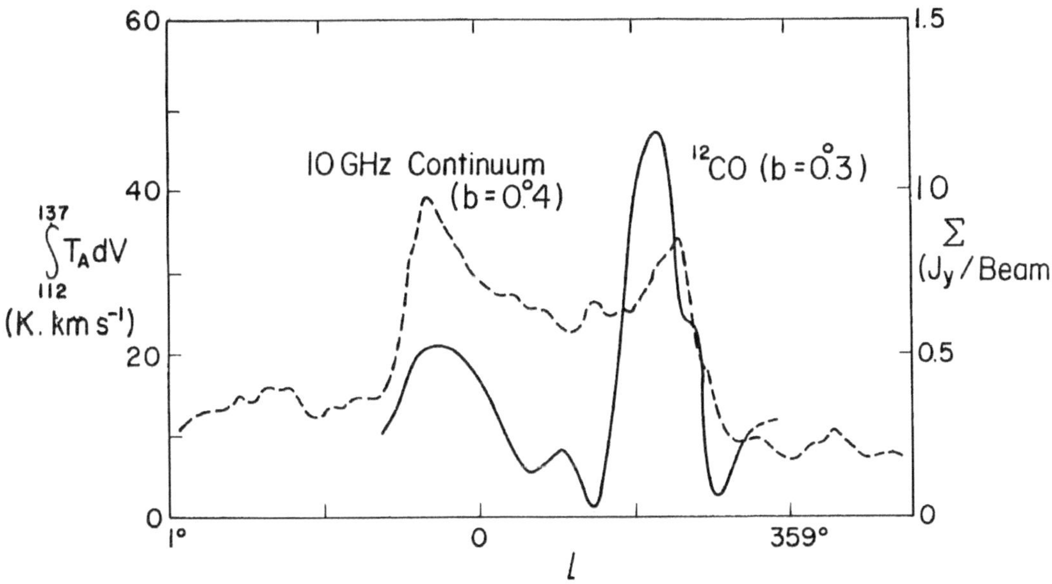

Fig. 3. Distribution of the CO line emission at a velocity range of from 112 to 137 km s^{-1}. The radio lobe ridges are associated with high-velocity CO gas.

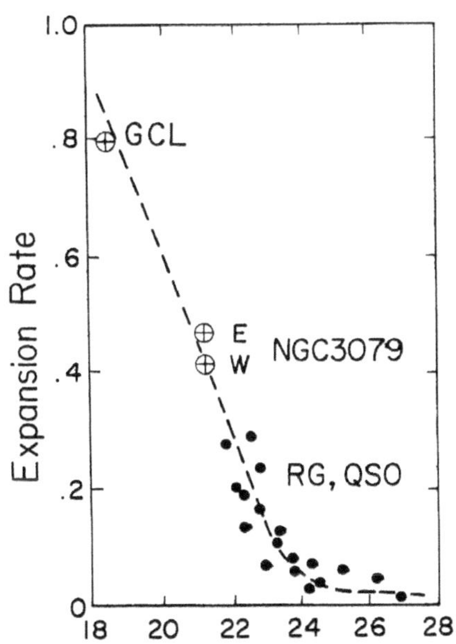

Fig. 4. A plot of lateral expansion rate, as defined by a ratio of the width to length of a jet, versus radio power at 5 GHz of the central radio source. Data for NGC 3079 were taken from Duric *et al.* (1983) and those for radio galaxies and QSOs were taken from Bridle (1984).

According to model B, which seems the most promising among the three, the radio lobe in our Galaxy may be a proto-type or mini-type of the cosmic jet phenomenon, which is frequently observed associated with active nuclei of radio galaxies and QSOs. In Figure 4 we plot the lateral expansion rate of radio jets in QSOs and radio galaxies (Bridle, 1984) versus the radio power of their central sources. In the figure we plot the same quantities determined for the galactic centre radio lobe and for radio lobes in the spiral galaxy NGC 3079. The same relationship obtained for radio jets at higher powers seems to apply for the radio lobes found in the spiral galaxy and in our own Galaxy center. This fact indicates that the radio lobe phenomenon in some spiral galaxies, including our own Galaxy, has common characteristics to the more powerful cosmic jet phenomenon observed in the active nuclei of radio galaxies and QSOs.

A full-description of this contribution will be given in a separate paper (Sofue, 1985).

References

Bridle, A. J.: 1984, *Astron. J.* **89**, 979.

Duric, N., Seaquist, G. G., Crane, P. C., Bignell, R. C., and Davis, L. E.: 1983, *Astrophys. J.* **273**, L11.

Sofue, Y.: 1984, *Publ. Astron. Soc. Japan* **36**, 539.

Sofue, Y.: 1984, *Publ. Astron. Soc. Japan* (submitted).

Sofue, Y. and Handa, T.: 1984, *Nature* **310**, 568.

Uchida, Y. and Shibata, T.: 1984, in M. Kundu (ed.), 'Unstable Current Systems and Plasma Instabilities in Astrophysics', *IAU Symp.* **107**.

TENMA OBSERVATION OF THE X-RAY PULSAR 4U1626–67*

T. KII, S. HAYAKAWA, and F. NAGASE

Department of Astrophysics, Faculty of Science, Nagoya University, Nagoya, Japan

(Received 2 July, 1985)

Abstract. Energy-dependent pulse profiles from the X-ray pulsar 4U1626–67, observed with Tenma in May 1983, were studied using an anisotropic radiation transfer model in an accretion column on a strongly magnetized neutron star. The result indicates that the strength of the magnetic field is about 8×10^{12} G at the surface of the neutron star.

1. Introduction

The X-ray pulsation of ~7.7 s of 4U1626–67 was discovered by SAS-3 (Rappaport *et al.*, 1977). However, the pulse period change due to the orbital motion of the binary system was not detected (Joss *et al.*, 1978). The optical counterpart was identified with a faint blue star ($m_v \sim 18$) with ~7.7. s optical pulsation (McClintock *et al.*, 1977; Ilovaisky *et al.*, 1978; Peterson *et al.*, 1980). From the optical timing observation Middleditch *et al.* (1981) confirmed that the system 4U1626–67 is a low-mass binary, made up of a neutron star and a dwarf star.

The pulse profile and the energy spectrum of 4U1626–67 was observed with HEAO-1 (Pravdo *et al.*, 1979). They measured the energy dependent nature of the pulse profile, and a change in the energy spectrum with pulse phase, and found that a continuum or broad emission feature around 19 keV appears at a particular pulse phase.

2. Observation and Result

The X-ray pulsar 4U1626–67 was observed with the Japanese X-ray astronomy satellite TENMA on 3 through 5 May, 1983. The gas scintillation proportional counters (GSPC's) on board TENMA have been used for the present observation. The total effective area of this system is 640 cm^2. The total exposure time was about 2×10^4 s.

During this observation a pulse period of $7.671\,330 \pm 0.000\,006$ s was obtained from pulse arrival time analysis, and the average X-ray luminosity was $(1.2 \pm 0.1) \times 10^{37}$ $(D/10\ \text{kpc})^2$ erg s^{-1} in the energy range 2 and 20 keV. In Figure 1(a) we show the light curves superposed modulo the pulse period in eight energy bands, after subtracting the off-source background. The pulse profile significantly changes with energy. The phase of the pulse profile reverses by ~180° around 15 keV.

* Paper presented at the IAU Third Asian-Pacific Regional Meeting, held in Kyoto, Japan, between 30 September–6 October, 1984.

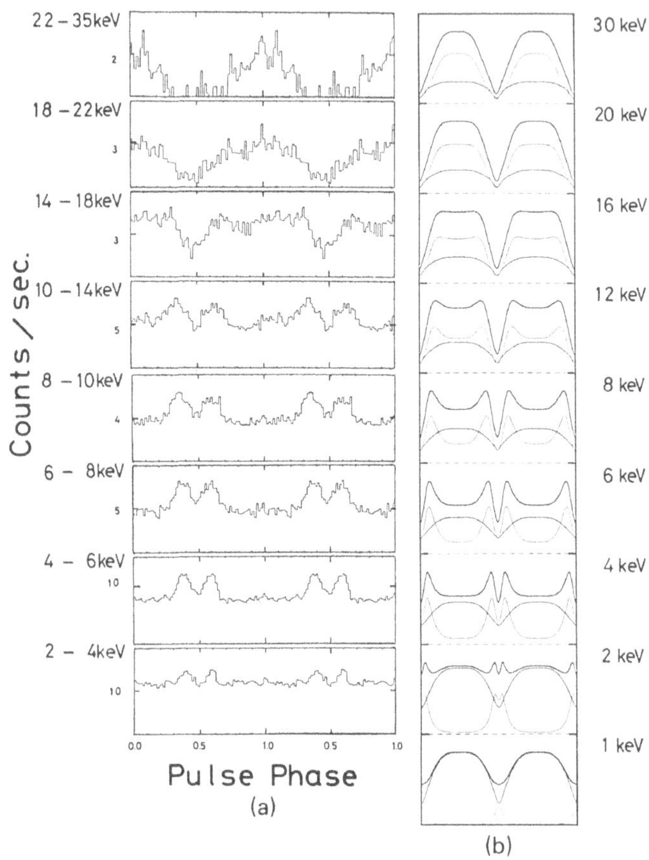

Fig. 1. (a) The pulse profiles observed by TENMA. (b) The pulse profiles calculated for cylindrical column with the plasma parameters; the strength of magnetic field of 8×10^{12} G, the temperature of 10 keV, the electron number density of 6.0×10^{22} cm^{-3}, and the column radius of 10^3 cm. The angles between rotation axis and magnetic pole and between rotation axis and view of sight are 40° and 50°, respectively. The boldest line is the total flux and others are ordinary and extraordinary polarized fluxes.

3. Discussion

This X-ray pulsar is characterized by a drastic change of pulse profile with energy. This effect cannot be interpreted by a simple geometrical model of a fan or a pencil beam. Radiative transfer in a strongly magnetized plasma should be taken into account to properly understand this feature of the pulse profile.

The method utilized in the present calculation is fundamentally similar to the Eddington approximation, and the 'two-stream' method (Nagel, 1981; Kaminker *et al.*, 1982). In an accretion column on a neutron star we must consider the influence of a strong magnetic field on the cross section. We assume the radiative region in the accretion column to be a cylinder parallel to the magnetic field. In this region the temperature and density are assumed to be uniform.

The drastic change of pulse profile is considered to occur in the radiative region below the cyclotron energy level. From the observed pulse profiles, the cyclotron energy is expected to be higher than 20 keV for 4U1626–67. We calculated the pulse profiles for varying strengths of magnetic field from 2 to 12 $\times 10^{12}$ G using the appropriate geometry of a cylinder. We are able to reproduce a change of pulse profile similar to the observed one. The critical energy, at which the pulse profile drastically changes, is found to be roughly proportional to the strength of magnetic field. And comparing the calculated profile with the observed one, we can estimate the field strength on the neutron star. The calculated profile with magnetic field of 8×10^{12} G is most similar to the observed one, as shown in Figure 1(b). HEAO-1 observations (Pravdo *et al.*, 1979) suggested that the pulse profile below 2 keV had the same sinusoidal pattern and phase as that above 15 keV. This profile is reproduced from the calculation for $E = 1$ keV in Figure 1(b). The present calculation indicates that the strength of the magnetic field of 4U1626–67 is $\sim 8 \times 10^{12}$ G.

References

Ilovaisky, S. A., Motch, C., and Chevalier, C.: 1978, *Astron. Astrophys.* **70**, L19.

Joss, P. C., Avni, Y., and Rappaport, S.: 1978, *Astrophys. J.* **221**, 645.

Kaminker, A. D., Pavlov, G. G., and Shibanov, Y. A.: 1982, *Astrophys. Space Sci.* **86**, 249.

McClintock, J. E., Canizares, C. R., Bradt, H. V., Doxsey, R. E., Jernigan, J. G., and Hiltner, W. A.: 1977, *Nature* **270**, 24.

Middleditch, J., Mason, K. O., Nelson, J. E., and White, N. E.: 1981, *Astrophys. J.* **244**, 1001.

Nagel, W.: 1981, *Astrophys. J.* **251**, 278.

Peterson, B. A., Wallace, P., Elliott, K. H., Hill, P. W., and Manchester, R. N.: 1980, *Monthly Notices Roy. Astron. Soc.* **190**, 33 p.

Pravdo, S. H., White, N. E., Boldt, E. A., Holt, S. S., Serlemitsos, P. J., Swank, J. H., Szymkowiak, A. E., Tuohy, I., and Garmire, G.: 1979, *Astrophys. J.* **231**, 912.

Rappaport, S., Markert, T., Li, F. K., Clark, G. W., Jernigan, J. G., and McClintock, J. E.: 1977, *Astrophys. J.* **217**, L29.

A MODEL OF THE GALACTIC CENTRE WITH MAGNETIC MONOPOLES*

WANG DE-YU

Purple Mountain Observatory, Nanjing, China

and

PENG QIU-HE and CHEN TING-YANG

Department of Astronomy, Nanjing University, Nanjing, China

(Received 19 July, 1985)

Abstract. A model of the galactic centre with magnetic monopole has been presented here. The positron can be produced continuously through magnetic monopoles to induce baryon decay (Rubakov catalytic reaction) and a lot of energy can be released as well. The calculation results show that even if the galactic centre contains only a few magnetic monopoles ($\zeta = N_M/N_B \approx 10^{-24}$), this massive object can not collapse into a black hole. This model can explain the observed intensities of the annihilation line and higher energy photons of $E_\gamma > 511$ keV from the galactic centre.

Recently, a surprisingly intense positron-electron annihilation line at the galactic centre has been detected by balloons and satellites (Leventhal *et al.*, 1978, 1980, 1982). In general, the model of a massive black hole can explain the production rate of positrons, however, it is rather difficult to account for the intensity of an $E_\gamma > 511$ keV gamma photon continuum spectrum (Matteson, 1982) and the luminosity of the galactic centre (Townes *et al.*, 1983). From another point of view, physicists have been interested widely in the problem of magnetic monopoles M should induce proton decay (catalytic reaction; cf. Rubakov, 1982)

$$\begin{aligned} P + M &\to M + e^+ + \pi^0 + \text{debris} \\ &\to M + e^+ + \mu^+ + \mu^- + \text{debris}, \end{aligned} \tag{1}$$

and this catalytic reaction should be a strong interaction $\sigma \sim 10^{-25}$–10^{-27} cm^2. Thus, positron can be produced continuously though this catalytic reaction of magnetic monopoles, and a lot of energy can be released as well. In this article, we assume that magnetic monopoles exist in the Universe, however, the number of magnetic monopoles and the cross-section of the catalytic reaction have not been determined in physics, we only choose $\zeta\langle\sigma v\rangle$ as a parameter to establish galactic centre model with magnetic monopoles (ζ is the ratio of magnetic monopole number and baryon number).

In another paper (Peng *et al.*, 1985) we have discussed the possibility that if magnetic monopoles exist in early cosmological evolution and their content is larger than the

* Paper presented at the IAU Third Asian-Pacific Regional Meeting, held in Kyoto, Japan, between 30 September–6 October, 1984.

Astrophysics and Space Science **118** (1986) 379–382.

Newtonian saturation value $\zeta_n = GM_B m_M/g_m^2 \simeq 1.0 \times 10^{-24}$, where G is gravitational constant; M_B, the mass of baryon; m_M, the of mass of magnetic monopole; g_m, the magnetic charge of monopole; so that primordial nebulae in the stage of fast contraction, forming active galactic nuclei (including galactic centre), may contain a similar proportion of magnetic monopoles within their nuclei. In this case, the radial balance equation of the galactic centre becomes

$$-\nabla P_r - \nabla P_g = \frac{GM}{R^2}\,(R)\,, \tag{2}$$

where $Pg = nkT$ and P_r represent gas and radiation pressure, respectively, these are produced by the release of the energy from the magnetic monopole catalytic reaction. M is the mass of galactic centre. When the galactic centre is in equilibrium,

$$-\nabla P_r(R) = \frac{Kn_B(R)L_m(R)}{4\pi RC}\;; \tag{3}$$

where K is an absorption coefficient; $n_B(R)$, the number density of baryon; and $L_m(R)$ is the luminosity of galactic centre at R. $L_m(R)$ is given by

$$L_m(R) \simeq 4\pi m_B C^2 \int_0^R n_B(r) n_m(r) \langle \sigma v \rangle\, r^2\, \mathrm{d}r\,, \tag{4}$$

where we assume ζ is a constant within the galactic centre, and $\langle \sigma v \rangle$ is also constant in the released energy reaction. The density distribution of baryons in the galactic centre satisfies the equation

$$\rho(r) = \frac{\rho_C}{(1 + r^2/a^2)^{1/2(2n+1)}}\,. \tag{5}$$

We can now substitute (3), (4), and (5) into (2). For simplicity, the equation of balance can be written in a dimensionless form as

$$\begin{aligned} &B\,\frac{(2n+1)}{\mu_s}\,(\xi\eta_n)^{1/4} M_8^{-1/4} x^{1/4} y^{1/2} + [D\mu_s^{-1}(\xi\eta_n)^{-3/4}\,\times \\ &\times \beta_n^{(2n+1)} M_8^{-(4n/3-1/12)}\, x^{-(2n/3+1/12)}\, y^{-(2n-1/2)} + 1]F^{-1}\,\times \\ &\times f_s(\xi\eta_n)^{1/8}\, \beta_n^{(2n+1)} M_8^{-(4n/3-5/24)}\, x^{-(2n/3-19/24)}\, y^{-(2n-3/4)} = 1\,, \end{aligned} \tag{6}$$

where B, F, and D are constants; η_n and β_n are constants related to the density index n; while f_s and μ_s are constants related to atomic weight; $y = R/R_g$, $\xi = (\zeta \langle \sigma v \rangle)/(\zeta_n \times 10^{-27})$, $M_8 = M/10^8\,M_\odot$, $x = P_c/1.84$, and $R_g\ (= 2MG/C^2)$ is the Schwarzschild black-hole radius. Therefore, if the mass and luminosity of the massive object is known, the balance radius R could be obtained in the case of different density index n and magnetic monopole content parameter ξ. For our own galactic centre, the

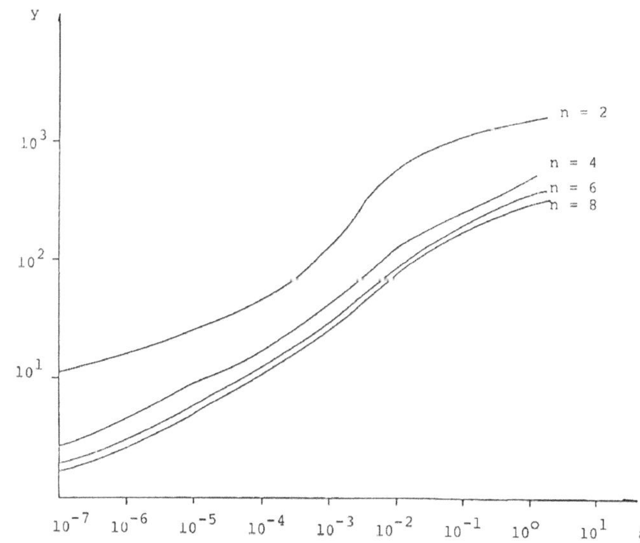

Fig. 1. The galactic center radius ($R = yR_g$) versus the parameter of magnetic monopole ξ, n is the density index.

mass is $M \simeq (3\text{–}10) \times 10^6\, M_\odot$ and $L_m \simeq (1\text{–}3) \times 10^7\, L_\odot$. The results calculated for this galactic centre are shown in Figure 1. It can be seen from this figure that the massive objects can not collapse into black holes even in the case of $\xi = 10^{-7}$. It corresponds to one of the two possibilities: (a) The content of magnetic monopoles ζ is at the Newtonian saturation value, whereas the catalytic reaction rate $\langle \sigma\beta \rangle \simeq 10^{-34}$, being a weak interaction. (b) The catalytic reaction is a strong interaction whereas $\zeta = 10^{-31}$, i.e., meaning that it contains only a few magnetic monopoles, in the galactic centre.

The positrons produced from the catalytic reaction (1) and high-energy photons decayed from π^0 and other unstable particles should undergo cascade processes for multiplication and slow-down within the galactic centre. The longest range $(R - R_a)$ of positrons within the galactic centre are given by the equation

$$\int_{R_a}^{R} n_B(R')\, \mathrm{d}R' = \int_{mC^2}^{E} \frac{\mathrm{d}E'}{|L(E')|}\,, \tag{7}$$

where the energy loss rate of positrons (mainly through bremsstrahlung) is determined by the formula

$$-\left(\frac{\mathrm{d}E}{\mathrm{d}R'}\right)_B = n_B(R')L(E) = n_B(R') \int_0^{E - mC^2} E_\gamma \sigma_\beta(E, E_\gamma)\, \mathrm{d}E_\gamma\,. \tag{8}$$

According to simple cascade shower theory, and considering positrons and photons through different thicknesses inside the galactic centre, the positron average multiplication number going outside the galactic centre surface is $(2\text{–}3) \times 10^2$.

If we take $\xi = 1$, $n = 3$, $M = 10^7 M_\odot$, and $L_m(R) = 3 \times 10^7 L_\odot$, then $R = 625 R_g \simeq 1.8 \times 10^{15}$ cm from (6), and the rate at which positrons leave the galactic centre is $\sim 10^{43} e^+ s^{-1}$. This agrees with the observed intensity of annihilation line.

The positrons will be slowed down further after they leave the galactic centre. The results calculated by Bussard *et al.* (1979) show that the Sgr A west warm clouds around the galactic centre, where temperature should be less than 5×10^4 K and the ionization of hydrogen larger than 10%, in order to conform with the observed spectrum line width. These results are available also for the model suggested in this paper. The higher energy photons of $E_\gamma \gtrsim 511$ keV also detected from balloons and satellites observation. The integral of the intensity from $E_\gamma = 511$ keV to 10 MeV is $(8\text{–}20) \times 10^{37}$ erg s^{-1} (Matteson, 1982), corresponding to about $\frac{1}{5}$ of the photon number of the annihilation line. According to the model suggested in this paper, there are always a fraction of photons with $E_\gamma > 511$ keV going out of the galactic centre surface after their slow down and multiplication within the galactic centre. Therefore, the higher energy photons of $E_\gamma > 511$ keV which have been detected are undoubtedly a strong support to the present model.

References

Bussard, R. W., Ramaty, R., and Drachmann, R. J.: 1979, *Astrophys. J.* **228**, 928.
Leventhal, M., MacCallum, C. J., and Stang, P. D.: 1978, *Astrophys. J.* **225**, L11.
Leventhal, M., MacCallum, C. J., and Stang, P. D.: 1980, *Astrophys. J.* **240**, 338.
Leventhal, M., MacCallum, C. J., and Stang, P. D.: 1982, *Astrophys. J.* **260**, L1.
Matteson, J. L.: 1982, in G. R. Reigler and R. D. Blandford (eds.), 'The Galactic Center', *AIP Conf. Proc.* **83**, 109.
Peng Qiuhe, Li Zongyun, and Wand Deyu: 1985, *Scientia Sinica* **A28**, 970.
Rubakov, V. A.: 1982, *Nucl. Phys.* **B203**, 311.
Townes, G. H., Lacy, J. H., Geballe, T. R., and Hollenbach, D. J.: 1983, *Nature* **301**, 661.

THE INJECTION OF ENERGY INTO THE INTERSTELLAR MEDIUM BY STARS (INVITED REVIEW)*

CHRISTOPHER F. McKEE
Departments of Physics and of Astronomy, University of California, Berkeley, CA, U.S.A.

(Received 2 July, 1985)

Abstract. Stars inject energy into the interstellar medium (ISM) by radiation, stellar winds, and supernova explosions. This energy injection causes the ISM to be inhomogeneous, which in turn alters the manner in which the energy is transferred through the ISM. A significant fraction of the energy is injected by massive stars, which form H II regions in the ISM. The structure and evolution of H II regions in a cloudy medium differs significantly from that in a homogeneous one. The strong stellar winds produced by massive stars form bubbles in the ISM, and the structure of these bubbles is often dominated by the structure of the H II region in which they are embedded. Finally, when the star explodes as a supernova, the evolution and appearance of the resulting remnant is determined by the structure of the bubble and H II region formed by the star during its lifetime.

1. Energy Injection and Inhomogeneity

The energy of the ISM in a spiral galaxy is derived from density waves on large scales and from stars on smaller scales. Stellar energy injection occurs by radiation and a wind during the star's life (both of which are particularly important for massive stars), and by a supernova explosion at its death. The effect of the radiation depends on the photon energy: FUV radiation between about 6 eV and the ionization limit of hydrogen at 13.6 eV ejects photoelectrons from dust grains, which may well be the dominant heating mechanism for neutral atomic gas (de Jong, 1980); EUV radiation, above 13.6 eV, ionizes the ambient gas and produces H II regions. A stellar wind can generate a large volume of hot gas around the star, as can a supernova explosion. To get an idea of the relative magnitudes of these energies, consider a B0 star, with an initial mass of about 20 $M_\odot$: over its lifetime it produces about $10^{52.1}$ erg of FUV radiation, $10^{51.7}$ erg of EUV radiation, $10^{49.6}$ erg of wind energy, and 10^{51} erg in the final supernova explosion. Ultimately, the energy injected into the ISM is reradiated into intergalactic space, so that the energization of the ISM can be viewed as an energy flow which originates in stars in several forms, moves outward through the ISM, and finally leaves the galaxy as radiation.

The ISM is observed to be highly inhomogeneous, with much of the mass concentrated in clouds occupying a small fraction of the volume. This inhomogeneity is a direct result of stellar energy injection into a thermally unstable medium. In the original two-phase model of the ISM, Field *et al.* (1969) showed that the coexistence of cold ($T \sim 10^2$ K) H I clouds in pressure equilibrium with warm ($T \sim 10^4$ K) intercloud

* Paper presented at the IAU Third Asian-Pacific Regional Meeting, held in Kyoto, Japan, between 30 September–6 October, 1984.

medium was possible, because the ISM is thermally unstable over a range of intermediate temperatures. When the gas is in thermal balance, the heating rate per particle Γ equals the cooling rate per particle $n\Lambda(T)$. Consider the example of photoelectric heating due to FUV absorption by dust grains so that Γ falls off as r^{-2}. For a collisionally ionized gas, the cooling function $\Lambda(T)$ rises sharply above 10^4 K because of Lα cooling. Hence, when the gas is close to the star, Γ is large, the temperature is of order 10^4 K, and the gas is homogeneous. At greater distances, Γ is lower and stable thermal balance is possible at either $T \sim 10^4$ K or at $T \sim 10^2$ K; there is an unstable equilibrium at intermediate temperatures. The gas then breaks up into cold clouds embedded in a warm intercloud medium. In principle, at yet greater distances from the star, Γ is so small that thermal balance is possible only at $T \sim 10^2$ K and the gas is again homogeneous, but, in practice, the density of stellar energy sources is too great for this to occur. Thus, as energy flows away from stars, the ambient medium necessarily goes through an unstable regime and becomes inhomogeneous.

It is now believed that there is a third phase of the ISM: hot ($T \sim 10^6$ K) gas in old supernova remnants (SNRs), which fill a significant fraction of the volume of the ISM (Cox and Smith, 1974; McKee and Ostriker, 1977). In an SNR the gas is heated impulsively by a shock and cools freely thereafter. The cooling is thermally unstable at all temperatures above about 10^5 K, but the instability cannot lead to inhomogeneity until the age of the remnant is comparable to the cooling time. Hence, in the three-phase case the injection of energy causes the ISM to become inhomogeneous over a range of distances from the site of the energy injection, just as in the two-phase case.

The inhomogeneity of the ISM has a profound effect on its structure and evolution, and on the mechanisms by which energy flows through the ISM. Here we discuss how the presence of clouds affects the nebulae around massive stars, which are responsible for much the energy injection into the ISM. We shall consider H II regions, stellar wind bubbles, and SNRs in turn.

2. H II Regions in a Cloudy Medium

The emission of ionizing radiation by massive stars tends to homogenize the ambient ISM. If the mean density of the ambient medium is n_m, then a star emitting S ionizing photons per second would produce a sphere of ionized gas of radius

$$R_{\rm St} = 66.9(S_{49}/n_m^2)^{1/3} \text{ pc}, \tag{1}$$

if the gas were homogeneous; where S_{49} ($= S/10^{49}$ s^{-1}) is about 0.04, 1, or 8 for a main-sequence B0, O6.5, or O4 star, respectively (Panagia, 1973). The pressure in the ionized gas exceeds that in the ambient medium, eventually leading to an expansion of the ionized gas obeying $R_i \propto t^{4/7}$ (e.g., Spitzer, 1978).

This simple picture is dramatically altered when the ambient medium is cloudy. Two cases must be considered, depending on the ratio of the cloud radius a to its distance r from the ionizing star. For $(a/r) > 1$, or for the star inside the cloud, one has an 'H II blister' (Icke, 1979) or the 'champagne effect' (Tenorio-Tagle, 1979); this case may occur near the beginning of the life of a massive star, when it is buried in its natal molecular cloud. Here we shall concentrate on the opposite case $(a/r) \ll 1$, corresponding to a star at a typical point in the ISM or a star in a molecular cloud which is itself inhomogeneous. In a classic paper, Oort and Spitzer (1955) discussed two important effects which occur in this second case: (1) the cloud is destroyed as gas at its surface is ionized and flows away (we term this process 'photo-evaporation'); (2) because the outflow of ionized gas is predominantly toward the star, the reaction drives the cloud away from the star (the 'rocket effect'). The photo-evaporation rate for a cloud of constant cross section is

$$\frac{dm}{dt} = -\frac{m_0}{t_{\rm ion}}, \tag{2}$$

where the characteristic photo-evaporation time for a cloud at a distance $r_{\rm pc}$ ($= r/1$ pc) from the star is

$$t_{\rm ion} = \frac{6300}{k_z}\left[\left(\frac{m_0}{M_\odot}\right)\left(\frac{0.03}{f_0}\right)\frac{n_m}{S_{49}}\right]^{1/2} r_{\rm pc}\ {\rm yr}, \tag{3}$$

where f_0 is the initial filling factor for the clouds; if most of the mass is in clouds of density n_{c0}, then the mean density is $n_m = n_{c0} f_0$. The factor k_z allows for the uncertainties associated with the cloud shape and the details of the outflow, and is unity in the Oort–Spitzer theory for a cloud which has been reduced to half its initial radius by the compression associated with the outflow. This theory assumes that the gas is opaque in the Lyman continuum and is at higher than ambient pressure; one or the other of these assumptions breaks down for $r \gtrsim R_{\rm St}$ for stars later than B0, or $M \lesssim 15\, M_\odot$.

If the star is stationary, the mass loss from the cloud drives the cloud away at a velocity

$$v = v_{\rm ex} \ln(m_0/m), \tag{4}$$

where we shall adopt 15 km s^{-1} for the exhaust velocity $v_{\rm ex}$. Equations (2) and (4) can be integrated to give the current position of the cloud in terms of its initial position r_0 and the ratio of its current mass m to its initial mass m_0: i.e.,

$$\frac{r}{r_0} = \exp\left[\beta\left(1 + \frac{m}{m_0}\ln\frac{m}{m_0} - \frac{m}{m_0}\right)\right], \tag{5}$$

where

$$\beta \equiv v_{\rm ex}(t_{\rm ion}/r) \tag{6}$$

is a constant. Since $t_{\rm ion} \propto m_0^{1/2}$, it is convenient to define a characteristic mass m_r such that

$$\beta \equiv \frac{1}{2}\left(\frac{m_0}{m_r}\right)^{1/2}, \tag{7}$$

where

$$m_r = 27\,\frac{S_{49}}{n_m}\left(\frac{f_0}{0.03}\right)k_z^2\,M_\odot\,. \tag{8}$$

When the cloud is completely destroyed, its distance is $r(m = 0) = r_0 \exp(\beta)$. Hence, for $m_0 \ll m_r$, $r(m = 0) \simeq r_0$: low mass clouds evaporate before they move. On the other hand, for $m_0 \gg m_r$, one has $r(m = 0) \gg r$: high mass clouds move away from the star before evaporating.

The effects of an ensemble of clouds in an H II region were analyzed by Elmegreen (1976). He showed that, in contrast to the homogeneous case, the radius $R_i > R_{\rm St}$ of the H II region initially contracts as the intercloud density builds up due to cloud photo-evaporation. The combined effects of photo-evaporation and the rocket effect imply that no clouds survive inside some radius R_{c1}, which is approximately (McKee *et al.*, 1984)

$$R_{c1} = (1/\beta)\,(1 + 2\beta)^{1/2}\,v_{\rm ex}t \tag{9}$$

for clouds of initial mass m_0. Hence, for $R < R_{c1}$ the gas is homogeneous and ionized; for $R_{c1} < R < R_i$ there are neutral clouds embedded in the H II region; and, for $R > R_i$, neutral H I clouds are embedded in a predominantly neutral intercloud medium.

For the more realistic case in which the clouds have range of masses, the clearing radius is generalized to the homogenization radius R_h, defined such that the expected number of clouds surviving inside R_h at time t is about 1 (McKee *et al.*, 1984; more precisely, they defined R_h such that the expected number of clouds initially inside R_h massive enough to survive until time t is 3; only about one of these remains inside R_h at t). Evaluation of R_h for a m^{-2} cloud distribution yields

$$\frac{R_h}{R_{\rm St}} = 1.05\left(\frac{t}{t_{\rm St}}\right)^{4/7}, \tag{10}$$

for $k_z = 1$ and $f_0 = 0.03$, where $t_{\rm St} = R_{\rm St}/C_i = 6.54 \times 10^6\,(S_{49}/n_m^2)^{2/3}$ yr is the characteristic Strömgren time, and $C_i = 10$ km s^{-1} is the isothermal sound speed of the ionized gas. The main-sequence lifetime of the star is related to its ionizing photon luminosity by $t_{ms} \simeq 4.5 \times 10^6\,S_{49}^{-1/5}$ yr for $100\,M_\odot \gtrsim M \gtrsim 10\,M_\odot$ (Van Buren *et al.*, 1985), so that at the end of the star's life the homogenized region extends out to a radius

$$R_h(t_{ms}) = 57 n_m^{-2/7}\,S_{49}^{1/35}\ {\rm pc}\,, \tag{11}$$

which is nearly independent of S. (If the star is moving fast enough that it traverses a distance comparable to R_h in its lifetime, then the size of the homogenized region is reduced somewhat.) The homogenized region has a low emission measure $EM \sim (n_m/2)^2 R_h \lesssim 15\,n_m^{12/7}$ cm^{-6} pc, so it is difficult to observe; most of the emission from the H II region emanates from the photo-evaporating clouds beyond R_h. Much of the mass initially inside R_h is in clouds which have rocketed outside R_h, producing an expanding H I shell of mass

$$M_h(t_{ms}) \simeq 2 \times 10^4\, n_m^{1/7}\, S_{49}^{3/35}\, M_\odot\,. \tag{12}$$

Note that this is only weakly dependent on the properties of the star or the ambient medium.

3. Stellar Wind Bubbles

In addition to being copious sources of ionizing radiation, massive stars are observed to produce powerful winds with mass loss rates

$$\dot{M}_w = 1.3 \times 10^{-16}\,(L/L_\odot)^{1.77}\, M_\odot\, \mathrm{yr}^{-1}\,, \tag{13}$$

and terminal velocities v_w of about 3000 km s^{-1}, where L is the bolometric luminosity (Abbott, 1982). The wind luminosity $L_w = \frac{1}{2}\dot{M}_w v_w^2$ is then about 10^{35} erg s^{-1} for a B0V star and 10^{36} erg s^{-1} for an O6.5V star.

The most thorough treatment of the interaction of a stellar wind with a homogeneous ambient medium has been given by Castor *et al.* (1975) and by Weaver *et al.* (1977). The wind expands at constant velocity until the ram pressure $\rho_w v_w^2$ drops to about the pressure of the surrounding gas, where a shock occurs. The shocked wind forms a bubble of hot gas which drives a shock into the ambient H II region, producing a shell of compressed gas outside the hot interior. The radius R_b of a nonradiative bubble can be estimated by equating the kinetic energy of the expanding shell, which is of order $\rho_m R_b^3 v_b^2$, to the total energy injected by the wind $L_w t$; since the expansion velocity v_b is of order R_b/t, this gives $R_b \sim (L_w/\rho_m)^{1/5}\, t^{3/5}$. In terms of the characteristic wind luminosity

$$L_{\mathrm{St}} \equiv 4\pi\rho_m R_{\mathrm{St}}^2 C_i^3 = 1.26 \times 10^{36}\,(S_{49}^2/n_m)^{1/3}\ \mathrm{erg\ s}^{-1}\,, \tag{14}$$

a precise evaluation gives

$$\frac{R_b}{R_{\mathrm{St}}} = 1.46 \left(\frac{L_w}{L_{\mathrm{St}}}\right)^{1/5} \left(\frac{t}{t_{\mathrm{St}}}\right)^{3/5}. \tag{15}$$

This predicts that quite large bubbles should exist around massive stars; for example, the radius of a bubble around an O6.5 star would reach $75 n_m^{-1/5}$ pc at the end of the star's life.

The evolution of a bubble is substantially altered in a cloudy medium because clouds which survive inside the bubble can 'poison' it (McKee *et al.*, 1984): photo-evaporation of the clouds injects dense gas into the bubble, which will mix with the hot gas and radiate away the internal energy of the bubble. Equations (10) and (15) give

$$\frac{R_b}{R_h} = 1.39 \left(\frac{L_w}{L_{\mathrm{St}}}\right)^{1/5} \left(\frac{t}{t_{\mathrm{St}}}\right)^{1/35} \tag{16}$$

for the ratio of the bubble radius to the homogenization radius. Since the factor $(t/t_{\mathrm{St}})^{1/35}$

is very nearly unity, the evolution depends primarily on L_w/L_{St}: For weak winds ($L_w \ll L_{St}$), the bubble radius is small compared to the homogenization radius, so that the bubble is confined to a homogeneous medium and the nonradiative theory of Castor *et al.* (1975) is valid. For moderate winds ($L_w \sim L_{St}$), the bubble expands to the edge of the cloud distribution; photo-evaporated gas from the clouds is injected into the bubble, inducing radiative losses, reducing the pressure, and thereby maintaining $R_b < R_h$. Finally, for strong winds ($L_w \gg L_{St}$), the bubble rapidly engulfs a number of clouds and radiates away most of its internal energy; its expansion then ceases, to resume only when R_h becomes larger than the bubble radius. Thereafter, R_b remains somewhat smaller than R_h. Thus, stellar wind bubbles are nonradiative only if the winds are weak; the dynamics of bubbles produced by strong winds are governed by photo-evaporating clouds. Note that since stellar motion reduces R_h, it would reduce the size of the bubble as well.

This theoretical model is in good agreement with observation. Chu (1981) introduced a classification for the nebulae around Wolf–Rayet stars which Lozinskaya (1982) has shown also applies to Of stars. The five categories are wind-blown bubbles (or, more precisely, wind-blown shells), amorphous H II regions, ring-like H II regions, stellar ejecta, and invisible. This can be interpreted as an evolutionary sequence (McKee *et al.*, 1984): young ($t \lesssim 10^5$ yr) O stars have a sufficiently strong wind that the shocked gas at the edge of the bubble appears as a sharp, bright shell (example: the Bubble nebula). Older O and B stars with $R_h < R_{St}$ have thick, clumpy shells of ionized gas and appear amorphous (example: the Rosette nebula). When R_h becomes comparable to R_{St}, the emission from the ionized gas is concentrated in a thin shell just beyond R_h, forming a ring-like H II region. Finally, in the late stages of evolution, the star ejects part of its envelope, forming a circumstellar nebula; in this interpretation, much of the nebulosity around Wolf–Rayet stars observed by Chu is circumstellar. For late O and B stars it is possible for the emission measure of the ring-like H II region to fall below the limit of observability as it expands outward, so that the H II region would become invisible for a time before the ejection of the circumstellar matter.

4. Type-II Supernova Remnants

The dynamical evolution and appearance of the SNR resulting from the explosion of a massive star as a Type-II supernova are determined by the processing of the ambient ISM by the progenitor star. As described above, when an isolated, stationary, massive star reaches the end of its main-sequence lifetime, it has converted the ambient ISM into a layered structure consisting of a bubble of hot shocked wind ($R < R_b$), a ring of homogenized, ionized gas ($R_b < R < R_h$), a ring of H I clouds undergoing photo-evaporation ($R_h < R < R_i$), and finally the nearly undisturbed ISM ($R_i < R$). There are several important caveats to this simple picture: the star must be nearly stationary (Elmegreen, 1976); it must be isolated from other massive stars (see Bruhweiler *et al.*, 1980, for a discussion of the evolution of an association of massive stars); and it must have a large ionizing flux (corresponding to $M \gtrsim 15\, M_\odot$) in order for the above description of photo-evaporation to be quantitatively correct. Nonetheless, this picture provides a useful framework for discussing how processing of the ambient medium by the progenitor star can alter the evolution of the SNR.

In a homogeneous medium, a SNR passes through four stages: (I) Ejecta dominated, or free expansion ($R \propto t$). The mass of the ejecta exceeds that swept up in the ambient medium so that most of the energy is in kinetic form; if there is circumstellar matter, as in Cas A, the remnant may be bright during its interaction with the SN ejecta. The interaction between the ejecta and the circumstellar matter may convert some of the kinetic energy into thermal energy, but, once the circumstellar matter has been swept up, the energy will return to kinetic form due to adiabatic expansion and the free expansion will continue. (II) Sedov–Taylor, or adiabatic ($R \propto t^{2/5}$). The swept-up mass exceeds the total ejected mass M_E (SN + circumstellar) for $R > 1.9(M_E/M_\odot)^{1/3}\, n_0^{-1/3}$ pc, and beyond this radius the evolution of the remnant approaches the Sedov–Taylor similarity solution (Sedov, 1969). (III) Radiative ($R \propto t^{2/7}$). Energy losses due to radiation lead to the formation of a cold dense shell surrounding a hot interior. (IV) Merge with ISM. When the shell velocity is comparable to the r.m.s. velocity in the ISM and the pressure is comparable to the ambient value, the SNR merges with the ISM, injecting cloud kinetic energy in the process.

In a three-phase ISM, much of the volume is filled by hot, low-density gas, so that an SNR remains in free expansion until $R \sim 13(M_E/M_\odot)^{1/3}$ pc for the typical density of the hot medium (McKee, 1983); the adiabatic stage is governed by cloud evaporation with $R \propto t^{3/5}$ (McKee and Ostriker, 1977); and the radiative shell forms in the interior rather than at the edge of the remnant (Cowie *et al.*, 1981). This evolutionary sequence is likely to apply to Type-I SNRs, both because they originate in an old stellar population at random points in the galaxy and because the progenitor star is unlikely to emit enough ionizing radiation to significantly alter the structure of the ambient medium (for example, if the progenitor is a 1 $M_\odot$ accreting white dwarf with an effective temperature of 10^5 K, then $S_{49} = 10^{-4}$, too small for effective photo-evaporation). This sequence may also apply to Type-II SNRs from relatively low mass or high velocity progenitors, since neither can strongly affect the ambient medium.

For the remaining Type-II SNRs, the early evolution is approximately independent of the nature of the undisturbed ISM since the progenitor star has processed the ISM into the layered structure described above. The low density of the central bubble of hot gas will prolong the free expansion stage, just as in the case of the early evolution of a remnant in a three-phase ISM (McKee, 1983; Mills *et al.*, 1984). After the interaction of the ejecta with the circumstellar matter is complete, the X-ray, optical, and radio emission from the SNR will be low until the blast wave hits the homogenized H II region beyond the bubble. Note that since the progenitor star goes through a red supergiant phase prior to exploding, the ionizing radiation is cut-off for a period of order 10^5 yr and the homogenized region may cool and form small-scale inhomogeneities due to thermal instability. If the wind-blown bubble is sufficiently large, the SNR may never be bright in X-rays after the initial interaction with circumstellar matter: the post-shock temperature of a Sedov–Taylor blast wave never reaches a minimal X-ray emission temperature of 10^6 K if the radius exceeds $R_x = 19(E_{51}/n_0)^{1/3}$ pc. Once the shock is well into the homogenized region, the evolution should approach the Sedov–Taylor form. Cloud evaporation becomes important only when the remnant expands beyond R_i, but by that time a significant amount of the energy may have been lost to radiation.

Several methods have been proposed for distinguishing Type-I and Type-II SNRs. For young remnants it is often possible to assign the type unambiguously by measuring the mass and composition of the ejecta. Young remnants, in which the optical spectrum is dominated by the Balmer lines, are taken to be Type I, based on their similarity to the known Type-I remnants Tycho and SN 1006. Remnants with central pulsars are generally regarded as Type II (Tammann, 1982), and, by extension, SNRs with radio morphologies resembling the Crab's ('plerions') are regarded as Type II as well (Weiler, 1983); however, since the Crab is an unusual remnant, this identification is not certain. MSH 15–52 may be an example of a Type-II remnant since it has young central pulsar with a spin-down age of 1600 yr embedded in a 17 pc shell (Seward and Harnden, 1982), which suggests that the remnant is in the free expansion stage. The X-ray morphology of this remnant can be explained in terms of the interaction of inhomogeneous ejecta with ambient clouds (McKee, 1983). The lack of proper motion in the optical filaments (van den Bergh and Kamper, 1984) shows that the filaments are shocked interstellar clouds and not ejecta; it does not rule out the free expansion hypothesis.

It is more difficult to assign a type to an SNR if there is no stellar remnant or observable ejecta, but the processing of the ambient medium by Type-II progenitors offers a method of doing this. In a discussion of N49, an SNR in the LMC, Shull *et al.* (1984) presented a detailed analysis of the pre-supernova condition of the ISM around a relatively low-mass B star with $S_{49} = 10^{-3}$ and a weak wind. They showed that photoevaporation of clouds in the H II region leads to an outflow of recombining gas from the H II region, forming a neutral shell. Assuming the shell is thin and homogeneous (which is questionable), they showed that the shell would expand back into the fossil H II region after the central star evolves into a red supergiant, and that the observed optical and X-ray morphology of the remnant could be explained by the emission from the shock as it decelerates in the expanded shell.

Closer to home, McCray and Snow (1979) have proposed that the Cygnus Loop is a Type-II SNR, because the clouds in its vicinity are arranged around the edge of the SNR rather than throughout the interior. Analysis of X-ray observations of this remnant has led Charles *et al.* (1985) to refine this argument and to suggest that the progenitor star had a mass between 10 and 20 $M_\odot$: a more massive star would have produced a bubble larger than 20 pc and would have destroyed the clouds observed at 20 pc, whereas a less massive star or a Type-I progenitor would not have eliminated the clouds inside that radius. Optical filaments in the Cygnus Loop are smaller and less dense than typical interstellar clouds, which can be attributed to the result of a thermal instability in the fossil H II region. Note that in this interpretation the Cygnus Loop may be substantially younger than the 1.5×10^4 yr based on the Sedov–Taylor assumption.

Other suggested examples of Type-II SNRs are W44 and HB21, which have large expanding H I shells around the shocked gas; these shells could have originated as H I clouds which were rocketed out of the progenitor H II region (McKee *et al.*, 1984).

On a global scale, indirect evidence for the processing of the ISM by progenitor stars comes from analysis of the observed number N of SNRs as a function of their radius. If all SNRs followed the same evolutionary path with $R \propto t^{\eta}$, then the number of SNRs smaller than R is simply the number younger than the age t corresponding to R, so that

$$N(<R) \propto R^{1/\eta} . \tag{17}$$

Remnants in the free-expansion stage should obey $N \propto R$, whereas Sedov–Taylor remnants should have $N \propto R^{5/2}$. Analysis of the SNRs in the LMC, where distances can be accurately measured, gives the remarkable result that $\eta \sim 1$ for $R \lesssim 20$ pc and $\eta \gtrsim 1$ for $R > 20$ pc (Clarke, 1976; Mathewson *et al.*, 1983; Mills *et al.*, 1984). The former result is consistent with free expansion, which we have argued is to be expected out to $R \sim 15$ pc for Type-I SNRs in a three phase ISM, and out to min $[R_b, 13(M_E/M_\odot)^{1/3}$ pc$]$ for Type-II SNRs. The observational result for $R \gtrsim 20$ pc demonstrates the incompleteness of the sample, since it is inconsistent with any model for SNR evolution; indeed, Hughes *et al.* (1984) have shown that the results for $R < 20$ pc can be explained by incompleteness as well, provided the number of remnants, as a function of (n_0/E), varies in the proper fashion. The correct explanation for the $N(<R)$ relation undoubtedly involves selection effects to some extent, but independent support for the importance of free expansion comes from the observation that the thermal energy of SNRs in the LMC and in M31 increases with radius up to about 15 pc (Dopita, 1979; Blair *et al.*, 1981), as would be expected from the gradual conversion of kinetic to thermal energy in the free-expansion stage (Mathewson *et al.*, 1983; McKee, 1983). Both interpretations of the $N(<R)$ relation predict that most adiabatic SNRs have yet to be detected, and it is a major challenge to find these missing remnants.

5. Conclusions

The complex and often beautiful morphology of the ISM is largely a manifestation of the processes by which stars inject energy into the ISM. Thermal and hydrodynamic instabilities accompanying this energy injection make the ISM inhomogeneous, or cloudy, and these inhomogeneities in turn dramatically affect the structure and evolution of the nebulae which mark the sites of the energy injection. Massive stars dominate the energy injection, forming H II regions, stellar wind bubbles, and finally supernova remnants. Photo-evaporation homogenizes the H II region out to a distance R_h, and the low mean density inside R_h implies that old H II regions should have a ring-like appearance with a surrounding shell of H I. Weak stellar winds blow adiabatic bubbles in the ISM, but stronger winds are constrained to remain inside R_h by radiative losses. When the star explodes as a supernova, it undergoes an extended free expansion in the low density bubble, and if the bubble is large enough, the remnant may never be visible in X-rays after the initial interaction with the circumstellar matter.

A variety of models have been proposed to explain the evolution of SNRs, including the classical Sedov–Taylor model for a homogeneous medium, expansion in a two-phase ISM, expansion in a three-phase ISM (with and without evaporation of embedded clouds), and expansion in a bubble/H II region around a massive star as described above. It is remarkable that, after years of study of SNRs, it has thus far proven impossible to make a definitive observational distinction among these models. One reason for this difficulty is that observations of the number vs radius relation for SNRs show that most SNRs in the Sedov–Taylor, or adiabatic, stage remain undetected. Ultimately, however, one should be able to relate the observed morphology of SNRs to both the character of the ambient ISM and to the nature of the progenitor star, thereby clarifying one of the major pathways by which stars energize the ISM.

Acknowledgements

I wish to thank the organizing committees of the Third Asian-Pacific Regional Meeting of the IAU for an opportunity to present these results. I am indebted to my colleagues B. Lazareff and D. Van Buren for their collaboration on the structure and evolution of the ISM around massive stars. This research is supported in part by NSF grant AST 83–14684.

References

Abbott, D. C.: 1982, *Astrophys. J.* **259**, 282.
Blair, W. P., Kirshner, R. P., and Chevalier, R. A.: 1981, *Astrophys. J.* **247**, 879.
Bruhweiler, F. C., Gull, T. R., Kafatos, M., and Sofia, S.: 1980, *Astrophys. J.* **238**, L27.
Castor, J., McCray, R., and Weaver, R.: 1975, *Astrophys. J.* **200**, L107.
Charles, P., Kahn, S., and McKee, C. F.: 1985, *Astrophys. J.* **295**, 456.
Chu, Y. H.: 1981, *Astrophys. J.* **249**, 195.
Clarke, J. N.: 1976, *Monthly Notices Roy. Astron. Soc.* **174**, 393.
Cowie, L. L., McKee, C. F., and Ostriker, J. P.: 1981, *Astrophys. J.* **247**, 908.
Cox, D. P. and Smith, B. W.: 1974, *Astrophys. J.* **189**, L105.
de Jong, T.: 1980, in P. A. Wayman (ed.), *Highlights of Astronomy*, D. Reidel Publ. Co., Dordrecht, Holland, Vol. 5, p. 301.
Dopita, M. A.: 1979, *Astrophys. J. Suppl.* **40**, 455.
Elmegreen, B. G.: 1976, *Astrophys. J.* **205**, 405.
Field, G. B., Goldsmith, D. W., and Habing, H. J.: 1969, *Astrophys. J.* **155**, L149.
Hughes, J. P., Helfand, D. J., and Kahn, S. M.: 1984, *Astrophys. J.* **281**, L25.
Icke, V.: 1979, *Astron. Astrophys.* **78**, 352.
Lozinskaya, T. A.: 1982, *Astrophys. Space Sci.* **87**, 313.
Mathewson, D. S., Ford, V. L., Dopita, M. A., Tuohy, I. R., Long, K. S., and Helfand, D. J.: 1983, in I. J. Danziger and P. Gorenstein (eds.), 'Supernova Remnants and Their X-Ray Emission', *IAU Symp.* **101**, 541.
McCray, R. and Snow, T. P.: 1979, *Ann. Rev. Astron. Astrophys.* **17**, 213.
McKee, C. F.: 1983, in I. J. Danziger and P. Gorenstein (eds.), 'Supernova Remnants and Their X-Ray Emission', *IAU Symp.* **101**, 87.
McKee, C. F. and Ostriker, J. P.: 1977, *Astrophys. J.* **218**, 148.
McKee, C. F., Van Buren, D., and Lazareff, B.: 1984, *Astrophys. J.* **278**, L115.
Mills, B. Y., Turtle, A. J., Little, A. G., and Durdin, J. M.: 1984, *Australian J. Phys.* **37**, 321.
Oort, J. H. and Spitzer, L.: 1955, *Astrophys. J.* **121**, 6.

Panagia, N.: 1973, *Astron. J.* **78**, 929.
Sedov, L.: 1969, *Similarity and Dimensional Methods in Mechanics*, Academic Press, New York.
Seward, F. D. and Harnden, F. R.: 1982, *Astrophys. J.* **256**, L45.
Shull, P., Dyson, J. E., Kahn, F. D., and West, K.: 1985, *Monthly Notices Roy. Astron. Soc.* **212**, 199.
Spitzer, L.: 1978, *Physical Processes in the Interstellar Medium*, Wiley, New York.
Tammann, G.: 1982, in M. J. Rees and R. Stoneham (eds.), *Supernovae: A Survey of Current Research*, D. Reidel Publ. Co., Dordrecht, Holland, p. 371.
Tenorio-Tagle, G.: 1979, *Astron. Astrophys.* **71**, 59.
Van Buren, D., Lazareff, B., and McKee, C. F.: 1985 (in preparation).
van den Bergh, S. and Kamper, K.: 1984, *Astrophys. J.* **280**, L51.
Weaver, R., McCray, R., Castor, J., Shapiro, P., and Moore, R.: 1977, *Astrophys. J.* **218**, 377.
Weiler, K. W.: 1983, in I. J. Danziger and P. Gorenstein (eds.), *Supernova Remnants and Their X-Ray Emission, IAU Symp.* **101**, 299.

THE SHAPE OF SPIRAL ARMS*

SØREN-AKSEL SØRENSEN

Department of Computer Science, University College, London, U.K.

(Received 2 July, 1985)

Abstract. The response by a gaseous disc to perturbations is investigated. Spectral analysis of the resulting pattern indicates that information about the nature of the perturber is maintained only if the perturbation is strong.

1. Introduction

The mathematical shape of the arms in spiral galaxies has frequently been used to discriminate between theories of their origin. Early investigations by Groot (1925) and Danver (1942) relied on manual identification of the spiral pattern, and their sample was restricted to galaxies with well defined large scale open arms which could be traced unambiguously. Only a small number of galaxies satisfy this requirement. In most systems the arms are bifurcated and it is often difficult clearly to identify a global pattern. The use of Fourier techniques provides an objective way of analysing the more complicated patterns. The technique can be used to analyse the distribution of spiral arm tracers (Kalnajs, 1973) as well as the luminosity distribution of galactic images (Grosbøl, 1979; Iye *et al.*, 1979, 1982). In contrast to the manual identification, spectral analysis provides an abundance of reliable information about the patterns present in a galaxy. The relation between this information and the dynamical properties of the galaxy is, however, still poorly known. The results discussed here are part of a series of model investigations aimed at determining how the dynamical properties of a galaxy influence the shape of the pattern, and whether information about any perturbation is retained. The latter is important if global gravitational instability creates a perturbation.

2. Bars and Companions

Global spiral structure is closely connected with the formation of large scale compression waves in the interstellar gas, which are excited by perturbations of the axisymmetric galactic potential. Most investigations have concentrated on how the gas responds to strong non-axisymmetric forces which are able to change the flow pattern radically. Several authors (Sanders, 1977; Matsuda and Isaka, 1980; Sørensen and Matsuda, 1982) have investigated the gas flow in galaxies with strong bars. The self-gravity of the gas phase was neglected in these models, and the potential from the stellar phase approximated by a Toomre (1963) disc with the bar superimposed.

* Paper presented at the IAU Third Asian-Pacific Regional Meeting, held in Kyoto, Japan, between 30 September–6 October, 1984.

Astrophysics and Space Science **118** (1986) 395–399.

Gradual introduction of a non-axisymmetric potential avoided transient patterns, and the models only investigated quasi-stationary solutions. Studies by Toomre and Toomre (1972) of interacting galaxies also concentrated on strong interactions, and the spiral pattern was formed by distortions of the disc. The spiral arms have in both cases a very slow growth rate, and the process requires large amounts of energy.

When the non-axisymmetric contribution to the potential is reduced to a small perturbation, the forces are too weak to distort the circular streamlines significantly, and the nature of the pattern changes. The patterns found by Sørensen (1983, 1985) in models of interacting galaxies were formed by epicyclic waves with low amplitudes. These waves have very fast growth rates, and only small amounts of energy are required to excite them. They were thus found in models where the perturber to galaxy mass ratio was as low as 0.03.

3. Spiral Perturbations

The results presented here are from a study of the expected response to spiral modes formed from a global gravitational instability in the stellar disc (Bardeen, 1975; Iye, 1978). The axisymmetric part of the potential is again approximated by a Toomre disc and the contribution by the gas to the potential is considered negligible compared to that

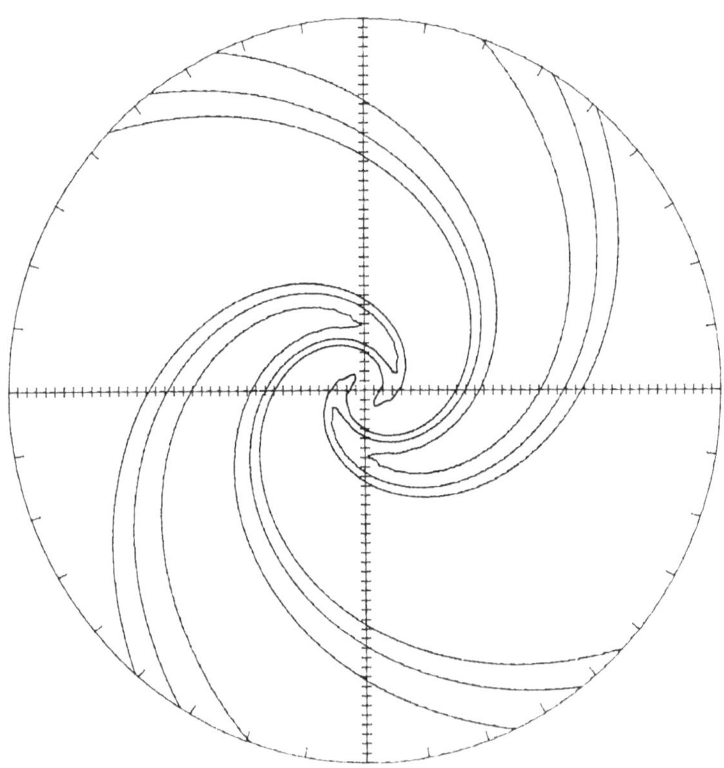

Fig. 1. Leading logarithmic spiral perturbation for the Toomre potential.

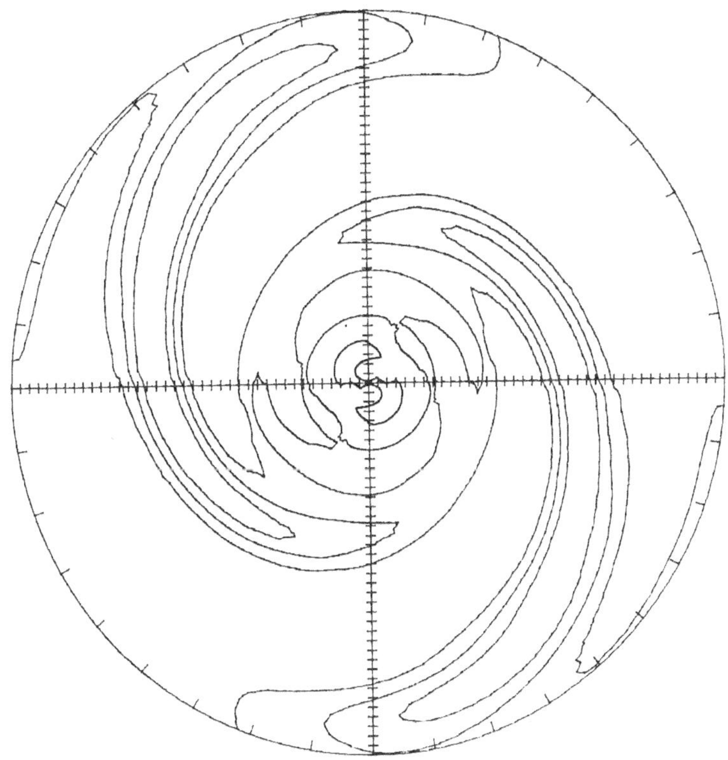

Fig. 2. Gas response to a 0.1% perturbation. Only positive contours are shown.

of the stellar phase. Superimposed on this potential is a leading spiral wave with a pitch angle of 30 degrees. The initial growth rate of this mode is assumed to be constant, but the allowed maximum amplitude varies for each model. The shape and phase of the perturbation is shown in Figure 1, which displays equi-potential lines for the non-axisymmetric potential only. In Figure 2 the response is shown for a weak perturbation, in which the spiral potential is 0.1% of the axisymmetric field. The figure shows equi-density contours for the positive response only; contours in the rarefied areas are disregarded to improve clarity. The response resembles closely that obtained when the system is perturbed by a low mass companion. From the power spectrum the only information about the nature of the perturbation is the bi-symmetry of the pattern. All other information is lost. When the maximum amplitude of the perturbation is increased the nature of the pattern changes, and the shape of the perturbation can be detected. Figure 3 shows the density contours for a model in which the perturbation is allowed to grow until it reaches 1% of the axisymmetric field. The global pattern in Figure 2 has been replaced by a whispy pattern, consisting of many small sections of arms. These individual arms are all trailing spirals with similar pitch angles to those found in Figure 2. The shape of the original pattern can, however, be detected by tracing the areas where the gas response is strongest. Although these areas are part of trailing spirals they all

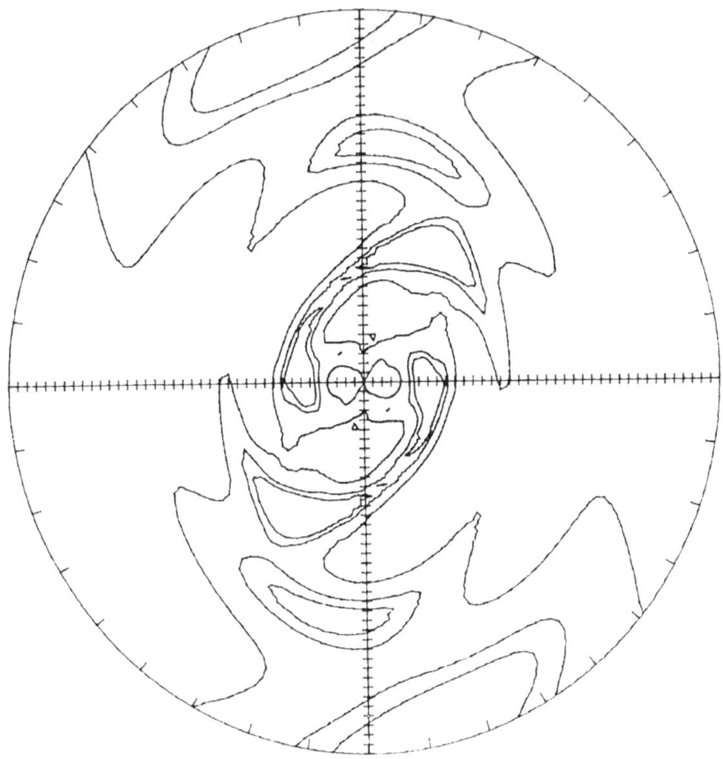

Fig. 3. Gas response to a 1% perturbation. Only positive contours are shown.

have their points of maximum response situated along the leading spiral potential. The power spectrum of this model is far more complicated than that of Figure 2. The radial wave number is again unrelated to the original perturbation, but also the azimuthal part of the spectrum is now contaminated with a number of higher modes. As the amplitude of the perturber is raised further the leading two-armed mode becomes more pronounced, and when the spiral contribution is allowed to reach a maximum value corresponding to 10% of the axisymmetric part this mode becomes dominant; only traces are left from the trailing modes.

4. Conclusions

The results presented above show that the weak perturbation model radically changes the interpretation of galactic power spectra. The information extracted by examining only the dominant mode may be of limited value. The dominant radial wave number can be determined either by the perturbation, or by the rotation curve, dependent on the strength of the perturbation. For weak perturbations the pattern furthermore depends on time. The information obtained from the azimuthal wave number of the dominant mode is also open to interpretation. The stronly distorted flow connected with strong non-axisymmetric forces will cause the dominant mode to reflect the azimuthal information about the perturbation well. The weak perturbation model will, however, allow a large number of waves with different amplitudes to exist in the gas disc simultaneously. The global pattern is likely to originate from interference between a number of waves, where bifurcations in the arms are tracers of the individual modes. It is thus necessary to interprete a large number of modes in each power spectrum in order to analyse the observed pattern.

References

Bardeen, J. M.: 1975, in A. Hayli (ed.), 'Dynamics of Stellar Systems', *IAU Symp.* **69**.
Danver, C.-G.: 1942, *Ann. Obs. Lund*, No. 10.
Groot, H.: 1925, *Monthly Notices Roy. Astron. Soc.* **85**, 535.
Grosbøl, P. J.: 1979, in D. S. Evans (ed.), *Photometry, Kinematics, and Dynamics of Galaxies*, University of Texas Press, p. 101.
Iye, M.: 1978, *Publ. Astron. Soc. Japan* **30**, 223.
Iye, M., Hamabe, M., Watanabe, M., and Okamura, S.: 1979, in D. S. Evans (ed.), *Photometry, Kinematics, and Dynamics of Galaxies*, University of Texas Press.
Iye, M., Okamura, S., Hamabe, M., and Watanabe, M.: 1982, *Astrophys. J.* **256**, 103.
Kalnajs, A.: 1973, in L. Weliachew (ed.), 'La Dynamique des Galaxies Spirales', *Colloque Int. CNRS*, No. 241.
Matsuda, T. and Isaka, H.: 1980, *Prog. Theor. Phys.* **64**, 1265.
Sanders, R. H.: 1977, *Astrophys. J.* **217**, 916.
Sørensen, S.-A.: 1983, in E. Athanassoula (ed.), 'Internal Kinematics and Dynamics of Galaxies', *IAU Symp.* **100**.
Sørensen, S.-A.: 1985, *Monthly Notices Roy. Astron. Soc.* **212**, 723.
Sørensen, S.-A. and Matsuda, T.: 1982, *Monthly Notices Roy. Astron. Soc.* **198**, 865.
Toomre, A.: 1963, *Astrophys. J.* **138**, 385.
Toomre, A. and Toomre, J.: 1972, *Astrophys. J.* **178**, 623.

HIGH-RESOLUTION OBSERVATIONS OF THE ORION BRIGHT BAR*[a]

TOSHIHIRO OMODAKA
College of Liberal Arts, Kagoshima University, Japan

MASAHIKO HAYASHI, SAEKO SUZUKI
Department of Astronomy, University of Tokyo, Japan

TETSUO HASEGAWA
Nobeyama Radio Observatory, Tokyo Astronomical Observatory, University of Tokyo, Japan

and

RYOSUKE MIYAWAKI
Hachioji Technical High School, Hachioji, Japan

(Received 2 July, 1985)

Abstract. High resolution strip maps of CS (J = 1–0) and H51α line emission across the Orion bright bar are presented. They reveal the existence of a high density molecular layer (*molecular sheet*) plane parallel to the ionization front. This molecular sheet is redshifted relative to the ambient molecular cloud by about 2 km s^{-1}. The rapid decrease of the CS emission at about 50 arc sec from the bar suggests that a shock front exists here and the sheet is a post shock layer.

1. Introduction

The Orion bright bar, a bar-shaped optically bright emission region, can be seen about 2′ southeast of the Trapezium stars. This bright bar is especially prominent in [O III] photographs, and seems to be a kind of bright rim; a high density region of ionized gas near an ionization front seen edge on (Pottasch, 1956). It has appeared in observations at all wavelengths (Becklin *et al.*, 1976; Keene *et al.*, 1982; Johnston *et al.*, 1983).

Schloerb and Loren (1982) and Omodaka *et al.* (1984) have observed molecular emission in this region. They suggested that the expansion of the H II region generated by the Trapezium stars has driven a shock wave propagating into the molecular cloud to the southeast of the bar.

* Paper presented at the IAU Third Asian-Pacific Regional Meeting, held in Kyoto, Japan, between 30 September–6 October, 1984.

[a] This work was carried out under the common use observation program at the Nobeyama Radio Observatory (NRO). NRO, a branch of the Tokyo Astronomical Observatory, University of Tokyo, Japan, is a cosmic radio observing facility open to outside users.

Astrophysics and Space Science **118** (1986) 401–404.

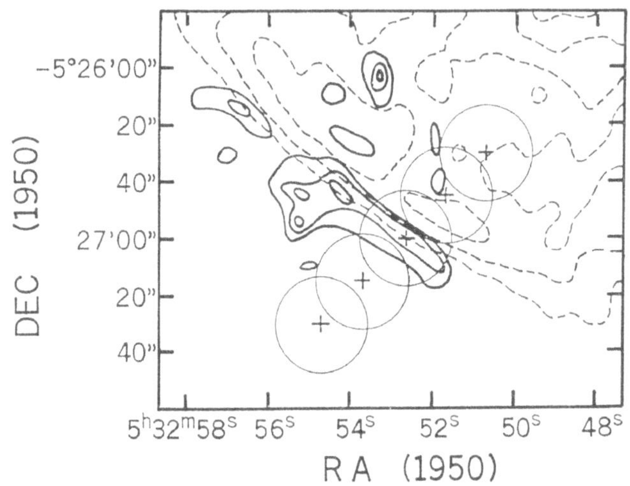

Fig. 1. The positions of the CS and H51α observations are indicated by crosses with the beam size superposed on the 10 μm (solid contours) and 6 cm (dashed contours) maps.

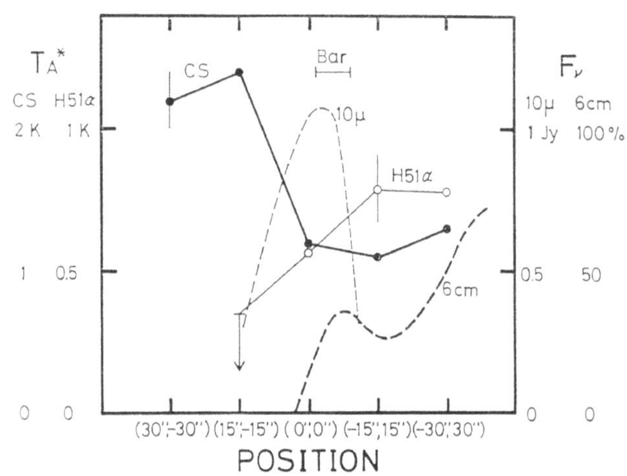

Fig. 2. The variation of the antenna temperatures of CS and H51α emission across the Orion Bright Bar. Also indicated are the 10 μm and 6 cm continuum flux densities. 'Bar' is the position of the optically bright rim. The vertical bars represent $\pm 1\sigma$ noise level for the CS and H51α antenna temperatures.

In this paper we present higher resolution strip maps across the bar in CS and H51α; they show the existence of a sheet of the molecular gas caused by the expansion of the H II region.

2. Observations and Results

The observations were made using the 45 m telescope of the Nobeyama Radio Observatory. The measured half power beamwidth was 33″ at 49 GHz.

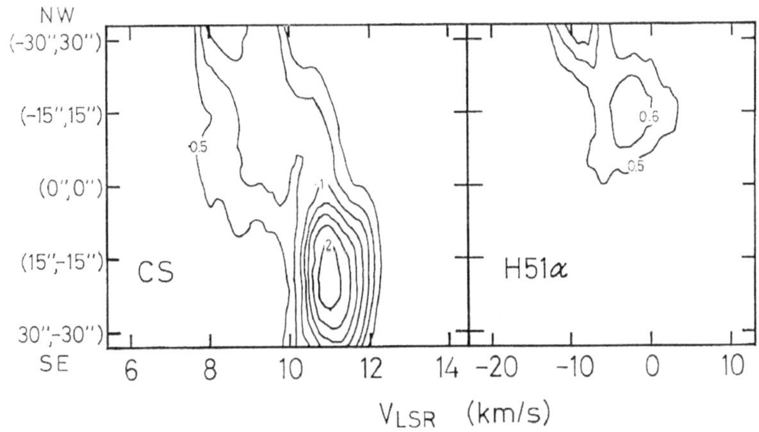

Fig. 3. The spatial-velocity maps for CS and H51α emissions.

Figure 1 shows the observed points superposed on the maps of 10 μm (Becklin *et al.*, 1976) and 6 cm continuum emission (Johnston *et al.*, 1976). The data were taken along a strip across the bar at a spacing of 21".2. The reference position is $\alpha(1950) = 5^h32^m52^s.7$ and $\delta(1950) = -5°27'00''$, which is on the ridge of the 10 μm emission. Offset angles are measured from this reference position.

Figure 2 shows the spatial variation of CS and H51α line intensities. For comparison, infrared (Becklin *et al.*, 1976) and radio continuum (Johnston *et al.*, 1983) intensities are also displayed. The H51α line and the 6 cm continuum are detected at the northwestern side and abruptly decrease at the bar, while the intensity of the CS emission is stronger at the southeastern side and decreases across the bar by a factor of two. It seems that a high density molecular cloud and the H II region are separated by this bar, which is probably the ionization front seen nearly edge on.

Figure 3 shows the spatial-velocity contour maps of CS and H51α. As noted above, H51α emission is detected northwest of the bar, while CS emission is stronger in the southeast. The peak velocity of the CS emission abruptly changes across the bar from 9 km s^{-1} on the northwest side to 11 km s^{-1} on the southeast side. The peak velocity of the H51α line is blueshifted by about 15 km s^{-1} with respect to the CS line, suggesting that the ionized gas is streaming toward us and that the bar is on the far side of the H II region.

In order to study whether these features are common to the whole bright bar region, we present three strip maps of CS and H51α emission across the bar. Figure 4 shows the spatial variation of CS and H51α integrated intensity. It is noted that the CS emission drops sharply at 50" distance from the bar. These tendencies are common to the three strip maps separated 21".2 from each other. As CS emission is a density tracer, this indicates the existence of a high-density layer (molecular sheet) plane parallel to the optically prominent ionization front.

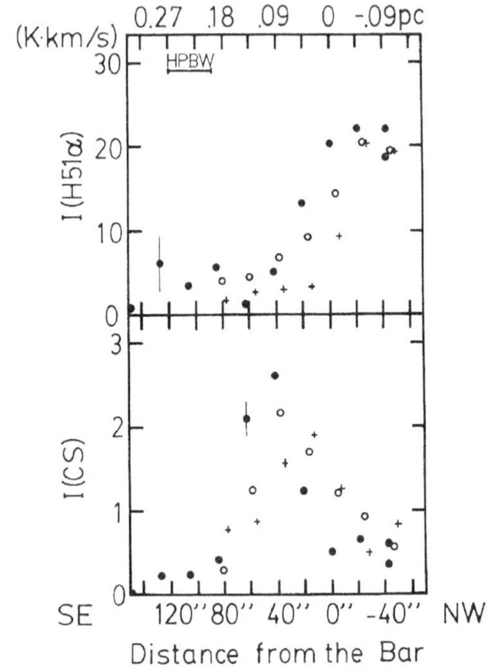

Fig. 4. The three strip maps of CS and H51α integrated intensity across the Bar (denoted by ○, ●, +, respectively). The spacing between each strip is 21ʺ.2.

3. Discussion

We have suggested in the previous section that the bright bar is the ionization front which separates the dense molecular cloud and the H II region. Also there exists a sheet of molecular gas plane parallel to the bar. The observed width of this sheet is $\sim 50''$. It is redshifted relative to the ambient molecular cloud by about 2 km s^{-1}. This shift can be due to an acceleration by thermal and kinetic pressures from the H II region; due to the passage of a shock. Thus the rapid decrease of the CS emission at about $50''$ from the bar suggests that a shock front exists here, and the sheet is the post-shock layer.

References

Becklin, E. E., Beckwith, S., Gatley, I., Mathews, K., Neugebauer, G., Sarazin, C., and Werner, M. W.: 1976, *Astrophys. J.* **207**, 770.

Johnston, K. J., Palmer, P., Wilson, T. L., and Bieging, J. H.: 983, *Astrophys. J.* **271**, L89.

Omodaka, T., Hayashi, M., and Hasegawa, T.: 1984, *Astrophys. J.* **282**, L77.

Keene, J., Hildebrand, R. H., and Whitcomb, S. E.: 1982, *Astrophys. J.* **252**, L11.

Pottasch, S.: 1956, *B.A.N.* **13**, 77.

Schloerb, F. P. and Loren, R. B.: 1982, in A. E. Glassgold, P. J. Huggins, and E. L. Schucking (eds.), *Symposium on the Orion Nebula to Honor Henry Draper*, New York Acad. Sci., New York, p. 32.

EXCITATION OF INTERSTELLAR MOLECULES IN THE ORI-KL SOURCE*

MASATOSHI OHISHI, NORIO KAIFU, HIROKO SUZUKI, and MASAKI MORIMOTO

Nobeyama Radio Observatory (NRO), Minamimaki, Nagano, Japan

(Received 2 July, 1985)

Abstract. A molecular spectral line survey of the title source detected 166 molecular lines from 18 interstellar molecules in the frequency ranges of 34.25–50.00, 83.50–84.50, and 86.00–91.50 GHz. For each molecule, gaussian decomposition of the velocity components in the transition profiles gave consistent radial velocity and linewidth. Rotation diagrams were drawn for each velocity component.

We have been performing a Molecular Spectral Line Survey using the 45 m radio telescope at Nobeyama with the 250 kHz resolution AOSs and the 37 kHz resolution AOSs. We have detected, toward the Ori–KL source, 170 molecular lines from 19 interstellar molecules in the frequency ranges of 34.25–50.00, 83.50–84.50, and 86.00–91.50 GHz. We tried gaussian decomposition to separate the velocity components in the profiles of various transitions. For each molecule the results of the decomposition for various transitions gave consistent radial velocities and linewidth. After this decomposition, we have drawn the rotation diagrams for each velocity component. The results are summarized in Table I.

TABLE I

Characteristics of velocity components

Component	V_{LSR} (km s^{-1})	ΔV (km s^{-1})	T_{rot} (K)	N_{tot} (cm^{-2})	Size (arc sec)
CH_3OH 9 km s^{-1}	9.1 ± 1.7	4.2 ± 0.7	105 ± 5	$(2.4 \pm 0.3) \times 10^{16}$	20
4 km s^{-1}	4.0 ± 1.7	3.9 ± 0.8	160 ± 10	$(4.2 \pm 0.6) \times 10^{15}$	20
SO_2 6 km s^{-1}	6.5 ± 1.4	14.4 ± 1.3	131 ± 5	$(6.2 \pm 0.6) \times 10^{16}$	25
20 km s^{-1}	20.3 ± 2.7	12.5 ± 3.1	146 ± 5	$(3.9 \pm 0.4) \times 10^{16}$	25
−3 km s^{-1}	-3.2 ± 4.2	16.5 ± 1.3	132 ± 8	$(7.2 \pm 1.0) \times 10^{16}$	25
HCO_2CH_3	9.0 ± 1.6	2.9 ± 1.1	24 ± 3	$(3.1 \pm 0.9) \times 10^{14}$	20
HNCO	7.9 ± 1.5	6.9 ± 3.3	126 ± 5	$(1.5 \pm 0.4) \times 10^{15}$	20
C_2H_5CN	6.8 ± 2.4	12.8 ± 4.9	133 ± 20	$(3.2 \pm 1.5) \times 10^{15}$	20
HCCN 9 km s^{-1}	8.9 ± 1.1	3.8 ± 0.7	35 ± 7	$(6.5 \pm 2.0) \times 10^{13}$	30
6 km s^{-1}	6.5 ± 2.6	15.9 ± 3.8	–	–	
$(CH_3)_2O$	8.2 ± 1.0	2.3 ± 1.1	–	–	
CCH	12.5 ± 0.3	3.4 ± 0.5	–	$\sim 0.8 \times 10^{15}$	
SO	7.2 ± 1.5	21.2 ± 5.6	–	–	

* Paper presented at the IAU Third Asian-Pacific Regional Meeting, held in Kyoto, Japan, between 30 September–6 October, 1984.

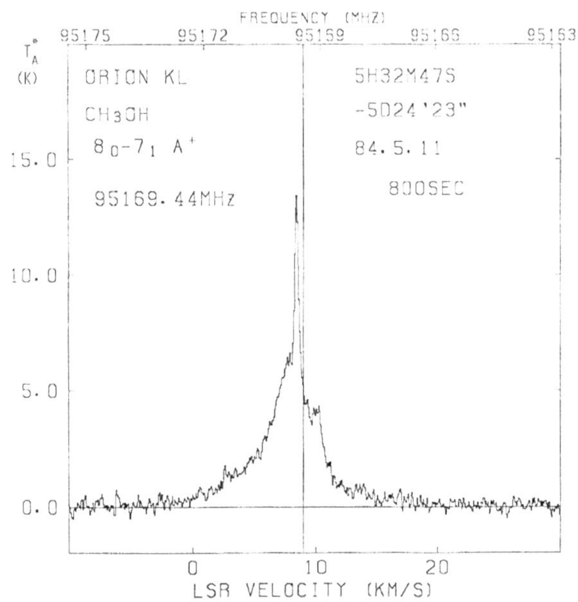

Fig. 1. Line profile of methanol 8_0–7_1 A^+ line. This spectrum was taken by AOS–H. A vertical line shows $V_{LSR} = 9$ km s^{-1}. A spike feature with $V_{LSR} = 8.5$ km s^{-1} and $\Delta V = 0.7$ km s^{-1} is the maser component.

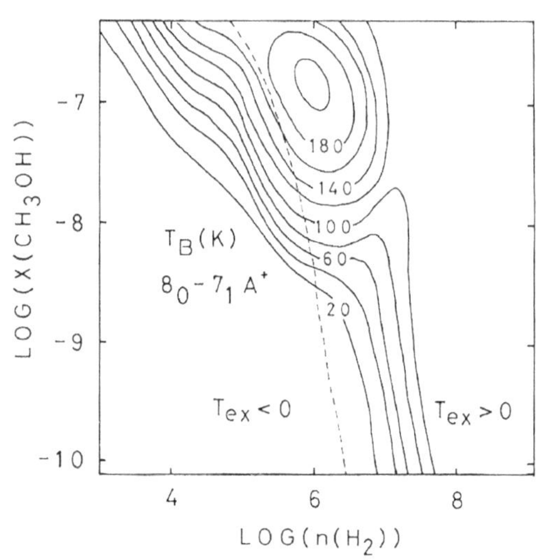

Fig. 2. Contour map of calculated brightness temperature of the 8_0–7_1 A^+ line. Parameters are $T_K = 100$ K and $dv/dr = 2$ km s^{-1} pc^{-1}. The dashed line shows $T_{ex} = \pm\infty$.

We also discovered two CH_3OH masering emissions, namely: 7_0–6_1 A^+ (44069 MHz) and 8_0–7_1 A^+ (95169 MHz) lines (Figure 1). These emissions have a linewidth of only 0.7 km s^{-1} and their spatial extents are less than 10″. These facts strongly support the view that these transitions are produced by masering. We analysed

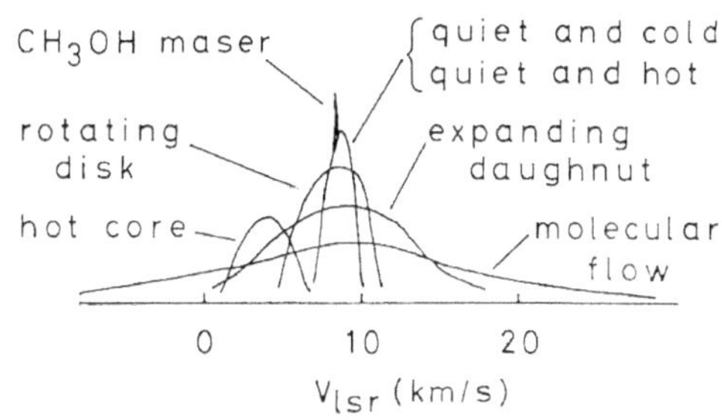

Fig. 3. Schematic presentations of velocity components toward the Orion–KL source.

these data to determine physical conditions of the masering region by comparing them with the results of our statistical equilibrium calculations (Figure 2). We estimate $T_K \sim 100$ K, $n(H_2) = 4 \times 10^5$ cm^{-3}, and $X(CH_3OH) \equiv n(CH_3OH)/n(H_2) \sim 1 \times 10^{-7}$.

From our data we can classify subregions in Ori–KL as follows:

(1) Quiet and cold region: $V_{LSR} \sim 9$ km s^{-1}, $\Delta V \sim (2\text{–}4)$ km s^{-1}, and $T \sim (20\text{–}30)$ K. This region corresponds to the emission region of the HCCCN 9 km s^{-1} component and HCO_2CH_3.

(2) Quiet and hot region: $V_{LSR} \sim 9$ km s^{-1}, $\Delta V \sim 4$ km s^{-1}, and $T \sim 100$ K. This region was previously called the 'ridge', and is observed in the CO, CS, HCN, HCO^+, and CH_3OH 9 km s^{-1} components. Our new CH_3OH masers may be associated with this subregion.

(3) Hot core: $V_{LSR} \sim (4\text{–}7)$ km s^{-1}, $\Delta V \sim (4\text{–}10)$ km s^{-1}, and $T \sim (120\text{–}160)$ K. The CH_3OH 4 km s^{-1} component, HNCO and C_2H_5CN are observed in this subregion.

(4) Molecular flow: $\Delta V \gtrsim 10$ km s^{-1}. This flow is observed in CO, HCO^+, H_2CO, and SO_2.

LARGE-SCALE STRUCTURES OF HIGH GALACTIC LATITUDE DARK CLOUDS*

YOSHIO TOMITA

Department of Astronomy, Faculty of Science, University of Kyoto, Sakyoku, Kyoto, Japan

(Received 2 July, 1985)

Abstract. The obscuring material distribution of an area of about 2000 square degrees was derived by analyzing about 130 plates, using the starcounts method.

We derived the obscuring material distribution over an area of about 2000 square degrees (covering $90° \leqq l \leqq 150°$, $+10° \leqq b \leqq +45°$). We observed this area using the 40 cm Schmidt telescope at Ohuda (Japan), with KODAK 103aD emulsions and a HOYA Y50 glass filter (colour is V band). About 130 plates have been obtained and analyzed by the starcounts method.

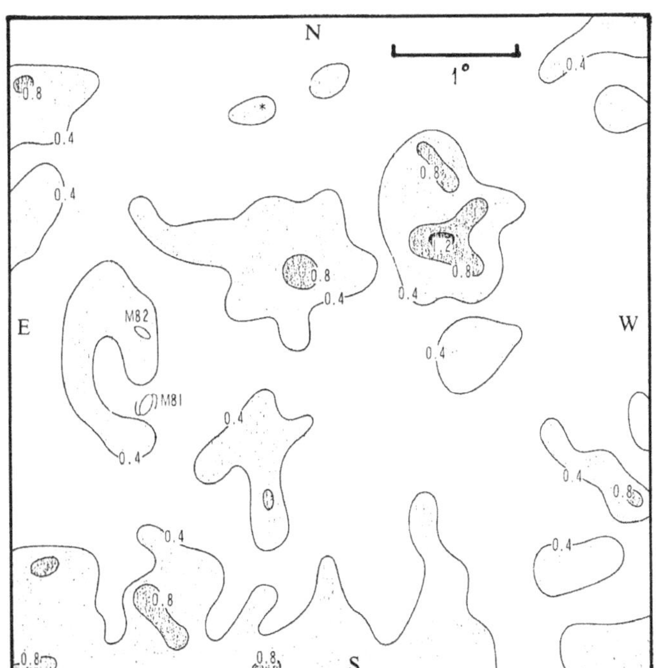

Fig. 1. Absorption map of the dark clouds in the complex *A*. Center position is RA = $9^h30^m.0$, DEC = $+70°.0$. The contours are 0.4 mag. interval in visual absorption units.

* Paper presented at the IAU Third Asian-Pacific Regional Meeting, held in Kyoto, between 30 September–6 October, 1984.

Astrophysics and Space Science **118** (1986) 409–410.

The first result is that there are three very large and diffuse dust complexes – *A*, *B*, and *C*. These are connected with each other sequentially, covering more than 50° in angular distance overall. This large-scale structures coincides well with that of the infrared 'cirrus clouds' discovered by the IRAS satellite. Figure 1 shows one of the dark clouds of the highest latitude dust complex *A* (the Messier 81/82 cloud).

The second result is some evidence of star-formation in the cloud complexes. Complex *A* does not show such evidence. On the other hand, there are small nebulosities, like Herbig–Haro objects, in complex *B*. And, the complex *C* includes A and B type stars accompanied by reflection nebulosities. Further, a remarkable feature is the linear chain of these star forming regions in complex *C*. Since these complexes may be situated near the Sun, they must provide important regions for detailed study of star-formation.

We also observed ^{12}CO and ^{13}CO molecular emission lines ($J = 1$–0) at several points of these complexes using the 4 m mm-wave telescope (Nagoya University). At a point in the Messier 81/82 cloud, the ^{12}CO antenna temperature is 1.6 K, the LSR velocity is 2.3 km s^{-1}, and the ^{13}CO column density is 7×10^{14} cm^{-2}. From this value, we derive a visual absorption $A_v = 0.3$ mag. by use of the general ratio of $N(^{13}\mathrm{CO})/A_v = 2.5 \times 10^{15}$. On the other hand the value of visual absorption at this point from starcounts is about 1 mag. From this result, we can say that diffuse high galactic latitude dark clouds may have larger dust/gas ratio than average.

A QUENCHED CARBONACEOUS COMPOSITE (QCC) GRAIN MODEL FOR THE INTERSTELLAR 220 NM EXTINCTION HUMP*

T. ONAKA, Y. NAKADA, and T. TANABE

Department of Astronomy, Faculty of Science, University of Tokyo, Bunkyo-ku, Tokyo, Japan

A. SAKATA

Laboratory of Applied Science, University of Electro-Communications, Chofu, Tokyo, Japan

and

S. WADA

Laboratory of Chemistry, University of Electro-Communications, Chofu, Tokyo, Japan

(Received 2 July, 1985)

Abstract. The dependence of the wavelength of peak absorption of dust grains on the grain size is investigated analytically by using an oscillator model for the absorption band. The peak wavelength of a weak absorption band is much less sensitive to the grain size than that of a strong band. This is explained by the fact that the surface mode, which is excited in the strong absorption band, is not raised in the weak absorption band. A quenched carbonaceous composite (QCC) synthesized from hydrocarbon plasma is found to have a weak absorption band at 220 nm. The absorption peak wavelength of the QCC grains falls well in the range of 217 ± 7 nm even if the grain size runs from 5 to 100 nm. This is compatible with the observed constancy of the 220 nm hump (217 ± 5 nm). By contrast, the absorption peak of graphite grains, which have a strong band around 280 nm and have been investigated as candidates for the hump, is very sensitive to the grain size. A quite narrow range of the grain size is required to account for the observed 220 nm feature. A weak absorption model, such as the QCC grains, is suggested to be a more likely candidate for the 220 nm extinction hump than a strong absorption model, such as graphite grains.

The 220 nm hump in the interstellar extinction curve has shown itself to be very uniform in interstellar space, and the position of peak absorption is remarkably constant (± 5 nm) (Savage, 1975; Savage and Mathis, 1979). The graphite grain model has been investigated intensively as a possible candidate to account for the hump (Stecher and Donn, 1965; Wickramasinghe and Guillaume, 1965; Donn, 1968; Gilra, 1971, 1972; Hecht, 1981). Indeed, bulk graphite has a strong absorption around 280 nm (Taft and Phillip, 1965). The surface plasmon mode for small particles shifts the absorption peak to shorter wavelengths (Gilra, 1972). However, because of high sensitivity of the surface mode to the particle size (Bohren and Huffman, 1983), the required range of particle sizes of graphite grains must be quite narrow to reproduce the constancy of the observed feature.

* Paper presented at the IAU Third Asian-Pacific Regional Meeting, held in Kyoto, Japan, between 30 September–6 October, 1984.

Astrophysics and Space Science **118** (1986) 411–413.

Recently, we have reported that a quenched carbonaceous composite (QCC), synthesized from a hydrocarbon plasma, has an intrinsic absorption feature around 220 nm (Sakata *et al.*, 1983). In this report, we investigate the dependence of the peak absorption on the grain size in a general manner, and apply the results to graphite and QCC grains for the 220 nm extinction hump.

We assume that the grain is spherical, and that the oscillator approximation is applicable to the absorption band considered. The complex dielectric constant ε at angular frequency $\omega(=2\pi c/\lambda)$ may then be written as

$$\varepsilon = 1 + \omega_p^2/(\omega_0^2 - \omega^2 + i\omega\gamma)\,,$$

where ω_p is the plasma frequency, ω_0 the peak absorption frequency, and γ the damping constant. In the case of a strong band ($\omega_p^2 \gg \gamma\omega_0$), ε usually has a region where the real part is negative, and where the surface mode can be excited (Bohren and Huffman, 1983). In this case, the absorption peak shifts to the wavelength where Real(ε) $\simeq -2$ for spherical grains. The dependence of peak frequency on the grain size can be obtained if the absorption cross-section is expanded around Real(ε) $= -2$ in terms of x, where $x = 2\pi a/\lambda$ (a, grain radius; λ, central wavelength), and is given by

$$\omega_{\text{peak}} = \omega_0[(1 + \tfrac{1}{6}(\omega_p/\omega_0)^2 - \tfrac{1}{15}(\omega_p/\omega_0)^2 x^2]\,,$$

when $x \ll 1$. The peak absorption changes with the grain size according to the term $\frac{1}{15}(\omega_p/\omega_0)^2 x^2$.

On the other hand, for a weak absorption band where the surface mode is not excited ($\omega_p^2 \ll \gamma\omega_0$), the cross-section should be expanded around $\omega \simeq \omega_0$ in terms of $x(\ll 1)$. The peak frequency is then obtained as

$$\omega_{\text{peak}} = \omega_0[1 + \tfrac{1}{6}(\omega_p/\omega_0)^2 - \tfrac{22}{75}(\omega_p/\omega_0)^2 x^2]\,.$$

For the 220 nm absorption, the dielectric constant of graphite can be approximated by the set of parameters; $\omega_0 = 4.4$ eV, $\omega_p = 7.9$ eV, and $\gamma = 1.35$ eV, when we fit the data of Taft and Phillip (1965) with the oscillator model. Then the peak shift due to grain size $\Delta\lambda$ is $60x^2$ nm. For QCC, the dielectric constant was obtained from the transmission spectrum (Sakata *et al.*, 1983) and the measurement of the thickness of the sample by means of the Kramers–Kronig relation. It was found to be approximated by parameters, such as $\omega_0 = 5.5$ eV, $\omega_p = 2.6$ eV, and $\gamma = 2.0$ eV, and turned out to be a weak absorption. The wavelength shift is given by $14x^2$ nm.

A Mie calculation shows that this analytical expression for the absorption peak is a good approximation when x is small (Figure 1). As seen from Figure 1, a possible range of the absorption peak wavelength of the QCC grains falls well around the observed peak wavelengths for the wide range of grain radius. By contrast, the size of graphite grains must be restricted in a quite narrow range in order to keep the peak wavelength within the observed values.

These results suggest that a weak absorption band model, such as QCC, is a more likely candidate for the 220 nm extinction hump than a strong absorption model, such as graphite, supporting an indication from the formation condition that material like

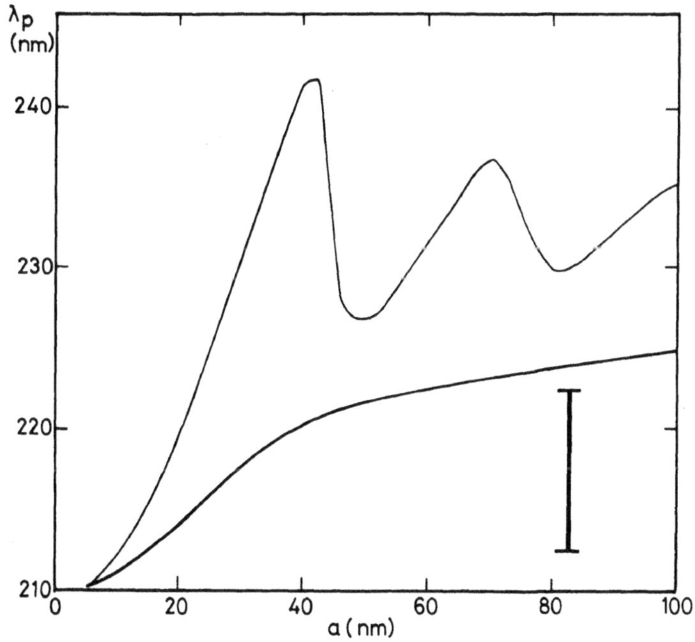

Fig. 1. The dependence of peak absorption wavelength λ_p on the grain size a for graphite grains (thin line) and for QCC grains (bold line). The observed range of the peak wavelength of the 220 nm hump (Savage, 1975) is also plotted in the lower corner by a straight line.

QCC is a more likely carbonaceous substance in space than graphite (Sakata *et al.*, 1983).

References

Bohren, C. F. and Huffman, D. R.: 1983, *Absorption and Scattering of Light by Small Particles*, Wiley, New York.

Donn, B.: 1968, *Astrophys. J.* **152**, L129.

Gilra, D. P.: 1971, *Nature* **229**, 237.

Gilra, D. P.: 1972, in A. D. Code (ed.), *The Scientific Results from the Orbiting Astronomical Observatory OAO-2*, NASA SP-301, p. 295.

Hecht, J.: 1981, *Astrophys. J.* **246**, 794.

Sakata, A., Wada, S., Okutsu, Y., Shintani, H., and Nakada, Y.: 1983, *Nature* **301**, 493.

Savage, B. D.: 1975, *Astrophys. J.* **199**, 92.

Savage, B. D. and Mathis, J. S.: 1979, *Ann. Rev. Astron. Astrophys.* **17**, 73.

Stecher, T. P. and Donn, B.: 1965, *Astrophys. J.* **142**, 1681.

Taft, E. A. and Phillip, H. R.: 1965, *Phys. Rev.* **A138**, 197.

Wickramasinghe, N. C. and Guillaume, C.: 1965, *Nature* **207**, 366.

PHOTOGRAPHIC AND SPECTROGRAPHIC OBSERVATIONS WITH A REDUCING CAMERA ON THE 61 CM NCU TELESCOPE

HSIN HENG WU

Institute of Physics and Astronomy, National Central University, Chung-Li, Taiwan, China

(Received 2 July, 1985)

Abstract. A Reducing Camera designed by Meinel has been constructed and tested on the 61 cm NCU Cassegrain telescope. The basic principle of RC is to reduce the reflector's *f*/13.5 to the equivalent *f*/2.7 such that the Cassegrain focus image can be reduced on a 35 mm film with almost no loss in sky coverage and in resolution but the exposure time is reduced by a factor of 25. The camera is able to obtain stellar spectra. With a grism and a standard lens the dispersion is 450 Å mm^{-1}, which is comparable to that required for the MK classification. RC constructure and testing results are reported. The method of a wide field spectrograph and that of narrow band differential photography of RC are also discussed.

1. Introduction

The concept of a reducing camera (RC) was developed in 1956 by Meinel (1956), in close collaboration with Morgan *et al.* (1953) at the Yerkes Observatory. The original goal was to reduce the 40 inch *f*/18 refractor to the equivalent of an *f*/3 prime-focus instrument in order to enable classification of the nuclear structure of galaxies and to explore the frequency of multiplicity of OB stars.

The idea of using an RC on the NCU 61 cm telescope was first suggested by Meinel during first visit to NCU in late 1978. The designing work of the RC for this particular 61 cm *f*/13.5 telescope was afterwards carried out.

The construction of the RC by P.I.D.C.–NSC dates from June 1980, and its major part was finished near the end of 1980. During this period many technical problems were solved by Dr and Mrs Meinel on their second visit to NCU in October 1980. Various tests were then performed and the results are reported as follows.

The purpose of this paper is to inform other astronomers who may be similar to us in the early developing stage of their observatory program, with the idea of this very versatile instrument.

2. Theory

The resolution of an astronomical observation is limited by the following four factors.

(a) The theoretical diffraction limit of the primary mirror: in the case of the 61 cm telescope this is 0.19 arc sec.

* Paper presented at the IAU Third Asian-Pacific Regional Meeting, held in Kyoto, Japan, between 30 September–6 October, 1984.

Astrophysics and Space Science **118** (1986) 415–418.

(b) The engineering limit due to the optical abberations of the mirrors and the mechanical mounting and driving of the telescope: the manufacture's specification is about 1 arc sec for this telescope.

(c) The atmospheric seeing: under typical observing condition the seeing disk of a point source (star) appears to be 1 to 4 arc sec in diameter.

(d) Resolving power of the detecting device: the resolving power of the most popular films used for astronomical purposes are (Kodak Publ. 315, 1973)

Kodak plates and films	Resolving power (line pairs/mm)
103a–F	80
IIa–F	100
IIIa–J	200
2415	320

Originally designated SO–115.

The Cassegrain focal length of the telescope is 8.2 m. If we take the type IIa–F film as an example then the resolving power turns out to be 0.25 arc sec.

It is quite clear that we need not consider the diffraction or engineering limits since they are smaller than the seeing disk, while this is about 5 to 12 times larger (in diameter) than the smallest angle the film can resolve at the Cassegrain focal plane. The concept of the reducing camera is to reduce the image at the Cassegrain focal plane by a factor of 5 and change the reflector's *f*/13.5 to the equivalent *f*/2.7.

We would not lose any practical resolution by making this reduction, only the linear image size is reduced by a factor of 5 and the intensity per unit area on the film is increased to 25 times that on the Cassegrain focus. So both the exposure time and the film size can be greatly reduced.

3. Observation Apparatus

The NCU 61 cm telescope installed in 1980 is a Cassegrain reflector manufactured by the Boller and Chivens. The mirror is Cervit and optical image quality is 80% in 0.5 arc sec. The focal ratio is *f*/13.5, with field size of 100 mm (42 arc min, flat field of view). The reducing camera consists of two main parts (Figure 1) (Meinel, 1980).

3.1. Star Guider Assembly

The star guider assembly (shown in the left part of Figure 1) attaches to the Cassegrain focus position of the NCU 61 cm telescope. The entire assembly can be rotated to any desired position. The eyepiece assembly can slide laterally and rack in and out permitting a rectangular field to be scanned until a guiding star is found.

3.2. Focal Reducer Assembly

The focal reducer assembly is shown in the center part of Figure 1. It consists of three subsystems: field lens, collimating lens, and the grism subassembly. The field lens is a plano convex lens of BK-7 glass with 10 cm diameter, 14 cm radius of curvature and about 28 cm focal length.

The term 'grism' denotes the combination of transmission grating and prims (Carpenter, 1957). The combination yields a dispersion in wavelength but no net deviation of the beam at 4800 Å. The grating is also blazed at about this wavelength for first order in transmission.

The collimating lens is a commercial *f*/4.5 lens selected to have the correct field image properties. When the camera is used in direct and filter photograph mode, the grism is replaced by a filter and the collimating lens is not needed. The filter slide unit has one clear opening and two filter holds for 50 × 50 mm filters.

3.3. Camera platform

The camera platform is shown at the right end of Figure 1. A 35 mm camera can be attached from here to the grism mounting plate or filter slide unit. The camera lens is a standard *f*/1.4 focal length lens of 35 SLR NIKON FM camera.

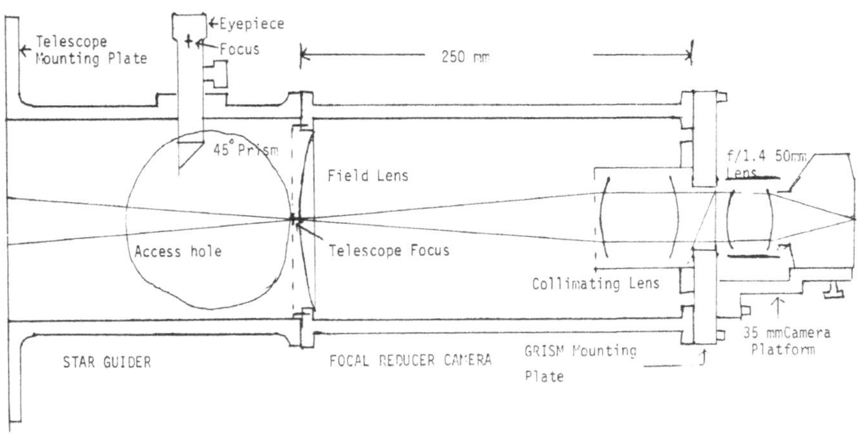

Fig. 1. Reducing camera.

4. Observations and Results

To test the direct photograph of the RC we chose the well-known Orion nebula (M42) as our object. Plus-X film was used. The best result was with a 1 min exposure. For shorter exposures, only the central stars and not the full shape of the nebula is shown. For longer exposures, the sky background gets 'foggy' and some of the faint structures of the nebula are washed out by the background.

For resulution, we compared the image diameters of the point source (stars) taken with the RC and a normal Cassegrain arrangement. We found that the resolution was not limited by film granularity but by the seeing disk.

For the spectrograph observation Kodak Tri-X PAN film was used. The spectra were dispersed in the east–west direction and widened by a north–south guide scan.

The stellar spectra we obtained cover the wavelength from 3600 to 5650 Å. Very clear hydrogen absorption lines were obtained on a spectrogram of Sirius (A1 V).

5. Other Proposed Research for the Reducing Camera

The results of prelimary tests of the RC are satisfactory both in direct photograph and objective 'grism' spectrograph modes. The research we propose using the RC concerns wide-field spectrography and narrow-band differential photography.

The ideal of wide-field spectrography is to photograph spectra of several stars in the same field simultaneously. Due to the presence of an intermediate focal plane (the Cassegrain focus) in the RC optical system, a special mask can be used in this plane to eliminate the sky background and the crowding of the star field. Appropriate holes can be drilled on such a mask.

The use of narrow passband filters to obtain direct photographs of emission objects and regions is a prime capability of the Reducing Camera. Several filter wavelengths will be selected, such as the strong lines in T Tauri stars or similar classes of objects.

Acknowledgements

The author should like to express his deep gratitude to Professor A. B. Meinel and Mrs M. P. Meinel. The work was supported in part by the National Science Council, Taipei, China through grant NSC 70–02040M008–01.

References

Carpenter, E. F.: 1957, *Publ. Astron. Soc. Pacific* **69**.
Meinel, A. B.: 1956, *Astrophys. J.* **124**, 652.
Meinel, A. B.: 1980, *Construction Notes on NCU Reducing Camera*, Arizona Scientific Research, 10121 Catalina Highway, Tucson, Az. 85715.
Morgan, W. W., Meinel, A. B., and Johnson, H. M.: 1953, *Astrophys. J.* **120**, 506.
Kodak Publication No. P-315: 1973, *Kodak Plates and Films for Scientific Photography*.

STAR COUNT OF THE NORTH GALACTIC POLE REGION IN THE *UBV* COLOUR BANDS*

R. S. STOBIE

Royal Observatory Edinburgh, Scotland, U.K.

K. ISHIDA and Y. YOSHII

Tokyo Astronomical Observatory, Japan

and

H. T. MACGILLIVRAY

Royal Observatory Edinburgh, Scotland, U.K.

(Received 2 July, 1985)

Star count data in the magnitude range $12 \leq V \leq 18$ covering 24 square degrees of the North Galactic Pole region have been obtained in the *UBV* colour bands. The photographic plates were taken with the Kiso Schmidt telescope, were subsequently scanned with the COSMOS machine, and were reduced at the Royal Observatory, Edinburgh. The merging of different plates successfully removed 'noise' images. Non-stellar images were segregated by an empirical relation between the magnitude and the area of an image. The stellar magnitudes were calibrated with the aid of photometric sequences, and were adopted only in those parts where very narrow calibration relations are obtained. The observed two-colour diagrams for each apparent magnitude bin were simulated by simple galaxy models. The present study places a new constraint on the galaxy model, especially in relation to the metallicity of stars by the ultraviolet excess on the two-colour diagram. The observed colour frequency distributions are well simulated by a slow metallicity gradient from the disk plane: $[Fe/H] = -0.5z$ for $0 \leq z \leq 0.56$ kpc, $[Fe/H] = -0.2-0.14z$ for $0.56 \leq z \leq 7.14$ kpc, and $[Fe/H] = -1.2$ for $7.14 \text{ kpc} \leq z$.

* Paper presented at the IAU Third Asian-Pacific Regional Meeting, held in Kyoto, Japan, between 30 September–6 October, 1984.

Astrophysics and Space Science **118** (1986) 419–420.

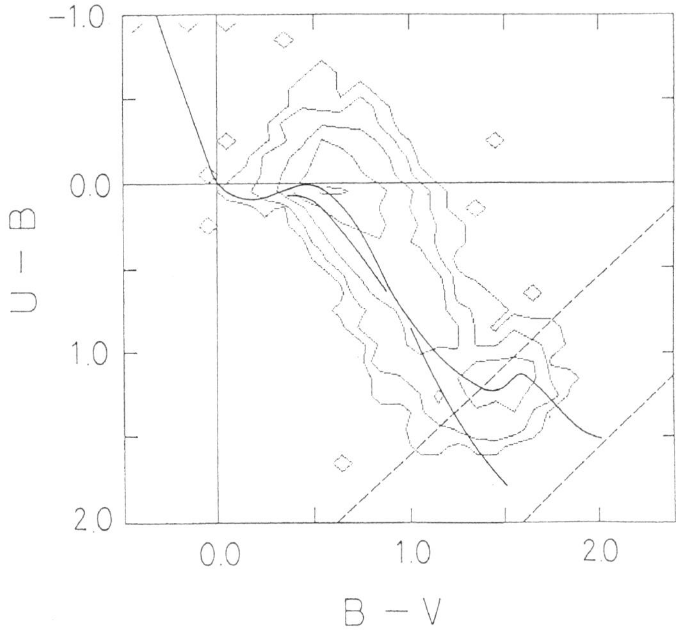

Fig. 1. An example of the observed two-colour diagrams. It shows the density function $A(V, (B-V), (U-B))$, which gives the numbers of stars counted for $15 \leq V \leq 16$ for each mesh of $d(B-V) = d(U-B) = 0.1$, on the $(B-V)$–$(U-B)$ two-colour domain. The equidensity contour lines stand for $A(V, (B-V), (U-B)) = 1, 3, 9, 27$, and 81. One of the curved lines crossing the origin stands for the Main-Sequence stars with [Fe/H] = 0.0 – the logarithmic metallicity in units of the solar abundance. Another one stands for giant stars with the same abundance. The two dotted straight lines show the limit of B and U mag corresponding to $V = 15$ and 16, respectively, and stars in the lower right part are missing. The following features are seen from Figure 1: (1) The overall contour lines envelope the two curved lines of AFGKM star sequences. (2) Some fraction of the stars are of too low metallicity to show ultraviolet excess. (3) There is a sharp peak ($A(V, (B-V), (U-B)) > 81$) of G-type stars. (4) There is another small peak ($A(V, (B-V), (U-B)) > 9$) of M-type Main-Sequence stars. The above mentioned features are seen in every magnitude range over $12 \leq V \leq 18$, and notably the sharp peak of G-type stars stands out nearly at the same colour range of $0.5 \leq (B-V) \leq 0.7$ and $0.0 \leq (U-B) \leq 0.1$ for each magnitude.

STAR FORMATION ASSOCIATED WITH HIGH-VELOCITY MASS OUTFLOWS (INVITED REVIEW)*

TETSUO HASEGAWA

*Nobeyama Radio Observatory**, Tokyo Astronomical Observatory, University of Tokyo, Nobeyama, Minamimaku, Nagano, Japan*

(Received 3 July, 1985)

Abstract. Energetic mass outflows have been detected in molecular line observations towards young stellar objects. In this review we take the Orion–KL as an example to discuss the overall structure of a high-velocity outflow and its environment. The kinematics of the high-velocity molecular emission show clear evidence of a bipolar jet which originates in the vicinity of IRc2, a massive protostar. Towards the ends of the jet, 0.05 pc away from the origin of the flow, the interaction between the high-velocity flow and the ambient molecular gas excites shocks. The protostar is encircled by a disc of dense molecular gas, the inner 0.04 pc of which is expanding while the outer part shows signs of rotation and contraction. A comparison between the dynamical timescales of the disk and the bipolar jet may suggest that the disk itself, or some mechanism of disk formation, is also responsible for the bipolar nature of the high-velocity flow.

1. The Outflow from IRc2

Of the regions of star formation associated with high-velocity flows, the Orion–BN/KL region has been the object of the most intense investigation (see also the review by Genzel and Downes 1982). The molecular 'plateau' emission was first recognized in the spectra of the 2.6 mm $J = 1–0$ line of CO (Zuckerman *et al.*, 1976; Kwan and Scoville, 1976). Subsequent observations of the $J = 2–1$, $J = 3–2$, and $J = 6–5$ lines of CO, and other molecular lines have been made (e.g., Wannier and Phillips, 1977; Waters *et al.*, 1980). Figure 1 shows a spectrum of the $J = 1–0$ line of CO from recent observations with the 45 m telescope of Nobeyama Radio Observatory (Hasegawa *et al.*, 1985). The line profile shows a broad wing of emission extending out to ± 90 km s^{-1} from the line center. The spatial resolution of the early molecular line observations was inadequate to resolve the source of the high velocity wing, so that they gave upper limits to the source size ranging from 25″ to 60″ (e.g., Knapp *et al.*, 1981; Phillips *et al.*, 1977). Several models for the kinematics of the high-velocity gas sampled by the wing emission have been tested against observations; gravitational collapse (Kuiper *et al.*, 1978), some explosive event – like a supernova (Kwan and Scoville, 1976), a rotating cloud (Clark *et al.*, 1979), and stellar winds (Zuckermann *et al.*, 1976). Each of the models has its own difficulties, and there was no decisive observational evidence to cause any one of them to be preferred at that time.

* Paper presented at the IAU Third Asian-Pacific Regional Meeting, held in Kyoto, Japan, between 30 September–6 October, 1984.

** NRO, a branch of the Tokyo Astronomical Observatory, is a cosmic-radio observing facility open to outside users.

Astrophysics and Space Science **118** (1986) 421–434.

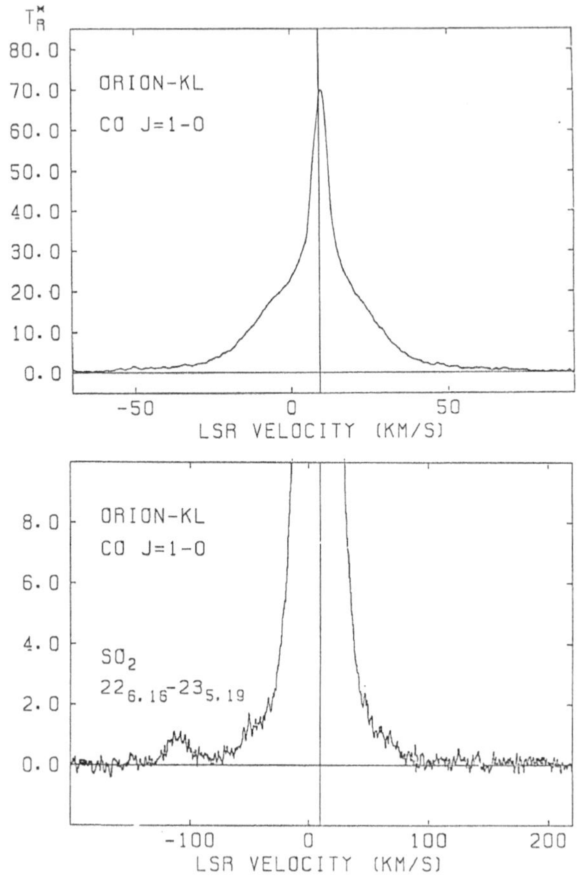

Fig. 1. The spectrum of the $J = 1$–0 CO emission observed toward IRc2 (Hasegawa *et al.*, 1985). The lower panel shows the same spectrum as the upper panel in a larger velocity range with an expanded vertical scale. The high-velocity emission is visible up to ± 90 km s^{-1} from the line center.

H_2O maser emission at 1.35 cm, covering a large velocity range, has been observed from the BN/KL cluster (Figure 2). The maser emission comes from a number of bright spots scattered over 60″ around the core of the infrared cluster. Genzel *et al.* (1981) have conducted a series of VLBI experiments to measure the proper motions of these maser spots. They found transverse motions of typically a few milli-arcsec per year, which are interpreted as kinematic motions of compact cloudlets of masering gas. The data suggest that the cloudlets are moving away from a common origin situated within 5″ of the compact infrared source IRc2. The velocity range and spatial extent of the H_2O masers are similar to those of the high-velocity CO emission, so that it is conceivable that the H_2O masers and the 'plateau' feature have the same kinematics; i.e., an outflow from an origin in the vicinity of IRc2.

IRc2 has been recognized by Rieke *et al.* (1973) as a bulge in the extended contours adjacent to stronger compact peaks on the 21 μm map. Downes *et al.* (1981) have made 8.7 μm and 20 μm maps with higher resolution (Figure 3) supplemented with spectrophotometric measurements in the range 4.8–20 μm. They revealed that IRc2 has a very

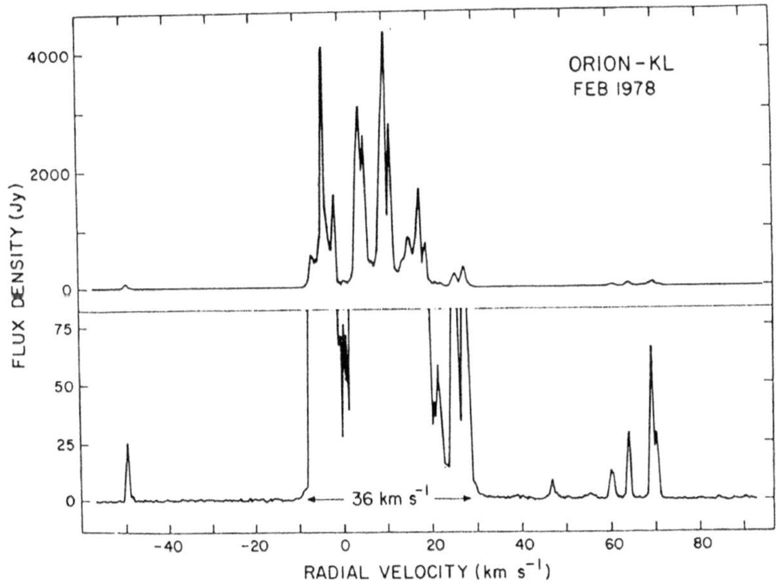

Fig. 2. The spectrum of the H_2O maser emission observed toward the Orion–KL (from Genzel *et al.*, 1981).

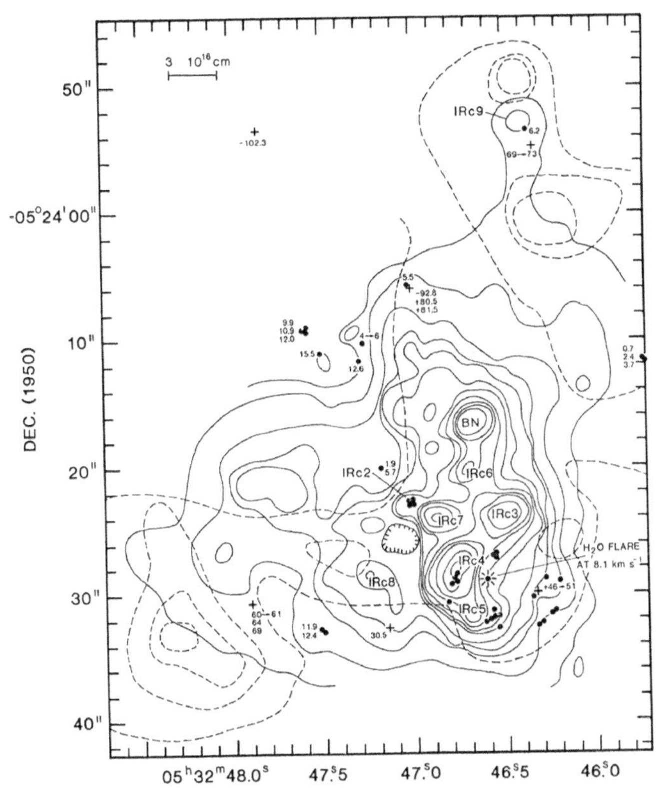

Fig. 3. Locations of H_2O masers in the KL nebula superposed on the 20 μm infrared contours (from Downes *et al.*, 1981). Labels are V_{LSR} in km s^{-1}. Dashed contours are those of the v = 1–0 S(1) transition of molecular hydrogen.

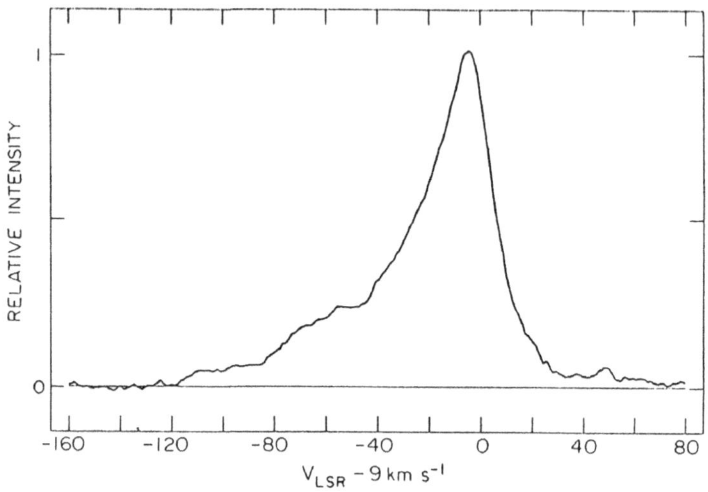

Fig. 4. The averaged spectrum of the v = 1–0 S(1) emission of H_2 observed toward the northern peak. The effects of the instrumental profile have been removed (from Nadeau and Geballe, 1979).

deep silicate feature, indicative of a very large extinction. The other infrared sources in the BN/KL region suffer from less extinction, so that IRc2, with an estimated luminosity of 10^5 $L_{\odot}$ is the most luminous object in this region. A polarimetric map at 3.8 μm, made by Werner *et al.* (1983), has demonstrated that the extended infrared nebula is due to scattered light emitted by two namely dominant sources BN and IRc2. From the maps of colour temperatures for 8–12.5 μm, 12.5–20 μm, and 20–30 μm, Wynn-Williams *et al.* (1984) have suggested that the geometry of the KL nebula is that of a clumpy cavity centered on IRc2, the source of essentially all the luminosity of the region and the most probable energy source of the high-velocity outflow. The peaks in the KL nebula other then BN and IRc9 may be irregularities in the material at the edge and surrounding the cavity, rather than individual self-luminous sources.

2. Shocked Molecular Gas

Other evidence for the high-velocity outflow has come from observations of infrared line emission from shocked molecular hydrogen (Shull and Beckwith, 1982; and references given therein). Gautier *et al.* (1976) pointed out a possibility that the H_2 molecules are excited by shock waves in their report on the initial discovery of the v = 1–0 vibration-rotation lines of H_2. Beckwith *et al.* (1978) made maps of the He_2 v = 1–0 S(1) emission, which show two dominant peaks 20″ (0.05 pc) northwest and southeast of the BN/KL region (Figure 5d). The line intensity ratios of the H_2 lines suggest that the level populations are characterized by a distribution of temperatures ranging from 1000 K for the lowest purely rotational levels to 3000 K for the higher vibrational levels (Beckwith *et al.*, 1983). The profiles of the 2.1 μm v = 1–0 S(1) line show prominent blueshifted

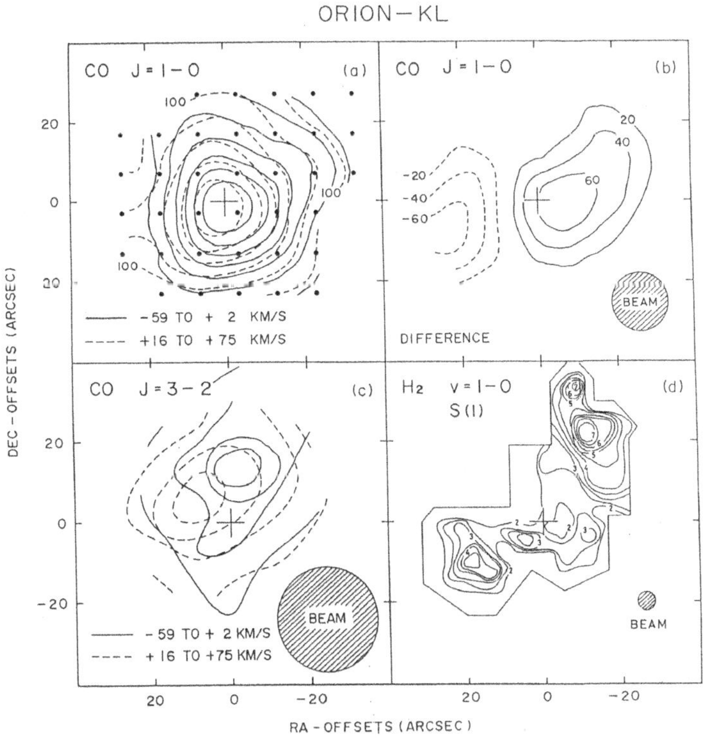

Fig. 5. (a) The distribution of the high-velocity $J = 1\text{–}0$ emission in the Orion–KL observed with the 45 m telescope at Nobeyama (Hasegawa *et al.*, 1985). The dots indicate observed positions and the positional offsets are taken from IRc2 indicated by the cross; R.A. = $5^h32^m47.0^s$, dec. = $-5°24'23''$ (1950.0). The integrated intensities are measured over the same velocity intervals as those of panel (c). Contours are at 50, 100, 150, 200, 250, 300, and 350 K km s^{-1} of T_R dV. (b) The asymmetry of the $J = 1\text{–}0$ CO line profile. The difference is calculated from the data of panel (a) by subtracting the intensity of the redshifted wing from that of the blueshifted wing. Contour labels are in K km s^{-1}. (c) The distribution of the high-velocity $J = 3\text{–}2$ emission observed by Erickson *et al.* (1982). (d) The distribution of the $v = 1\text{–}0$ S(1) emission of H_2 observed by Beckwith *et al.* (1978). Contour intervals are 6.1×10^4 ergs s^{-1} cm^{-2} sr^{-1}. The extent of the area mapped is enclosed by the straight-line boundary.

wings extending up to 100 km s^{-1} from the line center (Figure 4, Nadeau and Geballe, 1979; Scoville *et al.*, 1982). The asymmetry of the line profile is interpreted as due to differential extinction in a radially expanding source (Nadeau *et al.*, 1982), and indeed the spectrum of the 12.3 μm $v = 0\text{–}0$ S(2) line, which may be almost free from internal extinction, has a symmetric profile with a FWHM of 100 km s^{-1} centered near $V_{LSR} = 0$ km s^{-1} in both the NW and SE peaks (Beck, 1984).

The $J = 21\text{–}20$, $J = 27\text{–}26$, and $J = 30\text{–}29$ rotational transitions of CO have been detected from the region of H_2 emission (Watson *et al.*, 1980; Storey *et al.*, 1981). Also the 63 μm fine structure line of O I (Werner *et al.*, 1984) and the hydrogen recombination line H51α at 6 mm (Hasegawa and Akabane, 1984) have been observed, both of which may show characteristics of shocks.

Models of the shock have been constructed to help us to understand physical processes and energetics in the region of the H_2 emission. Earlier models assume that the magnetic field in the preshock molecular cloud is unimportant in determining possible sets of physical parameters for the shock (Kwan, 1977; Hollenbach and Shull, 1977; London *et al.*, 1977). In the absence of significant preshock magnetic field, a shock travelling into a molecular cloud with a minimum preshock particle density of 10^5 cm^{-3} at v_s = 10–24 km s^{-1} can excite the observed H_2 molecules, although it cannot be made to match the line intensities quantitatively. More recent studies recognize the role of a preshock magnetic field of the order of 1 milligauss. Provided that the ionization degree of the preshock gas is that of a typical molecular cloud (i.e., x_i less than 10^{-7}), the shock transition is fully mediated by collisions of neutral particles with streaming ions and dust grains accelerated in the magnetic field ('C-type' shock; cf. Draine, 1980). Models of a magnetohydrodynamic shock seem to succeed in explaining the observed emission from H_2 and other molecular and atomic species with certain physical parameters; the pre-shock density n_0 = (2–7) × 10^5 cm^{-3}, the shock velocity v_s = 36–38 km s^{-1}, and the transverse magnetic field B = 0.45–1.5 milligauss (Draine and Roberge, 1982; Chernoff *et al.*, 1982). However, the velocity range covered by the observed wing emission far exceeds that expected by shock models. Density inhomogeneities in the high-velocity wind and/or scattering off moving dust grains may have to be invoked to understand the line profiles (e.g., Nadeau *et al.*, 1982).

3. The Bipolar Flow

The anisotropy of the distribution of the high-velocity molecular gas was first demonstrated clearly by Erickson *et al.* (1982), who mapped the J = 3–2 CO emission with a 26″ beam using the Multiple Mirror Telescope as a phase coherent array (Figure 5c). The peaks in the emission strength of the redshifted and blueshifted high-velocity CO are offset with the redshifted emission peaking east and blueshifted emission peaking north of IRc2. Similar trends have been recognized in observations of the millimeter and sub-millimeter lines of CO, SiO, HCO^+, and SO (Phillips *et al.*, 1982; Olofsson *et al.*, 1981; Olofsson *et al.*, 1982; Friberg, 1984; Snell *et al.*, 1984) from single dish observations with similar or larger beamwidth.

Recent interferometric observations have presented a more detailed picture of the high-velocity flow. Maps of the v = 0, J = 2–1 SiO emission made at the millimeter-wave interferometer at Hat Creek delineates the denser part of the high-velocity gas with 6″ resolution (Wright *et al.*, 1983). The flow is bipolar and originates within a few arc sec of IRc2. The HCO^+ maps with 15″ resolution, made using the same interferometer, probe the interaction of an outflow from IRc2 with the quiescent gas (Vogel *et al.*, 1984). Masson *et al.* (1984) present results of a CO experiment made at the Owens Valley Millimeter Interferometer, which again suggests displacement between the intensity maxima of the redshifted and blueshifted emission.

The J = 1–0 CO emission has been observed using the 45 m telescope at Nobeyama (Figures 1 and 5; Hasegawa *et al.*, 1985). The beamwidth of 16″ coupled with a typical pointing accuracy of better than 3″ resulted in CO maps with the highest resolution. Figure 5(a) shows the distribution of the high-velocity emission integrated over the same velocity intervals as those of the $J = 3$–2 map by Erickson *et al.* (1982; Figure 5(c)). The $J = 1$–0 emission comes from a circular area centered on IRc2 and has a deconvolved size of 30″ (FWHM). The offset of the emission centroid 10″ north of IRc2, suggested by the former CO observations (e.g., Figure 5(c)), is not apparent. There is a difference between the distributions of the redshifted and blueshifted $J = 1$–0 emission, although it is not so clear as in the case of the SiO emission observed by Wright *et al.* (1983). Figure 5(b) shows this by subtracting the integrated intensity of the redshifted emission from that of the blueshifted emission. The region of excess blueshifted wing extends from IRc2 to 30″ northwest, while the excess of redshifted wing is observed 25″ east to southeast of IRc2 and extends farther. This distribution combines with that of the high-velocity SiO and HCO^+ emission observed at Hat Creek to make a strong point that there are a pair of well collimated jets of molecular gas flowing in the opposite direction from a common origin, most probably IRc2. However, the fraction of the CO emission showing kinematic properties of the bipolar jet is relatively small in the total intensity of high-velocity CO emission. Indeed, only 20% is sampled in the difference map of Figure 5(b). The remaining 80% of the high-velocity emission show no clear sign of asymmetry; this component may be interpreted as the expanding disk which is discussed in the next subsection.

The most reliable estimates of the mass, momentum, and kinetic energy of the high-velocity molecular flow have been given by Snell *et al.* (1984), who observed the $J = 1$–0 and $J = 2$–1 transition of CO and ^{13}CO. The high-velocity molecular gas amounts to 8 $M_\odot$ and its momentum and kinetic energy are 160 $M_\odot$ km s^{-1} and 4×10^{46} ergs, respectively. The dynamical timescale of 1.5×10^3 yr has been estimated from the separation of intensity maxima of the redshifted and blueshifted wings and the mean radial velocity. If the dynamical activity has been steady over this time-scale, the mechanical luminosity can be estimated as 230 $L_\odot$, which is about 0.2% of the bolometric luminosity of IRc2. The estimated momentum supply rate of 0.11 $M_\odot$ yr^{-1} km s^{-1}, however, far exceeds $L/c = 0.002$ $M_\odot$ yr^{-1} km s^{-1}, the momentum supply expected for single scattering of photons. Figures 6 and 7 show statistics of twenty sources with the luminosities of the young stellar objects ranging from 10 to 10^5 $L_\odot$. There are good correlations between the mechanical luminosity, or the momentum supply rate, and the stellar luminosity. The momentum supply required to continuously drive the flow far exceeds that supplied by photons suffering single scattering: the molecular flow should be driven in other ways, or the photons have to be scattered 10–100 times before they lose most of their energy (Bally and Lada, 1983).

4. The Disk Around IRc2

The existence of an expanding ring of dense molecular gas has been suggested by various observations. Plambeck *et al.* (1982) presented maps of the 86 GHz, 2_2–1_1 SO emission observed with an angular resolution of 6″ using the Hat Creek interferometer. They

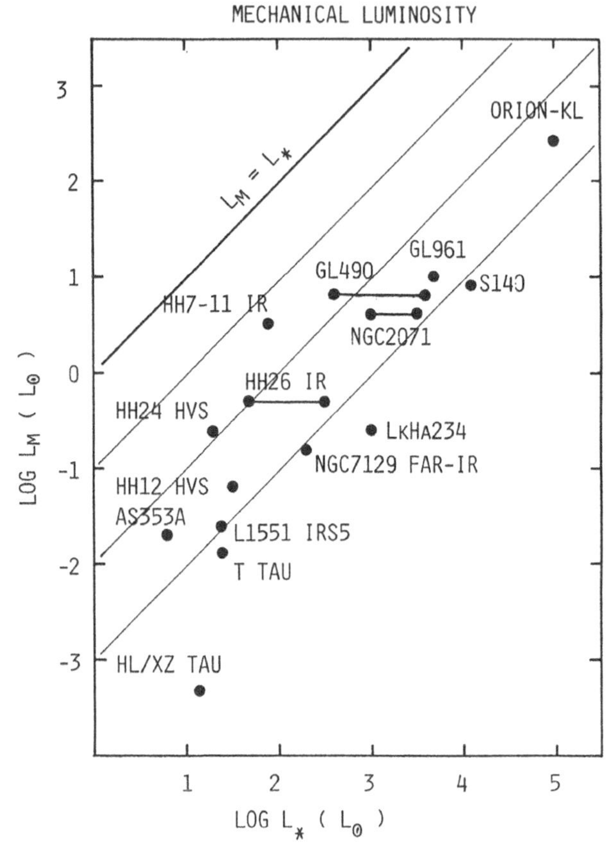

Fig. 6. The mechanical luminosity of the molecular flow compared with the luminosity of the young stellar object supposed to be driving the flow. The data are compiled from Snell *et al.* (1984) and Edwards and Snell (1983). The thick slanted line indicates $L_M = L_*$ and the thin lines are at $\frac{1}{10}$, $\frac{1}{100}$, and $\frac{1}{1000}$ of L_*.

show that the emission extending ± 20 km s^{-1} from the line center arises from a region of 19″ × 11″ (FWHM), centered within a few arc sec of IRc2 and is elongated in a northeast–southwest direction, perpendicular to the two lobes of the H_2 emission. A series of equivelocity maps of the SO emission show evidence for a systematic velocity structure: emission to the northwest of IRc2 tends to be redshifted, while emission to the southeast is blueshifted. The velocity shifts are consistent with a model of an expanding 'doughnut' of gas centered near IRc2.

More evidence for the expanding ring comes from the H_2O masers. Most of the maser spots within ± 20 km s^{-1} from $V_{LSR} = 9$ km s^{-1} are distributed to form a belt 30″ in length with a position angle of 30° (Figure 3). The three-dimensional velocity field implied by the VLBI experiments is that of an expansion at a constant velocity of 18 km s^{-1} from a common origin near IRc2 (Genzel *et al.*, 1981). The maser spots moving outward may be distributed in a ring seen nearly edge-on. The CO observations at Nobeyama show that the H_2O masers and the CO emission have similar spatial extent and velocity range, and it is plausible that the maser spots are dense (10^{9-10} cm^{-3};

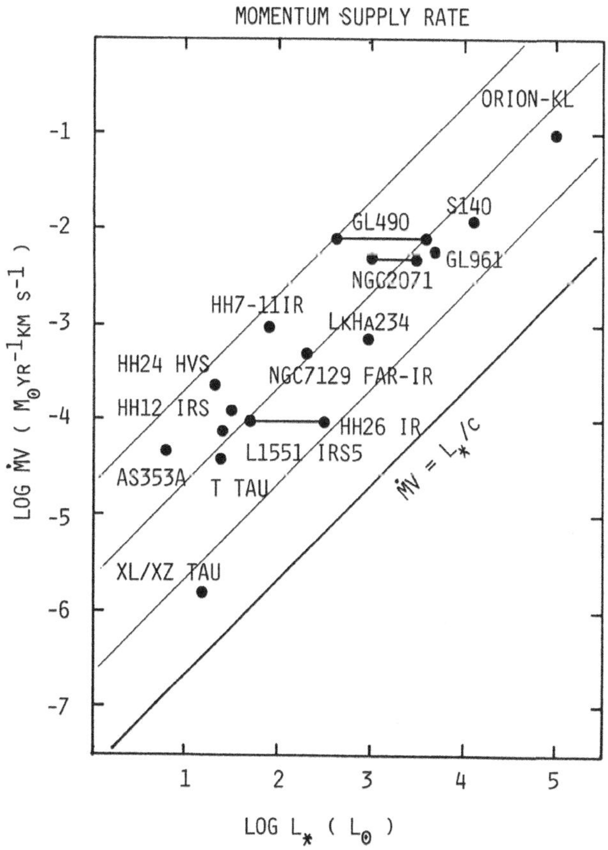

Fig. 7. The momentum supply rate compared with the stellar luminosity. The source of data are the same as Figure 6. The thick slanted line indicates the rate supplied via single scattering of photons from the star and thin lines are at × 10, × 100, and × 1000 of that rate.

Downes, 1984) clumps in the flow of less dense molecular gas sampled by the CO emission.

A more compact elongated feature appears in the maps of $v = 1$, $J = 2\text{–}1$ SiO emission (Wright *et al.*, 1983). Again within ± 20 km s^{-1} from $V_{LSR} = 9$ km s^{-1}, the emission has a deconvolved FWHM size of $10'' \times 5''$ and is elongated at a position angle of 65°. The mean density of this feature is estimated as $(0.3\text{–}3) \times 10^6$ cm^{-3}. OH masers at 1612 and 1665 MHz are observed with a similar spatial extent; $8'' \times 3''$ at a position angle of 70° (Hansen and Johnston, 1983; Norris, 1984). The OH masers show a systematic velocity change which may be consistent with a model that the stronger OH emission arises from a torus rotating about IRc2. It is possible that the SiO and OH emission samples the inner part of the expanding molecular disk compared to SO, H_2O, and CO. In that case of the position angle of elongation, from 30° for the outer part (30″ in diameter) to 65°–70° for the inner part (10″ in diameter), could be due to a warping of the inner disk, or to an optical depth effect in a systematic velocity field.

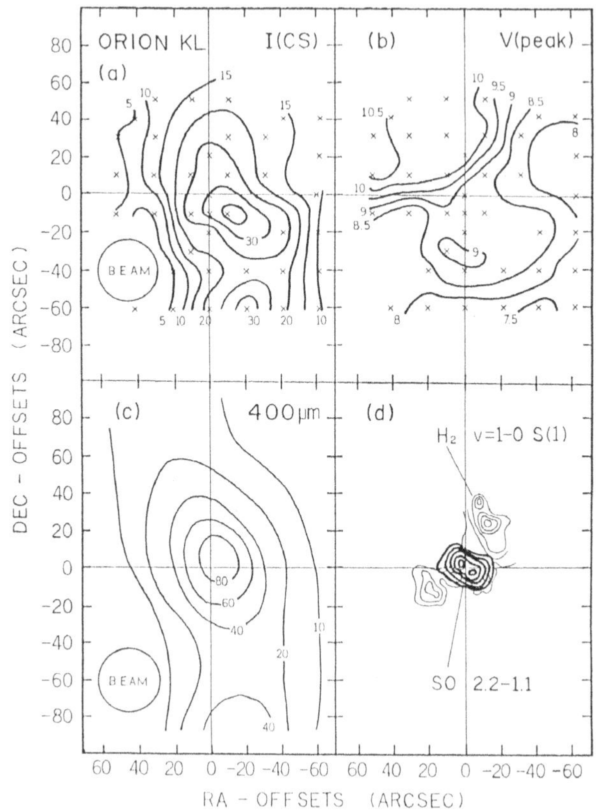

Fig. 8. (a) The distribution of the integrated intensity of the $J = 1$–0 CS emission observed at Nobeyama (Hasegawa *et al.*, 1984). The crosses indicate the observed positions and the positional offsets are taken from IRc2; R.A. = $5^h32^m47.0^s$, dec. = $-5°24'23''$. The contour interval is 5 K km s^{-1}. (b) Map of the peak V_{LSR} of the $j = 1$–0 CS line. (c) Map of the 400 μm continuum emission from warm dust taken from Keene *et al.* (1982). (d) Maps of the H_2 $v = 1$–0 S(1) emission (thin contours) taken from Beckwith *et al.* (1978) and the SO 2_2–1_1 emission (thick contours) taken from Plambeck *et al.* (1982).

There remains some controversy about the structure of the molecular cloud outside the expanding ring. A gradient in radial velocity has been recognized along the north–south ridge of molecular emission. Two models have been proposed; one is that the velocity gradient is due to rotation about an east–west axis of a disklike cloud (e.g., Liszt *et al.*, 1974; Evans *et al.*, 1975), and the other is that there are two separate clouds with the region of overlap being the KL nebula (Ho and Barrett, 1978; Bastien *et al.*, 1981). Recent high-resolution observations seems to support the former idea of a rotating cloud. Figure 6 shows a map of the $J = 1$–0 CS emission observed with a 33″ beam using the 45 m telescope at Nobeyama (Hasegawa *et al.*, 1984). The region of large CS integrated intensity has a single maximum near the KL nebula and is elongated northeast–southwest. The peak velocity shows a systematic shift along the axis of the elongated feature. Figure 7 shows strip scans taken along the major axis of the elongated feature. The appearance of the emission in the position vs velocity diagram is fairly symmetric about the central position and velocity, and there is no clear evidence for overlap of two clouds with different velocities.

These features are readily understood in terms of a coherent velocity field in a single object, rather than independent velocity structures of two clouds. The position vs velocity diagram of the $J = 1\text{–}0$ HCN emission made at Hat Creek shows an increase in the difference between the velocities NE and SW when we go closer to IRc2 (Vogel *et al.*, 1985). This is what we expect for a disk of gas rotating around a massive protostar; the rotation law approaches the Keplerian one in the vicinity of the central object, while the outer rotation curve reflects the mass distribution over the disk. Observations of the $J = 2\text{–}1$ CS emission with a 20″ beam at Nobeyama reinforces this interpretation. The CS observations show an indication that the disk could be still collapsing. The parameters of the outer disk derived from the CS $J = 1\text{–}0$ observations are summarized in Table I.

TABLE I

Physical parameters of the disk around Orion–KL

Parameter	Value
Diameter (defined by two $C^{34}S$ peaks)	0.20 pc or 4×10^4 AU
Systemic velocity (V_{LSR})	9.0 km s^{-1}
Rotational velocity	1.3 km $s^{-1}/\sin i$[a]
Rotational period	5×10^5 yr $\sin i$
Collapse velocity	0.4 km $s^{-1}/\sin i$
Collapsing time (radius/collapse velocity)	3×10^5 yr $\sin i$
Optical depth of the CS line	~1 (KL), 3–5 ($C^{34}S$ peak)[b]
$C^{34}S$ column density in the $J = 0$ level over the disk	~4×10^{12} cm^{-2}
H_2 column density over the disk	~2×10^{24} cm^{-2} $(T_{ex}/30\ K)\ ((H_2/C^{34}S)/(2 \times 10^{10})$
H_2 density in the disk	~1×10^7 cm^{-3} $(T_{ex}/30\ K)\ ((H_2/C^{34}S)/2 \times 10^{10})$[c]
Mass of gas in the disk	~300 $M_\odot$ $(T_{ex}/30\ K)\ ((H_2/C^{34}S)/2 \times 10^{10})$

[a] i = inclination of the disk.
[b] Equal excitation temperatures for CS and $C^{34}S$, and the terrestrial abundance ratio of $CS/C^{34}S = 22.6$ are assumed.
[c] The thickness of the disk along the line-of-sight is assumed to be $\frac{1}{3}$ of the diameter.

5. The Overall Structure of the Orion–KL Region

From the observations reviewed above, the overall structure of the Orion–KL region is illustrated in Figure 8. We are impressed by the highly symmetric nature of this region. The dynamical lifetime of the disk may be more than 3×10^5 yr, which is much longer than the estimated age of the high-velocity molecular flow (~ 10^3 yr). Probably the disk has formed first and collapsed to form a protostar which initiated the molecular flow.

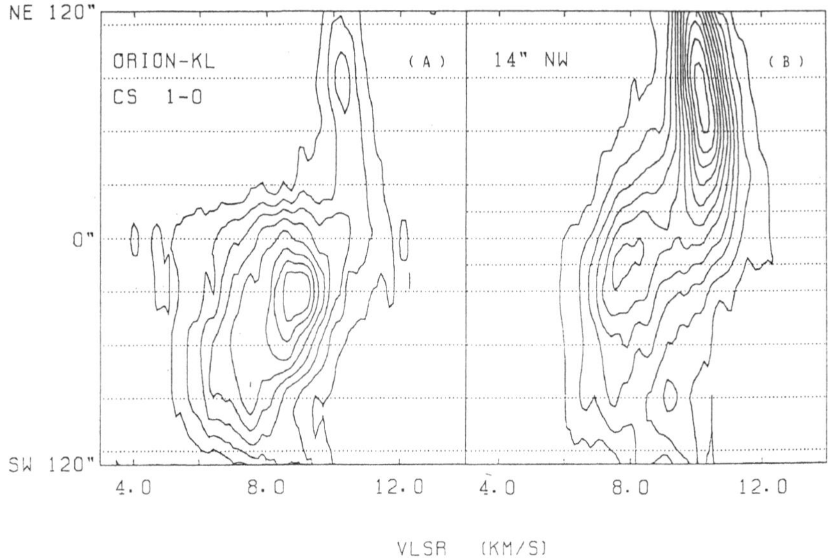

Fig. 9. Strip scans taken in NE–SW, an approximate direction of the major axis. (a) Scan through the center. (b) Scan along a line offset by 14″ toward NW.

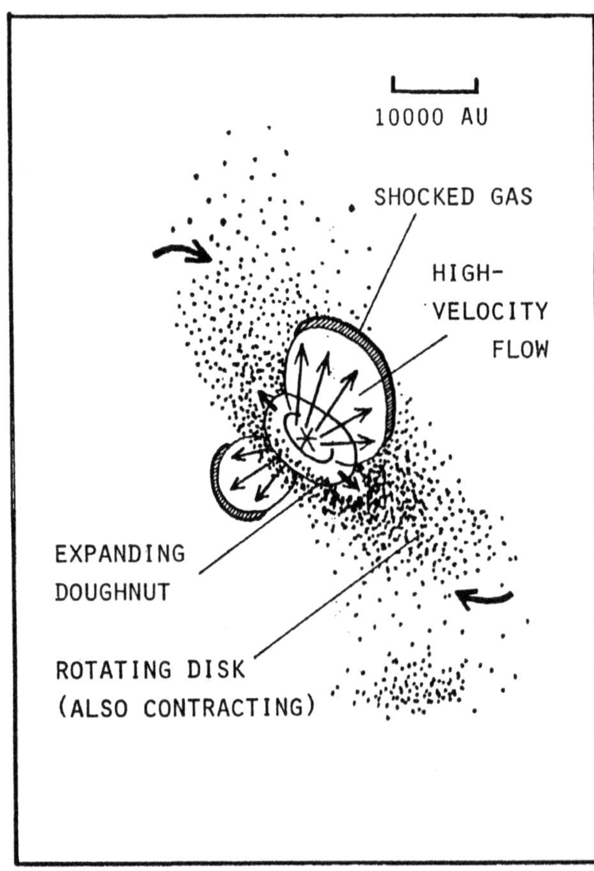

Fig. 10. The overall structure of the Orion–KL region (schematic).

The longer time-scale of the disk may argue for a hypothesis that the disk itself, or some mechanism of disk formation, is responsible for the bipolar structure of the high-velocity flow.

Two mechanisms of disk formation can be considered; the angular momentum and the magnetic field. The rotating disk discussed in the preceeding section fits smoothly in the global kinematics of the molecular cloud suggested by several authors (e.g., Linke and Wannier, 1974; Kutner *et al.*, 1977), so that it is possible that the axisymmetric nature of the Orion-KL region is somehow governed by the angular momentum. Alternatively, the magnetic field defines a preferred direction of gas motion. The flattened dense molecular cloud around the protostar and the collimated bipolar outflow can be formed under the influence of the magnetic field (e.g., Nakano, 1984; Shibata and Uchida, 1985; Pudritz, 1985). Far-infrared polarization at a position angle of 23° almost perpendicular to the axis of the disk was measured by Hildebrand *et al.* (1984), which is interpreted as due to emission from aligned dust grains. Hough *et al.* (1985) have made polarimetric mapping observations of the H_2 $v = 1$–0 S(1) line emission, which demonstrate an existence of extinction by alined grains between us and the shocked gas. These are evidence for the magnetic field penetrating the disk in the direction of its axis. With these observational supports for *both* of the mechanisms, we can say only that the rotation and/or the magnetic field may be the origin of the axisymmetric nature of the Orion-KL region. Further quantitative studies of their relative importance is needed to reveal the grand design of this region.

References

Bally, J. and Lada, C. J.: 1983, *Astrophys. J.* **265**, 824.
Bastien, P., Bieging, J., Henkel, C., Martin, R. N., Pauls, T., Wamsley, C. M., Wilson, T. L., and Ziurys, L. M.: 1981, *Astron. Astrophys.* **98**, L4.
Beck, S. C.: 1984, *Astrophys. J.* **281**, 205.
Beckwith, S., Persson, S. E., Neugebauer, G., and Becklin, E. E.: 1978, *Astrophys. J.* **223**, 464.
Beckwith, S., Evans, N. J., II, Gatley, I., Gull, G., and Russell, R. W.: 1983, *Astrophys. J.* **264**, 152.
Chernoff, D. F., Hollenbach, D. J., and McKee, C. F.: 1982, *Astrophys. J.* **259**, L97.
Clark, F. O., Biretta, J. A., and Martin, H. M.: 1979, *Astrophys. J.* **236**, 922.
Downes, D., Genzel, R., Becklin, E. E., and Wynn-Williams, C. G.: 1981, *Astrophys. J.* **244**, 869.
Draine, B. T.: 1980, *Astrophys. J.* **241**, 1021 (erratum: 1981, *Astrophys. J.* **246**, 1045).
Draine, B. T. and Roberge, W. G.: 1982, *Astrophys. J.* **259**, L91.
Edwards, S. and Snell, R. L.: 1983, *Astrophys. J.* **270**, 605.
Erickson, N. R., Goldsmith, P. F., Snell, R. L., Berson, R. L., Huguenin, G. R., Ulich, B. L., and Lada, C. J.: 1982, *Astrophys. J.* **261**, L103.
Evans, N. J., I, Zuckermann, B., Sato, T., and Morris, G.: 1975, *Astrophys. J.* **199**, 383.
Friberg, P.: 1984, *Astron. Astrophys.* **132**, 265.
Gautier, T. N., III, Fink, U., Treffers, R. R., and Larson, H. P.: 1976, *Astrophys. J.* **207**, L129.
Genzel, R., Reid, M. J., Moran, J. M., and Downes, D.: 1981, *Astrophys. J.* **244**, 884.
Genzel, R. and Downes, D.: 1982, in R. S. Roger and P. E. Dewdney (eds.), *Regions of Recent Star Formation*, D. Reidel Publ. Co., Dordrecht, Holland, p. 251.
Hansen, S. S. and Johnston, K. J.: 1983, *Astrophys. J.* **267**, 625.
Hasegawa, T. and Akabane, K.: 1984, *Astrophys. J.* (in press).
Hasegawa, T., Kaifu, N., Inatani, J., Morimoto, M., Chikada, Y., Hirabayashi, H., Iwashita, Y., Morita, K.-I., Tojo, A., and Akabane, K.: 1984, *Astrophys. J.* **283**, 117.

Hasegawa *et al.*: 1985, in preparation.
Hildebrand, R. H., Dragovan, M., and Novak, G.: 1984, *Astrophys. J.* **284**, L51.
Ho, P. T. P. and Barrett, A. H.: 1978, *Astrophys. J.* **224**, L23.
Hollenbach, D. J. and Shull, J. M.: 1977, *Astrophys. J.* **216**, 419.
Hough, J. H. *et al.*: 1986 (in preparation).
Keene, J., Hildebrand, R. H., and Whitcomb, S. E.: 1982, *Astrophys. J.* **252**, L11.
Knapp, G. R., Phillips, T. G., Huggins, P. J., and Redman, R. O.: 1981, *Astrophys. J.* **250**, 175.
Kuiper, T. B. H., Rodriguez Kuiper, E. N., and Zuckermann, B.: 1978, *Astrophys. J.* **219**, 129.
Kutner, M. L., Tucker, K. D., Chin, G., and Thaddeus, P.: 1977, *Astrophys. J.* **215**, 521.
Kwan, J.: 1977, *Astrophys. J.* **216**, 713.
Kwan, J. and Scoville, N. Z.: 1976, *Astrophys. J.* **210**, L39.
Linke, R. A. and Wannier, P. G.: 1974, *Astrophys. J.* **193**, L41.
Liszt, H. S., Wilson, R. W., Penzias, A. A., Jefferts, K. B., Wannier, P. G., and Solomon, P. M.: 1974, *Astrophys. J.* **190**, 557.
London, R., McCray, R., and Chu, S.-I.: 1977, *Astrophys. J.* **217**, 442.
Masson, C. R., Berge, G. L., Claussen, M. J., Heiligman, G. M., Leighton, R. B., Lo, K. Y., Moffet, A. T., Phillips, T. G., Sargent, A. I., Scott, S. L., Wannier, P. G., and Woody, D. P.: 1984, *Astrophys. J.* **283**, L37.
Nadeau, D. and Geballe, T. R.: 1979, *Astrophys. J.* **230**, L169.
Nadeau, D., Geballe, T. R., and Neugebauer, G.: 1982, *Astrophys. J.* **253**, 154.
Nakano, T.: 1984, *Fundam. Cosmic Phys.* **9**, 139.
Norris, R. P.: 1984, *Monthly Notices Roy. Astron. Soc.* **207**, 127.
Olofsson, H., Hjalmarson, A., and Rydbeck, G.: 1981, *Astron. Astrophys.* **100**, L30.
Olofsson, H., Ellder, J., Hjalmarson, A., and Rydbeck, G.: 1982, *Astron. Astrophys.* **113**, L18.
Phillips, J. P., White, G. J., and Watt, G. D.: 1977, *Monthly Notices Roy. Astron. Soc.* **199**, 1033.
Phillips, T. G., Huggins, P. J., Neugebauer, G., and Werner, M. W.: 1977, *Astrophys. J.* **217**, L161.
Plambeck, R. L., Wright, M. C. H., Welch, W. J., Bieging, J. H., Baud, B., Ho, P. T. P., and Vogel, S. N.: 1982, *Astrophys. J.* **259**, 617.
Pudritz, R.: 1985, *Astrophys. J.* **293**, 216.
Rieke, G. H., Low, F. J., and Kleinmann, D. E.: 1973, *Astrophys. J.* **186**, L7.
Scoville, N. Z., Hall, D. N. B., Kleinmann, S. G., and Ridgway, S. T.: 1982, *Astrophys. J.* **253**, 136.
Shibata, K. and Uchida, Y.: 1985, this issue.
Shull, J. M., and Beckwith, S.: 1982, *Ann. Rev. Astron. Astrophys.* **20**, 163.
Snell, R. L., Scoville, N. Z., Sanders, D. B., and Erickson, N. R.: 1984, *Astrophys. J.* **284**, 176.
Storey, J. W. V., Watson, D. M., Townes, C. H., Haller, E. E., and Hansen, W. L.: 1981, *Astrophys. J.* **247**, 136.
Vogel, S. N., Bieging, J. H., Plambeck, R. L., Welch, W. J., and Wright, M. C.: 1985, *Astrophys. J.* **296**, 600.
Vogel, S. N., Wright, M. C. H., Plambeck, R. L., and Welch, W. J.: 1984, *Astrophys. J.* **283**, 655.
Wannier, P. G. and Phillips, T. G.: 1977, *Astrophys. J.* **215**, 796.
Waters, J. W., Gustincic, J. J., Kaker, R. K., Kuiper, T. B. H., Roscoe, H. K., Swanson, P. N., Rodriguez Kuiper, E. N., Kerr, A. R., and Thaddeus, P.: 1980, *Astrophys. J.* **235**, 57.
Watson, D. M., Storey, J. W. V., Townes, C. H., Haller, E. E., and Hansen, W. L.: 1980, *Astrophys. J.* **239**, L129.
Werner, M. W., Dinerstein, H. L., and Capps, R. W.: 1983, *Astrophys. J.* **265**, L13.
Werner, M. W., Crawford, M. K., Genzel, R., Hollenbach, D. J., Townes, C. H., and Watson, D. M.: 1984, *Astrophys. J.* **282**, L81.
Wright, M. C. H., Plambeck, R. L., Vogel, S. N., Ho, P. T. P., and Welch, W. J.: 1983, *Astrophys. J.* **267**, L41.
Wynn-Williams, C. G., Genzel, R., Becklin, E. E., and Downes, D.: 1984, *Astrophys. J.* **281**, 172.
Zuckermann, B., Kuiper, T. B. H., and Rodriguez Kuiper, E. N.: 1976, *Astrophys. J.* **209**, L137.

GLOBULAR CLUSTERS AND GALACTIC EVOLUTION (INVITED REVIEW)*

SIDNEY VAN DEN BERGH

Dominion Astrophysical Observatory, Herzberg Institute of Astrophysics, Victoria, B.C., Canada

(Received 2 July, 1985)

Abstract. Observations of the frequency with which globular clusters surround galaxies are reviewed. With a single exception the data do not favour the view that elliptical galaxies are formed from merging disc systems. Attention is also drawn to the unusually high frequency of globular clusters in central galaxies of rich clusters.

1. Introduction

One of the simplest questions that can be asked about a globular cluster system is: what is the total number N of globulars that it contains? Since, as we shall see in Section 2, the total number of clusters is roughly proportional to apparent galaxy luminosity the specific cluster frequency S, defined as

$$S = N10^{0.4(15 + M_V)}, \tag{1}$$

is a more useful parameter for the characterization of globular cluster systems than is the total cluster population N. Recently, we (Harris and van den Bergh, 1981; van den Bergh and Harris, 1982) have studied the globular cluster populations around a number of galaxies at intermediate distances on a prime focus IIIaJ plates obtained with the Canada–France–Hawaii Telescope. Following Harris and Racine (1979) we shall assume that globular clusters have a Gaussian luminosity function. For galaxies in the Virgo cluster the luminosity function of globulars should peak near $J = 25$, whereas the observations typically extend to a limiting magnitude of $J = 23$ or $J = 24$. Some extrapolation from the observed cluster counts is, therefore, required to obtain the total number of clusters N associated with a galaxy and, hence, the specific frequency S. At this point it should, perhaps, be emphasized that our derivation of S values implicitly assumes that all globular cluster systems have the same luminosity function. Although we shall interpret our observations in terms of variations in S values, it might equally well be assumed that the shape of the cluster luminosity function differs from galaxy to galaxy.

2. Cluster Frequency in Elliptical Galaxies

The observations obtained to date show the following:

(1) Real differences in the specific globular cluster frequency exist between different

galaxies. These differences are perhaps most convincing in the Virgo cluster where selection effects should be similar for all galaxies.

(2) Over an apparent galaxy luminosity range exceeding 1000 there is no clearcut dependence of S on apparent galaxy luminosity.

(3) Within the rather limited accuracy of available statistics there is no clear dependence of the specific globular cluster frequency S on apparent galaxy ellipticity ε.

(4) Presently available data provide marginal evidence in favour of a slight dependence of the specific globular cluster frequency on environment. Even after excluding M87, $\langle S \rangle = 6.6 \pm 0.5$ for the Virgo ellipticals. This appears larger than the value $\langle S \rangle = 3.1 \pm 1.2$ for the Local Group.

3. Cluster Frequency in Lenticular Galaxies

Selecting only those galaxies for which Sandage and Tammann (1981) have given classifications, we find that $\langle S \rangle = 5.3 \pm 0.6$ for SO galaxies compared to $\langle S \rangle = 5.0 \pm 0.7$ for ellipticals, i.e. the present data indicate that elliptical and SO galaxies have similar specific cluster frequencies. It would be particularly important to see whether this conclusion also holds for highly flattened lenticulars, such as NGC 3115.

4. Cluster Frequency in Spiral Galaxies

Only in edge-on spirals is it possible to count globular clusters without seriously contaminating the data with stars and cluster in the galactic disc. As a result of this limitation the number of spiral galaxies for which data on the specific cluster frequency are available is very small (see Table I). The data in this table clearly show a very strong dependence of the specific cluster frequency on Hubble-type. For bulge-dominated systems such as NGC 4596 and NGC 4595 (the Sombrero) the specific globular cluster frequency is seen to be similar to that in ellipticals and lenticulars. For the four disc-dominated systems in Table I, $\langle S \rangle = 0.4$ which is an order of magnitude lower

TABLE I

Specific globular cluster frequency in spirals

Galaxy	Type	S
NGC 4596	SBa (early)	5.5 ± 0.9
NGC 4594	Sa	3.0 ± 1.0
NGC 224	Sb	0.4 ± 0.1
NGC 4565	Sb:	0.3 ± 0.1
Galaxy[a]	Sbc	0.9:
NGC 891	Sbc:	0.04 ± 0.07
LMC	Ir	0.24 ± 0.10
SMC	Ir	0.19 ± 0.19

[a] $M_V = -20.6$ and $N = 150$ assumed.

than that in ellipticals. This result implies that elliptical galaxies cannot have formed from merging disc systems similar to those of present-day spiral galaxies. A possible exception is provided by NGC 3557 (Morbey and McClure, 1985). This fast-rotating giant elliptical has $S = 0.8$ ($H = 50$ km s^{-1} Mpc^{-1}) or $S = 0.5$ ($H = 100$ km s^{-1} Mpc^{-1}). These values are quite similar to those typically observed in disc systems suggesting that this giant elliptical, which is not situated in a rich cluster, might have been formed from merging spirals.

It is sometimes suggested that the high globular cluster frequency in ellipticals might be due to the fact that such globulars were formed during the violent collisions between the gaseous discs in merging spirals. If this hypothesis were correct than the globular clusters so formed would have the same composition as the (presumably metal-enriched) interstellar gas. This conclusion appears to conflict with the colour observations by Forte *et al.* (1981) which appear to indicate that the globular clusters associated with Virgo ellipticals are significantly bluer than the dominant population of their parent galaxies. If the measurements by Forte *et al.* (1981) are correct then the clusters associated with Virgo ellipticals must be either metal-poor or quite young. The latter conclusion appears unlikely because the Virgo velocity dispersion has, within the recent past, been too large to permit galaxy mergers to proceed at a significant rate. In summary it seems most likely that the majority of elliptical galaxies were not formed by merging spirals.

5. Globular Clusters in Central Galaxies of Rich Clusters

It has been known for many years that M87, which is situated at the bottom of the Virgo potential well, is unusually rich in globular clusters. Our data indicate that $S = 18.2 \pm 4.5$ for this object, which is about 3 times the $\langle S \rangle = 6.6 \pm 0.5$ average for Virgo ellipticals. It is of interest to note that M49, which has $S = 6.1 \pm 1.5$, is the most luminous galaxy in Virgo. The high-specific globular cluster frequency of M87 cannot, therefore, be attributed to its high liminosity. The fact that M87 is located at the physical center of the Virgo cluster is more likely responsible for its unusually high globular cluster frequency. This suspicion is confirmed by the observations of Harris *et al.* (1982) of NGC 3311, which is the central galaxy in the Hydra I cluster, and by observations of NGC 1399 (Dawe and Dickens, 1976) which is the central galaxy of the Fornax cluster. All three of these central galaxies in rich clusters are found to have much higher S values than do the other elliptical galaxies in these clusters. Recently, Harris and Hanes (1984) have shown that NGC 524, which is the central galaxy in the compact but poor cluster CfA13, has a relatively normal specific cluster frequency $S \simeq 6$. This result suggests that only the central galaxies in rich clusters are surrounded by a rich globular cluster population. Because such central galaxies in rich clusters are believed to have grown by cannibalising other cluster members (which have normal S values) the original specific globular cluster frequencies of these central galaxies must have been even higher than they are now.

Since globular clusters are the oldest constituents of galaxies this result implies that the central galaxies of rich clusters must already have been unique very early in the cluster history. This result argues strongly against a hierarchical clustering model.

References

Dawe, J. A. and Dickens, R. J.: 1976, *Nature* **263**, 395.
Forte, J. C., Strom, S. E., and Strom, K. M.: 1981, *Astrophys. J.* **245**, L9.
Harris, W. E. and Hanes, D. A.: 1986, preprint.
Harris, W. E. and Racine, R.: 1979, *Ann. Rev. Astron. Astrophys.* **17**, 241.
Harris, W. E., Smith, M. G., and Myra, E. S.: 1983, *Astrophys. J.* **272**, 456.
Harris, W. E. and van den Bergh, S.: 1981, *Astron. J.* **86**, 1627.
Morbey, C. L. and McClure, R. D.: 1986, preprint.
Sandage, A., and Tammann, G. A.: 1981, *A Revised Shapley-Ames Catalog of Bright Galaxies*, Carnegie Institution, Washington.
Van den Bergh, S. and Harris, W. E.: 1982, *Astron. J.* **87**, 494.

THE STRUCTURE OF THE R MONOCEROTIS, NGC 2261, AND HH39 NEBULAR COMPLEX*

J. R. WALSH

Anglo-Australian Observatory, Epping, NSW, Australia

(Received 3 July, 1985)

Abstract. NGC 2261 is a reflection nebula that shows variations in brightness and structural changes of up to $\frac{1}{4}$ arc sec day^{-1}. UB observation of R Mon and NGC 2261 revealed a strongly blue-shifted line profile variation of the reflected lines with distance from R Mon.

R Monocerotis (R Mon) is the variable stellar object located at the apex of the cometary nebula NGC 2261. There is a strong near-infrared source displaced from the optical position of R Mon, indicating that the optical object is not a true star but a bright reflection nebula. NGC 2261 is a reflection nebula that shows variations in brightness; changes in structure of up to $\frac{1}{4}$ arc sec per day have been observed, corresponding to the speed of light at the probable distance of 800 pc. It is assumed that the changes are caused by illumination effects and do not represent high translational velocities. CO mapping of NGC 2261 shows an associated molecular cloud with bipolar morphology and evidence of a disc around R Mon (Kaifu *et al.*, 1984). Some 2.5 pc to the north are the group of Herbig Haro objects, HH39, which have large proper motions directed away from R Mon.

R Mon has been observed in the ultraviolet with the International Ultraviolet Explorer (IUE) satellite and shows a lack of high-excitation lines (found in T Tauri-type stars), no obvious 2200 Å extinction feature and a continuum typical of a late-type star. R Mon is considered to be a young star still emerging from parent cloud material, but its high luminosity ($\sim 1400\,L_{\odot}$; Cohen *et al.*, 1984) distinguishes it from T Tauri stars ($< 100\,L_{\odot}$). It is most probably an F–G type giant star. The lack of any 2200 Å extinction signature implies that the extinction is not typical of the general interstellar medium, as considerable local obscuration must be present to give rise to the embedded star and the cometary reflection nebula. In other respects the spectrum of R Mon is similar to T Tauri-type stars, and, for example, the lower Balmer line profiles are variable with time.

The spectrum of the nebula is confused with local H II material, which belongs to the outer regions of the nearby H II region NGC 2264. If we plot the fall-off in fluxes from R Mon into NGC 2261, and out to the surroundings, we find that the continuum and the hydrogen emission lines follow one another, and thus arise through reflection in the

* Paper presented at the IAU Third Asian-Pacific Regional Meeting, held in Kyoto, Japan, between 30 September–6 October, 1984.

Astrophysics and Space Science **118** (1986) 439–441.

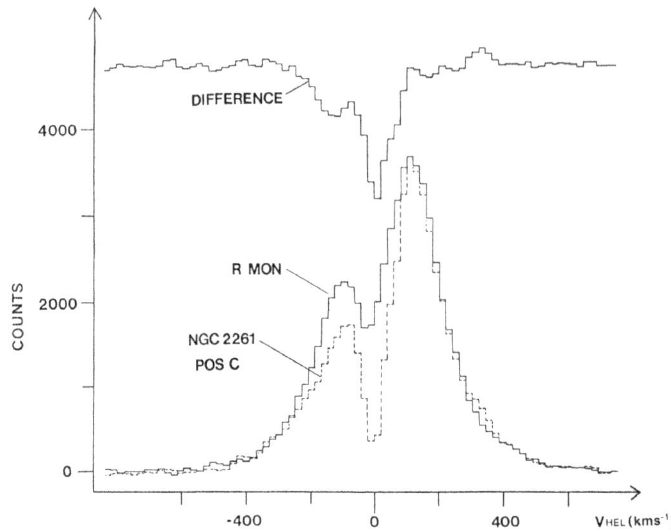

Fig. 1. The observed Hα profile of R Mon is shown compared to the observed Hα profile summed along a slit length at a position in NGC 2261 20 arc sec from R Mon. The observation of NGC 2261 was obtained 90 days after that for R Mon, to allow for the light travel time between R Mon and the region of the nebula. The difference between the two profiles (NGC 2261–R Mon) is also displayed. The ordinate is heliocentric radial velocity.

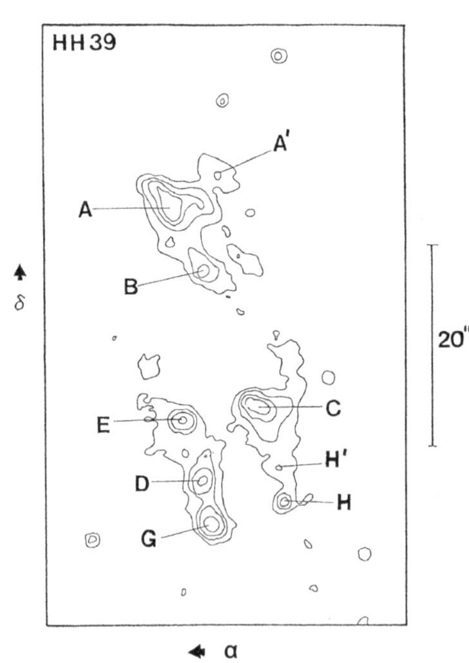

Fig. 2. A contour map of HH39 derived from an R band CCD frame taken at the prime focus of the 3.9 m AAT. The designation of the individual knots is indicated.

nebula. However, the forbidden lines are of similar brightness over the nebula and surroundings, suggesting an origin in the H II material only, although there could be some local [O II] emission in NGC 2261. The most unusual aspect of NGC 2261 is the line profile variation of the reflected lines with distance from R Mon. At 20 arc sec north of R Mon comparison of the nebular Hα line profile with that from R Mon, as obtained from photon counting spectroscopic observations with the 3.9 m Anglo-Australian Telescope (Figure 1), shows a strong blue-shifted absorption. Further north the absorption becomes stronger and systematically more blue-shifted. This phenomenon has been interpreted in terms of an azimuthally varying absorption close to the surface of R Mon, combined with reflection from different regions of the nebula (Jones and Herbig, 1982). The implication is that our view of varying mass loss over the surface of the star is being preferentially enhanced by reflection from the nebula. The direction of highest negative velocity absorption, however, points away from the receding Herbig-Haro objects; so that if this high velocity absorption is directly related to the driving mechanism of the Herbig-Haro objects, some directional change (precession?) is suspected.

The optical spectrum of the brightest knot in HH39, HH39A, can be fitted to an ~ 80 km s^{-1} shock model, even though its proper motion is 270 km s^{-1}. It appears that the velocity vector of the HH objects is out of the plane of the sky since a strong negative velocity component is observed on the Hα profile of HH39A. A recently obtained R band CCD exposure of HH39 (Figure 2 shows a contour map derived from the CCD image) reveals that one of the knots shows strong brightness variability (knot G in Figure 2). This knot was weakly present in 1960, had brightened in the early 1970s, and is now one of the brightest knots in the group. These brightness changes may be related to variability of the stellar wind from R Mon which excites the HH objects. In addition a loop has been detected extending from the eastern extremities of NGC 2261 to HH39 (Walsh and Malin, 1985), confirming the association of the Herbig-Haro objects with NGC 2261. The presence of such a loop has been predicted in the wind focussing model of Cantó and Rodriguez (1980), and in this model the HH objects form where the flows along the walls of a wind cavity collide. In such a case the presence of a single highly collimated flow (jet), as has been inferred for several HH regions, may not be required to explain the presence of HH objects.

References

Cantó, J. and Rodriguez, L. F.: 1980, *Astrophys J.* **239**, 982.
Cohen, M., Harvey, P. M., Schwartz, R. D., and Wilking, B. A.: 1984, *Astrophys. J.* **278**, 671.
Jones, B. F. and Herbig, G. H.: 1982, *Astron. J.* **87**, 1223.
Kaifu, N., Suzuki, S., Oishi, M., and Hasegawa, T.: 1984, Nobeyama Radio Observatory, preprint.
Walsh, J. R. and Malin, D. F.: 1985, *Monthly Notices Roy. Astron. Soc.* (in press).

FORMATION OF ASTROPHYSICAL JETS BY A CONTRACTING MAGNETIC ACCRETION DISK*

KAZUNARI SHIBATA

Department of Earth Science, Aichi University of Education, Japan

and

YUTAKA UCHIDA

Tokyo Astronomical Observatory, University of Tokyo, Japan

(Received 3 July, 1985)

Abstract. In the present paper, we discuss an MHD model for the formation of astrophysical jets, in which the directed flows are ejected along the rotation axis of an accretion disk formed from a cloud having a large scale magnetic field parallel to the angular momentum axis of the disk. The acceleration of jets is due to the $\mathbf{j} \times \mathbf{B}$ force in the relaxing magnetic twist which is produced by the rotation of the disk. The characteristic features of the jets, predicted by our mechanism and hopefully to be proven by observations, are the helical velocity and the hollow cylindrical shape of the jet, with a diameter of roughly the size of the region from which the accretion disk collected its mass. Justification for the assumption of the perpendicular orientation of the disk, or the parallelism of the jets, to the external magnetic field may be provided by the fact that the component of rotation whose axis is perpendicular to the field may have been damped in the earlier phase of the cloud contraction.

1. Introduction

Astrophysical jets form one of the categories of cosmic phenomena established recently. Well-known examples are jets with double radio lobes of active galaxies (Begelman *et al.*, 1984). The radio lobes above the nucleus of our Galaxy recently reported by Sofue and Handa (1984) may be related to these astrophysical jets in the galactic scale. It has also been revealed recently that there are jet-like phenomena having bipolar configuration in the stellar situation. Examples are the bipolar flows observed in star forming regions (Snell *et al.*, 1980; Bally and Lada, 1983), the jets from SS433 (Margon, 1982), and the double lobes of Sco X-1 (Fomalont *et al.*, 1983).

In spite of the progress in the observation, the origin of these astrophysical jets is not yet well explained theoretically (e.g., Begelman *et al.*, 1984). There are still several hypotheses as to the source of energy and the acceleration and collimation mechanisms. As for the collimation of the flow, mechanisms including the collimation of the strong 'wind' by polar funnels in a thick accretion disk (Koenigl, 1982; Fukue, 1982; Ferrari *et al.*, 1984), or collimation by blockading the isotropic flow in the direction of the high density disk (Sakashita *et al.*, 1984; Okuda and Ikeuchi, 1984) have been discussed; though the origin of the 'wind' remains unclear. It is, however, not easy to explain the high collimation of a flow by considering a funnel of small aspect ratio, and it is more

* Paper presented at the IAU Third Asian-Pacific Reginal Meeting, held in Kyoto, Japan, between 30 September–6 October, 1984.

Astrophysics and Space Science **118** (1986) 443–447.

difficult to explain the helical motion suggested in some observations. In this connection, we have proposed a mechanism in which both the acceleration and the collimation are caused by the action of a magnetic field. The magnetic field acts as an efficient converter for the gravitational energy of the contracting disk rotating around the center of gravity into that of the directed flow (Uchida and Shibata, 1984a, b, 1985; Shibata and Uchida, 1984).

We consider a cool gaseous disk rotating with a sub-Keplerian velocity around the central object. If there is a magnetic field penetrating the disk, the contraction of the rotating disk drags the magnetic field lines towards the center as well as towards the azimuthal direction. The magnetic field is thus twisted up around the axis of rotation, and after some time, the accumulated magnetic twists begin to relax along the external magnetic field and drive out the plasma by the $\mathbf{j} \times \mathbf{B}$ force helically towards the polar directions. In our model, the source of energy is the gravitational potential energy released in the contracting motion of the rotating disk. In the following we show that jets having a bipolar configuration can actually be created by our proposed mechanism.

2. Numerical Simulations

Assumptions in the calculations are (i) axial symmetry, but allowing B_φ and v_φ, (ii) ideal MHD, (iii) non-relativistic, and (iv) the self-gravity of the disk neglected. The basic equations are standard ideal MHD equations with the gravity source, the central star, located at the origin of cylindrical coordinates (r, φ, z).

As for the initial condition, we assume a cool accretion disk rotating around the z-axis with a velocity smaller than the Keplerian velocity in the gravitational potential of the central point mass. We further assume a hydrostatic corona and a uniform magnetic field which penetrates the disk vertically as the initial condition. These assumptions are for the sake of simplicity at this moment, but we know from observations that bipolar flows tend to be parallel to the external magnetic field, and the accretion disk lies perpendicular to both (Kaifu *et al.*, 1984).

Boundary conditions on the inner spherical boundary (for practical reasons in the calculation we cannot cover too wide a range of scales at a time) and on the top and side surfaces of the cylindrical region are all assumed to be free boundaries: i.e., waves and fluid can pass through these boundaries freely. Symmetrical conditions are imposed along the z-axis, and at the equatorial plane $z = 0$.

With an appropriate choice of dimensionless parameters, our problem becomes a scale-free one. The parameters are: β = the ratio of thermal and magnetic energies, ε_d = the ratio of thermal and gravitatinal energies, α = the ratio of rotational and Keplerian rotational velocities, and ρ_d/ρ_c = the ratio of the densities of the disk and corona across the interface. These ratios may be made functions of r, with a coefficient to be fixed, for example, at the inner edge of the disk. We have solved this problem numerically by using a modified Lax–Wendroff scheme with an artificial viscosity.

3. Results

We will now show a typical example of the numerical results (a full version will appear in Uchida and Shibata, 1985). $\alpha(r, z, t = 0)$, $\beta(r, z, t = 0)$, $\varepsilon_d(r, z, t = 0)$, and $\rho(r, z, t = 0)$ are prescribed by some simple suitable functions, as may be seen from the initial distributions in Figures 1 and 2, and parameters are specified as $\alpha(r, 0, 0) = 0.8$,

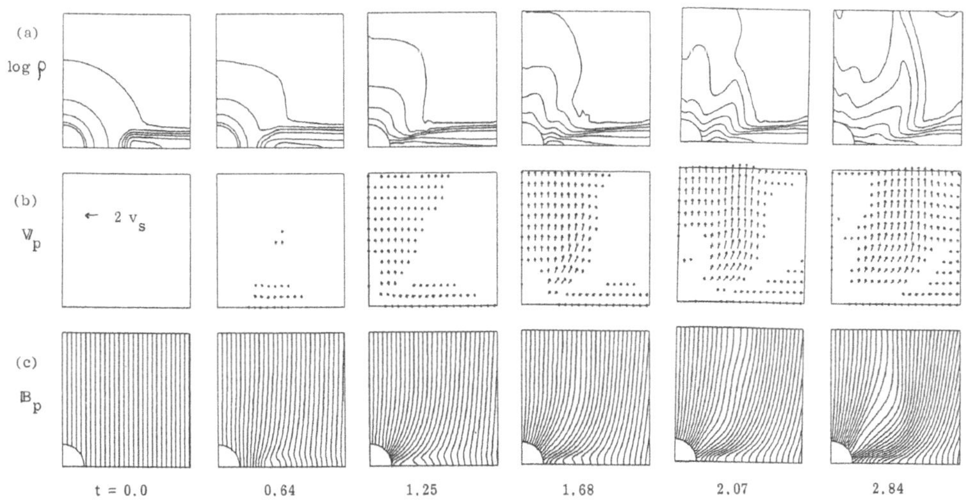

Fig. 1. Time variations of (a) $\log\rho$ (density), (b) the poloidal velocity vector $\mathbf{V}_p$, and (c) the poloidal magnetic field lines. The magnitude of the velocity vector is shown in the $t = 0.0$ figure of $\mathbf{V}_p$, where v_s is the sound velocity in the corona. Note that $\beta = 0.5$ in this case, and so V_A (Alfvén velocity in the corona) $= 1.55\, v_s$. Other initial parameters are: $v_\varphi/v_{\rm Kepler} = 0.8$, $\varepsilon_d = 3 \times 10^{-3}$, and $\rho_d/\rho_c = 400$. Times are in the unit of the free fall time at the inner edge of the disk.

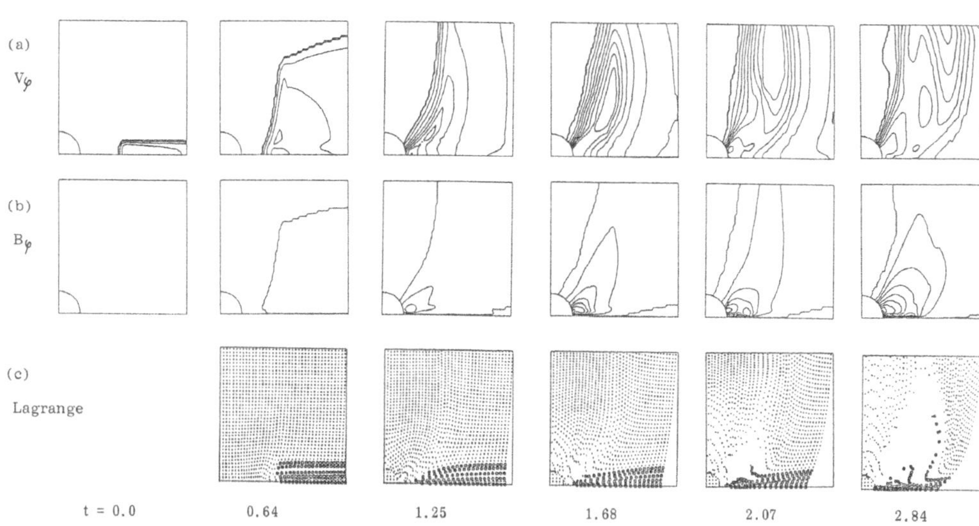

Fig. 2. Time variation of (a) the rotational velocity, v_φ, (b) the azimuthal component of the magnetic field, $\mathbf{B}_\varphi$, and (c) the Lagrangian description of the motion in the case of Figure 1. Other remarks are the same as in Figure 1.

$\beta(r, 0, 0) = 0.5$, $\varepsilon_d(r, 0, 0) = 3 \times 10^{-3}$ and $\rho_d/\rho_c = 400$, at $r = r_{ie}$, where r_{ie} is the radius at the innermost part of the initial disk. These values are based on an order-of-magnitude estimate of physical parameters in a typical stellar bipolar flow, L1551, for example.

Figure 1 shows the time variations of density, poloidal velocity vector, and poloidal magnetic field lines. The first one in each row shows the initial distributions of quantities as mentioned above. Times are in the unit of the free-fall time at the inner edge of the initial disk. The rotating disk starts to contract towards the center, because of its sub-Keplerian rotational velocity, and drags magnetic field lines towards the center as well as in the azimuthal direction. After some time of accumulation of the magnetic twist, relaxation of the magnetic twist starts, and a jet with a hollow cylindrical shape appears at around $t = 1.5$. The jet has a velocity which is a large fraction of the Alfvén velocity in the z-direction. Figure 2 shows the azimuthal velocity, the azimuthal component of the magnetic field, and the Lagrangian presentation of the motion for the same case as shown in Figure 1. At $t = 0.6$, we see the propagation of an Alfvén wavefront generated by the interaction between the rotating disk and the vertical field. Angular momentum of the disk is transported to polar directions along the vertical magnetic field. A helical jet is driven out when the gradient of the magnetic twists becomes large. The ejected gas comes from the layer near the surface of the disk.

4. Discussion

The main characteristic features of the jet in our model are (a) helical velocity and (b) hollow cylindrical configuration of the jet. In the galactic scale, the radio lobes observed above the galactic center (Sofue and Handa, 1984) actually seem to show hollow cylindrical jet structure. Although the helical motion has not yet been measured clearly in these lobes themselves, the rotating disk at their footpoints in the galactic plane may suggest the applicability of our model. In the stellar scale, some bipolar flows in the star forming regions, like GL490, also seem to show the hollow cylindrical jet structure (Fukui, private communication, 1984). We predict that helical motion in these structures will also be observationally proven in the near future.

Finally, it is remarked that the seemingly artificial assumption of perpendicular orientation of the plane of the disk relative to the direction of the magnetic field, and thus the parallelism of our jet to the external magnetic field is based on observations of some real cases, including L1551 (cf. Vrba *et al.*, 1976), and this may well have physical reasons. The component of the angular momentum perpendicular to the magnetic field may have been damped in earlier phases of the cloud contraction, due to some larger impediment by the action of the magnetic field on it than on the angular momentum parallel to the field (Mouschovias, 1976).

Acknowledgements

The authors thank Mrs H. Suzuki, Mr N. Shibuya, and Mr Y. Shiomi for their assistance in preparing the manuscript.

References

Bally, J. and Lada, C. J.: 1983, *Astrophys. J.* **165**, 824.

Begelman, M. C., Blandford, R. D., and Rees, M. J.: 1984, *Rev. Mod. Phys.* **56**, 255.

Ferrari, A., Habbal, S. R., Rosner, R., and Tsinganos, K.: *Astrophys. J.* **277**, L35.

Fukue, J.: 1982, *Publ. Astron. Soc. Japan* **34**, 163.

Fomalont, E. B., Gelzahler, Hjellming, R. M., and Wade, C. M.: 1983, *Astrophys. J.* **275**, 802.

Kaifu, N., Suzuki, S., Hasegawa, T., Morimoto, M., Inatani, J., Nagane, K., Miyazawa, K., Chikada, Y., Kanzawa, T., and Akabane, K.: 1984, *Astron. Astrophys.* **134**, 7.

Koenigl, A.: 1982, *Astrophys. J.* **261**, 115.

Margon, B.: 1982, *Science* **215**, 247.

Mouschovias, T.: 1976, *Astrophys. J.* **207**, 141.

Okuda, T. and Ikeuchi, S.: 1986, *Publ. Astron. Soc. Japan* **38** (in press).

Sakashita, S., Hanami, H., and Umemura, M.: 1985, *Astrophys. Space Sci.* **111**, 213.

Shibata, K. and Uchida, Y.: 1984, *Publ. Astron. Soc. Japan* **37**, 31.

Snell, R. L., Loren, R. B., and Plambeck, R. L.: 1980, *Astrophys. J.* **239**, L17.

Sofue, Y. and Handa, T.: 1984, *Nature* **310**, 568.

Uchida, Y. and Shibata, K.: 1985a, in M. R. Kundu (ed.), 'Unstable Current Systems and Plasma Instabilities in Astrophysics', *Proc. IAU Symp.* **107**, 287.

Uchida, Y. and Shibata, K.: 1985b, in A. B. Underhill (ed.), *Non-radiative Heating/Momentum in Hot Stars*, NASA Printing, Proc. NASA Workshop, p. 169.

Uchida, Y. and Shibata, K.: 1985c, *Publ. Astron. Soc. Japan* **37**, 515.

Vrba, F. J., Strom, S. E., and Strom, K. M.: 1976, *Astron. J.* **81**, 958.

CS, $C^{34}S$, AND CH_3OH OBSERVATIONS OF THE MOLECULAR CLOUD ASSOCIATED WITH NGC 7538*

OSAMU KAMEYA, TATSUHIKO HASEGAWA, NAOMI HIRANO, MUNEZO SEKI, MAKOTO TOSA,
YOSHIAKI TANIGUCHI, and KEIYA TAKAKUBO

Astronomical Institute, Tohoku University, Sendai, Japan

(Received 3 July, 1985)

Abstract. CS(J = 1–0), $C^{34}S$(J = 1–0), and CH_3OH(1_0–0_0 A and E) emission lines in the core region of molecular cloud associated with NGC 7538 have been surveyed at angular resolution of 33″. Distribution of CS column density shows two prominent peaks. Both blue and red wings in CS line are most prominent at 30″ northwest of IRS 11.

In order to make clear the physical conditions in the core of the molecular cloud associated with NGC 7538, some molecular emission lines in the 40 GHz band (CS(J = 1–0), $C^{34}S$(J = 1–0), CH_3OH (1_0–0_0 A and E)) have been surveyed using the high-spatial resolution (HPBW = 33″, beam interval = 30″) using NRO 45 m radio telescope.

Most of the CS profiles show narrow peaks (FWHM ~ 4 km s^{-1}) and wide wings (FWHM ~ 10 km s^{-1}). The size of the core is estimated to be 2.′5 × 2.′0 (2.0 pc × 1.6 pc at a distance of 2.7 kpc) from the distribution of $C^{34}S$ emission. The total mass of the core is of the order of $6 \times 10^3 M_\odot$.

Figure 1 shows the distribution of the CS column density, which is derived from CS and $C^{34}S$ intensities. The column density ranges over (2–4) × 10^{14} cm^{-2} in this region. Two prominent maxima are found at 30″ east of IRS 11 (position A) and at 40″ south–east of IRS 1 (position B). The molecular hydrogen density $n(H_2)$ and CS relative abundance $n(CS)/n(H_2)$ in these two positions, have been derived on the basis of the large velocity gradient approximation (Goldreich and Kwan, 1974). The velocity gradient dv/dr (km s^{-1} pc^{-1}) is about 1 km s^{-1} pc^{-1} over the observed region.

At position A $n(H_2) = 10^{4.5}$ cm^{-3} and $n(CS)/n(H_2) = 10^{-8.5}$. Position A is the densest part of the core and agrees with the distribution maxima of CS and CN (Churchwell and Bieging, 1982) which are excited in this dense region. The CH_3OH molecular peak antenna temperature T_A^* is also highest at position A. The analysis of T_A^* indicates that position A has the largest column density of CH_3OH molecule, and probably shows that the molecule distributes over a large pass length. Position A is

* Paper presented at the IAU Third Asian-Pacific Regional Meeting, held in Kyoto, Japan, between 30 September–6 October, 1984.

This work was carried out under the common use observation program at the Nobeyama Radio Observatory (NRO). NRO, a branch of the Tokyo Astronomical Observatory, University of Tokyo, is a cosmic radio observing facility open to outside users.

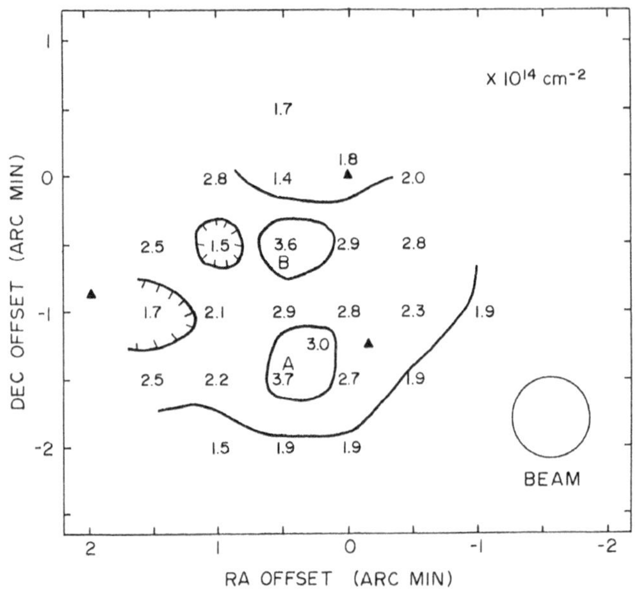

Fig. 1. The distribution of CS column densities. Triangles indicate the positions of IR sources; one of which at (0, 0) is IRS 1. Observed points are shown by numerals. Positions A and B are indicated.

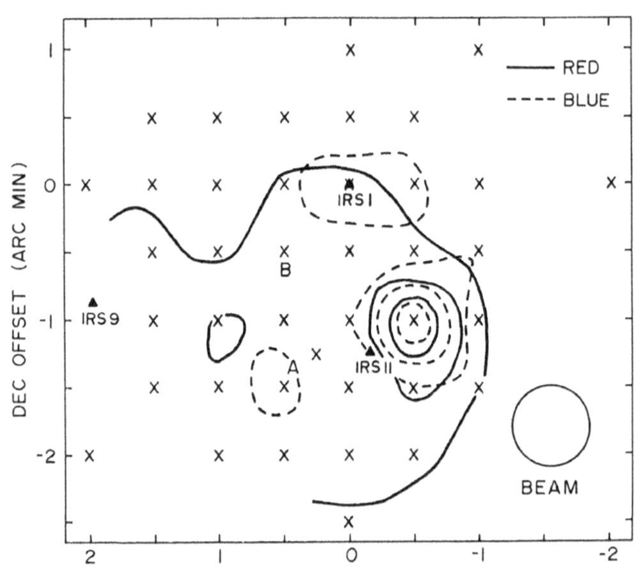

Fig. 2. A contour map of the $J = 1$–0 CS integrated antenna temperature in the blue and red-shifted high-velocity materials. The red-shifted one are integrated from -55 km s^{-1} to -41 km s^{-1}, while blue ones are from -66 km s^{-1} to -60 km s^{-1}. The contours are for 1, 2, and 3 K km s^{-1}, respectively.

possibly an undisturbed region, because around position A, not only the space and column densities but also the geometrical depth do not change so much.

At position B $n(H_2) = 10^{3.9}$ cm^{-3} and $n(CS)/n(H_2) = 10^{-7.9}$. The mean hydrogen density is low, but the relative abundance of CS is highest. Position B has the largest optical depth ($\tau \gtrsim 5$), and coincides with the distribution maxima of H_2S and H_2CO lines (Dickel *et al.*, 1981) which are high density indicators. Thus the region consists of relatively small but dense clumps. This idea is also supported by our CH_3OH line observations.

Figure 2 shows the distribution of high velocity wings of CS profiles. The ubiquity of weak red wings seems to show the presence of a cloud with $V_{LSR} = -50$ km s^{-1} (Campbell and Thompson, 1984; Israel *et al.*, 1973) which is situated in the same line-of-sight of the NGC 7538 region. Several maxima in the distribution of the CS wing intensity are found, indicating a large complicated gaseous motion, probably related to star-formation activity.

The most conspicuous of these maxima is situated at 20″ north–west of IRS 11, where both blue and red wings are very strong. Since these wings do not show a well-defined bipolar pattern, a contraction of this small region can explain the wide wings. However it may be difficult to consider that a region which has neither space nor column density maxima would contract. It is also difficult to consider that this high-velocity flow is related to IRS 1 directly (Campbell and Thompson, 1984), because of some definite difference in position. If the apparent disagreement in the positions of the flow and of IRS 11 arises from positioning errors, an energy source of the outflow may be IRS 11 itself. But if the disagreement of positions is real, we must think of a hidden energy source. Some higher resolution observations of the wings are very desirable, to clarify the relation between IRS 11 and the high velocity flow.

References

Campbell, B. and Thompson, R. I.: 1984, *Astrophys. J.* **279**, 650.
Churchwell, E. and Bieging, J. H.: 1982, *Astrophys. J.* **258**, 515.
Dickel, H. R., Dickel, J. R., and Wilson, W. J.: 1981, *Astrophys. J.* **250**, L43.
Goldreich, P. and Kwan, J.: 1974, *Astrophys. J.* **189**, 441.
Israel, F. P., Habing, H. J., and de Jong, T.: 1973, *Astron. Astrophys.* **27**, 143.

DEEP(EST) COLOUR-MAGNITUDE DIAGRAMS OF CLUSTERS*

C. A. CHRISTIAN

Canada–France–Hawaii Telescope Corp./Institute for Astronomy, University of Hawaii, Honolulu, Hawaii, U.S.A.

and

J. N. HEASLEY

Institute for Astronomy, University of Hawaii, Honolulu, Hawaii, U.S.A.

(Received 3 July, 1985)

Abstract. Color-magnitude diagrams are presented for the open cluster NGC 2158 and globular clusters Pal 4 and NGC 1904.

1. Introduction

It is now technically feasible to study colour-magnitude diagrams (cmds) to very faint levels to firmly establish cluster ages and the detailed relationships between cluster parameters and theoretical models. Deep, accurate photometry can be obtained on the Institute for Astronomy 2.2 m telescope on Mauna Kea using a 500^2 CCD camera (Hlivak *et al.*, 1984). This instrument has been used since 1981 to obtain broad-band photometry of intermediate aged and globular clusters. Some of this work is summarized here.

2. NGC 2158

Previous studies of NGC 2158 suggested that the cluster has a metallicity $[Fe/H] \sim -0.5$ and an age 8×10^8 yr (Arp and Cuffey, 1962; Janes, 1979; Hardy, 1981). The new colour magnitude diagram derived from CCD frames analyzed with a PSF fitting algorithm called DAOPHOT (Stetson, 1984) is shown in Figure 1. The 3.0 Gyr isochrone with $Z = 0.006$ and $Y = 0.25$ from van den Bergh (1984) were fitted using $E(B - V) = 0.45$. The best fitting isochrones have abundance $[Fe/H] = -0.46$, which makes the cluster similar in abundance to the globular 47 Tuc. Our new data are in agreement with the previous abundance measurements, but suggest an age substantially older than that quoted by Arp and Cuffey (1962). The large improvement in the photometric precision of the cmd allows us to determine the age of the cluster more accurately. We find that the cluster is a very good galactic analogy to the populous

* Paper presented at the IAU Third Asian-Pacific Regional Meeting, held in Kyoto, Japan, between 30 September–6 October, 1984.

Astrophysics and Space Science **118** (1986) 453–455.

clusters in the LMC and M33 (Hodge, 1983, and references quoted therein; Christian and Schommer, 1982).

A striking feature of the cmd is the gap in the turnoff region at $V = 17.5$ mag. This feature is sensitive to metallicity and age, and should not appear in objects older the 3 Gyr and/or more metal poor than [Fe/H] = − 1.0. Considering the similarity between NGC 2158 and clusters of the Large Magellanic Cloud, it would be interesting to

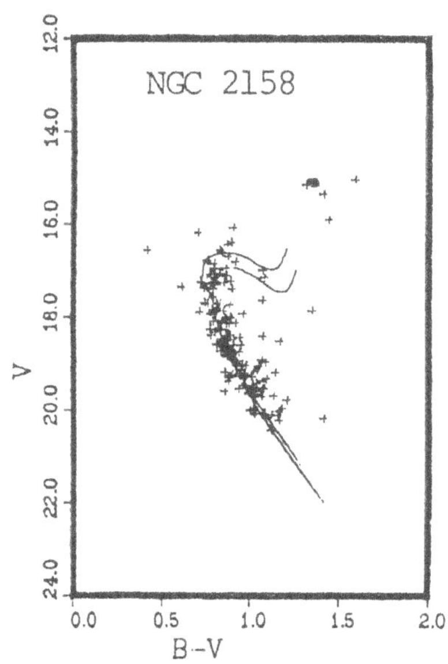

Fig. 1. Cmd of NGC 2158 fitted to the isochrones (van den Bergh, 1984). $Z = 0.006$, $Y = 0.25$, $t = 3$ Gyr (upper) and $Z = 0.01$, $Y = 0.25$, $t = 2.5$ Gyr (lower). Apparent $(m - M) = 14.3$ and 14.7, respectively.

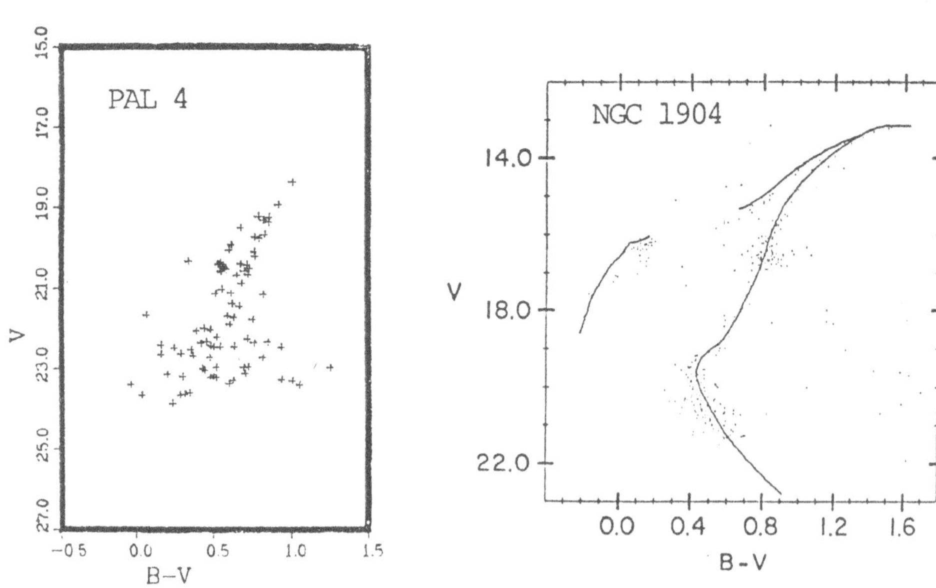

Fig. 2. Cmds of Pal 4 and NGC 1904.

determine whether the morphology of these clusters also show features (see for example the study of NGC 2190 and NGC 2162 by Schommer *et al.*, 1984).

3. Globular Clusters NGC 1904 and Pal 4

Figure 2 shows the cmd's for NGC 1904 and Pal 4, derived from CCD photometry, and, in the case of NGC 1904, the brighter photometry of Stetson and Harris (1977) was added. It is seen that the cmd of NGC 1904 is very similar to that of M13, as shown by the ridgelines in Figure 2(a). The characteristics of Pal 4 are just beginning to be unveiled with the 2.2 m data, but it appears to be a metal poor cluster with only a red horizontal branch. The cluster has a distance modulus of 19.9 mag., given a reddening of $E(B - V) = 0$ (Harris and Racine, 1979) and an $M_v = 0.6$ for the horizontal branch. Metallicity estimates from $(B - V)_{0.9} = 0.7$ and $\Delta V_{1.4} = 3.4$ (cf. Zinn and West, 1984), suggest the cluster has $[Fe/H] = -2.1$.

References

Arp, H. and Cuffey, J.: 1962, *Astrophys. J.* **136**, 51.
Christian, C. A. and Schommer, R. A.: 1982, *Astrophys. J. Suppl.* **49**, 405.
Hardy, E.: 1981, *Astrophys. J.* **86**, 217.
Harris, W. E. and Racine, R.: 1979, *Ann. Rev. Astron. Astrophys.* **17**, 241.
Hlivak, R. J., Henry, J. P., and Pilcher, C. B.: 1983, *Proc. S.P.I.E.* **331**, 96.
Hodge, P. W.: 1983, *Astrophys. J.* **264**, 470.
Janes, K. A.: *Astrophys. J. Suppl.* **39**, 135.
Landolt, A.: 1983, *Astron. J.* **88**, 439.
Schommer, R. A., Olszewski, E., and Aaronson, M.: 1984, preprint.
Stetson, P.: 1984, DAOPHOT manual.
Stetson, P. and Harris, W. E.: 1977, *Astron. J.* **82**, 954.
Van den Bergh, D.: 1984, preprint.
Zinn, R. and West, M. J.: 1984, preprint.
Zinn, R. and West, M. J.: 1984, *Astrophys. J. Suppl.* **55**, 45.

ELLIPTICAL GALAXIES UNDER PERTURBATION*

GUO-XUAN SONG

Shanghai Observatory, Academia Sinica, Shanghai, China

(Received 3 July, 1985)

Abstract. Three experiments on the initially slightly flattened spheroids ($\varepsilon = 0.45$) have been made. The first one is run without perturbation and the others with perturbation. The results have shown that the spheroid without perturbation will evolve to nearly sphere and the other two will almost keep the originally projected shape after perturbation and relaxation.

1. Introduction

In ellipticals the values of V_{max}/σ_{peak} from observations, where V_{max} is the maximum rotation velocity in the equatorial plane and σ_{peak} is the central velocity dispersion in one dimension, are lower than the values required for an oblate spheroid of the same projected shape and an isotropic distribution of stellar velocities. This leads to the suggestion by Binney (1976) that these galaxies are deformed by anisotropic stresses rather than by rotation. Nearly all the ellipticals studied to date are bright, with absolute magnitude between -20 and -23.

Eleven faint ellipticals have been observed, and the values of V_{max}/σ_{peak} are larger than those of bright ellipticals (Fall, 1982).

The question can be posed – what are the differences of formation between the bright and faint ellipticals?

In the simulations made by Aarseth and Binney (1978) it was shown that c/a should be very small in initial conditions. The observational results indicate that $\varepsilon_{contour}$ are between 0.25 and 0.70. If we take c/a larger than those adopted by Binney, it is clear the the galaxy mass will be less than Binney's values. What will happen with this kind of initial condition in the violent relaxation process? Especially, if there is a perturbation from another galaxy, will the result show some significant effects? This is the problem to be attacked, and has motivated us to do the following three simulations.

* Paper presented at the IAU Third Asian-Pacific Regional Meeting, held in Kyoto, Japan, between 30 September–6 October, 1984.

2. Models

Model	A	B	C
Axial ratios	10 : 10 : 5.5	10 : 10 : 5.5	8 : 8 : 4.4
Velocity dispersion (gaussian random isotropics)	4.10	4.10	4.65
Rotation velocity	0	0	0
Particle number (total mass)	10 000	10 000	10 000
Perturbing galaxy mass (as point particle)	0	10 000	10 000
The initial position of perturbing particle relative to the centre of mass of perturbed elliptical		13, 0, 17	13, 0, 17
The initial velocity of perturbving particle		0, 30, 0	0, 30, 0

These three models were set up.

3. Results

The results can be summarized as follows:

	Model A		Model B		Model C	
	Start	End	Start	End	Start	End
I_1/I_2	1	1.048	1	1.175	1	1.124
I_1/I_3	0.645	0.907	0.645	0.774	0.656	0.803
V_{max}/σ_{peak}		~ 0		0.38		0.41
$\bar{\varepsilon}_{contour}$		~ 0		0.49		0.42

References

Aarseth, S. T. and Binney, J.: 1978, *Monthly Notices Roy. Astron. Soc.* **185**, 227.

Binney, J. J.: 1976, *Monthly Notices Roy. Astron. Soc.* **177**, 19.

Fall, S. M.: 1982, in A. Wolfendale (ed.), *Progress in Cosmology*, D. Reidel Publ. Co., Dordrecht, Holland.

A PHOTOMETRIC STUDY OF POOR CLUSTERS OF GALAXIES*

TOMOHIKO YAMAGATA

Department of Astronomy, University of Tokyo, Japan

and

HIDEO MAEHARA

Kiso Observatory, Tokyo Astronomical Observatory, Mitaka, Tokyo, Japan

(Received 3 July, 1985)

Abstract. Sixteen poor clusters of galaxies with suspected cD galaxies are studied using *V*-band photographs taken with the 105 cm Kiso Schmidt telescope. Galaxian images were automatically detected using an image detection software developed for the present study. It is found from the luminosity functions that the richness of the clusters are well correlated with the optical properties of central galaxies and the X-ray luminosities of the clusters.

V-band photographs were taken with the 105 cm Schmidt telescope of the Kiso Observatory. The clusters studied in this work are poorer than rich Abell custers, but contain suspected cD galaxies. Twenty-three such clusters were listed by Morgan *et al.* (1975) and Albert *et al.* (1977). We have studied sixteen clusters out of their samples, as shown in Table I.

The cluster region on each plate was scanned with the PDS microdensitometer of Tokyo Astronomical Observatory. The scanning area was a 60 mm square region, setting the cD galaxies at the center. A two-dimensional cubic spline function was applied to determine the sky background level throughout the area. All the pixel data were converted into intensity values relative to the sky background.

We have developed an image detection system for counting stars and galaxies on photographic plates (Maehara and Yamagata, 1984). This is called the Kiso Image Detection System (abbreviated as KIDS). KIDS runs as follows: connections of pixels higher than some threshold value are picked up as celestial images. The moments of pixel distribution in images are calculated up to the second order, and several astronomical parameters are obtained for each image. The threshold value in this work was determined as the value corresponding to the surface brightness of 24 mag. arc sec^{-1}.

Stellar images were eliminated from the image data file with a newly discovered correlation between the sum of the second order moments; $M_{xx} + M_{yy}$ and peak intensity normalized by the square root of the area of the image; $I_{\max}/\sqrt{A}$. A field

* Paper presented at the IAU Third Asian-Pacific Regional Meeting, held in Kyoto, Japan, between 30 September–6 October, 1984.

Astrophysics and Space Science **118** (1986) 459–462.

TABLE I

Object	This work		Van den Bergh (1977) cD class	Kriss *et al.* (1983) $L \times (10 \text{ erg s}^{-1})$
	r (min.)	Type		
MKW1	12	II	A	<8.7
MKW2	16	II	A	<24
MKW3		III	A	–
MKW4	11	II	B	230
MKW6	11	I	B	54
MKW7	13	I	B	30
MKW8	14	I	B	–
MKW9	17	I	B	160
MKW10	9	II	A	11
MKW11	13	II	B	–
MKW12	12	III	A	4.9
MKW1s	9	III	A	5.9
MKW2s	16	II	B	18
MKW4s	14	I	–	70
AWM1	12	III	–	–
AWM2	13	II	–	–

correction was made statistically by using the region surrounding the cluster area. The luminosity function was defined as

$$n(m) = n_{\text{in}} - n_{\text{out}} \frac{S_{\text{in}}}{S_{\text{out}}} ,$$

where n_{in} and n_{out}, respectively, denote the number of galaxies having a magnitude between $m - \Delta m$ and m is the cluster area S_{in}, and in the surrounding area S_{out}. The magnitude bin of $\Delta m = 0.2$ was adopted in this study. The value of $n(m)$ can be negative, when the field correction is highly fluctuating.

After these procedures had been applied, we obtained logarithmic cumulative luminosity functions for each poor cluster. We classified the poor clusters into three types, based on the number of member galaxies and the shape of their luminosity functions. Clusters of type I are characterized by the luminosity functions similar in shape to those of rich clusters. Clusters of type III do not have any well-defined luminosity function because of scarce members. Clusters of type II have intermediate properties. The luminosity functions obtained in this work are plotted in Figure 1, where those of each type are collected together.

The results are summarized in Table I. The first column designates the cluster name. The second and the third columns, respectively, denote the radius and type determined in this work. The next column shows the class of central galaxy given by van den Bergh (1977); class A galaxies have faint symmetric envelopes while class B have bright asymmetrical envelopes, suggesting tidal encounters. The last column gives X-ray luminosities, observed by Kriss *et al.* (1983). The classification type proposed in this

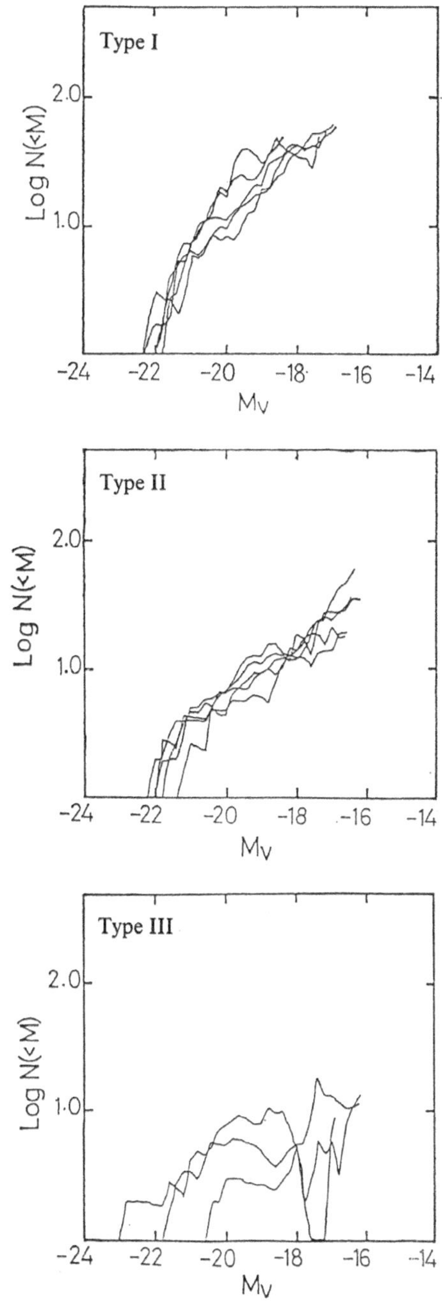

Fig. 1. Logarithmic cumulative luminosity functions.

study is well correlated with the optical properties of central galaxies and the X-ray luminosities of the cluster.

It is suggested that the proposed type represents a measure of clustering; type-I clusters are richer with prominent cD galaxies, while type-III clusters may not be real ones.

References

Albert, C. E., White, R. A., and Morgan, W. W.: 1977, *Astrophys. J.* **211**, 309.
Kriss, G. A., Cioffi, D. F., and Canizares, C. R.: 1983, *Astrophys. J.* **272**, 439.
Maehara, H. and Yamagata, T.: 1984, in M. Capacioli (ed.), 'Astronomy with Schmidt-type Telescopes', *IAU Colloq.* **78**, 169.
Morgan, W. W., Keyser, S., and White, R. A.: 1975, *Astrophys. J.* **199**, 545.
Van den Bergh, S.: 1977, *Publ. Astron. Soc. Pacific* **89**, 746.

N-BODY SIMULATION OF GIANT MOLECULAR CLOUDS IN A GALAXY*

MAKOTO TOSA and MASATAKA FUKUNAGA
Astronomical Institute, Tohoku University, Sendai, Japan

(Received 3 July, 1985)

Abstract. We made N-body simulation of Giant Molecular Clouds in a model galaxy and confirmed that their random motion is generated and accelerated by purely gravitational interaction among them in some 10^{7-8} yr. The results of the present simulation agree well with those of the previous theoretical calculation by Fukunaga (1984a, b).

One outstanding problem of Giant Molecular Clouds (GMCs) is that, in spite of their large mass, GMCs have significant random velocities, comparable to those of other components of the interstellar medium. Therefore, there must be some acceleration mechanism of GMCs in the Galaxy.

Fukunaga (1984a, b) proposed an acceleration mechanism based on purely gravitational encounters among GMCs in a differentially rotating galaxy. In the present paper we make an N-body simulation of a GMC system in a model galaxy to see how the random motion of the GMCs is accelerated through gravitational interactions, and compare the results with those of previous theoretical calculations.

As a model of the arrangement of GMCs in the Galaxy, we consider a differentially rotating thin disc consisting of gravitationally interacting particles. For simplicity, we adopt a spherically symmetric gravitational field for the model galaxy in which the velocity of the equilibrium circular rotation is given by $V(r) = V_0(r/r_0)/(1 + (r/r_0)^2)^{1/2}$, where r_0 and V_0 are the characteristic radius of a rigidly rotating region, and the maximum velocity of rotation, respectively. The three-dimensional motions of the particles, which are initially in centrifugal equilibrium with no random motion, are followed numerically by Aarseth's (1972) individual time step method.

We show a typical numerical example in Figure 1. The adopted parameters are as follows: the number of particles is 100; the mass of an individual particle is $5 \times 10^6\ M_\odot$; the inner cut-off radius of the gravitational field of the particle is 0.05 kpc; the radius of the disc of GMCs 1 kpc; $V_0 = 250$ km s^{-1} and $r_0 = 0.3$ kpc.

Figure 1 shows the time-evolution of the velocity dispersion of particles outside of 0.375 kpc from the galactic centre. σ is the total velocity dispersion, and σ_r, σ_θ, and σ_z are its radial, azimuthal, and vertical components, respectively. The unit of time is the galactic rotation period at $r = 1$ kpc, i.e., 2.6×10^7 yr.

The velocity dispersions σ, σ_r, and σ_θ steadily increase with time in a similar manner. In the early phase of the evolution, the velocity dispersions are small but show

* Paper presented at the IAU Third Asian-Pacific Regional Meeting, held in Kyoto, Japan, between 30 September–6 October, 1984.

Astrophysics and Space Science **118** (1986) 463–465.

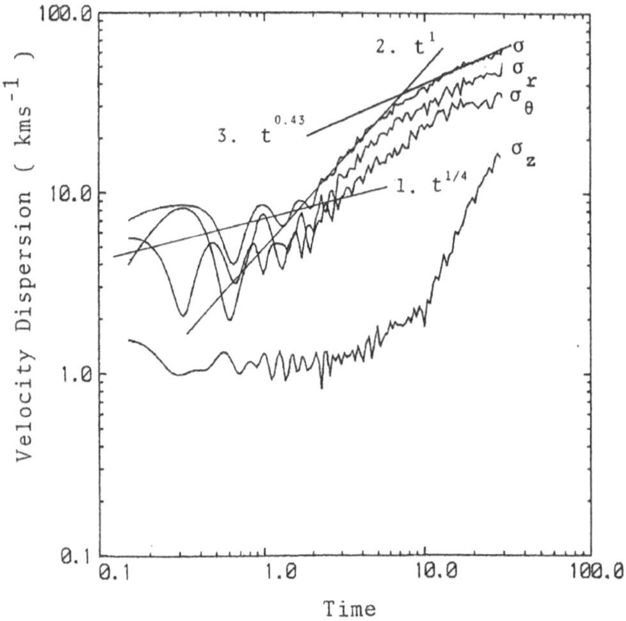

Fig. 1. The time-evolution of the velocity dispersion of GMCs at $r > 0.375$ kpc. σ is the total velocity dispersion; σ_r, σ_θ, and σ_z are the radial, rotational, and vertical components of velocity dispersion, respectively. The straight lines show the theoretically obtained time dependence of the velocity dispersion. The unit of time is the period of circular rotation at 1 kpc, 2.6×10^7 yr.

noticeable oscillations with epicyclic frequency. If these oscillations are averaged, the velocity dispersions increase monotonically with time. The straight line 1 indicates a power law of $t^{1/4}$ predicted for a thin disc of GMCs by Fukunaga (1984b) when the velocity dispersion is very small. In this stage the epicyclic radius is very small and the relative motion which causes encounters among particles is mainly due to shear motion of the system.

Thereafter comes a phase of rapid increase. The straight line 2 shows a power law of t^1 predicted by the theoretical calculation (Fukunaga, 1984a) when the random velocity of GMCs is significant. In this stage, the epicyclic radius becomes large and random motion dominates the relative velocity among particles.

As the velocity dispersion further increases, the rate of increase declines. The evolution in this stage fits to a power law $t^{0.43}$ as is indicated by a straight line 3. In this stage, the increase of the velocity dispersion eliminates large angle deflections of the particle orbits at encounters, because the cut-off radius exceeds the radius at which such deflections take place. This effect overwhelms other effects and reduces the rate of increase of the velocity dispersion.

The velocity dispersions σ, σ_r, and σ_θ follow similar evolution with time. However, the vertical component of the velocity dispersion σ_z first stays small and almost constant, and then starts to increase rapidly.

In conclusion, the present simulation confirms that a random motion of GMCs is generated and accelerated by purely gravitational interactions among GMCs in a differentially rotating galaxy in some 10^{7-8} yr.

The previous theoretical calculations for a thin disc of GMCs, based on an analytically derived heating rate, agree well with the present simulation and are proved to be valid and useful in the study of the dynamical and physical properties of the GMC system.

References

Aarseth, S. J.: 1972, in M. Lecar (ed.), 'Gravitational *N*-body Problem', *IAU Colloq.* **10**, 373.
Fukunaga, M.: 1984, in S. Aoki and Y. Yoshi (eds.), *Theoretical Aspects on Structure Activity, and Evolution of Galaxies, II*, p. 87.
Fukunaga, M.: 1984b, *Publ. Astron. Soc. Japan* **36**, 433.

A DESIGN FOR DIGITAL SIGNAL PROCESSING IN A LARGE FIELD RADIO PATROL CAMERA*

TSUNEAKI DAISHIDO, KUNIYUKI ASUMA, TOHRU OHKAWA, HIRAYOSHI OBARA, SHINICHI KOMATSU

Waseda University, Tokyo, Japan

and

KIYOSHI NAGANE

Nobeyama Radio Observatory, University of Tokyo, Minamimaki, Nagano, Japan

(Received 3 July, 1985)

Abstract. An FFT processor is being developed for the present pilot system of the large field radio patrol camera. A design of the processor is discussed.

1. Introduction

A phased array telescope containing eight elements has been developed at Waseda University as a pilot telescope for a large field radio patrol camera. The full system will be $64 \times 64 = 4096$ elements filled aperture at 10 GHz, with an overall size of 15 m × 15 m. The resolution will be 9 arc min, the field of view $9 \times 9°$, and the sensitivity is 200 mJy at 10 min integration (Daishido *et al.*, 1984). A 4096 points image will be obtained simultaneously, and the camera could map the whole sky once a week. This radio patrol camera will contribute to the search for radio supernovae in our Galaxy, and transient radio sources like Cyg X-3, Cir X-1, and SS 433 (Gregory, 1985). We discuss here a design for the digital signal processing of the image formed in the pilot telescope.

2. Digital Signal Processing

The telescope is a directional finding facility for electro-magnetic waves. Since the number of elements in the radio interferometer is finite, observable directions may be quantized according to the sampling theorem. In this discrete case, we can use a complex FFT processor as a directional finding system, i.e. a 'digital lens'. The benefit of an FFT processor lies in its capability for easy extension. Chikada *et al.* (1984) have developed an FFT processor for spectrum analysis with an image synthesizing telescope.

The FFT processor which is now being designed for use on the eight elements pilot telescope operates at 20 MHz bandwidth per channel (Figure 1). We make use of

* Paper presented at the IAU Third Asian-Pacific Regional Meeting, held in Kyoto, Japan, between 30 September–6 October, 1984.

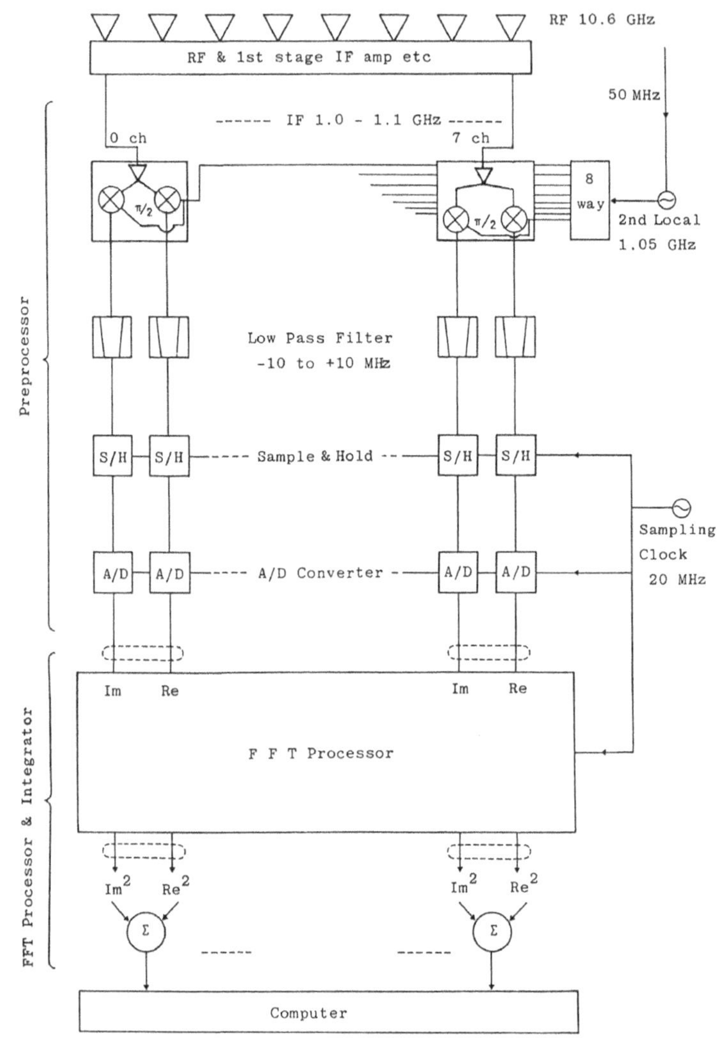

Fig. 1. Block diagram of digital signal processing. Sampling rate is 20 M samples s^{-1} in each real and imaginary part, i.e., the bandwidth is 20 MHz per channel.

consumer multipliers developed for digital VTR. Eight complex signals which have been down-converted to the range -10 to $+10$ MHz are sampled and A/D converted. The processor transforms these signals to the Fourier image, with a 20 MHz bandwidth speed per picture point. The total processing rate in the pilot system corresponds to a 160 MHz bandwidth, while in the full system it will be 80 GHz bandwidth. Twelve butterfly operations are needed in the present system, and 56 multiplier/adder LSI are used.

Acknowledgement

This work is being supported by a Grant-in Aid for developmental Scientific Research, from the Ministry of Education, Science and Culture (59840002).

References

Chikada, Y., Ishiguro, M., Hirabayashi, H., Morimoto, M., Morita, K., Miyazawa, K., Nagane, K., Murata, K., Tojo, A., Inoue, S., Kanzawa, T., and Iwashita, H.: 1984, in *Indirect Imaging*, Cambridge University Press, Cambridge, p. 387.
Daishido, T., Ohkawa, T., Yokoyama, T., Asuma, K., Kikuchi, H., Nagane, K., Hirabayashi, H., and Komatsu, S.: 1984, in *Indirect Imaging*, Cambridge University Press, Cambridge, p. 81.
Gregory, P. C.: 1985, in press.

A FAINT OBJECT SPECTROMETER FOR THE INFRARED*

ALAN T. TOKUNAGA and ROBERT G. SMITH

Institute for Astronomy, University of Hawaii, U.S.A.

(Received 3 July, 1985)

Abstract. A cryogenic astronomical spectrometer for the NASA Infrared Telescope Facility is described. This spectrometer will employ an array of at least 20 detectors and provide a resolving power of 100 to 1500 at 1–5 μm. The resolving power will be adjustable by changing gratings.

Low to moderate resolution near-infrared spectrometers make possible the study of very faint sources, for which little or no spectral data exists. Such a spectrometer has been constructed for the NASA Infrared Telescope Facility (IRTF) at Mauna Kea, Hawaii. Named the Cooled–Grating Array Spectrometer (CGAS), it has a resolving power of 100 to 1500 in the wavelength range 1–5 μm. The resolving power is adjustable by changing gratings. To reduce the thermal emission from the telescope and the instrument itself, the entire spectrometer is cooled with liquid nitrogen and the entrance aperture is small, only 3 arc sec. To maximize observing efficiency, the CGAS will have an array of at least twenty detectors. A spectrometer of this class represents an enormous advance over circular variable filter spectrometers, since not only are observations achieved 20 times faster, but systematic guiding and variable sky transparency errors are much reduced. A drawing of the instrument and its principal components are shown in Figure 1.

The astronomical problems which can be addressed by the CGAS span all fields of astronomy. These include the study of line ratios in QSOs; the energetics of active galactic nuclei; star formation in galaxies; the nature of protostellar objects; abundances in H II regions and the interstellar medium; composition of interstellar grains; composition of comets, asteroids, and satellites; and studies of planetary atmospheres. Of particular interest will be the 3–5 μm spectral range, since this is the spectral range in which the CGAS can make the greatest contribution because of its great sensitivity under conditions of high-background emission. Diagnostic absorption bands of ices such as H_2O and NH_3 exist at 3 μm, as do important emission features at 3.3, 3.4, and 3.5 μm. Also atomic emission lines of H, He, Mg, and N and molecular transitions of H_2 and CO occur at 3–5 μm.

* Paper presented at the IAU Third Asian-Pacific Regional Meeting, held in Kyoto, Japan, between 30 September–6 October, 1984.

Astrophysics and Space Science **118** (1986) 471–472.

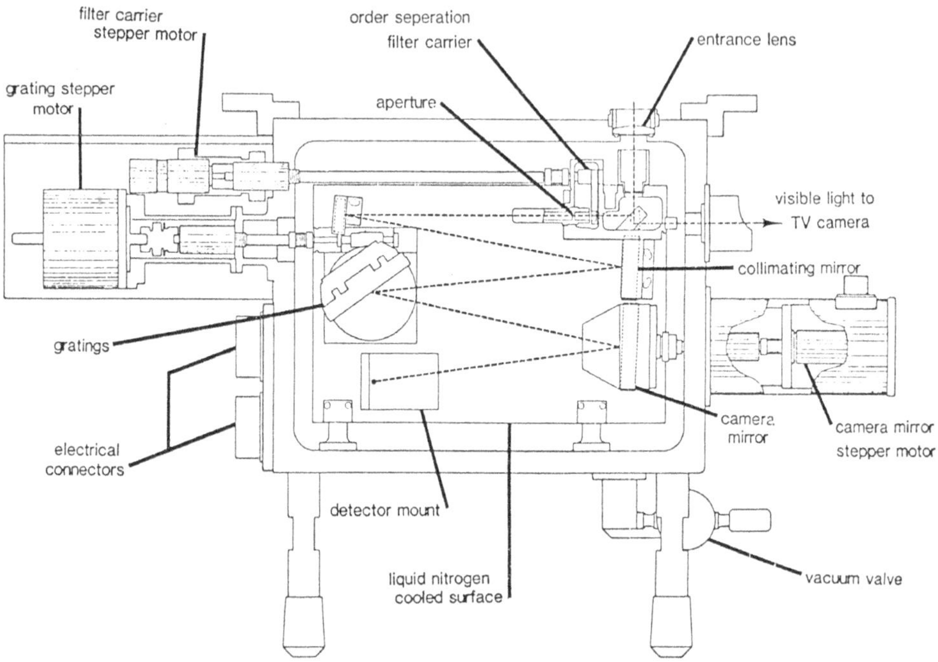

Fig. 1. Cross-section of the Cooled–Grating Array Spectrometer. All of the optical components are cooled to liquid nitrogen temperature (77 K). A cooled radiation shield (not shown) covers all the optics to prevent emission from the cryostat case to reach the detectors. The detectors are cooled to *solid* nitrogen temperature (46 K).

Acknowledgements

The support of NSF grant AST-8216217 and NASA contract NASW-3159 is greatly appreciated.

QSO ABSORPTION LINES (INVITED REVIEW)*

JIAN-SHENG CHEN

Beijing Astronomical Observatory, Academia Sinica, Beijing, China

(Received 19 July, 1985)

Abstract. Some techniques and selection effects in establishing uniform samples of absorption line systems are presented. Summaries of the basic properties of various absorbers are given. Finally, attention is paid to some special, interesting problems which are now widely discussed.

Many important review papers on this field have been published during last few years. Among them Weymann *et al.* (1981) gave the most comprehensive review on all aspects of this topic. Boksenberg and Sargent (1983), Sargent and Boksenberg (1983), and Weymann and Foltz (1983) gave up to date reviews on the subjects of narrow metal absorption line systems, the 'Lα forest' and the broad absorption lines, respectively. Bahcall (1979) reviewed the scientific needs of the ST for the studies of QSO absorption lines, and Chaffee (1983) reviewed the importance of the high resolution QSO absorption spectra. These papers are recommended here to readers who are interested in this field.

1. Techniques, Samples, and Selection Effects

1.1. Our knowledge of the QSO absorption lines is based upon the identifications of the absorption line systems, and the studies of the physical properties of these systems. The qualities of the spectra, including the resolution, signal to noise ratio and the wavelength coverage, are of most fundamental importance. Large increases in both the quality and the quantity of QSO absorption line data over the last few years have been in large measure due to the development of excellent high-resolution detectors, e.g., the IPCS (Boksenberg, 1972). To resolve completely the absorption lines, a spectral resolution as high as 10 km s^{-1} is needed. However, some balance must be taken into account between the observing times of the large telescopes and the qualities of the spectra. 21 cm observation can arrive at much higher resolution than the optical observations, but are limited to a few high-hydrogen column density systems. Most useful data come from the intermediate resolution spectra of 1–2 Å, with wavelength coverage from 3000 Å to 6000 Å, of high signal to noise ratio and which can be taken within a reasonably large telescope time (usually 1–2 clear nights for a 19 mag. QSO in 4 m telescope). These intermediate resolution spectra are now the most important basis for the studies of both the physical and statistical properties of the absorbers.

* Paper presented at the IAU Third Asian-Pacific Regional Meeting, held in Kyoto, Japan, between 30 September–6 October, 1984.

Astrophysics and Space Science **118** (1986) 473–486.

1.2. The line lists established from the spectra are the basis for the identifications of the systems. This is not difficult for strong narrow lines located at the long wavelength side of the Lα emission to be determined. However, it is not so reliable for weak lines merged each other in the Lα forest region. The continuum is suppressed in line crowded region. This, and the line blending, make the equivalent width of the lines uncertain, and even put the reality of the line in question. Important progress has been achieved by Young *et al.* (1979) invoking computer-based automatic continuum level determination, as well as actual line picking. How successful this has been in dealing with the line blending problem is not clear. In many cases some subjective choice, with a more or less artificial treatment, is unavoidable. All such uncertainties will finally impair the correctness of the identification of the redshift systems.

1.3. The dual difficulties of the determinations of redshift and ion line identification of observed systems have lead to the automatic computer searching method which has been described by Bahcall (1968). The principle of the method is to search by computer over a certain redshift interval for observed spectral lines which will coincide with a given set of standard lines, within a pre-set wavelength error requirement. The standard lines are selected on the basis of ultraviolet observations of interstellar absorption lines obtained by satellites (Morton, 1975). Many astrophysical constraints, such as the oscillater strengths within multiple lines have to be considered. The quality of the spectra, the blending and the uncertainty in equivalent width will definitely influence the identification. In effect, the distribution of the standard lines in wavelength and the differences in observed line densities from QSO to QSO, and also from wavelength region to region (e.g., blueward to redward of the Lα emission) will make the criteria of the identification of the systems heavily redshift and object dependent. Chen (1983) discussed selection effects in conventional computer searching for the absorption line systems. It was pointed out that unique criteria would tend to lead to identifications biased towards higher redshift, where more standard lines enter the observing window. Also, for high redshift QSOs, the Lα forest will greatly influence identifications, in the sense that the most important double lines, like C IV, Mg II, Si IV, Al III, will be ignored due to difficulties in distinguishing genuine from chance coincidences.

2. The Basic Properties of the Absorbers

2.1. The Classification of Absorption Line Systems

The systems identified from various QSO spectra can be classified by the degree of ionization, i.e., low (containing Lα and lower ionization ion lines), intermediate (containing typically C IV, Si IV) and high ionization (containing N V, O VI). There are also some 'mixture' systems, in which ion lines from different ionization stages occur together. This classification scheme correlates with that of Weymann (1980), who classified absorption line systems according to line shape and the location of the system relative to some emission redshift. Types *A* and *B* in the Weymann scheme always have high ionization, due to the proximity of the absorber to the QSO, type C is always shown to correspond to the intermediate ionization system. The typical self-damped hydrogen systems which have high H I column density $\log N(\text{H I}) > 19\ \text{cm}^{-2}$, can contain the subclass of narrow metal line systems, usually having lower ionization. The frequency of such systems, the high H I column density, as well as the lower ionization, all suggest that these systems are probably due to the absorption from discs of intervening galaxies distributed along the line of sight. The modified Weymann scheme, and a possible interpretation of the origin of the absorbers, can be suggested as shown in Table I.

TABLE I

Classification of absorption line system

Type	System	Ionization	Absorber
I	Broad metal absorption	High	Ejected from QSO
IIa	Narrow metal line	High	Close to QSO
IIb	Narrow metal line	Intermediate	Halo of galaxy $V > 0.01$ cm
IIc	Self-damped Lα	Low	Disc of galaxy
III	Lα forest	(Primordial?)	Intergalactic clouds?

2.2. Narrow Metal Line Systems

Brief summary of the basic physical properties of the absorbers:

(1) Velocity dispersion directly measured from the line widths is between 20 and 100 km s^{-1}.

(2) Kinematic temperature of the absorber $T = 1.8 \times 10^4$ K.

(3) Column density of the ions is 10^{13}–10^{16} cm^{-2}.

(4) Relative metal abundance is about 0.1 of the solar value.

(5) The neutral hydrogen column density $N(\text{H I})$ in the system (except the self-damped hydrogen system) is 10^{17}–10^{19} cm^{-2}.

(6) Electron density $n < 1$–2 cm^{-3}.

(7) The lines have velocity structure, from 40 km s^{-1} to a few thousand km s^{-1} with the peak at 150 km s^{-1}.

(8) Spin temperature from the 21 cm observations is about 100–500 K.

(9) Size of the absorbing region is 8 kpc $< D <$ 500 kpc.

(10) No variation in redshift of the absorption line systems has been found. Hunstead *et al.* (1983) found a time-scale of 11 days for the variation in intensity.

(11) The high-ionization line N V only appears for systems with a relative velocity from the QSO < 0.01 c (Hartquist and Snijders, 1982).

(12) The excited fine line C II* λ1335.7 is rare, and Si II* λ1265 has not been convincingly observed.

(13) The 21 cm absorption line is rare. So far it has only been detected in 3C286 (Wolfe *et al.*, 1976), PKS 1229 − 02 (Wolfe, 1980), PKS 1157 + 014 (Wolfe *et al.*, 1981), and 1331 + 170 (Wolfe and Davis, 1979). Its low occurrence is consistent with the absorber having a cross section of the order of the Holmberg radius of a galaxy.

(14) For the origin of the absorption line system, most data accumulated so far lean to the intervening galaxy hypothesis. Apart from statistical studies, evidences can also be provided by observations of the QSO-galaxy close pairs in the sky plane (3C232, Boksenberg and Sargent, 1978; Haschick and Burke, 1975; QSO 0046 – 208; Blades *et al.*, 1981; PKS 2020 – 370, Boksenberg *et al.*, 1980) and the close double QSO pairs (Shaver and Robertson, 1983a). The differences in ionization and frequency of the ion systems are likely to be due to differences in the cross section of the absorbers, as well as the distance of the absorber from the center of the intervening galaxy and the distance of the galaxy from the QSO.

2.3. The Lα Forest

The high contrast in line density, and the fraction of unidentified lines on both sides of the Lα emission line in high redshift QSOs lead to the Lynds (1971) hypothesis that most unidentified lines shortward of the Lα emission are due to Lα absorptions at different redshifts. Lα and Lβ correlation has been found (Young *et al.*, 1979; Chen *et al.*, 1981) and Lα and Lγ correlation in PKS 1442 + 101 (OQ172) was recently discovered by Peterson *et al.* (1985). Various statistical studies for a uniform Lα sample have been done (Sargent *et al.*, 1980; Zou *et al.*, 1982; Sargent and Boksenberg, 1983). These results are thought to favour the intervening galaxy hypothesis origin. The undetected metal abundance, and a different behaviour of the two point correlation function of Lα clouds from that of C IV systems have been considered by Sargent *et al.* (1980) as evidence for Lα clouds to have a particular population (primordial hydrogen clouds) in the Universe. Several different mechanisms for the origin of these clouds have been suggested, among which there are:

(1) That they are pressure confined by the hotter intergalactic medium (Sargent *et al.*, 1980; Ostriker and Ikeuchi, 1983).

(2) They come from an extended, diffuse, hydrogen halo in the outermost part of the intervening galaxy (Bahcall, 1979).

(3) They represent winds from dwarf irregular galaxies (Fransson and Epstein, 1982).

(4) They come from residual gas in a supercluster (Oort, 1981).

(5) They are due to an intergalactic shockwave (Chernomordik and Ozernoy, 1983).

The basic physical properties of the clouds can be summarized as:

(1) The dispersion velocity of the clouds is between 10 and 60 km s^{-1}.

(2) The kinematic temperature is about 10^4 K.

(3) The size of the clouds is between $10^{21} < D < 10^{23}$ cm (Sargent *et al.*, 1980, 1982; Weyman and Foltz, 1983; Shaver and Robertson, 1983c; Foltz *et al.*, 1984).

(4) The column density N(H I) is around 10^{14}–10^{15} cm^{-2} (Sargent *et al.*, 1981; Chen *et al.*, 1981).

(5) The ratio of H II to H I is about 1000.

(6) The total hydrogen density $n(\text{H I}) + n(\text{H II}) = 10^{-4}$ cm^{-3}.

(7) The average mass of a cloud is 10^6 solar mass.

The question of the evolution of such clouds in a co-moving coordinate system is a widely discussed problem, which will be considered in some detail later.

2.4. Broad absorption line (BAL) systems

The number of known BAL QSOs is presently about 40, and continues to increase rapidly. This is due to the discovery technique of low dispersion objective prism survey. Hazard *et al.* (1984) has discussed the selection effect of using this technique to find BAL QSOs. BAL QSOs can only be recognised when at least some of the region between C IV λ1550 and N V λ1240 lies within the response of the IIIaJ prism plates. To recognise a C IV trough, an appreciable stretch of continuum around Si IV λ1398 must be visible, that is $1.4 < Z < 3$. Among the QSOs satisfying this condition, the redshift interval $2.1 < Z < 2.3$ is particularly favoured, for then, all three troughs of C IV, Si IV, and N V are visible together. Four of the nine newly discovered BAL QSOs (Hazard *et al.*, 1984) lie in the very narrow redshift range $2.2 < Z < 2.31$. Relatively shallow absorption troughs are always going to be difficult to recognise. Broad absorption line stars, such as white dwarfs, M stars and carbon stars are also a source of confusion. Carbon stars, in particular, have spectra which on IIIaJ plates closely mimic a BAL QSO, having three absorption edges which, within the accuracy of measurements in the prism plates, correspond to C IV λ1550, Si IV λ1398, and N V λ1240 at $Z = 2.2$. These stars can, however, usually be eliminated from consideration on the basis of their colours.

Hazard *et al.* (1984) made a probability estimation that among bright (< 18 mag.) QSOs in the restricted redshift range 2.2–2.4 the fraction of 'extreme' BAL QSOs is at least 3%, and possibly as high as 10%. This fraction may be redshift and possibly luminosity-dependent, with BAL QSOs becoming more frequent at higher redshift. Turnshek (1984) summarized the basic statistical properties of BAL QSOs as follows:

(1) The total column density, under the assumption C IV/C I ~ 30%, is about 10^{20}–10^{21} cm^{-2}.

(2) The level of ionization in the BAL clouds is high – H II/H is $\sim 10^5$.

(3) Excluding QSOs with marginal BAL characteristics, none of the known BAL QSOs are strong radio sources.

(4) BAL QSOs can be X-ray sources.

The BAL QSOs exhibit a variety of structure in the absorption troughs. Outflow velocities near 65000 km s^{-1} have been reported, but 0–20000 km s^{-1} is more typical. The structure ranges from complexes of individual absorption lines to seemingly smooth, continuous absorption. Absorption can be adjacent to the emission or detached from it. Multiple troughs are often present. The C IV emission lines in BAL QSOs have smaller peak central intensities relative to the continuum when compared with non-BAL QSOs, and vary among themselves. A relation between BAL profile complexity and the emission line profile may exist. The N V emission line in BAL QSOs is strong relative to the C IV emission line and the continuum; this is in contrast with the results for non-BAL-QSOs.

Two schemes of classification of the BAL QSOs have been suggested. According to the red edges of the broad troughs, the shape of the bottoms, the blue edges, and the velocity displacement of the red edges, Weymann and Foltz (1983) classified the

BAL QSOs into six types excluding some unclassified objects. The order of the classification from type I to type VI depends on the increase of the displacement of the red edge from the emission, the decreasing steepness of the blue edge, and the increasing complexity in the bottom of the trough. This order of classification is also correlated with the intensity of C IV emission relative to the continuum.

In a different approach, Hazard *et al.* (1984) classified the system in accordance with temperature, which is characterized by the degree of the ionization. They suggested that the BAL QSOs are indeed young QSOs and that many of the spectral differences can be understood in terms of an evolutionary sequence. Such a sequence corresponds to a temperature sequence as the ejected material expands until, with its eventual dissipation, we are left with the more normal emission line QSO. Of course, there is a very long way to go before these complex and fascinating objects are well understood.

3. QSO Absorption Lines as Tool Probing the Universe

The narrow absorption systems are most likely to arise in intervening galaxies and hydrogen clouds distributed cosmologically along the line-of-sight. The detailed study of QSO absorption is, therefore, the only method presently available for probing the composition and the physical conditions in the early Universe. A lot of progress has been achieved along this line.

3.1. Do Lα clouds evolve with the epoch?

Under the assumption of the intervening galaxy hypothesis, the distribution of the Lα absorption lines, produced by intergalactic clouds with invariant cross-sections and a uniform space distribution, is given as a function of redshift; thus

$$\frac{dN}{dz} = \frac{C}{H} \sigma \rho_0 \frac{(1 + z)}{(1 + 2q_0 z)^{1/2}} ,$$

where σ is the cloud cross section; ρ_0, the number of clouds per unit volumn at the present epoch; H_0 and q_0, the Hubble constant and the deceleration parameter at the present epoch; and C, the speed of light. Peterson (1978) counted the number of the absorption lines in four QSOs at various redshifts, and has pointed out that the number density of the clouds shows evolution with epoch. Using a different sample of five QSOs Sargent *et al.* (1980) reached a different result; namely, that the number counts of Lα clouds is consistent with a Friedmann cosmology. Since then many workers have followed to study this problem with more QSO samples. Table II summarizes all these works. After a few years of controversy a general conclusion has been reached that the counts of Lα absorption lines do show clear evidence for evolution with redshift.

However, Carswell *et al.* (1982) pointed out that the dN/dZ is, to some extent, method-dependent. Using these methods for their sample they obtained three extremely different γ-values, namely $\gamma = 1.4$ from Peterson's method; $\gamma = 0.6$ by the method of Sargent *et al.*, and $\gamma = -2.1$ when some other method and taking the weighted mean

TABLE II

Summary of the works of dN/dZ of Lα clouds

Sample	Method	γ	References
1, 2, 3, 4	A	(rapid increase)	Peterson (1978)
5, 6, 7, 8, 9	B	0.48 ± 0.54	Sargent *et al.* (1980)
5. 6. 7. 8. 9. 10, 11, 12, 23	C	No evolution	Zou *et al.* (1982)
5, 6, 7, 8, 9, 17, 18, 19, 21		1.81 ± 0.48	Young *et al.* (1982)
5, 6, 7, 8, 15, 16	A	1.4 ± 0.7	Carswell (1982)
	B	0.6 ± 0.6	
	D	-2.1 ± 1.5	
5, 6, 7, 8, 9, 10, 11, 12, 23, 17, 18, 19, 21		1.9	Chen *et al.* (1983)
5, 6, 7, 8, 9	A	1.6 ± 1.3	Peterson (1982)
13, 10, 12, 8, 5, 6, 9, 7, 14, 15, 16, 17, 18, 19	A	2.2 ± 0.4	Peterson (1982)
5, 6, 7, 8, 9, 10, 11, 12, 23, 19, 21, 18, 15, 16, 22, 24, 25, 26, 13, 20	A	2.1	Sargent and Boksenberg (1983)
5, 6, 7, 8, 9, 10, 12, 13, 15, 16, 17, 18, 19, 21, 24, 25	A	2.36 ± 0.36	Peterson (1983)
?	?	(evolution)	Hunstead *et al.* (1984)

1 PKS 1442 + 101 (Peterson, 1978)
2 PKS 0805 + 046 (Peterson, 1978)
3 PKS 0329 − 255 (Peterson, 1978)
4 PKS 1448 − 232 (Peterson, 1978)
5 PKS 2126 − 158 (Young *et al.*, 1979)
6 Q0042 − 422 (Sargent *et al.*, 1979)
7 PHL 957 (Sargent *et al.*, 1980)
8 Q0453 − 423 (Sargent *et al.*, 1979)
9 B1225 + 31.7 (Sargent *et al.*, 1980)
10 PKS 1442 + 101 (Peterson *et al.*, 1985)
11 PKS 0528 − 250 (Morton *et al.*, 1980)
12 PKS 0805 + 046 (Chen *et al.*, 1981)
13 PKS 2000 − 330 (Hunstead *et al.*, 1984)
14 Q0420 − 388 (Smith, 1978)
15 Q0122 − 380 (Carswell *et al.*, 1982)
16 Q1101 − 264 (Carswell *et al.*, 1982)
17 Q1115 + 080 (Young *et al.*, 1982)
18 Q0119 − 046 (Sargent *et al.*, 1982)
19 Q0002 + 051 (Young *et al.*, 1982)
20 Q0237 − 233 (Sargent *et al.*, 1980)
21 Q0421 + 019 (Young *et al.*, 1979)
22 Q1704 + 710 (Sargent and Boksenberg, 1983)
23 Q1448 − 232 (Chen *et al.*, 1984)
24 Q1623 + 268 (Sargent *et al.*, 1982)
25 Q1623 + 269 (Sargent *et al.*, 1982)
26 Q2206 − 199 (Sargent and Boksenberg, 1983)

A Peterson (1978) method
B Sargent *et al.* (1980) method
C Zou *et al.* (1982) method
D Carswell (1982) single QSO method

of each QSO separately. If line density increases with redshift then Peterson's method would make the shape of dN/dz flatter than that of Sargent *et al.*, because Peterson's method requires a larger z interval to calculate the value of dN/dz. One speculation is that the result from the third method (D) cannot be simply due to the selection effects discussed by Carswell (1982); namely, blending, the contamination of Lα, and the size of the sample. Recent work by Bian *et al.* (1985) using all sixteen published QSO absorption line spectra shows that the line density does increase with the emission line redshift of the QSOs, while the line density along each QSO spectrum is statistically random with the wavelength. Bian *et al.* (1985) examined four different wavelength regions (for each wavelength region the redshift of the clouds for the different QSOs is, of course, the same) the density of pure Lα lines does indeed increase with the emission redshift of the QSO. Before a larger uniform sample is available it is hard to do anything further than call attention to the fact that it is very difficult to understand how most of the absorbing clouds could be at cosmological distances and unconnected with the QSOs.

3.2. Are Lα clouds primordial?

To confirm observationally the metal abundance in Lα clouds is difficult, because according to the ionization calculation by Carswell (quoted from Weymann *et al.*, 1981), the column densities of highly ionized species such as C IV and N V will be at most 3% that of the neutral hydrogen component if the metal abundance is $\sim\frac{1}{10}$ of solar values. Even with N(H I) as high as 10^{16} cm^{-2}, which is rare in Lα clouds, a very moderate additional reduction in C/H would make a C IV line undetectable. In order to improve the detection sensitivity of metal line abundance in Lα clouds the same statistical method has been applied by Sargent *et al.* (1982) and Norris *et al.* (1983). They added the portion of the spectrum in which, for example, the C IV $\lambda\lambda$1548, 1550 doublet would be expected, corresponding to each Lα line after being reduced to the rest frame. With 100 Lα lines it is possible, in principle, to increase the effective signal to noise ratio of the spectra by a factor of 10 in the region of longward of the Lα emission. The strongest expected ions – C IV, Si IV, N V, C II, Si II – have not been detected in both analyses. Sargent *et al.* obtained the upper limit of the rest equivalent width of the C IV 1548 in Lα clouds in Q1623 + 267 (Z_e = 2.52) as $W_0(1548) < 0.003$ Å; about a factor of 100 weaker than the same ion lines seen in a typical heavy element redshift system. On the other hand, Norris *et al.* (1983) added the rest frame spectra of 65 clouds that produced both Lα and Lβ lines in the spectra of the two QSOs 0805 + 046 (Chen *et al.*, 1981) and PKS 1442 + 101 (Peterson *et al.*, 1985). They concluded that they were able to detect O VI at the 99% confidence level, and measured a column density of log$N$(O VI) = 13.8 in the Lα/Lβ clouds, with log$N$(H I) = 14.9, and an upper limit of 13.2 for log$N$(C IV) and 13.5 for log$N$(N V). Norris *et al.* (1983), therefore, concluded that, the abundances in the clouds are similar to those of the heavy elements found in isolated extragalactic H II regions, and in Population II material in our Galaxy. The clouds are thus not primeval, and the result can be understood in terms of Oort's hypothesis, that the Lα/Lβ clouds are remnants of material which formed superclusters, and the clouds have been enriched by matter driven out by winds from the clusters of galaxies which comprise the superclusters. However, in the plot of the composite spectrum of Norris *et al.* (1983) the line center of the suggested O IV does not agree well with the rest wavelength, and this O VI doublet is located in the Lα forest region, and so has a higher background noise than the C IV doublet. Chen and Morton (1984), using the same technique and the same data, summed all the rest frame spectra of 55 clouds that have both Lα/Lβ lines in 0805 + 046 and can not find convincing evidence for the O VI line. A later analysis of Q1623 + 269 also did not reveal any evidence for O VI (Sargent and Boksenberg, 1983). Using the Lyman limit system, which consists of twelve lines of Lyman series at Z_a = 2.9982, in the spectrum of Q2126 − 158 (Z_e = 3.28) Sargent and Boksenberg found a hydrogen column density of $\log N(\text{H I}) = 17.3$, and velocity dispersion $v = 23 \pm 1$ km s^{-1}. They found a possible O VI line, however, the strength of the doublet is inverse and the 1031 feature is itself blended with a strong single Lα absorption line. The rest equivalent width of the 1037 feature is $W_0 = 0.16$ Å giving a column density $N(\text{O VI}) \leq 2.6 \times 10^{14}$ cm^{-2}. The O VI/H I abundance ratio provided by this one system is $\log(\text{O VI/H I}) < -2.9$, considerably below the value found by Norris *et al.* (1983). Foltz *et al.* (1983) have found a Lα line toward 0014+81 ($Z_e = 3.42$) for which $N(\text{H I}) \sim 10^{17}$ cm^{-2}. They have set stringent upper limits for the column densities of C II and C IV in this cloud, and from plausible photoionization models suggest that $Z/H < 10^{-3}$ of the solar value. Concerning the definition of the 'primordial' cloud it would be interesting to point out the recent result for metal poor stars in our own Galaxy, obtained by Bessell and Norris (1984) from objects HD 2796 and CD–38°245. The spectrum of HD 2796 shows a star with a very low abundance of heavy elements – about $\frac{1}{250}$ that of the Sun, with [Fe/H] = –2.4. The spectrum of CD–38°245 shows even lower metal abundances. With [Fe/H] = –4.5 this is lower than any other by a clear factor of 10. It would be interesting to see if the Lα clouds have a metal abundance lower than that of these K giants in our own Galaxy.

Chen and Morton (1984) reobserved the spectrum of the QSO PKS 0528 – 250 with a much improved signal to noise ratio in the region of the longward side of the Lα emission. In the autocorrelation function of this new spectrum there are two peaks due to the C IV doublet, further demonstrating that weak lines of this ion are prevalent in the spectrum. However, the corresponding Lα absorption is often missing. Thus, they have evidence for two types of narrow-line absorption systems: toward Q0805 + 046 there are numerous Lα and Lβ lines without detected C IV, while toward PKS 0528 – 250 some of the C IV systems do not have a measurable H I line.

3.3. A Search for Deuterium in the Early Epoch of the Universe

If H I Lα clouds do contain unprocessed material, then we may attempt to determine the cosmologically important D/H ratio in them. Since the velocity difference between the Lα line of D I and H I is only 81 km s^{-1}, and the D/H in the Galaxy is probably in the range 10^{-4} to 10^{-5}, such a detection is very difficult. For the H I Lα forest in high redshift QSOs the Doppler width of lines is so large that a H I Lα line of sufficient optical depth to produce a measurable D I Lα would swallow the D I feature. Only a very narrow combination of plausible column density of neutral hydrogen ($> 10^{17}$ cm^{-2}) and velocity dispersion (~ 15 km s^{-1}) will permit D I Lα to be detectable. Even if the Doppler width is as low as 10 km s^{-1}, very few Lα clouds having the requisite $N(\text{H I}) > 10^{17}$ cm^{-2} could be found. Carswell *et al.* (1984) pointed out another difficulty in that since the Lα line density is rather high, there is a good chance of a pair being present with velocity separation of about 80 km s^{-1}, which will mimic a D/H separation. Their observation of Q1101–264, with a resolution of 0.25 Å (FWHM) leads to six systems with a formal D/H number ratio ranging from 0.05 to 4 (but the column density is too low at $\sim 10^{14}$ cm^{-2}). This likelihood will make future seach for deuterium, in general, complicated. In another QSO 0420 – 388 ($Z_e = 3.12$) (Atwood *et al.*, 1983) six separations in the range 70–90 km s^{-1}, with formal D/H ratio ranging from 0.2 to 3, existed.

3.4. The correlation between the absorption line spectra of close QSO pairs

There are several suitable pairs of QSOs in the published catalogue (see, for example, Hewitt and Burbidge, 1980). The studies of the absorption line spectra of these pairs provide a new approach for many important astrophysical problems. There are three kinds of correlation between absorption lines in close QSO pairs:

(i) Common absorption line systems, that is absorption in both spectra at the same redshift.

(ii) Associated absorption systems, that is absorption at the redshift of the foreground QSO in the spectrum of the other member of the pair.

(iii) Statistical studies of the common lines in the Lα forest between the spectra of the close QSO pair.

The studies of these correlations can address several issues, as pointed out by Shaver and Rorbertson (1983a) as follows:

(i) The cosmological interpretation of QSO redshifts.

(ii) The origin of the narrow absorption lines (intrinsic or intervening).

(iii) The lateral sizes of typical absorption regions and their clustering.

(iv) The absorbing matter near QSOs-clusters of galaxies and QSO halos.

According to the cosmological interpretation, the foreground QSO should have the lower redshift of the pair. This is true from all the observations so far obtained (Q0028,9 + 003; Shaver *et al.*, 1982; Q0254 – 334A, B; Wright *et al.*, 1982; Q1228 + 076,7; Robertson and Shaver, 1983), and no case has been found in which the foreground QSO has the higher redshift. Shaver and Robertson pointed out that it is the symmetry of QSO pairs – both members are QSOs which makes this test particularly powerful.

The associated absorption systems also provide direct evidence that at least some high redshift absorption systems are due to intervening matter. The galaxy-QSO pairs investigated previously only samples for low redshift and excitation. The absorption of the associated kind can come from the outer part of the QSO, or the matter in the cluster which contains the QSO. In QSO pairs Q0307 – 195A, B and Q0028,9 + 003 the difference between the absorption and emission redshift is so small (<200 km s^{-1}) that Shaver and Robertson suggested that the absorption arises in a halo of the foreground QSO itself. Therefore, QSOs may be surrounded by halos with diameters of order 1 Mpc. Nebulosity surrounding low redshift QSO has been strongly supported by many QSO imaging observations (see, for example, Wyckoff *et al.*, 1983; Hutchings *et al.*, 1983). Common systems, with redshift much smaller than those of the QSO themselves, occur in the gravitational lens pair Q0957 + 561A, B (Young *et al.*, 1981) for which the projected separation is very small, as well as in Q0307 – 195A, B (Shaver and Robertson, 1983b), for which the projected separation is 376 h^{-1} kpc. Common absorption is evidently not very prevalent at projected separations of a few hundred kpc or larger. Therefore, it seems that the typical heavy element absorption region is no larger than several hundred kpc and that if such regions are clustered together at all, the absorption cross-section of the clusters is small. This is consistent with the suggestion that heavy element absorption systems originate in the halos of intervening galaxies. By cross-correlating the Lα forest in the two spectra of a QSO pair Sargent *et al.* (1982) found no significant correlation for the pair Q1623 + 268,9. The linear separation is $\sim 1\ h^{-1}$ Mpc over the relevant range of redshift. A marginal correlation has been found for the pair Q0307 – 195A, B (Shaver and Robertson, 1983a) at a projected separation of $0.4\ h^{-1}$ Mpc. It is very desirable to use high redshift QSO pairs of relatively low separation to set the constrains on the size of the Lα clouds by cross-correlation studies.

3.5. Detection of Molecular Hydrogen Clouds at High Redshift

The detection of molecular hydrogen at high redshift will be of great interest for cosmology. Bahcall and Salpeter (1965) were the first to point out that the abundance of intergalactic molecular hydrogen can be determined by investigating the absorption which molecular hydrogen causes in the Lyman continua of objects with redshift $Z_e > 2$. Levshakov and Varshalovich (1979) have identified molecular hydrogen with two redshift systems at $Z = 2.651$ and 3.092 using a spectrum obtained by Baldwin *et al.* (1974). Peterson *et al.* (1985) have obtained a spectrum of OQ172 at higher resolution. Peterson (1982) and Cui *et al.* (1983) have examined the same spectrum separately and concluded that there is no molecular hydrogen at the two redshifts found by Levshakov and Varshalovich (1979). Peterson further concluded that there is no other redshift for molecular hydrogen in the temperature range 3 K to 2000 K. The method used by Peterson is to cross correlate the standard molecular hydrogen line in the Lyman and Warner band with the observed spectrum of OQ172. Since most of the important molecular hydrogen lines in these bands would be located shortward of the Lα emission, it is not clear how one can properly estimate the upper limit of the abundance of molecular hydrogen from the noise contributed by the crowded Lα forest. Cui *et al.* (1983) used a different method and found a marginally detectable system at $Z = 2.273$ for molecular hydrogen in the OQ172 spectrum. Therefore, this question is still open.

3.6. The ST Program for QSO Absorption Line Studies

The high resolution in angular separation and in wavelength, as well as faint limiting magnitude possessed by the ST will enable a great capability to explore the Universe using QSO absorption lines. A comprehensive account has been given by Bahcall (1979). Here only a few topics are listed:

(i) To determine the border of the Lα forest at lower redshifts (so far the smallest redshift for a Lα cloud is $Z = 1.5$, from observations of a BL Lac object by Hunstead *et al.*, 1984).

(ii) To find the helium abundance in the early epoch. In a high redshift QSO absorption line spectrum the helium resonance lines would be observable, i.e., He I λ584, 537, 522 Å at $Z > 1.1$ and He II λ304 Å at $Z > 2.6$. It would be especially interesting to see if any 'Helium forest' exists in the ultraviolet region where a Lα forest appeared in the visual wavelength for the same high redshift QSO.

(iii) QSOs having strong UV radiation distributed behind some rich Abell clusters can serve as a net of probers to study the distribution, chemical composition, extent and the physical characteristics of moderate temperature gas in clusters of galaxies.

(iv) To study the frequency of the redshift systems containing both C IV and Mg II, so as to reveal the different cross-sections of these two ions and, thereby, the location of the absorber in the intervening galaxies.

(v) To study the evolution of the absorption system. The evolution of the system can be measured by comparing the spectra of a representative sample of large-redshift quasars observed with ground-based telescopes in the visible region, with the spectra of a similar sample of low redshift quasars observed in the ultraviolet with ST.

(vi) To study the D/H ratio. The advantage of the ST to study D/H is that it is possible to use small redshift QSOs, which are bright and have less crowded spectra.

(vii) To study the gaseous content of large (> 100 kpc) galactic halos of nearby galaxies.

We expect that ST will definitely throw light on many of these questions, though also open many new exciting questions before astronomers.

Acknowledgements

I would like to express my thanks to Drs Morton, Hunstead, Murdoch, and Robertson for sending me unpublished data, as well as for numerous discussions.

References

Aaronson, M., Black, J. H., and McKee, C. F.: 1974, *Astrophys. J.* **191**, L53.
Atwood, B., Baldwin, J. A., and Carswell, R. F.: 1983, in 'Quasars and Gravitational Lenses', *Proc. 24th Liège International Astrophysical Colloquium*, p. 581.
Bahcall, J. N.: 1968, *Astrophys. J.* **153**, 679.
Bahcall, J. N.: 1979, 'Scientific Research with the Space Telescope', *IAU Colloq.* **54**, 215.
Bahcall, J. N. and Salpeter, E. E.: 1965, *Astrophys. J.* **142**, 1677.
Baldwin, J. A., Burbidge, E. M., Burbidge, G. R., Hazard, C., Robinson, L. B., and Wampler, E. J.: 1974, *Astrophys. J.* **193**, 513.
Bessel, M. S. and Norris, J.: 1984, *AAO Newsletter*, No. 29.
Bergeron, J. and Boisse, P.: 1983, in 'Quasars and Gravitational Lenses', *Proc. 24th Int. Astrophys. Colloq.*, p. 589.
Bian, Y.-L., Chen, J.-S., and Zou, Z.-L.: 1985, *Acta Astrophys. Sinica* (in press).
Blades, J. C., Hunstead, R. W., and Murdoch, H. S.: 1981, *Monthly Notices Roy. Astron. Soc.* *194*, 669.
Boksenberg, A.: 1972, in *Auxiliary Instrumentation for Large Telescopes*, Proc. ESO/CERN Conf., Geneva, p. 295.
Boksenberg, A. and Sargent, W. L. W.: 1978, *Astrophys. J.* **220**, 42.
Boksenberg, A. and Sargent, W. L. W.: 1983, in 'Quasars and Gravitational Lenses', *Proc. 24th Liège Intern. Astrophys. Colloq.*, p. 504.
Boksenberg, A., Danziger, I. J., Fosbury, R. A. E., and Goss, W. M.: 1980, *Astrophys. J.* **242**, L145.
Carlson, R. W.: 1974, *Astrophys. J.* **190**, L99.
Carswell, R. F., Whelan, J. A. J., Smith, M. G., Boksenberg, A., and Tyler, D.: 1982, *Monthly Notices Roy. Astron. Soc.* **198**, 91.

Carswell, R. F., Morton, D. C., Smith, M. G., Stockton, A. N., Turnshek, D. A., and Weymann, R. J.: 1984, *Astrophys. J.* **278**, 486.
Chaffee, F. W.: 1983, *Publ. Astron. Soc. Pacific* **95**, 700.
Chaffee, F. W., Weymann, R. J., Latham, D. W., and Strittmatter, P. A.: 1983, *Astrophys. J.* **267**, 12.
Chen, J.-S.: 1983, in 'Quasars and Gravitational Lenses', *Proc. 24th Liège Intern. Astrophys. Colloq.*, pp. 560.
Chen, J.-S. and Morton, D. C.: 1984, *Monthly Notices Roy. Astron. Soc.* **208**, 167.
Chen, J.-S., Morton, D. C., Peterson, B. A., Wright, A. E., and Jauncey, D. L.: 1981, *Monthly Notices Roy. Astron. Soc.* **196**, 715.
Chen, J.-S., Morton, D. C., Peterson, B. A., Wright, A. E., and Jauncey, D. L.: 1984, *Proc. ASA* **5**, 355.
Chen, J.-S., Zou, Z.-L., Cui, Z.-X., Bian, Y.-L., Tang, X.-Y., and Sun, S.-D.: 1983, in *Proc. Third Marcel Grossmann Meeting on General Relativity*, p. 855.
Chernomordik, V. V. and Ozernoy, L. M.: 1983, *Nature* **303**, 153.
Coleman, G., Carswell, R. F., Strittmatter, P. A., Williams, R. E., Baldwin, J., Robinson, L. B., and Wampler, E. J.: 1976, *Astrophys. J.* **207**, 1.
Cui, Z.-X., Chen, J.-S., and Sun, S.-D.: 1983, *Acta Astron. Sinica* **24**, 233.
Field, G. B., Solomon, P. M., and Wampler, E. J.: 1966, *Astrophys. J.* **145**, 351.
Foltz, C. B., Chaffee, F. H., Jr., and Latham, D. W.: 1983, *Publ. Astron. Soc. Pacific* **95**, 589.
Foltz, C. B., Weymann, R. J., Roser, H.-J., Chaffee, F. H., Jr.: 1984, *Astrophys. J.* **281**, L1.
Fransson, C. and Epstein, R.: 1982, *Monthly Notices Roy. Astron. Soc.* **198**, 1127.
Hartquist, T. W. and Snijders, M. A. J.: 1982, *Nature* **299**, 783.
Haschick, A. D. and Burke, B. F.: 1975, *Astrophys. J.* **200**, L137.
Hazard, C., Morton, D. C., Terlevich, R., and McMahon, R.: 1984, *Astrophys. J.* **282**, 33.
Hewitt, A. and Burbidge, G.: 1980, *Astrophys. J. Suppl.* **43**, 57.
Hunstead, R. W., Murdoch, H. S., Pettini, M., and Blades, J. C.: 1983, in G. O. Abell and G. Chincarini (eds.), 'Early Evolution of the Universe and its Present Structure', *IAU Symp.* **104**, 359.
Hunstead, R. W., Murdoch, H. S., Pettini, M., and Blades, J. C.: 1986, in *Proceedings of the Third Asian-Pacific Regional Conference of IAU Astrophys. Space Sci.* **118**, 505.
Hutchings, J. B., Crampton, D., and Campbell, B.: 1983, in 'Quasars and Gravitational Lenses', *Proc. 24th Liège Intern. Astrophys. Colloq.*, p. 494.
Levshakov, S. A. and Varshalovich, D. A.: 1979, *Astrophys. Letters* **20**, 67.
Lynds, C. R.: 1971, *Astrophys. J.* **164**, L75.
Morton, D. C.: 1975, *Astrophys. J.* **197**, 85.
Morton, D. C., Chen, J.-S., Wright, A. E., Peterson, B. A., and Jauncey, D. L.: 1980, *Monthly Notices Roy. Astron. Soc.* **193**, 399.
Murdoch, H. S., Hunstead, R. W., Blades, J. C., and Pettini, M.: 1986, in *Proceedings of the Third Asian-Pacific Regional Meeting of the IAU Astrophys. Space Sci.* **118**, 501.
Norris, J., Hartwick, F. D. A., and Peterson, B. A.: 1983, *Astrophys. J.* **273**, 450.
Oort, J.: 1981, *Astron. Astrophys.* **94**, 359.
Ostriker, J. and Ikeuchi, S.: 1983, *Astrophys. J.* **268**, 63.
Peterson, B. A.: 1978, in M. S. Langian (ed.), 'The Large Scale Structure of the Universe', *IAU Symp.* **79**, 389.
Peterson, B. A.: 1982, in 'Early Evolution of the Universe and its Present Structure', *IAU Symp.* **104**, 349.
Peterson, B. A.: 1983, in 'Quasars and Gravitational Lenses', *Proc. 24th Liège Intern. Astrophys. Colloq.*, p. 563.
Peterson, B. A., Chen, J.-S., Morton, D. C., Wright, A. E., and Jauncey, D. L.: 1985, *Monthly Notices Roy. Astron. Soc.* (in press).
Robertson, J. G. and Shaver, P. A.: 1984, *Monthly Notices Roy. Astron. Soc.* **204**, 69.
Shaver, P. A. and Robertson, J. G.: 1983a, *Mem. Soc. Astron. Ital.* **54**, 664.
Shaver, P. A. and Robertson, J. G.: 1983b, *Nature* **303**, 155.
Shaver, P. A. and Robertson, J. G.: 1983c, *Astrophys. J.* **268**, L57.
Shaver, P. A., Boksenberg, A., and Robertson, J. G.: 1982, *Astrophys. J.* **261**, L7.
Sargent, W. L. W. and Boksenberg, A.: 1983, in 'Quasars and Gravitational Lenses', *Proc. 24th Liège Intern. Astrophys. Colloq.*, p. 518.
Sargent, W. L. W., Young, P., and Boksenberg, A.: 1982, *Astrophys. J.* **252**, 54.
Sargent, W. L. W., Young, P., Boksenberg, A., Carswell, R. F., and Whelan, J. A. J.: 1979, *Astrophys. J.* **230**, 49.

Sargent, W. L. W., Young, P. J., Boksenberg, A., and Tytler, D.: 1980, *Astrophys. J. Suppl.* **42**, 41.
Smith, M. G.: 1978, *Vistas Astron.* **22**, 321.
Smith, H. E., Jura, M., and Margon, B.: 1979, *Astrophys. J.* **228**, 369.
Smith, H. E., Turnshek, D. A., and Wolfe, A. M.: 1983, in 'Quasars and Gravitational Lenses', *Proc. 24th Liège Intern. Astrophys. Colloq.*, p. 567.
Strittmatter, P. A. and Williams, R. E.: 1976, *Ann. Rev. Astron. Astrophys.* **14**, 307.
Turnshek, D. A.: 1984, *Astrophys. J.* **280**, 51.
Varshalovich, D. A. and Levshakov, S. A.: 1978, *Soviet Astron. Zh. Letters* **4**, 115.
Varshalovich, D. A. and Levshakov, S. A.: 1982, in 'Early Evolution of the Universe and its Present Structure', *IAU Symp.* **104**, 365.
Weymann, R. J.: 1980, in 'Objects of High Redshift', *Proc. IAU Symp.* **92**, 107.
Weymann, R. and Foltz, C.: 1983a, in 'Quasars and Gravitational Lenses', *Proc. 24th Liège Intern. Astrophys. Colloq.*, p. 538.
Weymann, R. J. and Foltz, C. B.: 1983b, *Astrophys. J.* **272**, L1.
Weymann, R. J., Carswell, R. F., and Smith, M. G.: 1981, *Ann. Rev. Astron. Astrophys.* **19**, 2177.
Wingert, D. W.: 1975, *Astrophys. J.* **198**, 267.
Wolfe, A. M., Broderick, J. J., Condon, J. J., and Johnston, K. J.: 1976, *Astrophys. J.* **208**, L47.
Wolfe, A. M. and Davis, M. M.: 1979, *Astron. J.* **84**, 699.
Wolfe, A. M.: 1980, *IAU Symp.* **92**, p. 153.
Wolfe, A. M., Briggs, F. H., and Jauncey, D. L.: 1981, *Astrophys. J.* **248**, 460.
Wright, A. E., Morton, D. C., Peterson, B. A., and Jauncey, D. L.: 1982, *Monthly Notices Roy. Astron. Soc.* **199**, 81.
Wyckoff, P. A., Wehinger, P. A., Gehren, T., Fried, J., Spinrad, H., and Tapia, S.: 1983, in 'Quasars and Gravitational Lenses', *Proc. 24th Liège Intern. Astrophys. Colloq.*, p. 483.
Young, P. J., Sargent, W. L. W., Boksenberg, A., Carswell, R. F., and Whelan, J. A. J.: 1979, *Astrophys. J.* **229**, 89.
Young, P., Sargent, W. L. W., Boksenberg, A., and Oke, J. B.: 1981, *Astrophys. J.* **249**, 415.
Young, P., Sargent, W. L. W., and Boksenberg, A.: 1982, *Astrophys. J.* **252**, 10.
Young, P., Sargent, W. L. W., and Boksenberg, A.: 1982, *Astrophys. J. Suppl.* **48**, 455.
Zou, Z.-L., Chen, J.-S., Bian, Y.-L., Tang, X.-Y., and Cui, Z.-X.: 1982, *Acta Astrophys. Sinica* **2**, 253.

THE ENVIRONMENT OF QSOs
(INVITED REVIEW)*

ALAN STOCKTON
Institute for Astronomy, University of Hawaii, Honolulu, Hawaii, U.S.A.
and
Royal Observatory, Edinburgh, Scotland, U.K.

(Received 19 July, 1985)

Abstract. The evidence for the association of QSOs with groups of galaxies and the available data on the global properties of such groups are reviewed. At least for redshifts less than 0.5, QSOs tend to occur in richer than-average environments. There is some evidence that groups around steep-radio-spectrum QSOs are systematically richer than those around radio-quiet and flat-radiospectrum QSOs. The groups associated with the former may sometimes reach central galaxy number densities similar to those of moderately rich clusters, but are significantly more compact, and there is no evidence that any such groups have velocity dispersions above a few hundred km s^{-1}.

A possible explanation for this environmental preference can be found in the recent evidence, from a variety of directions, that galaxy interactions play an important role in triggering nuclear activity. A significant fraction of low-redshift QSOs either have close companions at the same redshift, or show some indication of recent interaction. The results of a recent search for resolvable extended emission around a large sample of luminous low-redshift QSOs are described. There are now substantial reasons for believing that the presence of these extended emission-line regions, which are found around approximately 25% of all low-redshift QSOs, is often a signature of a recent interaction.

Observed correlations between radio properties, optical spectra, and environment are mentioned briefly. The present evidence tends to support suggestions that there are two classes of QSOs, one being the extension of the Seyfert population to higher luminosities and having spiral host galaxies, the other being closely related to the broad-line radio galaxies. The host galaxies of this latter class appear often to be morphologically ellipticals, but may in fact be products of recent mergers.

1. Introduction

The interest in studying the environment of QSOs on scales larger than a few kiloparsecs derives largely from the fact that the relationship is not completely one-sided. The QSO can have a far-ranging physical influence on its surroundings, as it does, for example, if it produces an extended radio source. But it also seems to be the case that the physical processes that produce the characteristic observed properties of the QSO, which operate on scales of a parsec or less, are in some way determined by (or at least cognizant of) the physical environment on a scale of tens of kiloparsecs and greater.

The principal avenues of investigation of the large-scale environment of QSOs have been, firstly, observations of their associated galaxy groups, secondly, broad band imaging and spectroscopy of the faint extended luminous material immediately surrounding QSOs, and thirdly, the use of the extended radio structure as a probe of

* Paper presented at the Third Asian–Pacific Regional Meeting, held in Kyoto, Japan, between 30 September–6 October, 1984.

Astrophysics and Space Science **118** (1986) 487–497.

the intergalactic medium. The excellent review of Balick and Heckman (1982) covers most of the relevant work up to 1981, to which I shall make only a few passing references. I shall further limit myself to work relating to three specific entities: groups or clusters of galaxies associated with QSOs, the galaxies in which QSOs are embedded, and extended ionized gas surrounding QSOs.

2. Groups of Galaxies Associated with QSOs

Early attempts to identify QSOs as events in giant elliptical galaxies had to face one particularly disconcerting fact: QSOs, unlike radio galaxies and elliptical galaxies generally, did not show any tendency to be associated with clusters of galaxies (Sandage and Miller, 1966; Roberts *et al.*, 1977). The lack of association between QSOs and clusters was even used as an argument against the cosmological interpretation of QSO redshifts (Arp, 1970).

The first indication that QSOs were not completely isolated was the discovery by Gunn (1971) of a companion galaxy to PKS 2251 + 113. Robinson and Wampler (1972) confirmed Gunn's redshift for this galaxy and found a second galaxy in the same field with a similar redshift. Oemler *et al.* (1972) showed that 3CR 231.1 belonged to the outskirts of Zwicky cluster 1445.1 + 2104.

More recent systematic surveys of galaxies in QSO fields have sought to answer two questions: firstly, is there a statistically significant excess of galaxies in QSO fields over that expected by chance? Secondly (given an affirmative answer to the first question), do these groups differ in any significant way from groups associated with normal galaxies? These surveys fall into two general classes: spectroscopic surveys (Stockton, 1978, 1980; Heckman *et al.*, 1984) which give great precision in identifying galaxies associated with a particular QSO, but are time-consuming and, therefore, have been restricted to moderate samples; and imaging surveys (French and Gunn, 1983; Green and Yee, 1984; Yee and Green, 1984; Hintzen, 1984), which permit large samples to be examined but give only statistical information on the association of galaxies with QSOs. Without exception, both the spectroscopic and the imaging surveys show a significant excess of galaxies around QSOs. Although there are apparent differences in the conclusions of the various authors concerning the character of these groups, a consistent picture seems to be emerging that satisfies all of the observational constraints.

Several investigations have attempted to compare the clustering properties of galaxies around QSOs with those of galaxies amongst themselves, using the two-point cross-correlation characteristics (see, e.g., Peebles, 1980) of the samples. The limitations of these samples generally do not give one enough information to determine the radial dependence of the spatial cross-correlation function ξ, so the usual approach has been to assume the dependence found for the galaxy–galaxy correlation function, $\xi_{gg} = B_{gg}r^{-1.77}$, so that for galaxy–QSO associations, $\xi_{gq} = B_{gq}r^{-1.77}$. Then it is possible to simply compare the coefficients B_{gq} and B_{gg} that measure the amplitude of the clustering; because of the uncertainty of the assumption that the radial dependence is the same as that for galaxies at large, this measurement should be carried out for a fixed metric diameter at the QSO redshift distance for all fields. Even so, there are potential sources of systematic error. In order to compare the QSO field surveys with published estimates of B_{gg} based on the Zwicky and Lick galaxy surveys (Groth and Peebles, 1977), one must have both an accurate galaxy luminosity function and a good estimate of the magnitude limit of the QSO field surveys.

Weymann *et al.* (1978), using the results from Stockton (1978), found $B_{gq}/B_{gg} \approx 6$, i.e., galaxies are significantly more strongly correlated with QSOs than they are, on the average, with other galaxies. French and Gunn (1983), on the other hand, from their own imaging survey found $B_{gq}/B_{gg} = 1.24 \pm 0.50$, while Yee and Green (1984) found $B_{gq}/B_{gg} = 2.76 \pm 0.57$ for their entire sample of QSOs and $B_{gq}/B_{gg} = 4.4 \pm 1.2$ for the radio-loud subset. Yee and Green found that part of the difference between their results and those of French and Gunn could be attributed to differences in the assumed luminosity functions; however, they also found evidence to support a suggestion by French and Gunn that such a difference might be due to a substantially steeper radial gradient for galaxies associated with QSOs than typical for galaxy–galaxy associations. If groups associated with QSOs are unusually compact, then averaging over larger fields will dilute their apparent density enhancement. If is, therefore, suggestive that the field sizes used by French and Gunn were considerably larger than those of Yee and Green, which in turn were larger than the fields for Stockton's spectroscopic survey.

A strong central concentration of groups associated with QSOs is also indicated by the spectroscopic survey of Heckman *et al.* (1984). All but one of a sample of twenty galaxies at projected distances of less than 50 kpc from a QSO or luminous Seyfert galaxy had redshifts close to that of the adjacent active object.

Stockton and MacKenty (in preparation) have obtained redshifts for 60 of a total of 73 galaxies in magnitude-limited samples in 0.6 Mpc ($H_0 = 75$, $q_0 = 0$) radius fields around four low-redshift QSOs. These QSOs were chosen without regard for their radio properties, and include one each of radio-quiet, steep-spectrum, and flat-spectrum types, the fourth object being a weak radio source with uncertain spectral characteristics but a D2 morphology. The radial distribution of the galaxies associated with the four QSOs gives $\xi \propto r^{-2.4}$, confirming the strong concentration of galaxies to QSOs inferred by French and Gunn (1983), Yee and Green (1984), and Heckman *et al.* (1984), although the distribution around the radio quiet QSO is consistent with the $r^{-1.77}$ dependence found for normal galaxies. The one strong steep-spectrum QSO both the greatest density enhancement and the strongest central concentration of galaxies.

Differences in the properties of groups associated with different classes of QSOs are on less firm ground, because of the smaller sample, but follow a consistent trend. In Stockton's (1978, 1980) spectroscopic survey of galaxies in small fields around 27 high-luminosity QSOs, with steep-spectrum and flat-spectrum radio sources roughly equally represented, 10 of the 14 galaxies found associated with QSOs were in fields of known steep-spectrum radio QSOs, and only one was associated with a flat-spectrum QSO (the remaining three were in fields of radio QSOs for which no spectral information was available). Thus, the steep-spectrum QSOs appear, on the average, to be found in richer environments than flat-spectrum QSOs. Yee and Green (1984) found marginally significant evidence that groups associated with radio-quiet QSOs were both poorer and comprised of generally less-luminous galaxies than were groups associated with radio-loud QSOs (virtually all of which were steep-spectrum sources). Since the flat-spectrum radio QSOs and the radio-quiet QSOs show many similarities, these two studies tend to support each other, although neither can be considered conclusive. A similar result for radio galaxies is on firmer ground: Prestage (1984) finds a significantly smaller clustering amplitude for flat-spectrum sources than for steep-spectrum sources.

The central concentration of the groups associated with at least the steep-spectrum QSOs gives us a fresh perspective from which to approach the old question of whether QSOs are found with any significant frequency in the centres of rich clusters. Recent work by Hintzen and his collaborators (Hintzen and Scott, 1978; Hintzen *et al.*, 1981, 1983; Hintzen, 1984) indicates that QSOs with distorted triple radio structure are in systematically denser galaxy groups than are QSOs with collinear triple structure, as would be expected if the distortions were induced by interaction with the comparatively dense gaseous medium found in the centers of rich clusters. The most impressive example, 3CR 275.1, was found by Hintzen *et al.* (1981) to lie in a faint, elliptically-shaped nebulosity in what appears to be a dense cluster. Hintzen (1984) confirmed that two of the galaxies had redshifts consistent that of 3CR 275.1.

Although some of the QSOs with distorted structure appear to have substantial numbers of associated galaxies, others do not (e.g., 4C 25.01, at $z = 0.284$), so the distorted structure may not always indicate the effect of a gaseous medium. The central question remaining, from the interpretative point of view, is whether those QSOs that appear to have the largest numbers of associated galaxies are actually in rich, symmetrical, relaxed clusters with large velocity dispersion. So far, there seems to be no evidence for QSOs in groups with velocity dispersions in excess of ~ 500 km s^{-1}, including 3CR 275.1, for which Hintzen's (1984) measurements of two galaxies give velocities relative to the QSO of -230 and -307 km s^{-1}. Therefore, it appears that even these results can be accommodated by a picture in which the steep-radio-spectrum QSOs frequently occur in small, compact groups; where the local galaxy density may inded approach that of rich clusters, but the velocity dispersion remains low.

3. Host Galaxies, Close Companions, and Extended Emission Regions

Because the first QSOs discovered were identified primarily from low-frequency radio surveys, their radio properties were considered to be generally similar to those of the powerful classical double radio galaxies. As those radio galaxies appeared morphologically to be ellipticals, it was natural to suppose that QSOs, too, were in giant elliptical galaxies and to seek ways of detecting these galaxies in spite of the overwhelming luminosity of their QSO nuclei. Kristian (1973) compared the predicted and actual presence of faint extensions to QSO images on available large-scale photographic exposures with a sample with a fairly wide range in luminosity, and showed that the results were consistent with QSOs being in first-ranked elliptical galaxies. Miller (1981) tried a spectroscopic approach, obtaining very high signal-to-noise ratio data for a number of QSOs and N-galaxies, and looking for strongly diluted late-type stellar absorption features. Assuming a stellar content like that of a normal giant elliptical, Miller found that the luminosities of galaxies underlying QSOs must be at least two magnitudes fainter than first ranked cluster ellipticals.

Over the past several years it has become clear that current techniques are capable of resolving extended optical structure for most QSOs with $z \leq 0.5$ (Hawkins, 1978; Hutchings *et al.*, 1981, 1984a, b; Wyckoff *et al.*, 1981; Gehren *et al.*, 1984; Malkan *et al.*, 1984). What is still unclear, in spite of a great deal of effort on the part of these groups, is what these results tell us about the host galaxies of QSOs. There is still disagreement among the different investigators concerning, for example, whether a Hubble-law or an exponential is a better fit to the residual radial profiles after subtraction of a scaled point-spread function to remove the excess luminosity of the QSO nucleus. To a large extent such disagreements may reflect real differences in the samples observed. Malken *et al.* (1984), who find exponential profiles (i.e., spiral-like host galaxies), observed an X-ray-selected, generally low-luminosity sample of QSOs. Hutchings *et al.* (1984a) observed a more heterogeneous sample, with significant representation from X-ray, optical, and radio surveys, but still weighted towards lower-luminosity objects. They argue that the radial surface brightness profiles generally cannot discriminate between a spiral-like and an elliptical-like light distribution, being sensitive to the chosen decomposition into unresolved and extend components, so they have used other morphological criteria instead. They find a high proportion of spiral features present in their optically and X-ray-selected samples, but a significantly lower proportion in their radio-selected sample; however, they emphasize that none of their objects is definitely in an elliptical galaxy. On the other hand, Wyckoff *et al.* (1981) and Gehren *et al.* (1984), who concentrated primarily on radio-loud QSOs, generally of moderate to high luminosity, found a clear preference for Hubble-law (elliptical-like) profiles.

In principle, one might resolve this question of QSO host galaxy types by spectroscopy of their stellar populations; however, such observations are extremely difficult, because of the faintness of the galaxies and because of scattered light from the QSO nucleus. Among classical, high-luminosity QSOs, the only objects in which stellar absorption lines have been reported are the atypically large host galaxies for 3C 48 (Boroson and Oke, 1982, 1984) and Mrk 1014 (MacKenty and Stockton, 1984). Both galaxies show absorption lines characteristic of a mixture of early and late-type stars – i.e., unlike those expected for a normal elliptical galaxy. However, the colours of the 3C 48 galaxy are bluer than those of the resolved regions of most other QSOs, so this result cannot be generalised.

One factor that may complicate attempts to identify unambiguous morphologies for QSO host galaxies is the recent evidence that a substantial fraction of QSOs may have had their activity triggered by violent encounters or mergers with other galaxies. Stockton (1982) noted three examples of low-redshift QSOs with very close compact companions and suggested that these galaxies had interacted strongly with the QSO host galaxies. Both Hutchings *et al.* (1982, 1983, 1984a) and Gehren *et al.* (1984) find a large fractino of QSOs with asymmetric extensions or other evidence of interaction in their imaging surveys, and the spectroscopic survey of Heckman *et al.* (1984) confirms that many QSOs have companion galaxies at projected distances of less than 50 kpc. Other evidence for interactions involving specific QSOs has been presented by Stockton and MacKenty (1984) and MacKenty and Stockton (1984). The rather specialized environment seemingly preferred by at least the steep-radio-spectrum QSOs, i.e., one of high galaxy density but low velocity dispersion, would tend to promote strong interactions. Particularly if mergers are important for a significant fraction of QSOs, part of the ambiguity in attempting to classify host galaxies of QSOs quite likely reflects the ambiguous nature of the galaxies. It may, in fact, make more sense to ask what kind of galaxies they *were*, rather than what kind they *are*: as de Vaucouleurs (1974) has pointed out, "After a collision a car is a wreck, not a new type of car". But as a careful examination of a wreck may allow one to infer the appearance of the original car, so we may, in cases of interactions or mergers of galaxies, look for clues to the nature of the precursor galaxies.

One promising approach for such a search, largely neglected until recently, is that of observing the morpholgy of extended emission-line gas around QSOs. Virtually all imaging surveys of QSOs to date have used standard broad-band filters, with characteristic widths of 1000 Å. The filter bandpass usually includes a mixture of line and continuum radiation; the line radiation is sometimes strong enough in the extended luminous regions of QSOs to dominate the total flux received through the filter, so errors of interpretation can arise if it is assumed that the observed distribution is due to stars. Three QSOs have been known for some time to have strong extended emission: 3C 48 (Wampler *et al.*, 1975), 4C 37.43 (Stockton, 1976), and 3C 2491. (Richstone and Oke, 1977).

We (cf. Stockton and MacKenty, 1984) have recently carried out an extensive search for extended emission for a sample of luminous QSOs by obtaining deep CCD images through 30 Å bandpass filters centred on the redshifted [O III] λ5007 line. The initial sample of 71 objects included all QSOs in the Hewitt and Burbidge (1980) catalogue and the Palomar BQS list (Schmidt and Green, 1983) with $Z \leq 0.45$, $\delta > -25°$, and listed $V < 19.62 + 5 \log z$. Three QSOs were deleted because the [O III] line fell too close to the night sky (NaD) and for an additional eight no filters were purchased, because each QSO would have required a separate one. For the remaining 60 QSOs spectra were recorded of the Hβ-[O III] region, both to define the character of the spectra and to obtain accurate line positions for the filter observations. The [O III] images were obtained with the Galileo/Institute for Astronomy CCD system, using a Texas Instruments 500 × 500 CCD, on the University of Hawaii 2.2 m and the Canada–France–Hawaii 3.6 m telescopes. Fairly early in the programme, it became clear that the presence of extended emission-line regions is correlated with the nuclear QSO spectrum, in the sense that moderate to strong narrow-line emission seems to be necessary, but not sufficient, for extended emission to be detected by this technique. Because of lack of time, 11 QSOs with little or no narrow-line emission were not observed, leaving a total of 49 for which interference filter observations were available. In discussing our results,

I shall assume that these 11 QSOs would not have shown extended emission, had they been observed.

Although the detailed quantitative analysis of these data is not yet complete, some conclusions can be stated now:

(1) As mentioned above, it appears to be necessary for a QSO to have a narrow-line region if it is to show extended emission. In our sample of 60 QSOs, 15 showed some extended emission, and all of these were from the sub-sample of 40 QSOs with at least moderately strong narrow-line emission.

(2) The extended emission tends to be highly structured, showing filaments and clumps. The morphologies strongly suggest that the ionized gas is mostly density-bounded rather than radiation-bounded: i.e., the actual distribution of the gas is very nearly what we see. This statement will be qualified somewhat in the discussion in the following section.

(3) We have obtained line-free continuum exposures of most of the QSOs that show strong extended emission-line regions. Any extended continuum emission visible is always more smoothly distributed than the line emission, and any structure in the continuum distribution is rarely correlated with the emission-line morphology.

4. The Abundances Appear to be Approximately Solar

This extended ionized gas is therefore processed material, but it no longer shares the distribution of any detectable stellar component. Its correlation with the narrow-line gas much closer to the QSO nucleus makes plausible a common origin. Two possibilities come to mind: either the gas has been ejected from the QSO host galaxy, or it is debris from a companion galaxy that has been disrupted during an interaction or merger. In the first case, there are at least two further possibilities: the gas may have been ejected as a consequence of either an external event, such as an encounter, or as a result of internal processes, such as radiative acceleration.

Sometimes individual examples of extended emission regions give suggestive indications of their origin. Stockton and MacKenty (1983) have argued that the morphologies and velocity fields of the emitting gas around 3CR 249.1 and Ton 202 can be most easily explained in terms of very close interactions or mergers. In each case, two tail-like structures are seen, which are interpreted as tidal tails from both the QSO host galaxy and its interacting companion. If this explanation is correct, both galaxies possessed extensive gaseous disks prior to their encounter. For the emission-line gas in 3CR 249.1, no velocities relative to the QSO significantly greater than 100 km s^{-1} were observed, indicating (in the absence of unusual projection effects) that the two galaxies involved were relatively low-mass systems. Other cases exist (e.g., PKS 2251 + 113) where only one tail is seen, and at least one (4C 25.40) with an off-center gaseous ring. All of these can plausibly be attributed to interactions, and it seems reasonable that least some of the remaining examples that do not show any of the obvious signatures may be cases for which the accidental features such as interaction parameters or viewing angles are unfavourable.

One of the more puzzling objects is 4C 37.43, which has the most luminous, most highly structured, and most extensive of the emission-line regions studied. Although the observed structure consists of large-scale filaments, some of which are 40 kpc long, there is no obvious way of understanding it as resulting from an interaction. Yet there almost certainly was an interaction: some 10″ ($\sim$ 50 kpc) away is a compact galaxy with a radial velocity (in the QSOs rest frame) about 300 km s^{-1} higher than that of the QSO (Stockton, 1973, 1978). Deep line-free continuum images shown an apparent bridge, presumably of stars, between the two. But the emission-line gas on the same side of the SO as the galaxy (i.e., putative 'bridge-like' material) has a *negative* velocity with respect to the QSO, as does most of the gas generally. The velocity field of the gas has local systematic gradients corresponding to individual filaments but is fairly chaotic on a global scale. This object offers a tantalizing wealth of detail, but so far remains an enigma.

While I believe that the qualitative evidence that many of these extended emission-line regions owe their origins to interactions or mergers is quite strong, a detailed demonstration that the observed morphologies are consistent with such an interpretation will be difficult, involving consideration of the hydrogas during the action. What is more important than any specific interpretation put forward at the moment is the potential that these striking and often beautiful structures have as probes of the recent physical history of the regions around QSOs, once they are understood in detail.

4. A Speculative Overview

For most of the remainder of this paper, I shall put aside any qualifications or qualms concerning the significance of some of the trends mentioned in the preceding sections and assume the following picture to be substantially correct for QSOs near the present epoch:

(1) Steep-spectrum radio-loud QSOs are typically found in regions of substantially higher galaxy density than is the average galaxy, ranging up to dense but compact groups with low to moderate velocity dispersions.

(2) Flat-spectrum radio-loud QSOs and radio-quiet QSOs are typically found in regions of only slightly higher galaxy density than is the average galaxy.

(3) The presence an easily-detected, highly structured, extended emission-line region around a QSO can always be taken as evidence of a recent interaction or merger involving at least one galaxy with a pre-existing cold gaseous disk.

Boroson and Oke (1984) have emphasized that the correlation between the presence of extended emission and the ordinary QSO narrow-line emission is part of a broader picture of correlations among observed properties of QSOs (Setti and Woltjer, 1973, 1977; Miley and Miller, 1979; Steiner, 1981; Blumenthal *et al.*, 1982; Bergeron and Kunth, 1984). Briefly, steep-spectrum radio-loud QSOs with classical double radio structure tend to have very broad permitted lines with irregular profiles, weak Fe II emission in the optical region, and strong narrow-line emission, while flat-spectrum core-dominant radio-loud QSOs and radio-quiet QSOs usually have less broad and smoother permitted lines, stronger Fe II emission, and weak narrow lines. To these differences we add the apparent disparity in average local galaxy density mentioned above.

QSOs as a class have always been in a somewhat ambiguous position, being viewed on the one hand as related to the powerful extended-double radio galaxies, and on the other, as a continuation of the Seyfert galaxies to higher luminosities. It has always been difficult to visualize these two relationships merging into a smooth sequence, since virtually all Seyfert galaxies are morphologically spirals and all powerful radio galaxies appear to be ellipticals. The position that the two types of QSOs outlined in the previous paragraph correspond, respectively, to extensions of the radio galaxy and the Seyfert galaxy sequences is, therefore, extremely attractive (see, e.g., Yee and Green, 1984; Peacock *et al.*, 1986). As Boroson and Oke (1984) have noted, however, the correlation of extended emission regions with the steep-spectrum radio QSOs (i.e., elliptical host galaxies) produces an apparent inconsistency, since it would be much more likely that a spiral would have the required gas. They suggest that for those QSOs showing neither conventional narrow-line emissions nor extended emission (which on the present picture would have spiral host galaxies), the gas lies in a plane coincident with the plane of the optically thick broad-line-emitting gas, which shields the lower density gas from the photoionizing continuum. The implication is that, if encounters are important in triggering QSOs in the Seyfert-like sequence, they generally do not perturb significant amounts of gas into the unshielded double cone.

For the steep-spectrum radio QSOs, like 3CR 249.1 can point the way to a speculative interpretation I have hinted at earlier. If, as Stockton and MacKenty (1984) have argued, the two gaseous tails of this object imply that the two galaxies involved in the supposed merger each possessed a substantial cold gaseous disk, then it is likely that they were morphologically spirals prior to their encounter. Since 3CR 249.1 has a classical asymmetric double radio structure, it may be that the elliptical morphology of some radio sources in this class has been assumed relatively recently, and that the steep-spectrum radio-loud QSOs may generally be the results of mergers. In these violent encounters, the large-scale distribution of gas will not be confined to a plane, and the gas can be photoionized by the central source.

This picture, while painted with a very broad brush, is consistent with the apparent differences in group environment, the different conclusions reached by various groups doing broad-band imaging regarding host galaxies of QSOs, and at least some of the observed correlations among various observed properties of QSOs. It presently rests on a rather shaky observational foundation, which needs to be firmed up by (1) observations of the properties of groups associated with carefully selected samples of steep-spectrum radio-loud, flat-spectrum radio-loud, and radio-quiet QSOs, (2) high-resolution line-free continuum imaging of similar samples to determine galaxy morphologies and colours, and (3) spectroscopy of stellar populations in the host galaxies of steep-spectrum radio-loud QSOs, to differentiate, if possible, between galaxies that have a genuinely old stellar population and those that have one that is characteristic of recently merged spirals. To make substantial further progress on these last two programmes almost certainly will require the Space Telescope.

Acknowledgements

Much of the original work reported here has been done in collaboration with John MacKenty. I have benefitted from discussions with Lance Miller, John Peacock, and Adrian Webster. This work has been supported in part by grants from the US National Science Foundation.

References

Arp, H.: 1970, *Astrophys. J.* **162**, 811.
Balick, B. and Heckman, T. M.: 1982, *Ann. Rev. Astron. Astrophys.* **20**, 431.
Bergeron, J. and Kunth, D.: 1984, *Monthly Notices Roy. Astron. Soc.* **207**, 263.
Blumenthal, G. R., Keel, W. C., and Miller, J. S.: 1982, *Astrophys. J.* **257**, 499.
Boroson, T. A. and Oke, J. B.: 1982, *Nature* **296**, 397.
Boroson, T. A. and Oke, J. B.: 1984, *Astrophys. J.* **281**, 535.
de Vaucouleurs, G.: 1974, in J. R. Shakeshaft (ed.), 'The Formation and Dynamics of Galaxies', *IAU Symp.* **58**, 1.
French, H. and Gunn, J. E.: 1983, *Astrophys. J.* **269**, 29.
Gehren, T., Fried, J., Wehinger, P. A., and Wyckoff, S.: 1984, *Astrophys. J.* **278**, 11.
Green, R. F. and Yee, H. K.: 1984, *Astrophys. J. Suppl.* **54**, 495.
Groth, E. J. and Peebles, P. J. E.: 1977, *Astrophys. J.* **217**, 385.
Gunn, J. E.: 1971, *Astrophys. J.* **164**, L113.
Hawkins, M. R. S.: 1978, *Monthly Notices Roy. Astron. Soc.* **182**, 361.
Heckman, T. M., Miley, G. K., and Green, R. F.: 1984, *Astrophys. J.* **281**, 525.
Hewitt, A. and Burbidge, G.: 1980, *Astrophys. J. Suppl.* **43**, 57.
Hintzen, P.: 1984, *Astrophys. J. Suppl.* **55**, 533.
Hintzen, P. and Scott, J.: 1978, *Astrophys. J.* **224**, L47.
Hintzen, P., Boeshaar, G., and Scott, J.: 1981, *Astrophys. J.* **246**, L1.
Hintzen, P., Ulvestad, J., and Owen, F.: 1983, *Astron. J.* **88**, 709.
Hutchings, J. B. and Campbell, B.: 1983, *Nature* **303**, 584.
Hutchings, J. B., Crampton, D., Campbell, B., and Pritchet, C.: 1981, *Astrophys. J.* **247**, 743.
Hutchings, J. B., Crampton, D., and Campbell, B.: 1984a, *Astrophys. J.* **280**, 41.
Hutchings, J. B., Crampton, D., Campbell, B., Duncan, D., and Glendenning, B.: 1984b, *Astrophys. J. Suppl.* **55**, 319.
Kristian, J.: 1973, *Astrophys. J.* **179**, L61.
MacKenty, J. W. and Stockton, A.: 1984, *Astrophys. J.* **283**, 64.
Malkan, M. A., Margon, B., and Chanan, G. A.: 1984, *Astrophys. J.* **280**, 66.
Miley, G. K. and Miller, J. S.: 1979, *Astrophys. J.* **228**, L55.
Miller, J. S.: 1981, *Publ. Astron. Soc. Pacific* **93**, 681.
Oemler, A., Jr., Gunn, J. E., and Oke, J. B.: 1972, *Astrophys. J.* **176**, L47.
Peacock, J. A., Miller, L., and Longair, M. S.: 1986, in preparation.
Peebles, P. J. E.: 1980, *The Large-Scale Structure of the Universe*, Princeton Univ. Press, Princeton.
Prestage, R. N.: 1984, in F. Mardirossian *et al.* (eds.), D. Reidel Publ. Co., Dordrecht, Holland, p. 559.
Richstone, D. O. and Oke, J. B.: 1977, *Astrophys. J.* **213**, 8.
Roberts, D. H., O'Dell, S. L., and Burbidge, G. R.: 1977, *Astrophys. J.* **216**, 227.
Robinson, L. B. and Wampler, E. J.: 1972, *Astrophys. J.* **171**, L83.
Sandage, A. and Miller, W.: 1966, *Astrophys. J.* **144**, 1238.
Schmidt, M. and Green, R. F.: 1983, *Astrophys. J.* **269**, 352.
Setti, G. and Woltjer, A.: 1973, *Ann. N.Y. Acad. Sci.* **224**, 8.
Setti, G. and Woltjer, A.: 1977, *Astrophys. J.* **218**, L33.
Steiner, J. E.: 1981, *Astrophys. J.* **250**, 469.
Stockton, A.: 1973, *Nature Phys. Sci.* **246**, 25.
Stockton, A.: 1976, *Astrophys. J.* **205**, L113.
Stockton, A.: 1978, *Astrophys. J.* **223**, 747.

Stockton, A.: 1980, in G. O. Abell and P. J. E. Peebles (eds.), 'Objects of High Redshift', *IAU Symp.* **92**, 89.
Stockton, A.: 1982, *Astrophys. J.* **257**, 33.
Stockton, A. and MacKenty, J. W.: 1983, *Nature* **305**, 678.
Wampler, E. J., Robinson, L. B., Burbidge, E. M., and Baldwin, J. A.: 1975, *Astrophys. J.* **198**, L49.
Weymann, R. J., Boroson, T. A., Peterson, B. M., and Butcher, H. R.: 1978, *Astrophys. J.* **226**, 603.
Wyckoff, S., Wehinger, P. A., and Gehren, T.: 1981, *Astrophys. J.* **247**, 750.
Yee, H. K. C. and Green, R. F.: 1984, *Astrophys. J.* **280**, 79.

HEAVY ELEMENT ABUNDANCES IN ABSORPTION LINE SYSTEMS TOWARDS Q2206–199N

J. G. ROBERTSON
Anglo-Australian Observatory, Epping, NSW, Australia

R. F. CARSWELL
Institute of Astronomy, Cambridge, U.K.

and

P. A. SHAVER
European Southern Observatory, Garching, F.R.G.

(Received 19 July, 1985)

Abstract. Spectroscopy of the $z = 2.0763$ absorption line system in Q2206–199N shows that it has a very large hydrogen column density (5×10^{20} cm^{-2}) but extremely weak absorption due to heavy elements. We discuss the possibility that the gas cloud causing this absorption system may be exceptionally metal-poor.

The degree of heavy element enrichment in the clouds which produce the Lα absorption lines in QSO spectra is of great interest. Sargent *et al.* (1980) showed that these clouds are not spatially associated with galaxies, and it has been suggested that they are primordial material which has undergone little or no heavy element enrichment through stellar nucleosynthesis. Certainly such systems generally show only the hydrogen Lα (and sometimes Lβ) absorption line(s).

However, as Weymann *et al.* (1981) pointed out, the lack of absorption lines due to heavy elements in typical Lα forest absorption systems does not exclude the heavy elements at a significantly low abundance. Metal abundances equal to those of extreme Population II stars (~ 0.003 solar) would easily escape detection in a typical cloud. More sensitive searches have been made by Norris *et al.* (1983), Chaffee *et al.* (1985), and Sargent and Boksenberg (1983). The latter two works obtained increased sensitivity to low metal abundances by examining systems with unusually large H I column densities.

Here we report a very sensitive search for heavy elements in one of the absorption systems of the quasar Q2206–199N. The system in question has by far the highest H I column density of any known metal-poor absorption cloud and is, therefore, an excellent candidate for demonstrating very low metal abundances. However, it should be noted that such a high column density system is not a typical Lα forest absorber, and is likely to have a different origin (e.g., galaxy or protogalaxy; see below).

* Paper presented at the IAU Third Asian-Pacific Regional Meeting, held in Kyoto, Japan, between 30 September–6 October, 1984.

Q2206–199N (Savage and Bolton, 1979) is a 17.6 mag. QSO with an emission redshift of 2.55. It was observed with a spectral resolution of 1.5 Å FWHM at the Anglo–Australian Telescope (Robertson *et al.*, 1983). As well as several metal-rich absorption systems, it has a very metal-poor system at $z_a = 2.0763$. The Lα line in this system is very strong, with a profile indicating that damping wings dominate. Profile fitting gives a column density $N(\text{H I}) = 5 \times 10^{20}$ cm^{-2}. However, no certain detections of metal lines in this system were made by Robertson *et al.* (1983) at this spectral resolution. A very important feature of this cloud is that the H I column indicates that it is extremely optically thick to the Lyman continuum. Therefore, in the body of the cloud H will be predominantly neutral, C will be C II, and Si will be Si II. The uncertainties due to ionization models, which have affected other studies of metal-poor absorption clouds, are considerably reduced.

Higher resolution (0.3 Å) observations of the metal-poor system were obtained at the Anglo–Australian Telescope in September 1984. Lines due to Si II 1193, Si III 1206, Si II 1304, Si IV 1402, and C IV 1548, 1550 were detected, with equivalent widths ranging from 0.07 to 0.40 Å (observed frame). These observations will be reported in more detail elsewhere. The column densities (and, hence, abundances) which these detections imply depend critically on the Doppler velocity dispersion (*b* parameter) of the cloud. The equivalent width ratio for Si II 1193 and Si II 1304 indicates that *b* is probably close to 10 km s^{-1}, in which case $N(\text{Si II}) = 1.4 \times 10^{13}$ cm^{-2} and the Si abundance is ~ 0.0009 solar. However, if *b* could be as low as 3 km s^{-1} then the Si abundance would be ~ 0.025 solar. A study of the uncertainties in the equivalent widths of the detected lines will be made in order to assign confidence limits to the value of *b* and, hence, the Si abundance. If an abundance as low as ~ 0.001 solar is confirmed, the absorbing cloud may be a protogalaxy or young galaxy in which very little star formation has taken place.

References

Chaffee, F. H., Foltz, C. B., Röser, H. J., Weymann, R. J., Latham, D. W., and Schaeffer, J.: 1985, *Astrophys. J.* **292**, 362.
Norris, J., Hartwick, F. D. A., and Peterson, B. A.: 1983, *Astrophys. J.* **273**, 450.
Robertson, J. G., Shaver, P. A., and Carswell, R. F.: 1983, 'Quasars and Gravitational Lenses', *Proc. 24th Liège Astrophys. Colloq.*, p. 602.
Sargent, W. L. W. and Boksenberg, A.: 1983, 'Quasars and Gravitational Lenses', *Proc. 24th Liège Astrophys. Colloq.*, p. 518.
Sargent, W. L. W., Young, P. J., Boksenberg, A., and Tytler, D.: 1980, *Astrophys. J. Suppl.* **42**, 41.
Savage, A. and Bolton, J. G.: 1979, *Monthly Notices Roy. Astron. Soc.* **188**, 599.
Weymann, R. J., Carswell, R. F., and Smith, M. G.: 1981, *Ann. Rev. Astron. Astrophys.* **19**, 41.

AN ABSORPTION LINE STUDY OF GALAXIES AT HIGH REDSHIFT*

H. S. MURDOCH, R. W. HUNSTEAD
School of Physics, University of Sydney, Australia

J. C. BLADES
Space Telescope Science Institute, Baltimore, Md., U.S.A.

and

M. PETTINI
Royal Greenwich Observatory, Herstmonceux Castle, Hailsham, Sussex, England

(Received 19 July, 1985)

Abstract. Complex velocity structure is revealed in several absorption systems in the spectrum of 0215 + 015 (z = 1.715) at resolutions of 20–30 km s^{-1} FWHM. Striking differences are found in the relative strengths of low to high ions among the components, reinforcing our interpretation of these systems as intervening galaxies.

The bright BL Lac object 0215 + 015 (z = 1.715) provides a unique opportunity for a detailed study of absorption by intervening galaxies in the range z = 1.2 to 1.7. The results described below have been obtained at the Anglo–Australian Telescope (AAT), supplemented by a low resolution UV spectrum obtained with IUE. Over the period of our observations, we have observed the apparent magnitude to vary between the extremes of V = 14.5 in August 1981 (and again in August 1984) to V = 18.3 in July 1983. The redshift of 0215 + 015 quoted above is based on weak Lα and C IV emission observed in October 1983.

Absorption systems have been detected at z_{abs} = 1.719, 1.686, 1.649, 1.549, 1.491, 1.345, and 1.254. A small number of unidentified lines form a sparse Lα forest. Most of the prominent lines occurring between 3400 and 7400 Å in the above systems have been observed at high signal-to-noise ratio with a resolution of 20–30 km s^{-1}.

The 1.649 and 1.549 systems exhibit strong, complex C IV absorption with 9 and 7 components covering 900 and 300 km s^{-1} respectively (Pettini *et al.*, 1983). The 1.491 system has two components separated by 92 km s^{-1}. The average b value for these 18 components is 12 km s^{-1} with some as low as 6 km s^{-1}. The 1.686 system has a single weak C IV with $b \leq 5$ km s^{-1}. The b values set stringent upper limits to the temperatures of the absorbing gas and suggest that photoionization of carbon may well be the predominant mechanism. Higher resolution data are needed to settle this question.

* Paper presented at the IAU Third Asian-Pacific Regional Meeting, held in Kyoto, Japan, between 30 September–6 October, 1984.

Astrophysics and Space Science **118** (1986) 501–503.

We interpret the 1.491 system as an intervening galaxy and the 1.549 and 1.649 systems as intervening rich clusters (Pettini *et al.*, 1983). On the alternative ejection hypothesis, the minimum velocity for the 1.549 and 1.649 systems would be 19000 and 7000 km s^{-1}, respectively. We are unaware of any ejection model which can adequately explain the complexity of these systems. Our interpretation that the components arise in separate galaxies in a cluster is reinforced by the striking variation in the relative proportion of low to high ions among the components (Blades *et al.*, 1985).

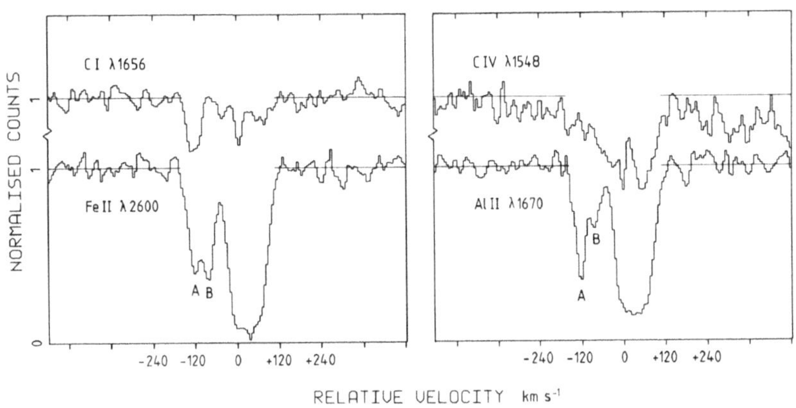

Fig. 1. Absorption for representative species in the $z_{abs} = 1.345$ system in 0215 + 015 at resolutions ~25 km s^{-1} FWHM.

There is an even stronger case that the 1.345 system represents an intervening galaxy (or group of galaxies). In Figure 1 we present, on a common velocity scale, the absorption profiles for representative singly ionized species Fe II λ2600 and Al II λ1670, together with C I λ1656 and C IV λ1548. There is a striking difference between the velocity distributions for the three ionization stages represented. The main component is strong in C II, Al II, Mg II, Si II, and Fe II and evidently represents a predominantly galactic disk sightline. The two minor components *A* and *B* may represent companion galaxies. Component *A* is particularly strong in the neutral species C I λλ1560, 1656 and Mg I λ2852, and is, therefore, unlikely to be a halo cloud.

Component *B*, in Fe II, was found to almost double in strength over a 3 week period in November–December 1981 (Hunstead *et al.*, 1983). This change was detected in both members of Fe II λλ2586, 2600. We have observed these lines on many subsequent occasions and have not detected any comparably large variation. The difference of the December 1981 result from the mean remains, however, highly significant and further observations are planned.

We have identified 30 separate lines (not counting sub-components) in the 1.345 system and have calculated column densities for many elements. The values relative to hydrogen are typically an order of magnitude less than in the solar neighbourhood. Since this result includes nitrogen and oxygen, we believe it represents genuine underabundance rather than depletion onto dust; nitrogen and oxygen are lightly depleted in dusty regions in the Galaxy. It is an open question of considerable interest whether this underabundance is an evolutionary effect, or whether it results from radial abundance gradients within galaxies and the impact parameter of the particular sightline (Blades *et al.*, 1985).

References

Blades, J. C., Hunstead, R. W., Murdoch, H. S., and Pettini, M.: 1985, *Astrophys. J.* **288**, 580.
Hunstead, R. W., Murdoch, H. S., Pettini, M., and Blades, J. C.: 1983, in G. O. Abell and G. Chincarini (eds.), 'Early Evolution of the Universe and its Present Structure', *IAU Symp.* **104**, 459.
Pettini, M., Hunstead, R. W., Murdoch, H. S., and Blades, J. C.: 1983, *Astrophys. J.* **273**, 436.

QSOs AS PROBES OF THE EARLY UNIVERSE*

R. W. HUNSTEAD, H. S. MURDOCH
School of Physics, University of Sydney, Australia

M. PETTINI
Royal Greenwich Observatory, Hailsham, Sussex, England

and

J. C. BLADES
Space Telescope Science Institute, Baltimore, Md., U.S.A.

(Received 19 July, 1985)

Abstract. The spectrum of the highest redshift QSO 2000−330 ($z = 3.78$) contains four heavy-element absorption systems with $z_{abs} > 3.0$. Interesting features include velocity structure at $z_{abs} = 3.552$ which suggests a cluster origin and a purely low ionization system at $z_{abs} = 3.1881$ typical of a galactic disk sightline.

The nature of the absorbing material responsible for the narrow absorption lines seen in the spectra of high redshift QSOs is a subject which receives wide attention. On the commonly-held view of QSOs as highly-luminous, very distant objects, this absorption is most likely to arise in intervening galaxies and hydrogen clouds distributed cosmologically along the line-of-sight. The detailed study of QSO absorption is, therefore, the only method we have at present for probing the composition and physical conditions in the early Universe.

Studies of the interstellar medium in our Galaxy, using halo stars or extragalactic objects as probes, have been valuable in mapping the spatial distribution of various ionic species. Highly-ionized species such as C IV $\lambda\lambda$1548, 1550 and Si IV $\lambda\lambda$1393, 1402 are shown to occur mainly in the Galactic halo, whereas singly-ionized and neutral species are predominant in the disk. By analogy, many absorption systems seen in QSOs show a diversity of ionization states which can most simply be interpreted as arising in a sightline which intercepts both disk and halo gas in an intervening galaxy or galaxies. Our high-resolution observations of various systems in the BL Lac object 0215 + 015 (Pettini *et al.*, 1983; Murdoch *et al.*, 1985) provide strong support for this interpretation.

The identification of heavy-element systems in QSO spectra commonly includes the recognition of a C IV doublet. In many cases the system is defined by C IV (and maybe Lα) alone, due to the difficulty in claiming weak metal lines which mostly fall in the Lα forest. In the case of the highest-redshift QSO 2000 − 330 ($z = 3.78$; Peterson *et al.*, 1982), which we have now observed at the AAT over the past three seasons, our initial data obtained in 1982 did not extend past Lα emission. Hence, the search for high-

* Paper presented at the IAU Third Asian-Pacific Regional Meeting, held in Kyoto, Japan, between 30 September–6 October, 1984.

redshift absorption systems had to be conducted entirely within the Lα forest (Hunstead *et al.*, 1985). We adopted a cross-correlation procedure, together with detailed checking for plausibility, and identified four heavy-element systems with $z_{abs} > 3.0$. All of these systems have recently been confirmed by spectra redward of Lα.

A particularly interesting region in 2000 – 330 is shown in Figure 1. The highest-redshift system in this object at $z_{abs} = 3.552$ shows a clear Si IV doublet with probable velocity structure. This structure may extend over the full width of the corresponding Lα (indicated by the bar labelled *A*), equivalent to a velocity range ~ 900 km s^{-1}. A velocity range of this magnitude suggests a cluster origin. If confirmed by our planned C IV observations this would be the earliest epoch detection of galaxy clustering, with important implications for theories of cluster formation and evolution.

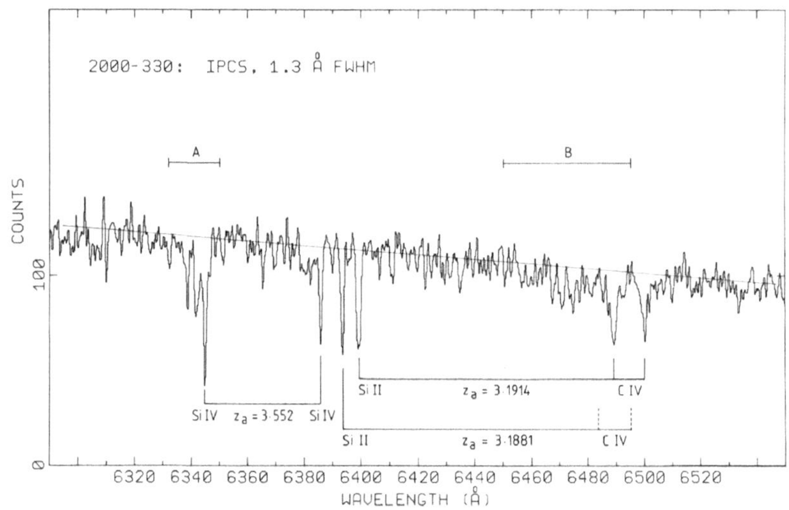

Fig. 1. A red-region AAT spectrum of 2000 – 330 at 1.3 Å FWHM showing metal lines in three high-redshift absorption systems. Note the absence of C IV in the $z_{abs} = 3.1881$ system. The bar labelled *A* indicates the extent of Lα in the $z_{abs} = 3.552$ system at the wavelength of Si IV λ1393. Similarly, *B* refers to the Lα trough at $z_{abs} = 3.18$ translated to C IV λ1548.

Also shown in Figure 1 are lines from two further systems, close together in redshift, which exhibit markedly different ionization structure. The higher-redshift system at $z_{abs} = 3.1914$ shows the expected C IV doublet together with Si II λ1526, while additional lines due to Si IV, Si III, Si II, C II, and O I have been identified. This mixed ionization structure is typical of a halo + disk line-of-sight. On the other hand, the system at $z_{abs} = 3.1881$ appears to contain no C IV (or Si IV) but still has a rich low-ionization spectrum typical of a galactic disk. This is probably the first reported case of an absorption system without the signature doublet of C IV. While this interesting system may have important ramifications for studies of galaxy evolution, its detection certainly impinges directly on present techniques for identifying QSO absorption systems.

The Lα corresponding to the systems at $z_{abs} \simeq 3.19$ is a very broad trough with a velocity width of 2100 km s^{-1}. The corresponding region of C IV λ1548 is indicated by the bar labelled *B* in Figure 1. There is suggestive evidence for weak C IV components spanning an appreciable fraction of the trough width, once again implying a sightline through a cluster of galaxies.

In conclusion, our preliminary analysis of the heavy-element absorption systems in 2000 – 330 has shown that the ionization structure in these very young galaxies is remarkably similar to that seen in Galactic halo and disk sightlines. In addition we have found tentative evidence for clusters of galaxies at these early epochs, although further observations are needed before this can be firmly established.

References

Hunstead, R. W., Murdoch, H. S., Peterson, B. A., Blades, J. C., Jauncey, D. L., Wright, A. E., Pettini, M., and Savage, A.: 1985, *Astrophys. J.*, submitted.

Murdoch, H. S., Hunstead, R. W., Pettini, M., and Blades, J. C.: 1985, *Astrophys. Space Sci.* **118**, 501 (this issue).

Peterson, B. A., Savage, A., Jauncey, D. L., and Wright, A. E.: 1982, *Astrophys. J.* **260**, L27.

Pettini, M., Hunstead, R. W., Murdoch, H. S., and Blades, J. C.: 1983, *Astrophys. J.* **273**, 436.

THE BARYON CLUMP WITHIN AN EXTENDED DARK MATTER REGION*

SATORU IKEUCHI

Tokyo Astronomical Observatory, University of Tokyo, Mitaka, Japan

(Received 19 July, 1985)

Abstract. As a model of the Lyman α absorbing systems of quasar lights, the gravitational equilibrium of a baryon clump within an extended dark matter composed of collisionless particles is examined. There exists the critical mass below the clump is stable. It is highly probable that these baryon clumps may be the Lyman α absorbers.

Various models have been presented to account for the origin of the Lyman α forests of QSOs (Sargent *et al.*, 1980, 1982). They are: pressure confined intergalactic clouds (Sargent *et al.*, 1980; Ostriker and Ikeuchi, 1983; Ikeuchi and Ostriker, 1986); self-gravitating intergalactic clouds (Melott, 1980; Black, 1981), supernova remnants in galactic halos (Sarazin, 1981), winds from dwarf galaxies (Fransson and Epstein, 1982), and shells of cosmologically expanding shock waves (Ikeuchi *et al.*, 1983; Chernomordik and Ozernoy, 1983a, b).

In the present paper, we investigate the gravitational equilibrium of a baryon cloud within an extended dark matter region, which is composed of collisionless particles, and we examine whether these clouds can be a candidate of Lα absorbing systems of quasar light or not.

The distributions of baryons and dark matter are assumed to be spherically symmetric with respect to their common center. The velocity dispersion σ of collisionless particles (dark matter) is assumed to be spatially constant. For baryons, we assume the barotropic relation $P_b = K\rho_b^\gamma$.(Black, 1981). The gravitational equilibrium of baryons within the dark matter is described by

$$\frac{1}{\rho_b}\frac{\mathrm{d}P_b}{\mathrm{d}r} = -\frac{G}{r^2}(M_{b,r} + M_{d,r}), \tag{1}$$

where we use the subscripts b and d to denote the baryon and dark matter. The mass, for either case, within radius r is written as

$$M_{i,r} = \int_0^r 4\pi r^2 \rho_i(r)\,\mathrm{d}r \quad (i = b, d). \tag{2}$$

* Paper presented at the IAU Third Asian-Pacific Regional Meeting, held in Kyoto, Japan, between 30 September–6 October, 1984.

Astrophysics and Space Science **118** (1986) 509–514.

We assume that the gravitational equilibrium of dark matter is effectively described in the same form as for the baryons by using the velocity dispersion

$$\frac{\sigma^2}{\rho_d}\frac{\mathrm{d}\rho_d}{\mathrm{d}r} = -\frac{G}{r^2}(M_{b,r} + M_{d,r}). \tag{3}$$

Equations (1) and (3) give the relation

$$\rho_d(r)/\rho_d(0) = (\rho_b(r)/\rho_b(0))^\eta \quad \text{for} \quad r = 1, \tag{4a}$$

$$\frac{\rho_d(r)}{\rho_d(0)} = \exp\left[\frac{\eta}{\gamma - 1}\left(\left(\frac{\rho_b(r)}{\rho_b(0)}\right)^{\gamma-1} - 1\right)\right] \quad \text{otherwise}, \tag{4b}$$

where we define $c^2 = \mathrm{d}P_b/\mathrm{d}\rho_b = \gamma K\rho_b^{\gamma-1}(0)$ and $\eta = (c/\sigma)^2$. Then, if $\eta \ll 1$ $(c \ll \sigma)$ we have $\rho_d \simeq$ constant, and if $\eta \gg 1$ we have $\rho_b \simeq$ constant. The velocity dispersion of collisionless particles depends upon the mass of a particle, and the epoch when these particles decoupled from the cosmic expansion by their own self-gravity.

On the other hand, the sound speed of baryons is determined by the heating and cooling processes. We assume, in the present paper, $\eta \ll 1$, that is, the dark matter extends uniformly around a baryon clump. Then, we have

$$\frac{1}{\rho_b}\frac{\mathrm{d}P_b}{\mathrm{d}r} = -\frac{GM_{b,r}}{r^2} - \frac{4\pi}{3}\rho_d G r. \tag{5}$$

We express the baryon density $\rho_b(r)$ as

$$\rho_b(r) = \rho_b(0)f(r/R_B) = \rho_b(0)f(x), \tag{6}$$

with $x = r/R_B$ for an arbitrary function $f(x)$, which satisfies $f(0) = 1$ and $f(1) = 0$. Defining the function $F_{n,m}(x)$ as

$$F_{n,m}(x) = \int_0^x x^n f^m(x)\,\mathrm{d}x, \tag{7}$$

the baryon mass within the radius r is

$$M_{b,r} = 4\pi R_B^3\rho_b(0)F_{2,1}(x), \tag{8}$$

and the total mass of a baryon clump with the radius R_B becomes $M_B = M_{b,r} = R_B = 4\pi R_B^3\rho_b(0)F_{2,1}(1)$.

Equations (5) and (8) are easily integrated for various sets of parameters, $X = \rho_b(0)/\rho_d$, ρ_d and the baryon temperature T_b. In this section, we examine the baryon clump within a region of extended dark matter for a model of Lα absorbers. The cloud is assumed to be ionized and heated by the diffuse UV flux, which is taken as (Ikeuchi and Ostriker, 1986)

$$J(z) = 10^{-21}J_{21}\left(\frac{1+z}{3.5}\right)^4 \mathrm{erg\,cm^{-2}\,s^{-1}\,Hz^{-1}}. \tag{9}$$

When this UV heating balances the radiative cooling the cloud temperature is $T_b \simeq 3 \times 10^4$ K (Black, 1981; Ostriker and Ikeuchi, 1983).

In Figure 1, we illustrate the relation between the equilibrium mass M_B and the density ratio $X = \rho_b(0)/\rho_d$ for the case of $T_b = 3 \times 10^4$ K and $\rho_d = \rho_{cls}(1 + z = 5) = 2.5 \times 10^{-27}$ g cm^{-3} (Case a). Two other cases of $\rho_d = 10 \times \rho_{cls}(1 + z = 5)$ (Case b) and $\rho_d = 0.1 \times \rho_{cls}(1 + z = 5)$ (Case c) are also illustrated for comparison. Since the

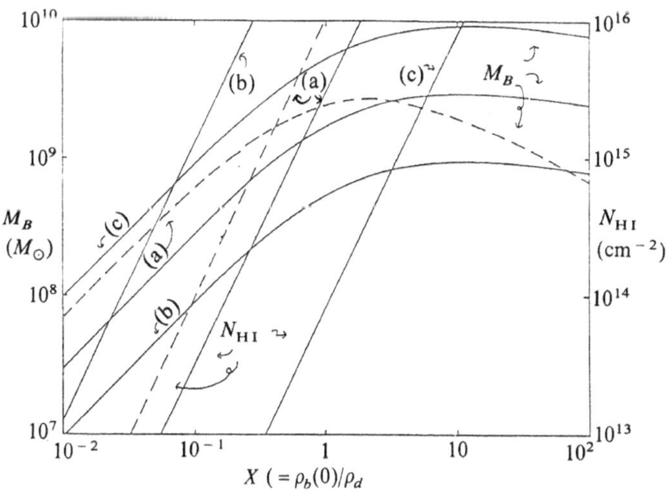

Fig. 1.

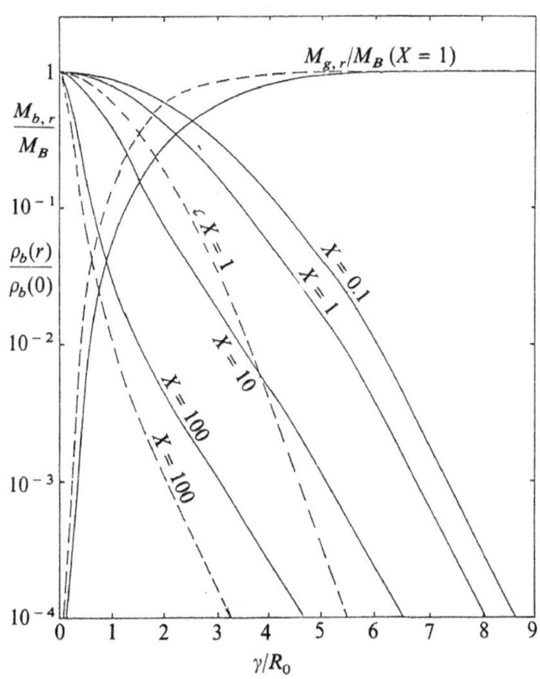

Fig. 2.

isothermal cloud within an external gravity extends infinitely, we cut it at the radius R_B, where the density decreases to $\rho_b(R_B)$, such that $\rho_b(R_B) = 5 \times 10^{-4}\, \rho_b(0)$. As long as $\rho_b(R_B) < 10^{-2}\, \rho_b(0)$, the baryon mass is effectively independent of the definition of the surface of the cloud as seen in Figure 2.

In Figure 2 we illustrate the distribution of baryon density $\rho_b(r)/\rho_b(0)$ and the mass fraction within the radius r, $M_{b,r}/M_B$, for Case a by solid lines and Case b by dashed lines. Formally, the density distribution is written, from Equation (5), as

$$\frac{\rho_b(r)}{\rho_b(0)} = \exp\left[-\frac{G}{c^2}\int_0^r \frac{GM_{b,r}}{r^2}\,\mathrm{d}r - \frac{2\pi G}{3c^2}\,\rho_d r^2\right]. \tag{10}$$

For a fixed ρ_d, the degree of central condensation increases with increasing the density ratio $X = \rho_b(0)/\rho_d$. At a small $X(= 0.1, 1)$, we can neglect the self-gravity and the density distribution becomes gaussian. With increasing $X(= 10)$, the first term in the right-hand side dominates the second term and the baryon density decreases rapidly with radius r. At the limit of large X ($= 100$), this clump nearly corresponds to the self-gravitating isothermal cloud, which is highly condensed to the center, with the polytropic index $N = \infty$. In other words, the degree of central condensation reflects the importance of self-gravity. On the other hand, for a fixed X the baryon density decreases more steeply with increasing the density of dark matter. The core radius, at which the gas density decreases to e^{-1} of the center, represents the ratio of the internal energy to the gravitational energy near the center. With increasing ρ_d this core radius decreases, though not for the same physical reason as the preceding case. As is seen, 90% of the mass is inside the radius $\sim R_B/2$, at which point the density decreases to one hundredth of the central one.

The ionization equilibrium of hydrogen atoms illuminated by the diffuse UV flux is described by Black (1981)

$$JG_H n_{\mathrm{H\,I}} = n_e n_{\mathrm{H\,II}} \alpha_H\,, \tag{11}$$

where J is the UV flux defined in (9), G_H is the photo-ionization parameter and α_H is the recombination coefficient. Here, we take a H/He plasma with the particle number ratio 9/1. When the line-of-sight passes through the center of a cloud, the H I column density for this line-of-sight should be

$$N_{\mathrm{H\,I}} = 2\int_0^{R_B} n_{\mathrm{H\,I}}\,\mathrm{d}r = 0.26\left(\frac{3.5}{1+z}\right)^4 J_{21}^{-1}\int_0^{R_B} n_b^2(r)\,\mathrm{d}r\,, \tag{12}$$

where $n_b(r)$ is the number density of the cloud medium. In this expression, we take the UV flux as in Equation (9) and the cloud is assumed to be wholly ionized, $n_{\mathrm{H\,II}} = n_b$ and $n_e = 1.1n_b$. In Figure 1, the H I column density is also plotted, which is represented by a nearly straight line. The dashed line shows a homogeneous model, which well reproduces the numerical one. We have the relation $N_{\mathrm{H\,I}} \propto X^2(X+d)^{-1/2}\,\rho_d^{3/2}$, where

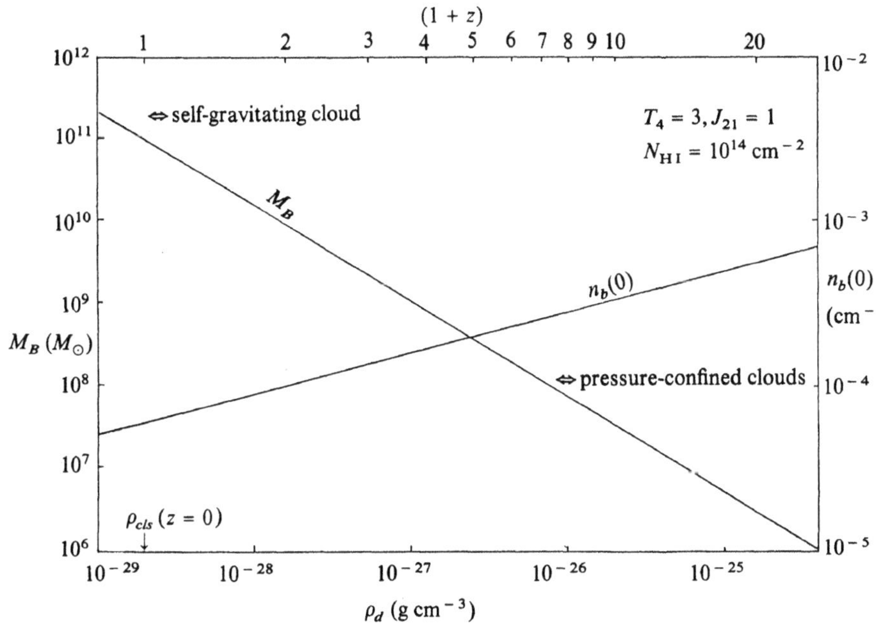

Fig. 3.

d is a structure-dependent numerical factor. The results in Figure 1 well agree with this relation.

The typical H I column density of Lα absorbers is $N_{\rm HI} \sim 10^{14}\ \rm cm^{-2}$ (Sargent *et al.*, 1980). In Figure 3, the baryon mass M_B and central density $n_b(0)$ of the cloud with $N_{\rm HI} = 10^{14}\ \rm cm^{-2}$ are shown with respect to ρ_d. On the upper horizontal axis, the corresponding redshift $1 + z$ is plotted if $\rho_d = \rho_{cls}(z)$. The cloud within the dark matter $\rho_d \gtrsim 2 \times 10^{-27}\ \rm g\ cm^{-3}$ is very similar to a pressure-confined cloud in the mass and density (Ostriker and Ikeuchi, 1983), as is expected. Both models attempt to confine a hot cloud by some external force: the gravity of dark matter or the pressure of the ambient medium. At the limit $\rho_d \sim 0$, the present model reduces to the self-gravitating cloud presented by Black (1981). Figure 3 shows simple relations $M_B \propto \rho_d^{-5/4}$ and $n_b(0) \propto \rho_d^{1/4}$ (for a fixed $N_{\rm HI}$). These relations can be easily derived.

References

Black, J. H.: 1981, *Monthly Notices Roy. Astron. Soc.* **197**, 553.
Chernomordik, V. V. and Ozernoy, L. M.: 1983a, *Nature* **303**, 153.
Chernomordik, V. V. and Ozernoy, L. M.: 1983b, *Astrophys. Space Sci.* **97**, 19.
Fransson, C. and Epstein, R.: 1982, *Monthly Notices Roy. Astron. Soc.* **198**, 1127.
Ikeuchi, S. and Ostriker, J. P.: 1986, *Astrophys. J.* **302**, No. 2, in press.
Ikeuchi, S., Tomisaka, K., and Ostriker, J. P.: 1983, *Astrophys. J.* **265**, 583.
Melott, A. L.: 1980, *Astrophys. J.* **241**, 889.
Ostriker, J. P. and Ikeuchi, S.: 1983, *Astrophys. J.* **268**, L63.

Sarazin, C. L.: 1981, *Astrophys. Letters* **21**, 23.
Sargent, W. L. W., Young, P. J., and Schneider, D. P.: *Astrophys. J. Suppl.* **42**, 41.
Sargent, W. L. W., Young, P. J., Boksenberg, A., and Tytler, D.: 1982, *Astrophys. J.* **256**, 374.

THE EFFECTS OF SELF-GRAVITY ON THE SOLAR NEBULA*

CHI YUAN**, CHIH-KANG CHOU

Institute of Physics and Astronomy, National Central University, Chung-Li, Taipei, China

and

TA-JEN LEE

Department of Atmospheric Physics, National Central University, Chung-Li, Taipei, China

(Received 19 July, 1985)

Abstract. We study the distribution and transport of angular momentum in a self-gravitating accretion disk formed during the collapse of a rotating gas cloud. Using the surface density for the low-viscosity models and minimum-mass models presented by Cassen and Summers, Poisson's equation is solved explicitly to determine the effects of self-gravitation of the protostellar disk. Analytic expressions for the angular momentum of the central star and other relevant quantities of interest during the formation stage are presented.

1. Introduction

In a recent paper, using the models previously developed by Cassen and collaborators (Cassen and Moosman, 1981; Cassen and Summers, 1983) for the formation of a protostar and disk from a rotating protostellar cloud, Yuan and Cassen (1984) calculated the angular momentum, the rate of accretion of angular momentum and the mass of the central star as functions of time. They showed that the accretion rate of angular momentum is orders of magnitude smaller than the surrounding disk through density waves, provided that the central star becomes slightly distorted into a triaxial shape. This suggests that the angular momentum (nondimensionalized by $(GM_*^3 R_*)^{1/2}$ where M_* and R_* denote, respectively, the mass and radius of the central star) cannot exceed some critical value J_s at which the central star becomes triaxially distorted. Furthermore, their results showed that the nondimensional angular momentum $J_*/[GM_*^3 R_*]^{1/2}$ of the central star tends to a constant asymptotic limit J_c with time, independent of the initial rotation and sound speed of the original protostellar cloud. Moreover, for the low-viscosity models considered by Cassen and Moosman (1981), this limit J_c equals $\frac{2}{7}$ (0.2857). Consequently, in terms of nondimensional angular momentum, the central star has two possible angular momentum final states as the rotating protocloud collapses, namely, it reaches the constant limit J_c if this constant limit is smaller than the critical value for triaxial stars ($J_c < J_s$); it reaches the critical value J_s if this value is smaller than the constant limit ($J_s < J_c$).

Yuan and Cassen (1984) have emphasized that a slightly triaxial distortion of the central star can lead to extremely effective transport of angular momentum from the star

* Paper presented at the IAU Third Asian-Pacific Regional Meeting, held in Kyoto, Japan, between 30 September–6 October, 1984.

** On leave from the City College of the City University of New York, U.S.A.

Astrophysics and Space Science **118** (1986) 515–521.

to the surrounding disk of the nebula by means of exciting density waves at the resonance regions. The central star would maintain its angular momentum at the critical value J_s by transporting the excessive amount to the outer Lindblad resonance and beyond if the dimensionless angular momentum $J_*(GM_*^3 R_*)^{-1/2}$ ever becomes greater than the critical value J_s. However, the results obtained by Yuan and Cassen as delineated above are based on the following assumptions:

(1) The mass and angular momentum of the central star and its surrounding disk are those determined by the low-viscosity models (Cassen and Moosman, 1981).

(2) The angular velocity in the disk may be approximated by Keplerian rotation.

In addition, the self-gravitation of the disk was entirely neglected in their analysis. Therefore, the main purpose of this study is to consider how the effects of self-gravity in the surrounding disk would modify their results.

Using the low-viscosity models for the nebula, we formulate the general problem in Section 2. We then present our solution in closed form and explain the physical significance of our results in Section 3. The role played by self-gravity in the disk with the minimum-mass models (Cassen and Summers, 1983) is also briefly mentioned in Section 3. A summary of our results follows in Section 4.

2. Description of the Problem

One of the main results of Cassen and Moosman (1981) was that the size of the nebula produced from the collapse of a rotating cloud would be the larger of $R_{CF} \equiv J^2/k^2GM^3$ and $R_V \equiv (GM\nu/2a^3)^{1/2}$, where J, M, and a are the total angular momentum, total mass, and sound speed of the protostellar material; G is the gravitational constant; k, a number of order unity; and ν, the effective viscosity in the nebula. From this result it can be deduced that low-mass nebulae are produced if $P \equiv (R_V/R_{CF})^2 \gg 1$; massive nebulae result if $P \lesssim 1$. Disks for which $P \gg 1$ are very viscous and spread rapidly beyond R_{CF} whereas $P \ll 1$ implies relatively little viscous redistribution of angular momentum. Thus a high-viscosity disk would retain less mass than a low-viscosity disk because the viscous torque can remove angular momentum from the gas closer to the center, which then flows in to become part of the protostar.

In the low-viscosity models developed by Cassen and Summers (1983), the internal viscous torques are assumed to operate only in a relatively small region of radius r_* near the center of the disk. Outside of r_* the $P = 0$ solution of Cassen and Moosman (1981) applies. This solution does not neglect the viscous mixing of the infalling gas with disk gas. The infalling gas produces a drag on the disk gas that causes the latter to flow inward to become part of the protostar. The first, low-angular-momentum material falls directly on the growing protostar, whereas higher-angular-momentum gas collapses to form a disk around the protostar. Most of the collapsing gas impinges directly on the disk rather than on the protostar, and expansion of the disk is due entirely to the addition of material with ever-increasing angular momentum. Most of the mass within r_* will be in the protosun; this is given by

$$M = \int_0^{r_*} 2\pi r\sigma(r, t)\,\mathrm{d}r\,, \tag{1}$$

where the surface density $\sigma(r, t)$ can be cast in the convenient form (Cassen and Summers, 1983)

$$\sigma(r, t) = \frac{\dot{M}t}{16\pi[r_d(t)]^{1/4}}\, r^{-7/4} \int_u^1 S(x)\,\mathrm{d}x \equiv \overline{\alpha}(t)S(u)r^{-7/4}\,; \tag{2}$$

where

$$u = r/r_d\,, \qquad r_d(t) = \Gamma^2(t)/GM_c\,, \qquad S(u) = \int_u^1 S(x)\,\mathrm{d}x\,,$$

$$S(x) = \frac{(1 - X^{1/2})(3 - 2X)}{X^{1/4}(1 - X)^{3/2}}\,, \qquad M_c = \dot{M}t\,,$$

M_c, mass of the core; and $\Gamma(t)$ is the specific angular momentum of the gas just arriving at the outer edge of the disk from the collapsing cloud. The radius of the disk is given by $r_d(t)$. For a given collapse model $\Gamma(t)$ is a known function; for Shu's (1977) expansion-wave collapse, $\Gamma(t) = \Omega_0(0.488at)^2$ where Ω_0 is the cloud's original angular velocity, and $\dot{M} = 0.975a^3/G$.

In order to study the effects of self-gravity of the protostellar disk we adopt a heliocentric cylindrical coordinate system (r, θ, z) where the equatorial plane is described by $z = 0$. The nebula is disk-like and rotating around the protosun with negligible pressure effects. The governing equations for this axisymmetrically rotating system reduce simply to

$$r\Omega^2(r) = \frac{u_\theta^2}{r} = \left.\frac{\partial\phi}{\partial r}\right|_{z=0}, \tag{3}$$

where $u_\theta(r)$ is the velocity component in the θ-direction and $\Omega(r)$ is the angular velocity. The gravitatinal potential ϕ is related to the surface density $\sigma(r, t)$ by the Poisson equation

$$\frac{\partial^2\phi}{\partial r^2} + \frac{1}{r}\frac{\partial\phi}{\partial r} + \frac{\partial^2\phi}{\partial z^2} = 4\pi G\sigma(r, t)\,\delta(z)\,, \tag{4}$$

where $\sigma(r, t)$ is given by Equation (2) and $\delta(z)$ is the Dirac delta function. In the above expression, we have implicitly assumed that the nebula has an infinitesimally thin disk.

The angular momentum of the central star is given by

$$J_*(t) = \int_0^{r_*} 2\pi r\sigma(r, t) r^2 \Omega(r)\, \mathrm{d}r\,. \tag{5}$$

For a given model of the solar nebula, $\sigma(r, t)$ is a known function; for the low-viscosity models, $\sigma(r, t)$ is given by Equation (2). Using this equation in Equation (4), the gravitational potential ϕ is obtained by solving Poisson's equation. The angular velocity $\Omega(r)$ is then computed by substituting ϕ into Equation (3) and the angular momentum of the central star is calculated from Equation (5). Thus, the effects of self-gravitating disks are completely determined through the gravitational potential ϕ.

3. The Effects of a Self-Gravitating Disk

In order to study the effects of a self-gravitating disk, we have to solve the Poisson equation explicitly. Because the axisymmetry of the problem, this equation can be conveniently solved in terms of Hankel transform. We adopt the following notation and formulae for the Hankel transforms:

$$g_n(\xi) = \int_0^{\infty} r g(r) J_n(\xi r)\, \mathrm{d}r\,, \tag{6a}$$

$$g(r) = \int_0^{\infty} \xi g_n(\xi) J_n(\xi r)\, \mathrm{d}\xi\,, \tag{6b}$$

$$g_n'(\xi) = (n-1) \int_0^{\infty} g(r) J_n(\xi r)\, \mathrm{d}r - \xi g_{n-1}(\xi)\,; \tag{6c}$$

where $J_n(x)$ is the Bessel function of order n.

We now apply the method of Hankel transformation to solve Equation (4). After some transformations, it becomes

$$\frac{\mathrm{d}^2\phi}{\mathrm{d}z^2} - \xi^2\phi = 4\pi G\sigma\delta(z)\,, \tag{7}$$

where

$$\phi \equiv \phi_0(\xi, z) = \int_0^{\infty} r\phi(r, z) J_0(\xi r)\, \mathrm{d}r\,, \qquad \sigma \equiv \sigma_0(\xi, t) = \int_0^{\infty} r\sigma(r, t) J_0(\xi r)\, \mathrm{d}r\,.$$

Equation (7) may be integrated to yield

$$\phi(\xi, z) = -\frac{2\pi G}{\xi} e^{-|z|\xi} \sigma(\xi, t) = -\frac{2\pi G}{\xi} e^{-|z|\xi} \int_0^{\infty} r\sigma(r, t) J_0(\xi r)\, \mathrm{d}r\,. \tag{8}$$

Using Equation (2) in Equation (8) and setting $z = 0$, we obtain

$$\phi(\xi, t) = -\frac{2\pi G}{\xi} \int_0^\infty r\,\mathrm{d}r\, J_0(\xi r) \left\{ \bar{\alpha}(t) r^{-7/4} \left[\mathscr{S}_0 - \int_0^u S(x)\,\mathrm{d}x \right] \right\}, \tag{9}$$

where $\mathscr{S}_0 \equiv \int_0^1 S(x)\,\mathrm{d}x = 2.72$, and $u = r/r_d$. For small u, Equation (9) may be written as

$$\phi(\xi, t) = -\frac{2\pi G}{\xi} \int_0^\infty r J_0(\xi r)\,\mathrm{d}r \,\{\bar{\alpha}(t) r^{-7/4} [\mathscr{S}_0 - \mathscr{I}(u)]\}, \tag{10}$$

with

$$\mathscr{I}(u) \equiv \int_0^u S(x)\,\mathrm{d}x = 4u^{3/4} - \tfrac{12}{5}u^{5/4} - \tfrac{10}{7}u^{7/4}.$$

Equations (9) and (10) cannot be evaluated, even if we retain only the lowest order term in u. It is seen that these equations can be formally integrated only if $\mathscr{I}(u)$ is entirely neglected. Thus, in the case of very small u ($r \ll r_d$), Equation (10) becomes

$$\phi(\xi, t) = -\frac{2\pi G}{\xi} \bar{\alpha}(t) \mathscr{S}_0 \int_0^\infty r^{-3/4} J_0(\xi r)\,\mathrm{d}r. \tag{11}$$

The inverse Hankel transform of Equation (11) then yields the solution

$$\phi(r, t) = -2\pi G \bar{\alpha}(t) \int_0^\infty J_0(\xi r)\,\mathrm{d}\xi \int_0^\infty r^{-3/4} J_0(\xi r)\,\mathrm{d}r = -\pi G \bar{\alpha}(t) \beta \mathscr{S}_0 r^{-3/4}, \tag{12}$$

where $\beta \equiv \Gamma(\frac{1}{8})\Gamma(\frac{3}{8})\,[\Gamma(\frac{5}{8})\Gamma(\frac{7}{8})]^{-1}$ (cf. Gradshteyn and Ryzhik, 1980).

We now turn to calculate the angular momentum of the central star. This is done by substituting Equation (12) into Equation (3) and using the resulting expression for the angular velocity in Equation (5), we have

$$J_* = \tfrac{8}{7}(2\pi\alpha)\,(\tfrac{3}{4}\pi G\alpha\beta)^{1/2}\, r_*^{7/8},$$

where

$$\alpha \equiv \alpha(t) \equiv \mathscr{S}_0 \bar{\alpha}(t) = \frac{\mathscr{S}_0 \dot{M} t}{16\pi[r_d(t)]^{1/4}}, \qquad \mathscr{S}_0 = 2.72. \tag{13}$$

The mass of the central star may be obtained from Equations (1) and (2) with the result

$$M_* = 8\pi\alpha(t) r_*^{1/4}. \tag{14}$$

The non-dimensional angular momentum of the central star is then

$$J_c \equiv \frac{J_*}{\sqrt{GM_*^3 r_*}} = \frac{2}{7}\,\zeta\,, \qquad \zeta = \frac{1}{4}\sqrt{\frac{3}{2}}\,\beta = \frac{1}{4}\sqrt{\frac{3}{2}\,\frac{\Gamma(\frac{1}{8})\Gamma(\frac{3}{8}}{\Gamma(\frac{5}{8})\Gamma(\frac{7}{8})}} = 1.038\,. \quad (15)$$

The effects of self-gravity of the disk may be represented by the factor ζ, which is only slightly greater than unity. Thus the asymptotic angular momentum of the central star for a self-gravitating disk is slightly increased.

Similar results are obtained for the minimum-mass models. For instance, if the nebular viscosity is assumed to increase linearly with radius, then the surface density in the inner part of the nebula can be shown to vary as $r^{-3/2}$ (Cassen and Summers, 1983; Hayashi, 1981). For this case the asymptotic ceiling J_c may be cast in the form

$$J_c = \frac{J_*}{\sqrt{GM_*^3 r_*}} = \frac{2}{5}\,\omega\,, \qquad \omega = \frac{\sqrt{2}\,\Gamma(\frac{1}{4})}{4\Gamma(\frac{3}{4})} = 1.046\,. \quad (16)$$

The factor ω represents the effects of self-gravitation of the nebular disk. The details of this calculation will be given in a forthcoming paper.

From the above discussion it can be seen that the effects of self-gravity on the asymptotic ceiling J_c are, in general, very small in the inner part of the nebula ($r \ll r_d$, $r_* \ll r_d$) and that the existence of a constant ceiling for the central star is a general result, although J_c is model-dependent.

The rate of accretion of angular momentum of the central star may be computed by differentiating Equation (13) with respect to t. The ratio of this accretion rate to the torque Υ exerted on the triaxial star by spiral density waves for Keplerian rotation has been estimated to be (Yuan and Cassen, 1985; Goldreich and Tremaine, 1979)

$$\frac{1}{\Upsilon}\frac{dJ_*}{dt} = 1.19\varepsilon^{-2} \times 10^{-9}\,, \quad \varepsilon \sim 10^{-2}\text{–}10^{-3}\,, \quad \text{(low-viscosity model)} \quad (17)$$

where ε is the ratio of the triaxial field to the mean field at the outer Lindblad resonance. For a self-gravitating disk, with $P = 0$, this ratio turns out to be

$$\frac{1}{\Upsilon}\frac{dJ_*}{dt} \simeq 2.88\varepsilon^{-2} \times 10^{-9}\,; \qquad \text{(for the low-viscosity model)} \quad (18)$$

whereas, for the minimum-mass models with linear viscosity variation $\nu = \nu_0(r/r_0)$, this ratio is given by

$$(1) \quad \frac{1}{\Upsilon}\frac{dJ_*}{dt} \simeq 5.53\varepsilon^{-2} \times 10^{-8} \quad \text{for a Keplerian disk}\,,$$

(19)

$$(2) \quad \frac{1}{\Upsilon}\frac{dJ_*}{dt} \simeq 6.10\varepsilon^{-2} \times 10^{-8} \quad \text{for a self-gravitating disk}\,.$$

Therefore, the transport of angular momentum from a triaxial star to the surrounding disk by density waves is extremely effective for both the low-viscosity model and minimum-mass models. Moreover, the effects of self-gravity are relatively small in the inner parts of the nebula. The details of these calculations will be presented elsewhere.

4. Summary

In this paper we have investigated the properties of self-gravitating disks that are formed by the accretion of gas from a rotating cloud onto a gravitational source. The effects of self-gravitation of the protostellar disk can be studied by solving Poisson's equation with the aid of nebular models. For the low-viscosity models and minimum-mass models discussed by Cassen and Summers (1983), it is found that analytic solutions can be obtained only in the inner part of the nebula, and that the effects of self-gravity are relatively small.

Our results can be summarized as follows:

(1) There exists an asymptotic ceiling J_c for $J_*/(GM_*^3 r_*)^{1/2}$, as time increases. For the low-viscosity model, $J_c = \frac{2}{7}\zeta$ ($\zeta = 1.038$); for the minimum-mass model with linear viscosity variation $\nu = \nu_0 r/r_0$, $J_c = \frac{2}{5}\omega$ ($\omega \simeq 1.046$), where the constant factors ζ and ω represent the effects of self-gravity of the protostellar disk. Thus the existence of a constant ceiling is a general result, although J_c is model-dependent.

(2) A relatively slight triaxial distortion of the central star can excite outgoing spiral density waves at the outer Lindblad resonance. These waves exert a torque on the triaxial star and, as a consequence, carry its angular momentum to the outer parts of the protostellar disk. The transport of angular momentum from a triaxial star to the surrounding disk by density waves is extremely effective.

Acknowledgement

This work has been supported by the National Science Council of the Republic of China under Grant NSC74–0208–M008–04.

References

Alpar, M. A., Cheng, A. F., Ruderman, M. A., and Shaham, J.: 1982, *Nature* **300**, 728.
Cassen, P. and Moosman, A.: 1981, *Icarus* **48**, 353.
Cassen, P. and Summers, A.: 1983, *Icarus* **53**, 26.
Goldreich, P. and Tremaine, S.: 1979, *Astrophys. J.* **233**, 857.
Gradshteyn, I. S. and Ryzhik, I. M.: 1980, *Tables of Integrals, Series, and Products*, Academic Press, New York, p. 684.
Hayashi, C.: 1981, *Prog. Theor. Phys. Suppl.* **70**, 35.
Lynden-Bell, D. and Pringle, J. E.: 1974, *Monthly Notices Roy. Astron. Soc.* **163**, 603.
Shu, F.: 1977, *Astrophys. J.* **214**, 488.
Yuan, C. and Cassen, P.: 1985 (to appear in Icarus).

ECHELLE SPECTROSCOPY OF NARROW LINE REGIONS OF SEYFERT GALAXIES*

M. IYE

European Southern Observatory, Garching, Freistaat, Bayern, F.R.G.
and Tokyo Astronomical Observatory, Mitaka, Tokyo, Japan

and

M.-H. ULRICH

European Southern Observatory, Garching, Freistaat, Bayern, F.R.G.

(Received 19 July, 1985)

Abstract. Preliminary results of spectroscopic studies of selected Seyfert galaxies employing a new Cassegrain echelle spectrograph completed at ESO are presented. Subjects reported are the asymmetric profiles of narrow emission lines, spatial variation of narrow emission line profiles, and the interstellar absorption lines.

A new Cassegrain echelle spectrograph (CASPEC) coupled with a cooled CCD was completed at ESO (D'Odorico *et al.*, 1983) and was successfully operated at the ESO 3.6 m telescope during its first scientific observing run in April 1984.

We obtained a number of echellograms of one hour exposure for two type-1 Seyferts IC 4329A and ESO 141–G55, a type-2 Seyfert NGC 4388, and a high excitation galaxy Tololo 74. The spectral resolution is as high as $\lambda/\Delta\lambda = 1.4 \times 10^4$ or $\Delta V_r = 22 \text{ km s}^{-1}$ at Hα. Although one pixel of CCD corresponds to 0.″72 in the sky, the spatial resolution is limited by the seeing size of typically 2″. The entrance slit was aligned to the major and the minor axes of the galaxy. The spatial extent covered by the slit was up to 10″ across the nucleus. A single echellogram covers a spectral region of 1000 Å width around Hα or Hβ.

Figure 1 shows a CASPEC/CCD image of the Hα spectral region of the type-2 Seyfert galaxy NGC 4388. Spectra extracted from respective orders in this image are calibrated and merged to form a reduced spectrum (Figure 2). The emission lines He I λ5876, [O I] λ6300, [O I] λ6364, [N II] λ6548, Hα λ6563, [N II] λ6584, [S II] λ6717, and [S II] λ6731 are clearly resolved. All the echellograms were reduced using a CASPEC reduction package run in MIDAS (Ponz, 1984).

Many interesting features are revealed in these high resolution spectra of Seyfert galaxies, among which three subjects are reported here.

* Paper presented at the IAU Third Asian-Pacific Regional Meeting, held in Kyoto, Japan, between 30 September–6 October, 1984.

Astrophysics and Space Science **118** (1986) 523–528.

Fig. 1. A CASPEC/CCD image of the type-2 Seyfert NGC 4388. A spectral region from 5850 Å (lower left) to 6900 Å (upper right) is covered. Note the red wings of emission lines, and a small tilt of lines due to rotational motion. The [O I] λ6300 emission is seen at pixel coordinates around (228, 176) with a redshift of 0.0082. The corresponding atmospheric [O I] λ6300 is seen at (95, 160) as a very sharp line.

1. Asymmetric Profiles of Narrow Emission Lines

The line profiles of the type-2 Seyfert NGC 4388 are asymmetric and have extended red wings as shown in Figure 3(b). On the other hand, ESO 141–G55, a type-1 Seyfert, shows blue wings in its narrow emission lines (Figure 3(a)). Predominance of blue wings in the narrow emission lines has been reported by Osterbrock and De Robertis (1984) for type-1 Seyferts and by Turtle and Pence(1985) for their sample galaxies. If the

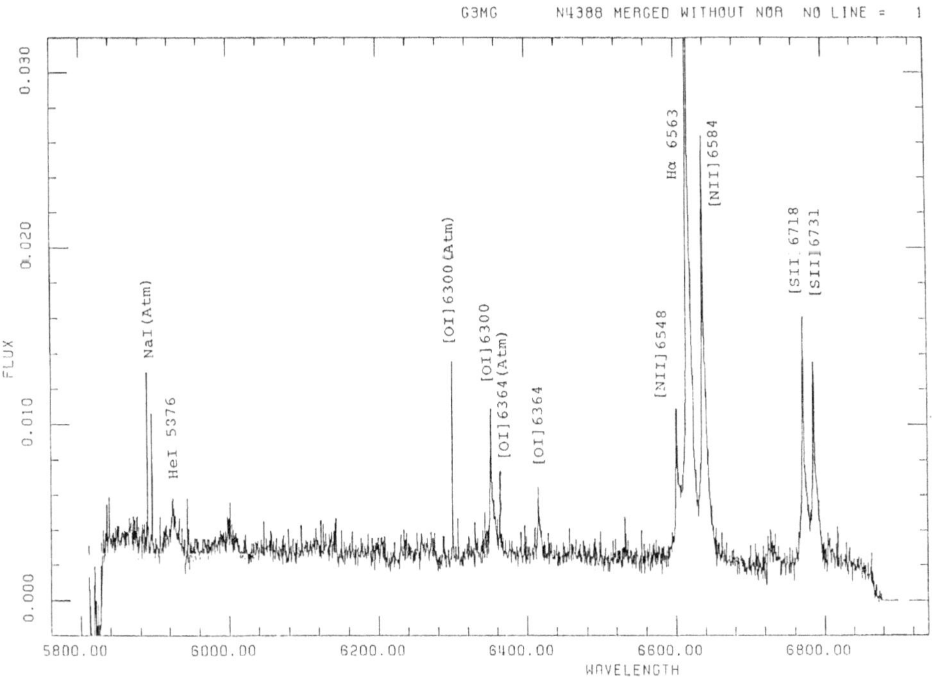

Fig. 2. Reduced spectrum of the nucleus of NGC 4388.

narrow line regions of these active galaxies are to be explained in a unified model, e.g., outflow/inflow obscured by dusts, distribution of a few discrete emission clouds, or accelerated/decelerated flow with varying line widths, then the presence of a red as well as blue emission wings in Seyferts places a strong constraint in modelling them.

2. Spatial Variation of Narrow Emission Line Profiles

Figure 4 gives the reduced spectrum of the nuclear region of the high excitation galaxy Tololo 74. The inlet figure shows the spatial variation of emission line profiles along the slit, set at position angle 45°. Each spatial element covers 1''.44, and the element 1 is to the NE of the nucleus, which is at element 5. Two components of emission clouds are clearly resolved. The sharp redward component extends from element 6 to element 1. The corresponding spatial extent is no less than 1.0 kpc, even if a seeing size of about 2'' is taken into account (H_0 = 75 km s^{-1} Mpc^{-1} is assumed). Since the galaxy is almost face-on, the relative velocity of 110 km s^{-1} between the two components, and the constancy of velocity over a few spatial elements, suggest a bi-polar nature of these emission line clouds. However, some alternative configuration of clouds in the disk plane, revolving with a flat rotation curve, cannot be ruled out from the present data alone.

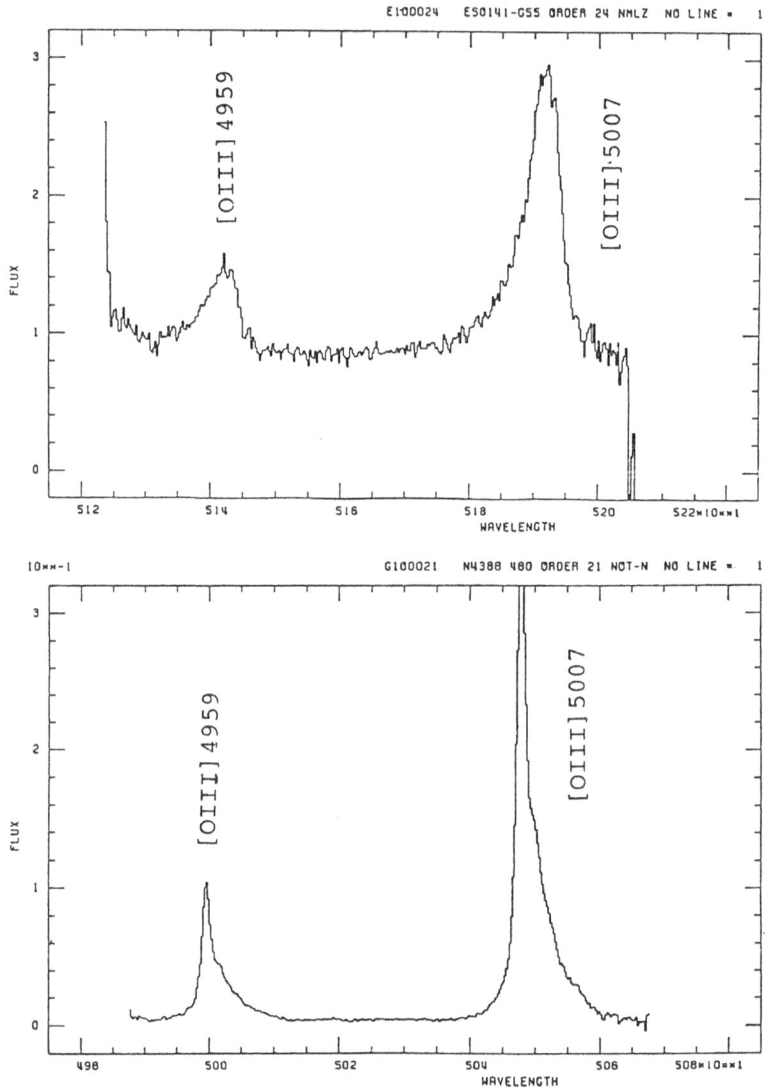

Fig. 3. Asymmetric profiles of narrow emission lines. Note that the type-1 Seyfert ESO 141–G55 shows conspicuous blue wings ((a) above), while the type-2 Seyfert NGC 4388 exhibits red wings ((b) below).

3. Interstellar Absorption Lines

Another topic of interest is the interstellar absorption lines. Figure 5 shows the interstellar absorption of Na I *D* lines in the disk of the type-1 Seyfert IC 4329A. The disk of this galaxy is edge-on and the column density of Na I is about 10^{15} cm^{-2} (Wilson and Penston, 1979). The FWHM of the *D* lines is 65 km s^{-1}, and the residual intensity at the line center is less than 20% of the continuum level.

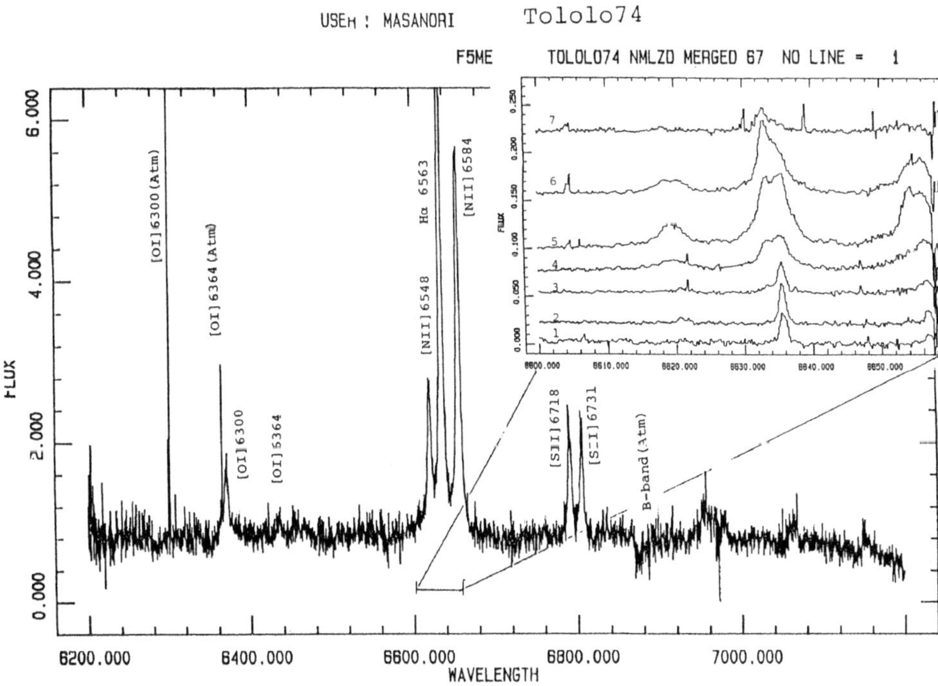

Fig. 4. Reduced spectrum of the nucleus of the high excitation galaxy Tololo 74. Inlet figure shows the spatial variation of emission line profiles. Seven profiles are plotted with an interval of 1".44 along the slit. Note the sudden change of profiles between the spatial elements 3 and 4, and the spatial extent of emission cloud components.

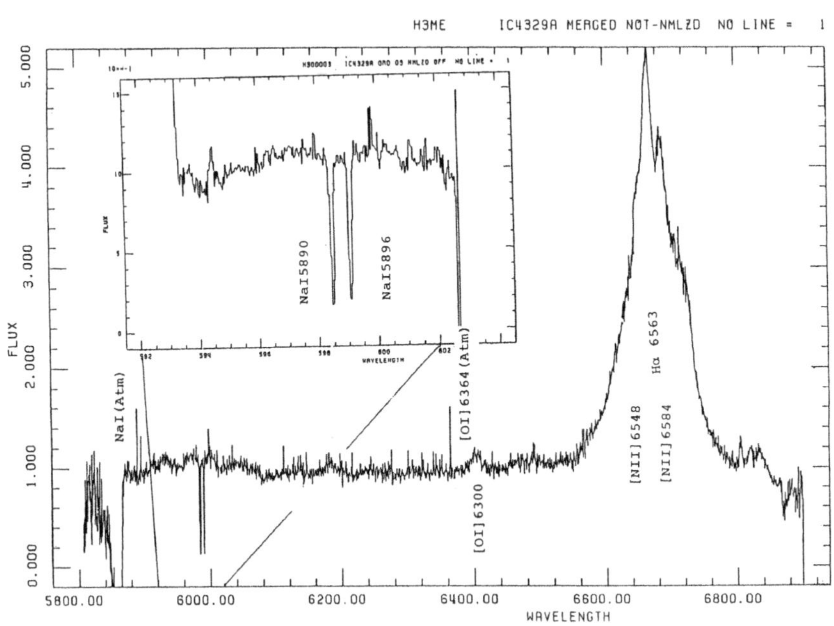

Fig. 5. Reduced spectrum of the type-1 Seyfert IC 4329A. Note the resolved interstellar absorption of Na I *D* lines at the redshift of the galaxy, $z = 0.016$.

Acknowledgements

CASEC has proved to be well suited to the study of narrow line regions of active galactic nuclei. The authors are grateful to S. D'Odorico for his introduction to CASPEC and to D. Ponz for his patient help in data reduction.

References

D'Odorico, S., Enard, D., Lizon, J. L., Ljung, B., Nees, W., Ponz, D., Raffi, G., and Tanne, J. F.: 1983, *Messenger* **33**, 2.
Osterbrock, D. E. and De Robertis, M. M.: 1984, *AAS Bull.* **16**, 440.
Ponz, D.: 1984, in *The MIDAS Users Guide,* ESO Operating Manual, No. 1.
Turtle, A. J. and Pence, W. D.: 1985, *Astrophys. Space Sci.* (this issue).
Wilson, J. C. and Penston, M. V.: 1979, *Astrophys. J.* **232**, 389.

NEW POLAR RING GALAXIES IN RICH CLUSTERS OF GALAXIES*

YOSHIAKI TANIGUCHI, KATSUNORI SHIBATA

Astronomical Institute, Tohoku University, Sendai, Japan

and

KEN-ICHI WAKAMATSU

Junior Technical College, Gifu University, Gifu, Japan

(Received 30 July, 1985)

Abstract. Four polar ring galaxies discovered in rich clusters of galaxies are presented. Brief comments on their structural properties are given.

We present four polar ring galaxies discovered in the rich clusters of galaxies A548, A1631, and A1644 on the plates taken by Dressler (1980). All of them are included in Dressler's (1980) catalogue. Their basic quantities are summarized in Table I, and direct photographs are reproduced in Figure 1.

TABLE I

Basic quantities of the presented polar ring galaxies

	A548–17	A548–71	A1631–14	A1644–105
Position: α(1950.0)[a]	$5^h42^m33\overset{s}{.}2$	$5^h43^m03\overset{s}{.}5$	$12^h49^m54\overset{s}{.}3$	$12^h54^m58\overset{s}{.}7$
δ(1950.0)[a]	−26 06 34	−25 52 45	−15 36 08	−16 55 06
Apparent photographic magnitude[a]	15	15	16	16
Absolute photographic magnitude[b]	−22.3	−22.3	−21.5	−21.3
Central component				
Angular size (″)	9.4 × 6.4	6.4 × 5.7	9.1 × 5.3	3.7 × 3.3
Linear size (kpc)	10.9 × 7.5	7.5 × 6.0	14.0 × 7.0	5.2 × 4.8
Ellipticity	0.3	0.1	0.4	0.1
Ring				
Angular size (″)	22″	21″	33″	9″.8
Linear size (kpc)	25.6	24.4	50.8	14.0
Inclination angle[c]	90	55	55	90

[a] Data are taken from Dressler (1980).

[b] Distance is estimated with the redshift of the cluster and $H_0 = 50$ km s^{-1} Mpc^{-1}.

[c] Inclination angle between the central and ring components.

* Paper presented at the IAU Third Asian-Pacific Regional Meeting, held in Kyoto, Japan, between 30 September–6 October, 1984.

Astrophysics and Space Science **118** (1986) 529–530.

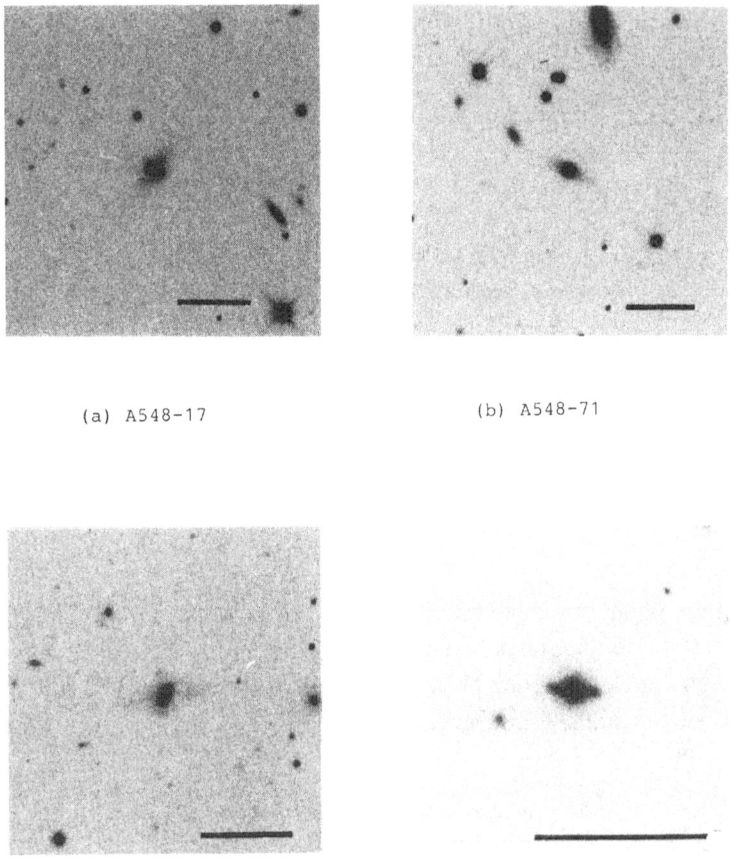

(a) A548-17 (b) A548-71

(c) A1631-14 (d) A1644-105

Fig. 1. Photographs of the new polar ring galaxies: (a) A548–17, (b) A548–71, (c) A1631–14, and (d) A1644–105. North is top and east is left. Scale is shown by a horizontal bar (= 30″).

We give brief comments on their structural characteristics. (1) All but A1644–105 have fainter polar components than previous well-known examples such as NGC 4650A (Laustsen and West, 1980), MCG5–29–86 (Wakamatsu and Arp, 1983), and A0136–0801 (Schweizer *et al.*, 1983). (2) The ring structures of both A1631–14 and A1644–105 can be discerned by the presence of a brightness dip near the central components. However, the other two show no evident ring feature. (3) In the cases of A548–71 and A1631–14, the inclination angles between the ring and central component are only 55°, which is significantly different from polar (90°).

References

Dressler, A.: 1980, *Astrophys. J. Suppl.* **42**, 565.
Laustsen, S. and West, R. M.: 1980, *Astron. Astrophys.* **1**, 177.
Schweizer, F., Whitmore, B. C., and Rubin, V. C.: 1983, *Astron. J.* **88**, 909.
Wakamatsu, K. and Arp, H. C.: 1983, *Astrophys. J.* **273**, 167.

DISTANCE TO THE ANTI-CENTER SHELL*

SHRINIVAS R. KULKARNI

Dept. of Astronomy, Univ. of California, Berkeley, U.S.A.

and

R. MATHIEU

Center for Astrophysics, Cambridge, Mass., U.S.A.

(Received 30 July, 1985)

Abstract. Heiles (1984) has identified a large angular diameter ($\sim 30°$), anomalous velocity H I structure in the anti-center region of the Galaxy. The distance to this structure is the key to understanding of the energetics and, hence, the origin. Distance estimates based on indirect arguments have ranged from less than 100 pc to a few kiloparsecs. We report the absence of any anomalous velocity Ca K absorption features in high sensitivity echelle spectra towards 17 stars located within the boundaries of this structure. We conclude that the structure is definitely beyond 500 pc and barring unusually low Ca II abundance is more distant than 2.5 kpc. This results supports the hypothesis that the anti-center structure is the result of collision of high-velocity clouds with the galactic H I disk.

1. Introduction

Recently, Heiles (1984) has identified a remarkable H I structure in the galactic anti-center region. This structure is impressive on two counts:

(1) It appears as a nearly complete shell with a diameter of $\sim 30°$ (see Heiles, 1984, for further details).

(2) It is moving at high negative velocities. The structure can be recognized at velocity, v as high as -90 km s^{-1} – the negative-velocity limit of Heiles' (1984) H I data. For $0 > v > -40$ km s^{-1} the structure is confused with normal-velocity H I, and it is not visible in any positive velocity map. Since Galactic rotation does not contribute significantly to radial velocity in this region of the sky, the observed velocities represent real motion in the Galactic standard of rest.

The morphology of the structure resembles a shell and for only this reason we will henceforth refer to this structure as the Anti-Center Shell (ACS). The origin of the ACS is unclear. In fact, from the H I data alone, it is not clear whether the ACS is a moving H I-ring or an expanding H I-shell. However, the absence of high-velocity polar caps, the absence of the shell at positive velocities and the high shell velocity are inconsistent with the expectation of an expanding shell model (see Kulkarni, 1983; for more details). Mirabel (1982) finds a perturbation in the kinematics of the low-velocity gas in the vicinity of $(l, b) = (185°, -11°)$ and he argues for a direct physical association of the anti-center group of high velocity clouds (ACI, ACII, etc.) and the intermediate velocity gas ($-40 > v > -100$ km s^{-1}). It is precisely this gas that forms the ACS.

* Paper presented at the IAU Third Asian-Pacific Regional Meeting, held in Kyoto, Japan, between 30 September–6 October, 1984.

All the data discussed so far seem to be consistent with the hypothesis that the ACS is a result of the collision of high-velocity cloud(s) with the galactic H I disk. The distance of 2 to 5 kpc implied by the 'cosmic splatter' model (Mirabel, 1982) implies that

TABLE I

Star name	m_v	Sp. type	A_v	D^* (pc)	W_λ (Ca K) (ACS) m Å	τ	σ_τ	Notes
(1)	(2)	(3)	(4)	(5)	(6)	(7)	(8)	(9)
Summary of Optical Observations toward nearby SAO stars								
SAO 78259	6.3	A0	~0	132	70	0.5	0.018	
SAO 78275	7.1	A0	0.06	190	117	1.2	0.05	
SAO 78319	8.5	B8	>0.4	550	117	1.2	0.02	
SAO 78322	8.3	B?	>0.3	–	164	1.2	0.05	A
SAO 78333	6.5	B9	0.09	173	94	0.7	0.10	
SAO 78349	6.0	A0	0:	114	140	1.0	0.018	B
SAO 78384	8.7	A0	0:	398	94	0.7	0.10	B
SAO 78393	8.7	B9	0.2	436	117	0.9	0.045	
SAO 78047	8.7	A0	0.09	398	234	1.7	0.04	
SAO 78415	8.2	A0	–	316	140	1.0	0.045	C
SAO 78524	5.0	A0	~0	72	47	0.17	0.02	
SAO 95819	7.6	A0	0.09	549	140	0.34	0.025	
Summary of Optical observations towards distant Cruz-Gonzales stars								
HD 45314	6.64	OPE	–	< 1000	140	0.34	0.03	D, F
HD 248894	9.29	O8.0:V:NN	–	2680	105	0.26	0.05	E
HD 255055	9.38	O9.0VP(E)?)	–	2480	117	0.17	0.05	E
HD 256035	9.16	O9.0VP	–	1500	105	0.26	0.05	E
HD 256725	9.7	O	–	<2500	105	0.26	0.05	D, F

Notes:

Column 4 is the visual extinction estimated by assuming that $R = N(\text{Ca II})/N(\text{HI}) = 1.3 \times 10^{-8}$, $N(\text{H I})/E(B - V) = 6 \times 10^{21}$, and finally $R_v = Av/E(B - V) = 3$.

Column 5 is the deduced photometric distance to the star, uncorrected for extinction.

Columns 6 and 7 are the expected Ca II K equivalent width and the maximum optical depth for the ACS gas.

Column 8 is the noise in the continuum and is simply $n^{-1/2}$ where n is the number of photons detected in the continuum (per resolution element).

Column 9 refers to additional notes (below).

A, Spectral sub-type not known.

B, The ISM line is confused with the stellar line; however, the ISM line (the narrower line) appears to be quite weak.

C, The ISM line is confused with the stellar line.

D, Upper limit to the distance derived from the observed equivalent width of the Ca K absorption line arising from the normal ISM.

E, Distances are photometric determination from Cruz-Gonzales *et al.* (1974).

F, HD 45314, controversy in the literature as to nature, varying from subdwarf to $M_v = -6.8$. HD 256725, luminosity class not known.

the ACS is (a) large ($\sim$1 kpc) and (b) very energetic ($\sim 10^{52}$ to 10^{53} ergs in kinetic energy). For these and other such indirect reasons the distance to the ACS has been argued to be as small as 100 pc (Weaver, 1979; Kulkarni, 1983). Thus it is clear that the determination of the distance to the ACS is crucial to the acceptance of any model.

2. Optical Observations and Results

During 1983 and 1984 we have obtained high-resolution (Δv = 10 km s^{-1}), high-sensitivity ($\sigma_\tau \sim 0.01$) echelle spectra towards early-type stars located within the boundaries of the ACS. The echelle spectrograph was at the Cassegrain focus of the Whipple Observatory's 1.5 m telescope and provided a 50 Å spectrum centered on the Ca K line. The first set of observations included 12 SAO stars no later than A0 with a maximum distance of less than 500 pc; the second set consists of 5 O stars from the Catalog of Galactic O stars (Cruz-Gonzales *et al.*, 1974) with distances between 1 and 2.5 kpc. Both sets of observations are summarized in Table I; the essential result is that no anomalous velocity Ca II absorption features were detected.

The conclusion drawn from our optical work is that the shell is certainly more distant than 500 pc, and quite possibly more distant than 2500 pc. Crucial to this conclusion, however, is our estimation of the expected optical depth of the Ca K absorption line (columns 7 and 8 in Table I) arising from the ACS gas. The expected optical depth has been calculated by assuming that $R = N(\text{Ca II})/N(\text{H I}) = 1.3 \times 10^{-8}$ (Habing, 1969). However, fluctuations by a factor of 3 are not uncommon – thus possibly weakening the results for the more distant stars in Table I. However, it is well known that R increases with the peculiar velocity of H I clouds (Spitzer, 1978); this is attributed to the return of metals to the gas phase from the destruction of grains in high-velocity shocks. Kulkarni *et al.* (1985) have studied the H I emission and absorption properties at two points in the shell and conclude that the ACS H I (a) appears to have properties similar to that of the classic high velocity clouds and (b) has anaomalous spin temperature properties suggestive of a post-shock cooling region. Thus we conclude that the shell is certainly more distant than 500 pc, and beyond 2.5 kpc unless the Ca II abundance is much lower than normal.

References

Cruz-Gonzales, C., Recillas-Cruz, Costero, R., Peimbert, M., and Torres-Peimbert, S.: 1974, *Rev. Mex. Astron. Astrofisika* **1**, 211.

Habing, H. J.: 1969, *Bull. Astron. Inst. Neth.* **20**, 177.

Heiles, C.: 1984, *Astrophys. J. Suppl. Ser.* **55**, 585.

Kulkarni, S. R.: 1983, Ph.D. Thesis, Univ. of Calif., Berkeley.

Kulkarni, S. R., Dickey, J. M., and Heiles, C.: 1985, *Astrophys. J.* (submitted).

Mirabel, I. F.: 1982, *Astrophys. J.* **256**, 112.

Spitzer, L.: 1976, *Comm. Astroph.* **6**, 177.

Weaver, H. F.: 1979, in W. B. Burton, *The Large Scale Structure of the Galaxy*, D. Reidel Publ. Co., Dordrecht, Holland, p. 295.

INDEX OF CONTRIBUTORS

(Volume 118)

ANNOUNCEMENT

Third Asian-Pacific Regional Meeting of the International Astronomical Union. Part 1. September 30– October 5 1984, Kyoto, Japan

Editors: M. Kitamura and E. Budding

Please note that a hardbound edition of this special issue of *Astrophysics and Space Science,* Vol. 118, Nos. 1/2 (January 1986), is available from the publishers.

ISBN: 90–277–1867–9

Prices: Dfl. 230,– / \$98.00 / £63.75
(for participants: \$34.00)

GPSR Compliance

The European Union's (EU) General Product Safety Regulation (GPSR) is a set of rules that requires consumer products to be safe and our obligations to ensure this.

If you have any concerns about our products, you can contact us on ProductSafety@springernature.com

In case Publisher is established outside the EU, the EU authorized representative is:

Springer Nature Customer Service Center GmbH
Europaplatz 3
69115 Heidelberg, Germany

Batch number: 10607564

Printed by Printforce, the Netherlands